FALCO PEREGRINUS, LINN.
Wanderfalke.

Altes Weibchen   Junges Männchen.

The
Peregrine Falcon

Paul 10 7 v4

Thanks for providing all your superb photographs over the last few years.

I hope you enjoy reading & referring to it.

All best,
Steve

The Peregrine Falcon

Richard Sale

Steve Watson

SNOWFINCH PUBLISHING

Snowfinch Publishing
The Beeches
Coberley Road
Coberley
Gloucestershire GL53 9QY
UK

© in the text: Richard Sale and Steve Watson, 2022.
© in the figures: Richard Sale and Steve Watson, 2022.
© in the drawings: Kerry Jane, 2022.
© in all unattributed photographs: Richard Sale.
Contact the authors at raptoraidSteve@outlook.com

Every effort has been made to make contact with the authors of the paper from which the image on p12 (lower) was taken, but without success.

ISBN 978-0-9571732-6-2

All rights reserved. No parts of this publication may be reproduced, stored in a retrieval system, or transmitted in any form or by any means, electronic, mechanical, photocopying, recording or otherwise, without the prior written permission of the Publisher.

The left-hand front endpaper shows an image entitled *Rhynchodon submelanogenys*, the Western Black-cheeked Falcon, now more correctly termed *Falco peregrinus submelanogenys*, the Western Black-cheeked Peregrine which some experts believe to be a separate sub-species rather than a darker form of the Australian sub-species *Falco peregrinus macropus*. This image is an original, hand-coloured lithograph by the Danish natural history artist Henrik Gronveld from a drawing by G.E. Lodge for Gregory Macalister Matthews' *The Birds of Australia* published in London between 1910 and 1927.

The right-hand front and left-hand rear endpapers are engravings of nominate Peregrines by Otto von Riesenthal for his folio of images of Birds of Prey (raptors and owls) sold in a slip case entitled *Die Raubvögel Deutschlands und des angrenzenden Mittel-Europas* published by Cassel Th. Fischer in 1876. The right-hand front endpaper shows, left, an adult female (exact translation 'old lady') and, right, a young male. The left-hand rear endpaper shows, above, a 'middle-aged lady' and, below, an 'old man'.

The right-hand rear endpaper is by Johannes Gerardus (John Gerard) Keulemans, a Dutch artist who spent much of his adult life in England. This particular image of an adult pair of African Peregrines, *Falco peregrinus minor*, was used as an illustration in the *Catalogue of the Birds in the British Museum*, Volume 1 (*Catalogue of the Accipitres or Diurnal Birds of Prey, in the Collection of the British Museum*), by R. Bowdler Sharpe published in 1874.

Acknowledgements

The production of this book would not have been possible without the assistance of people from across the globe who have willingly given their time, their knowledge and, in many cases, their photographs to aid the authors in fulfilling the dream of a book which covered both all sub-species and all aspects of the behaviour of this most remarkable of birds. The limitations of space and the risk of omitting a name by accident means we offer our sincere thanks to everyone who has aided us. We also thank all the photographers named against their individual images for use of those photographs.

RS thanks Steve Watson without whose ~~persistent bullying~~ gentle persuasion, this book would probably not have been written. I thank him for his friendship and for stimulating conversations throughout the time this book was being prepared. I also thank him for acting as chauffer during a long period of poor health which meant I was limited in my ability to drive myself for an extended period. RS thanks Arjun Amar, Rob Biljsma, Nick Dixon, David Ellis, Malcolm Henderson, Neil McDonald, Nathalie Mahieu, Simon Potier, Remo Probst, Nathan Sale, George Smith, Iñigo Zuberogoitia and the folk at Evesham Bell Tower for help with information and in the field. Other individuals are thanked specifically against the figures for which they have given permission for use. For their help with the construction of the IMUs flown on falconry birds for the collection of data set down in *Chapter 3* RS thanks Waldo Cervantes, Natan Morar and Seb Madgwick, and thanks Nick Havemann-Mart and Oliver Tomlinson for allowing him to fly the IMUs on their Peregrines. RS also thanks Joanne Cooper of the Natural History Museum, Tring for assisting in the loan of Peregrine bones for CT scanning to underpin work set down in *Chapter 3*, and Adam Bloch and Tom Dutton for aid with the CT scanning and analysis of the data produced. Thanks also to the library services of the University of Oxford, particularly Sophie Wilcox of the Alexander Library of Ornithology and Marija Babic of the Vere-Harmsworth Library.

Finally, I thank my wife Susan for her unfailing support during the production of the book. Poor health during early 2022, and the unremitting nature of both the research for, and the work on, the book meant that support was instrumental in the completion of the project.

SW thanks Richard Sale for his valued friendship and for allowing me to 'gently persuade' him to write this book. I consider it an honour and a privilege to have been able to support him in this project and would like to pay tribute to his extraordinary ability to synthesise complicated material and data and resolve it into a highly intelligible format. It would also be remiss of me not to acknowledge Richard's steadfast assiduousness, in the face of ill health at times, together with his open-minded approach to all hypotheses and suggestions put to him in the course of writing this book. I have learnt so much from him.

SW acknowledges the inspiration provided by his mentor and friend, Richard (Dick) Treleaven MBE (dec'd), who taught me so much in the field, and the seminal literary work of Derek Ratcliffe, Tom Cade and Prof Ian Newton, all of whom have inspired me with their writings on Peregrines and raptors generally.

SW would like, in particular, to acknowledge the help of Mark Wilson of BTO who answered a long series of queries on UK population with unfailing courtesy and promptness. Nathalie Mahieu of FaB Peregrines provided meticulously produced breeding data, images and notes for our Charing Cross Hospital study and also answered many information requests with good grace and humour, as did Stuart Harrington regarind London Peregrines.

SW would like to thank the following for their friendship and field discussions over the period of his 40-year Peregrine studies; Jon Watson, Russ (Hawkeye) Peacey, Nick Dixon, Frank Williams, Andrew Bluett, Richard Tyler, Dave Pearce, Jacky Morrell, Peter Welsh, Roger Finnamore, Greg Curno, Dan Williams, Jimmi Hill, Wayne Middlemist (dec'd), Chris Jones, Richard Scantlebury, James Aldred, Rob Husbands, Gordon Kirk, Natalie Roberts, Gareth Jones, Anna Field, Ed Drewitt, Annette Evans, Bill Williams (dec'd), Gavin and Jude Black, Ken Eames, Phil and Chris Andrews, Mike (the bike), Brian and Linda Woodroffe, Richard and Jenny Cross, Malcolm Stokes and Polly Goldsworthy, Chris Griffin, Tony Whitehead, Jarrod Sneyd, Denis Jackson, Ed Hutchings, 'Welsh' Clive, Chris Ridler, Martin Buckley, Rob Lewis, Dom Blake, Vince Stanford, Julian Wylie, Tina Freeman and Catherine Inglis.

And finally I thank my wife, Mary, for her unfailing support and forbearance whilst I was single-mindedly preoccupied both physically and mentally with the demands of producing this book.

Above *Falco peregrinus peregrinus*, photographed in the snow in Poland. *Ondřej Prosicky.*

Below *Falco peregrinus macropus*, photographed on the coast near Sydney, Australia. *Ken Griffiths.*

Contents

Introduction

1	**The Falcons**	10
2	**The Peregrine Falcon**	34
3	**Flight Characteristics** *Flight, Hunting Techniques and Strategies*	92
4	**Diet**	162
5	**Food Consumption and Energy Balance**	246
6	**Breeding Part 1** *Pair Formation to Nest Sites*	256
7	**Breeding Part 2** *Eggs to Fledglings*	292
8	**Movements and Winter Grounds**	364
9	**Friends and Foes**	394
10	**Population** *Survival and Population Numbers*	416

References 492

Index 526

Introduction

Peregrine Falcons are aerial apex predators, the subject of copious volumes of purple prose attesting to the sometimes mystical regard in which they have been held through the ages, on occasions being elevated to the status of myth and legend. Their spectacular hunting flights attract mankind's admiration for their athleticism and jaw-dropping turns of speed. Matchless and unmatchable, they have been, and are, revered, to such an extent that they are known as Shaheen, the 'king of birds'.

The Peregrine is one of the most recognisable birds on the planet, breeding on six continents, a range shared by only one other bird, the Barn Owl. This range is testament to the Peregrines' adaptability, as shown in their recent move to urban breeding areas, and a highly plastic prey spectrum. Yet despite long history of the Peregrine as a falconry bird and its being one of the most famous and photographed of all the world's birds, studies of it and its behaviour are more limited than might be anticipated. But studies do exist, covering all of the numerous sub-species found across the globe, though not always to the same depth and extent. In this book we have attempted to collate that information and so create a pen picture of the various aspects of behaviour that make the Peregrine iconic in the eyes of many bird-watchers. Our aim is to provide a readable, well-illustrated text defining the current knowledge of all aspects of *Falco Peregrinus* and its sub-species. The book starts with the evolution of falcons, then looks at the history of the relationship between Man and Falcon. That interaction has not always been of the awestruck kind, as such interactions can have consequences that are benign or malign, direct or indirect, witting or unwitting and clearly humans now, as never before, have a major impact on the density, range and distribution of this species throughout the world. Man has, through both accident and design, had a radical, negative effect on Peregrine numbers and in some parts of its range continues to do so. The book concludes by looking at those aspects of mankind's influence on the Peregrine population. Filling the pages between those topics we include a scientific understanding of the Peregrines flight characteristics as well as breeding behaviour, migration and the falcon's interaction with other species.

The authors come from two very different backgrounds and specialisms, RS primarily a scientist with a current interest in avian flight dynamics, and an established raptor author. SW is an obsessive (some might say) Peregrine observer and student of their behaviour with a determination to understand everything there is to know about the species. He is also a lecturer on raptors. But despite those inherent differences, we share an enthusiasm for falcons in general and Peregrines in particular.

We have both strived to provide a detailed understanding of the Peregrine. We hope you enjoy reading the book.

Above   *Falco peregrinus babylonicus*, Little Rann of Kutch, Gujarat, India, *Nirav Bhatt*.

Below   *Falco peregrinus anatum*, Point Lobos, California, USA. *Beth Hamel*.

1   THE FALCONS

The origin of birds as a surviving line from the dinosaurs is now well-established aided, in part, by a recent startling discovery of an in-ovo articulated embryo fossil of a non-avian oviraptorid therapod dinosaur of the late Cretaceous in southern China (Xing *et al.*, 2022). The fossil and an X-ray of the skeletal remains are illustrated opposite, with other relevant images on pages overleaf.

But, while the evolutionary line of birds is now accepted, the lineage of individual bird families is still debated. Early studies of the way various avian species were related used fossils and current morphology to piece together an overall family tree, but more recently the development of genetic analysis has been added to the tools scientists may use to investigate ancestry. Carole Griffiths used a combination of morphological differences (specifically of the syrinx of various bird families) and genetics (sequences based on mitochondrial cytochrome *b*) – Griffiths, 1994a, 1994b and 1997 – to establish a view of falcon evolution, suggesting that the Falconidae originated in South America, an idea supported by the number of genera and endemics found there. This proposal is still debated, though the Falconidae do appear to have evolutionary roots in the Neotropics (Griffiths, 1999) where many species are still found.

The Family Falconidae of the Order Falconiformes is now most often divided into two sub-families, the Polyborinae, which are largely confined to neotropical South America and include the caracaras and the forest-falcons, and the Falconinae, which are more widespread and include the falconets and

Above One of the fossil eggs found in the Hekou Formation, southern China. With thanks to Darla Zelenitsky and the remaining team of Xing *et al.* (2022).

Below X-ray of the above fossil showing the position of the embryo. With thanks to Darla Zelenitsky and the remaining team of Xing *et al.* (2022).

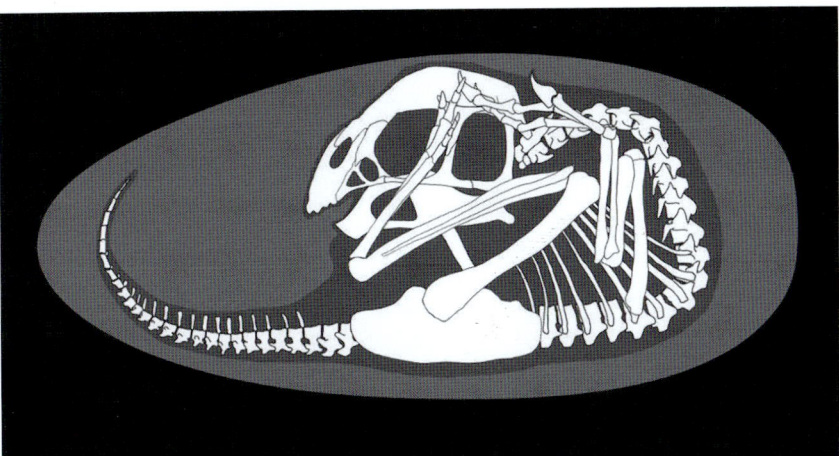

the pygmy falcons, as well as the 'True' falcons (species of the genus *Falco*[1]). As the behaviour of the caracaras resembles that of vultures and, to a lesser extent, buzzards, while the forest-falcons are hawk-like, this, unsurprisingly, led to the early assumption of a close evolutionary link between the Falconidae and the Accipitridae (hawks) families. That assumption seemed supported by the fossil record, which indicated that both families had origins in the Eocene (some 35 million years ago) but was contradicted by the known

---

[1] Some authorities define three families, adding the Herpetotherinae (usually including the Laughing Falcon (*Herpetotheres cachinnans*) alone, but sometimes also including the *Micrastur* forest-falcons).

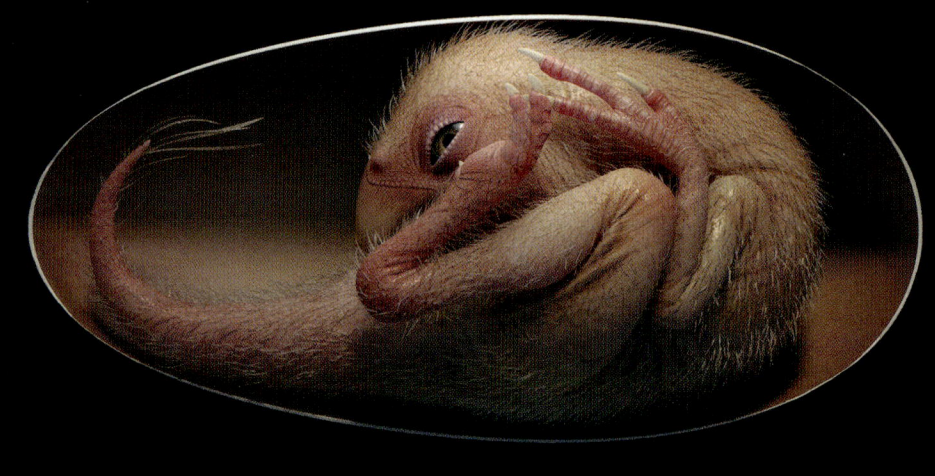

Above Reconstruction of the embryo of the fossil on the previous page. With thanks to Darla Zelenitsky, Lida Xing and Shoulin Animations.

Below Photograph of the embryo of a domesticated Rock Dove removed from its shell at 17 days from laying. Reproduced from Łukasiewicz *et al.* (2014) - see Note on p4.

morphological differences between 'falcons' and 'hawks', as well as differences in the moult sequence. While it is also occasionally proposed that the main difference between the two families is the obvious one – that 'falcons' do not build nests whereas 'hawks' do – that suggestion ignores the fact that caracaras construct untidy piles of sticks.

Studies of molecular phylogenetics across 169 species of bird by Hackett *et al.* (2008) suggested that the Passeriformes (perching birds, species which include more than half of all avian species) were all related. The work suggested

Above Fossil eggs, as found in southern China. With thanks to Kohei Tanaka and Darla Zelenitsky.

Below Artist's impression of the dinosaur and nest of eggs based on the above find and advice of dinosaur specialists. With thanks to Masato Hattori and Darla Zelenitsky.

that the Falconiformes had branched from an ancestral line which also included corvids, finches and thrushes among others. Equally interesting, the work of Hackett and co-workers suggested that the long-held belief that falcons and hawks were related, and part of the same order (the Falconiformes, which also included the vultures, the Osprey (*Pandion haliaetus*) and the Secretary Bird (*Sagittarius serpentarius*), the latter a seemingly unlikely cousin, but one which has often been referred to as the 'pedestrian eagle'), was incorrect, and that the two were distinct. Hackett *et al.* noted that the Falconidae and the Accipitridae formed different clades in all their analyses, confirming the suggestion of those who had previously noted differences both of morphology, moult sequence and behaviour.

Hackett *et al.* (2008) also confirmed an ancestral link which had been noted in earlier work by Ericson *et al.* (2006) that falcons were closely related to parrots. That suggestion had been met with scepticism by some who considered the link was more likely to be an artefact of the Ericson *et al.* study. But the work of Hackett *et al.* (2008) required a re-appraisal of the apparently (and surprisingly close) ancestral link between parrots and falcons, the suggestion being reconfirmed by Suh *et al.* (2011) whose findings implied that parrots were the closest link, and falcons the second closest link, to passerines.

Wright *et al.* (2008) had also noted a similar link between falcons and parrots, suggesting that an ancestral form of both had originated on the supercontinent of Gondwanaland during the Cretaceous period (65-145 million years ago). Further confirmation came from Pyle (2013) who examined 4500 museum specimens of 65 falcon species and 375 species of parrot to investigate moult sequences and noted similarities in the two families (Fig. 1.1), similarities which are not shared by other avian groups. The exception was the Kakapo (*Strigops habroptilus*) whose moult pattern resembles that of most other birds.

The link between parrots and falcons was further strengthened by results from the work of the Avian Phylogenomics Consortium, a group of some 200 scientists from 20 countries. Late in 2014 the Consortium produced two-dozen or so papers across the entire spectrum of ornithological scientific journals suggesting an ancient link between falcons, parrots, the seriemas of South America and all Passeriformes. The Consortium suggests these four groups form the 'Australaves' clade, one of two clades which comprise core landbirds, the other being the 'Afroaves' which includes, among others, the remaining diurnal birds of prey, owls and woodpeckers. For more information on the Consortium's work see, for instance, Jarvis *et al.* (2014).

However, while the relationship between falcons and parrots has merit, genetics is an evolving subject (a somewhat ironic situation given its importance to evolution) and there are discrepancies between the results of molecular methods and data based on both palaeo-environmental and palaeontological (particularly morphological) evidence, which is perhaps not surprising as there is no reason to suppose that the rates of evolution of morphology and molecules are similar.

The Falcons

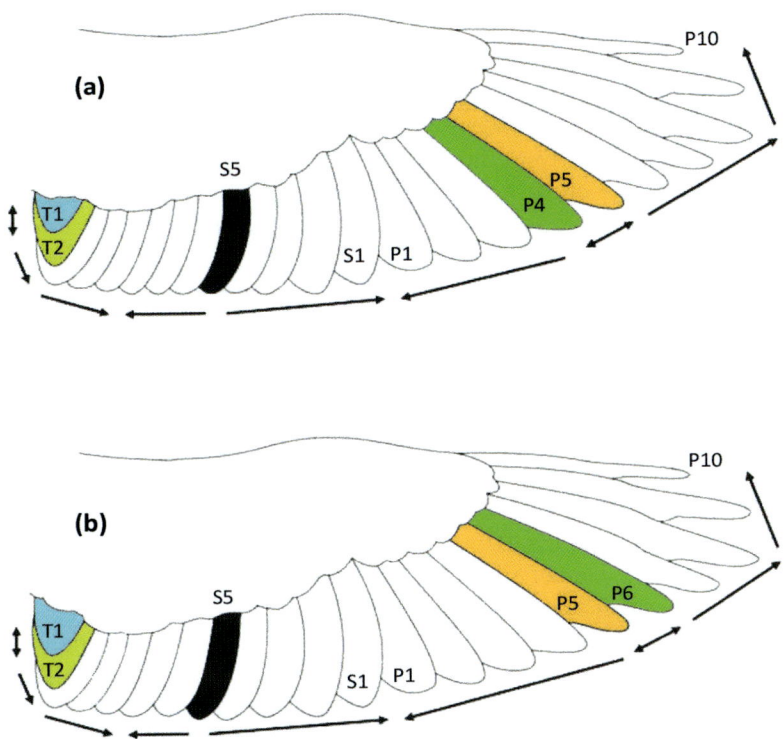

Figure 1.1 Moult sequences of the Primaries (P), Secondaries (S) and Tertials (T) of falcons (a) and parrots (b). Falcons have 10 primaries and 12 or 13 secondaries (including tertials): 13 are shown here. Parrots have 10 primaries and 12 secondaries/ tertials. Moulting starts from nodal feathers which are illustrated in colour. In falcons the nodal primary is P4 or P5, in parrots it is P5 or P6. S5 is the nodal secondary in both. The arrows indicate the direction of moult. Redrawn from Pyle (2013).

It is, therefore, worth noting the cautions expressed by some experts. Ballard and Whitlock (2004) note that '*it is not safe to assume a priori that mtDNA evolves as a strictly neutral marker because both direct and indirect selection influence mitochondrial*', while Galtier et al. (2009), when considering the application of mtDNA analysis to the evolutionary reconstruction of species, noted that mtDNA is '*far from neutrally evolving and certainly not clock-like*'. More recently Cadena and Zapata (2021) note that '*evolutionary theory...calls for an integration of genomics with phenomics in avian species delimitation.*'

But despite early misgivings, in the 53rd supplement to the American Ornithologists' Union check-list (Chesser et al., 2012) the linear sequence of orders has been changed, with Falconiformes (which comprise a number of orders and genera, one of the latter being Falco) and Psittaciformes being placed immediately preceding Passeriformes, a change 'reflecting the close relationship among these orders'. This remains the position, supported by the most recent study of the genomes of over 90% of bird families (Feng *et al.*, 2020).

While the use of genetics to explore the ancestry of birds has aided an understanding of the development of orders and families, its use has not been limited to that field, the work of Ng and Li (2018) having used genetics to examine feather diversity in an exquisitely illustrated article. In another study of genomic sequencing, Zhan *et al.* (2013) looked at the evolutionary basis of predatory adaptations, taking the Peregrine and the Saker Falcon as their models. The analysis noted the evolution of the falcon's bill, and also the changes in the Saker genome that allowed the species to thrive in an arid environment.

But whatever their origins, the True falcons have proved extremely successful and have colonised all Earth's ecozones except Antarctica, the failure to colonise that last continent probably due to its isolation, though whether a falcon could survive Antarctica's climate is another matter: the Gyrfalcon (the world's largest True falcon) is an Arctic dweller, but the southern winter is far harsher than that of the northern polar region, and it is debatable whether any falcon could successfully migrate across the open water that separates Australia, New Zealand or South America in order to avoid the southern polar winter. In Chapter 2 *Distribution* it will be noted that Peregrines have been seen on South Georgia, which is 1500km from the Falkland Islands where the falcons breed. As it is less than 1000km to the Antarctic Peninsula from Tierra del Fuego where Peregrines also breed it is possible the falcons have reached (but not been seen on) Antarctica though it is widely assumed that those spotted on South Georgia were ship-assisted in the transit from South America/Falkland Islands. Of course, reaching and long-term surviving/breeding in the remote south are very different matters.

In their genetic analysis of the Falconidae (the largest order of the Falconiformes), Fuchs *et al.* (2015) considered that species divergence probably began about 16 million years ago, but that the rapid divergence of the True Falcons came much later, around the time of transition from the Miocene to the Pliocene, coinciding with the rapid expansion of the C4 grasslands (so-called because the plants had evolved a new method of fixing carbon from the atmosphere) primarily in Africa. That the divergence of the True Falcons and the expansion of the C4 grasslands are linked would then explain why the falcons are grassland/savannah rather than woodland species. In his classic book on falcons, Tom Cade (Cade, 1982) noted that the expansion of open grasslands had stimulated not only the development of falcons, but of mankind, as early fossil hominids are also found around the transition of the Miocene and Pliocene, and wondered whether those early men, gazing at the speeding birds in the sky, heralded the start of the association between man and falcon.

The timing of the emergence of the True Falcons is consistent with the earliest known fossil in Europe which dates from the late Miocene/early

Pliocene (about 2 million years ago – Mlíkovský, 2002). Fossil records of other falcons are also known from the Pleistocene, both from the Nearctic – *e.g.* Becker, 1987, Brodkorb, 1964 and Emslie, 1985 in North America, and *e.g.* Campbell (1976) in South America – in Australasia (*e.g.* Vickers-Rich *et al.*, 1991), and in the Palearctic (*e.g.* Tyrberg, 2008).

The True Falcons do not build nests, using nests constructed by other birds in which to lay their eggs, making a scrape on the ledge of a cliff or on the ground, or utilising convenient holes in trees or cliffs. But that characteristic (and others) does not aid the grouping of the species within the genus, groupings which, initially, were based on geography (where the falcons occurred in the world) and morphology (form and structure, *e.g.* size, plumage colour, hunting strategies *etc.*). These groupings were contentious, with all suggested groups seemingly having as many adherents as critics. However, ultimately four groups were most-commonly suggested, these according with the current best estimate for the evolution of the falcons which saw the hierofalcons branch off the ancestral line first, followed by the kestrels, the hobbies and then the peregrines. There were still complications with this four-group structure, not least the number of species which did not readily fit into any of the groups. These, basically morphological groups, though modified by early genetic studies, are organised here with the proviso that they are presented as much as a way of listing the Falcos as of examining the way in which the genus is truly organised.

The four groups comprise:

The Hierofalcons
Gyrfalcon (*Falco rusticolus*), which has a circumpolar Arctic distribution;
Lanner Falcon (*Falco biarmicus*), of southern Europe and Africa;
Saker Falcon (*Falco cherrug*), of central Asia;
Laggar Falcon (*Falco jugger*), of India.

The name Hierofalcon was coined by Otto Kleinschmidt (1870-1954)[2], a German pastor and ornithologist who developed the theory of 'form circles' (self-contained, stable units of 'hidden nature'). Kleinschmidt probably considered 'hiero' to derive from the Greek for 'hawk' because of the favoured, hawk-like, hunting strategy of the larger falcons as opposed to the stooping technique favoured by peregrines. However, in view of the pastor's creationist views it has been suggested he might have favoured a derivation from the Greek for 'sacred'.

An alternative name for the Hiero group is 'Desert Falcons' because of the species' preferred habitat: the Gyrfalcon is a true Arctic species, so the definition of desert needs expanding to include the northern polar desert. Despite most specialists now considering that the Hierofalcons include just

---
[2] For more information on Kleinschmidt and his ideas see, for instance, Potapov and Sale, 2005.

four species, unscrambling their evolution has not been easy (see Potapov and Sale, 2005, and references therein) and is still debated: the Prairie Falcon has also been included on occasions, although that species would appear much more closely associated with the peregrines (Wink and Sauer-Gurth, 2000; Nittinger *et al.*, 2005). In another complication, Wink *et al.* (2004) consider that the Brown Falcon of Australia is associated with the Hiero group (but see below).

Nittinger *et al.* (2005) considered that Africa was the probable origin of the Hierofalcons, while in a subsequent paper (Nittinger *et al.*, 2007) it is suggested that the Lanner is the closest species to the African ancestor of the group, subsequent phases of emigration splitting off into Eurasia and south Asia to form the Gyr, Saker and Lagger falcons. Nittinger *et al.* (2005) suggest that the Saker emerged as a species only about 34,000 years ago, a time consistent with the view of Potapov and Sale (2005) who suggested that the Gyr and Saker falcons were initially a single species, the growth of the taiga separating the population and forming the two species we see today.

The Peregrines
Peregrine Falcon (*Falco peregrinus*), of all continents apart from Antarctica[3];
Barbary Falcon (*Falco pelegrinoides*), of North Africa, the Middle East and west central Asia;
Prairie Falcon (*Falco mexicanus*), of North America.

In their studies of genome sequencing of the Peregrine and the Saker falcons, work that noted the evolutionary development of bill design and olfaction of the two species, Zhan *et al.* (2013) considered that the demographic history of the species involved a bottleneck approximately 20,000 years ago when the last Ice Age limited the available habitat in Eurasia. Although the work of Zhan *et al.* considered Eurasia specifically, the same bottleneck would have occurred in North America, though Peregrine populations in the remote south of the planet would not have been as habitat-constrained (though the genetic differences between the nominate, *F.p. brookei* and *F.p. minor* and the Barbary Falcon are considered 'slight' (Helbig, 2000)). For more specific consideration of genetic differences between sub-species see Nesje *et al.* (2000), who noted significant differences between Scandinavian and Scottish Peregrines, and between the falcons of northern and southern Scandinavia. Interestingly, the study did not find significant differences between the cliff-nesting Peregrines of northern Sweden and the bog-nesting Peregrines of northern Finland/Sweden, suggesting that nesting habitat is not genetically differentiated. See also Brown *et al.* (2007) who considered the effect of the North American population crash induced by the DDT crisis and the potential genetic bottleneck it caused between *F.p. anatum*, *F.p. pealei* and *F.p. tundrius*.

---
[3] Although the Peregrine Falcon and Barn Owl (*Tyto alba*) breed on six continents, they do not show a uniform pattern of breeding in those continents. The reason for the curious pattern is unknown, but is likely to be due to ecological factors or competition by established species which have prevented occupation.

A Barbary Falcon on Fuerteventura, Canary Islands. The position of these falcons within or without the Peregrine family is still discussed. *Tomas Grim*.

Nevertheless, the overall range of the Peregrine and the speciation since the last Ice Age has allowed a large number of sub-species to be identified. Nineteen are identified in the superb book on the Peregrine by White *et al.* (2013a). The debate about the authenticity of some sub-species, and the sheer majesty of the bird, have not surprisingly produced copious material in the scientific press regarding the Peregrine's evolution and family tree. White *et al.* sensibly declined to get involved with the various pros and cons of the sub-species in favour of dealing with the currently 'agreed' status, though they do present a brief, but cogent, assessment of the position and reference the relevant papers. The debate on sub-species is not aided by the fact that Peregrines readily breed (when in captivity) across sub-species' 'barriers' and with Hierofalcons.

A more obvious debate is the position of the Barbary and Prairie falcons. Are these a sub-species of the Peregrine rather than full species? Barbary-Peregrine crosses are fertile, but the Barbary is a semi-desert species, the ranges of the two rarely overlapping. However, since there are small areas where the two falcons may be found during the breeding season, and since Peregrine-Prairie hybrids definitely occur naturally in the wild (Oliphant, 1991[4]) the debate on the position of Barbary, Prairie and Peregrine is likely to be ongoing. The genetic analysis of White *et al.* (2013b) has suggested that an ancestral falcon may have evolved into both the Prairie Falcon and another line, which itself then branched, with one branch being ancestral to the Hiero and Peregrine falcons, while another was ancestral to the Kestrels, Hobbies and, presumably, those falcons which do not comfortably fit the four-group pattern.

In Chapter 2 of this book, where the sub-species of the Peregrine are considered, the Barbary Falcon is listed as a possible sub-species, but the Prairie Falcon is not included. This decision is in line with majority opinion at this time.

The current 'best estimate' list of sub-species and the breeding/hunting/wintering range they are believed to occupy is considered in more detail in Chapter 2.

The Kestrels
Common (or Eurasian) Kestrel (*Falco tinnunculus*), of Eurasia and Africa;
Lesser Kestrel (*Falco naumanni*), of southern Europe and central Asia;
Grey Kestrel (*Falco ardosiaceus*), of central Africa;
Dickinson's Kestrel (*Falco dickinsoni*), of central Africa;
Fox Kestrel (*Falco alopex*), which ranges across central Africa;
Greater or White-eyed Kestrel (*Falco rupicoloides*), of southern and eastern Africa;

---

[4] Oliphant (1991) reports the pairing of a male Peregrine and female Prairie Falcon in southern Saskatchewan in 1985. Two eggs were laid, both of which hatched. Both youngsters were male. They looked very similar to hybrids bred in captivity, being very different from Prairie Falcons and much more similar to their Peregrine father.

Rock Kestrel (*Falco rupicolus*) of southern Africa;
Seychelles Kestrel (*Falco araeus*), endemic to the Seychelles;
Mauritius Kestrel (*Falco punctatus*), endemic to Mauritius;
Madagascar or Malagasy Kestrel (*Falco newtoni*), of Madagascar, Mayotte and Comores, together with an Aldabran sub-species;
Madagascan Banded Kestrel (*Falco zoniventris*), endemic to Madagascar.
American Kestrel (*Falco sparverius*), of North and South America;
Moluccan or Spotted Kestrel (*Falco moluccensis*), of Indonesia;
Australian or Nankeen Kestrel (*Falco cenchroides*), of Australia and New Guinea.

Some authorities also include the following species within the Kestrel group:

Eastern Red-footed or Amur Falcon (*Falco amurensis*), of China;
Western Red-footed Falcon (*Falco vespertinus*), of eastern Europe and west-central Asia;
Red-necked or Red-headed Falcon (*Falco chicquera*), of India.

That said, it is also worth noting that the work of Wink *et al.* (2007) suggests the kestrels to be rather more loosely associated, their work suggesting that while the bulk of the group branched from a single ancestral form, the American Kestrel is more closely related to the Aplomado Falcon (*Falco femoralis*) those two sharing an ancestral link to the red-footed falcons. Furthermore, Wink and co-workers consider Dickinson's Kestrel to be far-removed from the Kestrel group, being genetically closer to the New Zealand Falcon (*Falco novaeseelandiae*).

The kestrels are clearly a complex group, with both rufous and grey forms[5]. Although many kestrels are to be found in Africa, which Groombridge *et al.* (2002) considered the evolutionary home of the species, the Common Kestrel is the most widespread form, breeding from the Atlantic coasts of Europe and above the Arctic Circle in Fennoscandia, as well as being found in Japan and central Africa. Groombridge *et al.* used genetic analysis to study not only kestrel evolution, but the interdependence of various African/near African sub-species. The work supported the idea of an African origin for kestrels with an ancient move to the Nearctic and a more recent move from Africa towards Madagascar and on to Mauritius and the Seychelles, though the timing of the movement to Australia is ambiguous. However, while the idea that kestrels evolved in Africa fits with the distribution of species, it is not the only interpretation, complicating factors being the position of the red-footed falcons which may be related to the American Kestrel, and the fact that the Lesser Kestrel is so distinct from the other Palearctic kestrels that it seems basal to all.

---

[5] The inclusion of the Dickinson, Grey and Madagascan Banded Kestrels is contentious as these are grey (or mostly grey in the case of the Madagascan Banded) and so have been grouped by some authorities as a sub-genus, *Dissodectes*.

Adult *F.p. anatum* with three chicks, San Pedro, Los Angeles, USA. *Gabriel Gruia*.

The Hobbies
Occasionally referred to as a sub-genus, Hypotriorchis. The sub-genus consists of four tree-nesting species:

Eurasian Hobby (*Falco subbuteo*), of central Eurasia;
African Hobby (*Falco cuvierii*), of central Africa;
Oriental Hobby (*Falco severus*), ranging from north-east India to Indochina;
Australian Hobby (*Falco longipennis*), of Australia.

Some experts consider that the following two, cliff-, rather than tree-nesting, species are also hobbies:
Sooty Falcon (*Falco concolor*), of the Middle East;
Eleonora's Falcon (*Falco eleonorae*), of the Mediterranean coasts of Europe and Africa.

The behavioural and morphological similarities between these two and the Eurasian Hobby are supported by molecular data (Wink *et al.*, 1998) and certainly imply a close relationship. Wink and co-workers suggest that an ancestral Falco branch gave rise to the Merlin and a branch which then split, with the Sooty Falcon taking one path, while the second path branched again to form the Eurasian Hobby and Eleonora's Falcon.

The Bat Falcon (*Falco rufigularis*), of Central and South America is also very hobby-like: it has similar plumage, having red legs and vent feathers, a characteristic shared by the 'true' hobbies and by Eleonora's Falcon, though not by the Sooty Falcon, and a similar hunting technique and diet, with males specialising in hunting insects. However, it nests in tree holes rather than the stick nests favoured by the Hobbies. The Aplomado Falcon of South and Central America and the southern USA has also been allied to the Hobbies by some. Olsen *et al.* (1989) made this link based on feather protein similarities (but suggested other falcon linkages which are not favoured by more recent genetic studies). The Aplomado has red legs and vent feathers, and there are also behavioural similarities: Fiuczynski and Sömmer (2000) noted the parallel between the prey and urban breeding habits of the Aplomado (in Rio de Janeiro) and the Eurasian Hobby (in Berlin). However, it is worth noting that the mere existence of red leg feathering as an indicator of a close relationship with Hobbies would mean having to consider the red-footed falcons and the New Zealand falcon as possible cousins which would be contrary to the story told by genetics.

The four-group classification presented above leaves several species adrift, these forming distinct continental or country groups:

Orange-breasted Falcon (*Falco deiroleucus*), Aplomado Falcon and Bat Falcon of Central and South America, if the Bat and Aplomado falcons are not grouped with the hobbies. The Orange-breasted Falcon is a medium-sized predator of avian prey which has occasionally been considered part of the Peregrine group.

Taita Falcon (*Falco fasciinucha*), a rare species of east and southern Africa which has occasionally been placed with the hobbies by some experts, but more recently has been suggested as being part of the Peregrine group based on mtDNA sequencing (White *et al.*, 2013b, Bell *et al.*, 2014).

Black Falcon (*Falco subniger*), Brown Falcon (*Falco berigora*) and Grey Falcon (*Falco hypoleucos*), all of which are endemic to Australia. As noted above the Brown Falcon has occasionally been placed in the Hiero group by some experts, while others suggest it belongs with the hobbies.

The above groupings leave two further outliers, the New Zealand Falcon and the Merlin (*Falco columbarius*). The former is endemic to New Zealand. There it is one of only two diurnal birds of prey (the other being the Australasian, or Swamp, Harrier (*Circus approximans*)) and has characteristics which suggest it fills the role of both falcon and sparrowhawk. The Merlin is found in both

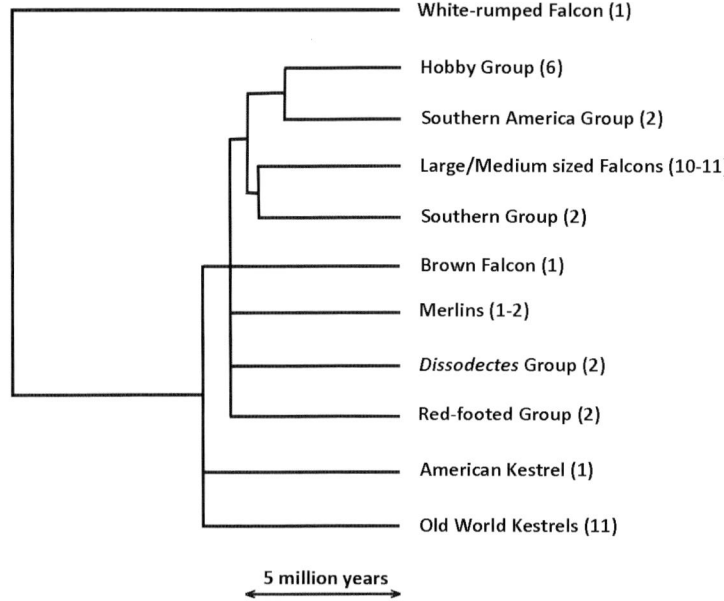

Figure 1.2 The evolution of groups of True Falcons based on phylogeny of Fuchs et al. (2015). The numbers in each bracket pair are the species within each grouping.

Hobby Group (6)
African Hobby, Australian Hobby, Eurasian Hobby, Oriental Hobby, Eleonora's Falcon, Sooty Falcon.

South American Group (2)
Bat Falcon, Orange-breasted Falcon.

Large/Medium Falcons (10-11)
Hiero falcons (Gyr, Lagger, Lanner, Saker), Peregrine, Barbary Falcon, Prairie Falcon, Black Falcon, Grey Falcon, Red-necked Falcon, Taita Falcon.

The number depends on whether the Barbary Falcon is, or is not, considered a subspecies of the Peregrine. The current genomic position is that the Barbary is a species.

Southern Group (2)
Aplomado Falcon, New Zealand Falcon.

Brown Falcon (1)

Merlins (1-2)
American Merlin, Eurasian Merlin.

The number depends on whether the New World and Old World Merlins are separate species or are sub-species of a single species. The current genomic position favours that there are two species.

Dissodectes Group (2)
Dickinson's Kestrel, Grey Kestrel.

Red-footed Group (2)
Amur Falcon, Red-footed Falcon.

American Kestrel (1)

Old World Kestrels (11)
Common Kestrel, Fox Kestrel, Greater Kestrel, Lesser Kestrel, Madagascan Banded Kestrel, Malagasy Kestrel, Mauritius Kestrel, Nankeen Kestrel, Rock Kestrel, Seychelles Kestrel, Spotted Kestrel.
Redrawn from Wilcox et al. (2019).

northern Eurasia and the northern reaches of North America. Merlins share characteristics with the hobbies and the peregrines but are distinct from both. The position of the Merlin is complicated further by the fact that while most falcons are birds of temperate climates (warm and damp maritime, or warm and dry continental) the Merlin is a cold climate species.

The groupings set down above, and the species which are outliers of the four groups, have not survived more recent genetic work which has produced a family tree with 10 groups, a family tree which also suggests that the Hiero falcons, the Peregrines and several outliers actually form a 'super group' of large/medium sized falcons distinct from their smaller cousins. The fundamental work to prepare this family tree was carried out by Fuchs *et al.* (2015), but more recently an impressive article by Wilcox *et al.* (2019) has amalgamated the genetic work of Fuchs *et al.* (2015) with a history of falconry, integrating the evolution of falcons with human culture and the needs of conservation. What is apparent from these fine bodies of work is that the True Falcons (the Falcos) as a genus are an evolutionary success, both in terms of their global distribution and species diversity – each of which is significantly greater than not only the Falconidae family, but birds as a whole. Fig. 1.2 (above, with details on opposite page) illustrates the present best estimate of the True Falcon family tree and how it has evolved over the last 15 million years since the currently assumed separation of the White-rumped Falcon (*Polihierax insignis*) – now considered the closest extant relative of the genus Falco – and the True Falcons from the other members of the family Falconidae.

General Characteristics of Falcons
Falcons are diurnal raptors, adapted for predation. The eyes are enlarged, to such an extent that they cannot be manoeuvred in the manner of human eyes, movement being highly restricted. The bird must therefore move its head to change its line of sight and field of view. Falcons also have two foveae (from the Latin fovea for a pit) or positions of visual acuity, rather than the one in the human eye. One fovea is positioned to give forward binocular (stereo) vision, the other being positioned at an angle of about 40° to the axis of the bird which is probably associated with the tracking of prey. All aspects of Peregrine vision are considered further in Chapter 3 *Vision*. The eyes are protected by a distinctly developed superciliary ridge which extends above and in front of the eye. The ridge and its feathers give the birds their famous piercing stare. The ridge is assumed to protect the eye from wind and debris. Falcons also have a region of dark feathers in front and below the eye which create an equally distinctive malar stripe (more commonly called the 'moustache'). The stripe was long assumed to reduce glare from the sun when the bird is hunting, a theory that was finally proven in a study by Vrettos *et al.* (2021) who studied photographs of Peregrine sub-species across the world and compared the size of the stripe to local solar radiation, temperature and rainfall. The only positive correlation was with annual solar radiation.

The sickle-wing shape is well-illustrated in this shot of a Peregrine in flight. *Ray Gilbert.*

The falcon wing is the classic design for rapid flight, the leading-edge outboard of the wrist being swept back and tapering to a point: the falcon's wing shape defines the bird as a hunter, rather than a soaring raptor which searches for prey on the wing. Interestingly, the wing shape may be the origin of the family name, deriving from the Latin *falx* (the genitive of *falcis*) a curved blade or sickle. However, while that is a majority opinion, some maintain that the name is equally good at describing the shape of the talons and bill, so either of those aspects of the falcon's design would work well as the basis of the name.

The upper mandible (maxilla) of the bill is decurved and hooked at the tip for tearing at the flesh of prey[6]. The tomia (cutting edges) of the maxilla have a distinct notch which creates the tomial tooth: this tooth is used to sever the spinal cord of the prey by biting the base of the neck. This manner of killing differs from that of hawks which tend to kill by squeezing their prey with their talons, killing by either asphyxiation or by penetrating the skin and reaching vital organs, thus causing death by multiple body traumas, or a combination of the two. The muscles which control a falcon's bill are large, giving the falcons a powerful bite, allowing them to deal with prey larger than themselves. Use of the bill to kill is inherited rather than learned, appearing to be an automatic reflex when faced with live prey. Adult falcons overcome this auto-function during the fledgling stage of their offspring's development as they occasionally deliver live prey for the young to deal with themselves. During the study at London's Charing Cross Hospital which is dealt with in Chapter 7 there were two observed instances of this behaviour. On each occasion the single, near-fledgling chick did not dispatch the prey, the adult male stepping in to to complete the task.

---

[6] It would be assumed that the shape of the Peregrine's (indeed all raptors') bill was an evolutionary adaptation to diet. But work by Bright *et al.* (2016) has shown that about 80% of the shape variation in raptor bills arises primarily from the integration of bill and skull, *i.e.* changes to bill shape result in predictable changes in the skull and vice versa. The major adaptive difference in terms of diet is size, though the falcon's tomial tooth is, of course, an adaptive feature.

The Peregrine's bill, and tomial tooth, are beautifully illustrated in this image. *Anne Hay*.

In a study of the use of live prey Spofford and Amadon (1993 and references therein) investigated known instances of live prey delivery and other occasions when live prey has been seen in nests, falcons and other raptors. They noted instances where prey had blundered into raptor nests and also where prey had been brought in alive but had escaped before being killed and fed to the chicks. In most instances provision of live prey is avoided because, Spofford and Amadon suggest, with elevated nests any struggle between nestlings and prey might result in the nestlings falling to their deaths. There is also no evidence that raptor chicks 'play' with prey in the manner of, for instance, domestic cats, so there is no specific advantage in the practice. Spofford and Amadon also report instances where raptors had adopted chicks which they had brought to their nests and not killed, mentioning several instances in which Bald Eagles (*Haliaeetus leucocephalus*) had adopted Red-tailed Hawks (*Buteo jamaicensis*). In another instance a pair of American Kestrels whose brood had apparently been predated, ousted adult (Eurasian) Starlings (*Sturnus vulgaris*) from a nest box and reared the young Starlings on a diet of mice. Spofford and Amadon conclude that only Peregrines are definitely known to use the dropping of live prey as a teaching method: dropping in this case refers not to dropping into the nest, but dropping from flight of captured, possibly injured, prey so that recently fledged chicks can catch the prey in mid-air. This behaviour has been observed on multiple occasions by one of the authors (SW) at a long-term English (Gloucestershire) study site and is also referenced by Treleaven (1998). The behaviour has also been seen in New Zealand Falcons and is suspected (anecdotally) in other falcons.

The talons and underfoot structure of an *F.p. anatum* photographed in Alpine, New Jersey. *John Spreitzer.*

Falcons have hooked claws (talons) on their toes which are used to seize prey in an unremitting grip which may actually be fatal (causing death as in hawks), even if the main purpose is to ensure, and maintain, capture. The toes also have a ratchet-like tendon which means that once prey is gripped no muscular effort is required in maintaining the grip, a useful attribute when prey must be carried away to a nest or secure place for consumption.

The talons may also be used to deliver a blow to prey during a stoop: if the blow causes the talons (particularly the rear talon or hallux) to rake across the prey, this may cause severe damage, perhaps even death. Prey stunned by the blow, or injured by the talons, will usually fall towards the ground having lost flight control. After a sharp turn the falcon will reach the falling prey and deliver a final fatal bite. Further details on prey strikes is given in Chapter 4 *Hunting Strategies*.

Falcons share other design details common to all birds – the fusion and elimination of some bones, the hollowing of others to reduce weight (but with the use of strut-reinforcing to increase strength), and a respiratory system which maximises oxygen uptake to aid flight.

While the characteristics noted above are directly observable, the internal structures of falcons have also evolved. In Chapter 3 Stooping modifications to the skeleton of Peregrines will be investigated. But other studies have noted that as well as skeletal changes, the internal organs of Peregrines have also been modified as a result of the Peregrine's predatory lifestyle. Barton and Houston (1996) compared the weight of the organs of several raptors and owls, their work including a very interesting comparison between these weights for a nominate male Peregrine and a male Common Buzzard (*Buteo buteo*) which are roughly comparable in size. The weights are set down in Table 1.1.

Species	Mass	Liver	Heart	Kidneys	Stomach
Peregrine	760	2.76±0.43 (0.36%)	1.81±0.05 (0.24%)	0.86±0.20 (0.11%)	1.24 (0.16%)
Common Buzzard	850	2.15±0.47 (0.25%)	0.86±0.25 (0.10%)	0.92±0.05 (0.11%)	1.48±0.41 (0.17%)

Table 1.1 Comparison of the weight (in grams, including ± standard deviation across number of samples, and as percentage of mean body weight) of certain internal organs of Peregrines (*F.p. peregrinus*) and Common Buzzards. Data from Barton and Houston (1994).

The marked difference is in the size of the heart. As chasers, Peregrines need a larger heart to fuel the muscles that allow prey pursuit. In their study, Barton and Houston found that the Peregrine's heart was significantly heavier than not only Buzzards and Red Kites (*Milvus milvus* – soaring feeders), but of both Eurasian Sparrowhawks (*Accipiter nisus*) and Northern Goshawks (*Accipiter gentilis*) which are also pursuit hunters. The heart of the Common Kestrel, which primarily hunts mammals by flight-hunting (the approved technical term for 'hovering') was significantly smaller, but that of the Merlin, a much smaller falcon was actually larger (1.2% .v. 0.24%) *i.e* five times larger as a percentage of body weight. The difference is due to the usual hunting

Continuing 'evolution' means that Peregrines are becoming an urban species. Nominate pre-fledgling hatched and reared on a church roof in England.

technique: Merlins are persistent, open-country pursuit hunters, a life-style which requires a large energy output, often over an extended period, and, therefore, a larger heart.

In previous work, Barton and Houston (1994) showed that other soft internal organs of falcons had also evolved to cater for a predatory life. Barton and Houston analysed the bodies of 583 raptors, including all four British breeding falcons, together with other species ranging in size from the Eurasian Sparrowhawk to the Golden Eagle (*Aquila chrysaetos*). They also looked at several owl species. Barton and Houston defined species as 'attackers' and 'searchers', analogous to the 'hunters' and 'soarers' mentioned above.

When comparing morphological parameters of species that differ in body mass, it is often not possible to simply correlate the parameter with body mass, a scaling factor being needed. Such scaling factors require data accumulated between species: the more data available the less approximate the scaling factor is. For their comparisons, Barton and Houston measured aspects of the birds' skeleton such as keel length (measured from the base of the sternum to its anterior edge) and diagonal (the distance between the base of the sternum and the distal point of the coracoid) and defined variables such as:

$$\text{Body} = \sqrt{(\text{keel} \times \text{diagonal})} \text{ mm}$$

$$\text{Linearised wing loading} = ((\text{body mass})^{1/3}/(\text{wing area})^{1/2}) \text{ g/cm}^2$$

to allow comparison between species.

The Falcons

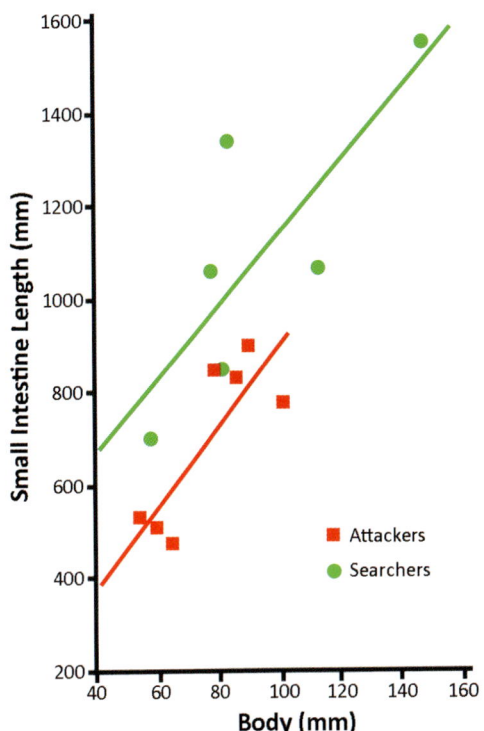

Figure 1.3 Variation of digestive tract with body size (as defined in the text). In each case the points are means across species with male and female data combined. Redrawn from Barton and Houston (1994).

Barton and Houston also measured the length of the small intestine of the studied birds, then predicted the length according to their scaling factors to allow a 'residual length', the difference between the actual and the calculated lengths, to be determined.

Barton and Houston note that attackers feed predominantly on avian prey, the capture of which depends on agility, acceleration and speed, while searchers hunt slow-moving, predominantly mammalian (and stationary, *i.e* carrion) prey. They note that attackers need to reduce the weight of all body parts except musculature and so would be expected to have shorter, lighter digestive tracts which process prey quickly, than searchers, whose hunting technique can afford the extra weight and volume of larger, slower processing tracts which gain maximum energy from prey. They note that the hunting success rate of attackers is significantly lower than that of searchers as would be expected if the digestive tracts differed in length.

The results of the comparison are shown in Figs. 1.3 and 1.4. The comparisons for owl species have been omitted from Fig. 1.3, but are set down in Fig. 1.4 (overleaf) where the inclusion is instructive.

As predicted, Fig. 1.3 indicates that the length of the key section of the digestive tract is consistent with the theory of attackers having a smaller tract.

The Peregrine Falcon

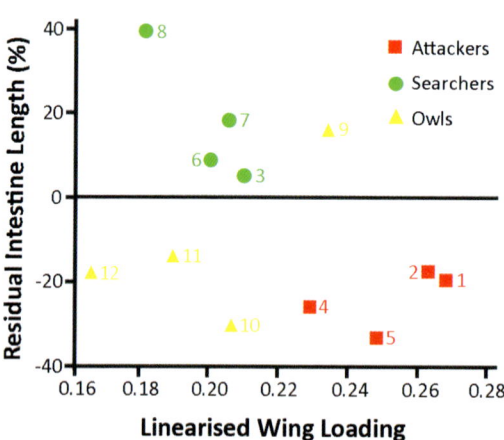

Figure 1.4 Variation of residual intestine length with linearised wing loading (as explained in the text). The studied owls have been added for completeness. The numbered species are: 1 Peregrine; 2 Merlin; 3 Kestrel; 4 Sparrowhawk; 5 Goshawk (*Accipiter gentilis*); 6 Hen Harrier (*Circus cyaneus*); 7 Buzzard (*Buteo buteo*); 8 Red Kite (*Milvus milvus*); 9 Tawny Owl (*Strix aluco*); 10 Barn Owl; 11 Long-eared Owl (*Asio otus*); 12 Short-eared Owl (*Asio flammeus*). Redrawn from Barton and Houston (1994).

Fig. 1.4 is equally instructive. The attackers – Peregrine, Merlin, Eurasian Sparrowhawk and Northern Goshawk – having intestines which are, statistically, significantly smaller than those of searchers (which include the perch-hunting or 'hovering' Common Kestrel which is primarily a mammal feeder). The owls appear, at first glance, to offer a contradiction, but as Barton and Houston point out, the hunting technique of the three attacker species differs, those owls having very low wing-loading and using a quartering flight and audio signals to locate mammalian prey, while the hunting technique of the Tawny Owl (*Strix aluco*) is closer to that of the Kestrel. All four studied owls had lighter muscle mass because they do not need fast acceleration and high-speed flight to hunt.

One final interesting aspect of the Barton and Houston study was addressing why attackers actually chose an attack strategy which is both energy-demanding and high risk, as the prey might fight back. Given the relatively high chance of finding carrion, why not concentrate on it? They suggest that carrion deteriorates with time and so offers less energy content; that it might be contaminated with bacteria which make it potentially unpalatable, but also difficult to digest or toxic; and that it is cold, much lower than the hunter's body temperature, and so requires energy input to warm as well as consume. Peregrines (and other attackers) have evolved not only to seek and obtain highly nutritious prey and to have weaponry to minimise the risk of conflict, but to have evolved digestive tracts to aid the hunt. The issue of the efficiency of the digestive tract of avian-hunting falcons is mentioned again in Chapter 5 where the food consumption and energy balance of the Peregrine is considered.

Further adaptations of the Peregrine will be dealt with in Chapter 4 *Hunting Strategies* where their influences on hunting techniques are considered.

Nominate Peregrine, England. *Paul Hayes*.

2 THE PEREGRINE FALCON

As noted in Chapter 1, Tom Cade wondered whether the association of humans and falcons began when falcons first evolved in Africa and were watched in awe by early humans. It is certainly the case that from the start of recorded history there are stories of gods in the form of raptors (particularly eagles and falcons). In Egypt the solar god Horus had the head of a falcon (Fig. 2.1) and falcons were often captured, mummified and placed in sacred sites. The practice of mummification of animals, not just falcons, in ancient Egypt was astonishing in its scale. Cooper (2020) notes that hundreds of thousands of animal mummies have been found at sites at Saqqara and Abydos. Initially the mummies were considered to be of little cultural value – Copper notes that in 1890 over 20 tonnes of cat mummies were shipped to England and sold as fertiliser. Though animal mummification was practised from about 3000BCE, it peaked in the centuries prior to and after the beginning of the Christian era (*i.e* 0AD), the reasons apparently varying from the mundane – the preservation of pets and providing food for humans in the afterlife – to the spiritual, as votive offerings to the gods. The mummification was big business – over 8 million dogs and the same number of ibises – have been found, with similar figures being likely for other species. Ibises and falcons were favoured birds because they were the familiars of the gods Thoth (god of wisdom)

The Peregrine Falcon

Figure 2.1 A watercolour copy by Howard Carter of the god Horus from the mortuary temple of Queen Hatshepsut at Deir-el-Bahari, Egypt. Carter (1874-1939) is famous as the discoverer of the tomb of Tutankhamun. Reproduced with permission of the Griffith Institute, University of Oxford.

and Horus respectively, but the size of the trade meant that occasionally corners were cut in the production process. Mummification meant that the purchasing individual was presented with a linen-wrapped bundle whose shape was identifiable, but whose contents might contain a complete bird, or an assortment of bones, bulked out with sticks or reeds (Fig. 2.2 overleaf).

The Egyptians were not alone in deifying falcons or other raptors, later civilisations also had raptorial winged or headed gods. The Greek legend of Icarus reflects man's envy of birds' ability to fly and so connect with a different world, a link with more modern examples such as the Thunderbird of native Americans to the familiars of Siberian shamans (see photograph overleaf). An echo of that link between humans and raptors, and falcons in particular, can still be seen in falconry where the preferred bird is the Peregrine.

While all sources agree that the name Peregrine derives from the Latin peregrinus, wanderer, the first use of the name is disputed. One possible source was Albertus Magnus (Albert the Great), also known as Albert of Cologne. Born in Bavaria sometime in the 1190s (no exact date is known, but he was over 80 when he died in 1280), Albert was educated at Padua University in Italy, became a Dominican Friar and returned to Cologne as a teacher. He also continued his studies, becoming the first German Master of Theology, a position which allowed him to travel to France where he taught at the University of Paris, one of his pupils being Thomas Aquinas. Albert's fame as a scholar and teacher resulted in his being made a bishop, but he resigned

The Peregrine Falcon

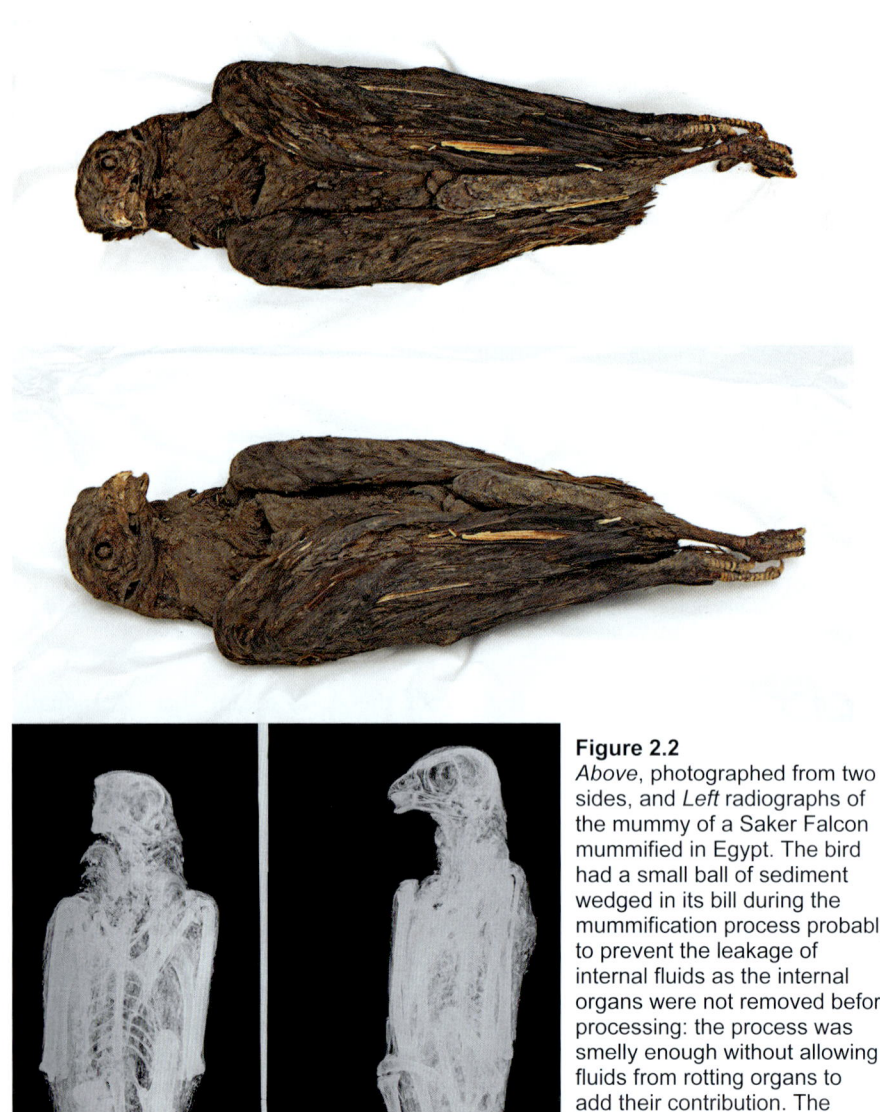

Figure 2.2
Above, photographed from two sides, and *Left* radiographs of the mummy of a Saker Falcon mummified in Egypt. The bird had a small ball of sediment wedged in its bill during the mummification process probably to prevent the leakage of internal fluids as the internal organs were not removed before processing: the process was smelly enough without allowing fluids from rotting organs to add their contribution. The bird also had a third femur between its legs, so in this case the purchaser received more than had been paid for. With permission of the Natural History Museum, Tring, England.

The Peregrine Falcon

Shaman 'familiar' figure from the Koryak people of Kamchatka, Russia. The figure was formed around a wooden tube delicately created from a series of narrow laths of wood glued together. The clothing is reindeer hide and hair (the Koryaks are Reindeeer herders), the falcon's bill being the claw of a Kamchatka Brown Bear. The shaman drum was created from a Reindeer bone. The figure was finished with modern beedwork.

The falcon local to the Koryaks was *F.p. japonensis*.

the position to return to monasticism. His studies and writings covered all sciences from astronomy to zoology, and also included law, music and morality, as well as alchemy, which at the time (and for many years after his death) was considered a legitimate science. Albert produced several books, one of which, on falcons and falconry, was later incorporated into De Animalibus, a vast book on all aspects of the natural world. In his writings he refers to *Falco peregrinus*, the name being given to a large falcon, favoured by falconers, but a species whose eyries were often on the ledges of vertical cliffs, too difficult and dangerous to climb. Falconers would therefore bide their time, waiting for a fledgling to 'wander' from the nest, or for an adult or sub-adult 'wandering' back to their natal area in later years. The alternative source for the name is *De Arte Venandi cum Avibus* (The Art of Hunting with Birds) by Frederick II (1194–1250), who became King of Germany at the age of two, lost that title after one year, but was crowned King of Sicily the following year, aged four. Frederick became King of Germany again in 1212 and Holy Roman Emperor in 1220, though his position did not prevent a series of conflicts with the

Figure 2.3 *left*

An image from *De Arte Venandi cum Avibus* which shows a bird of prey (a Peregrine?) hitching a ride on the stern of a ship. It has been suggested that *peregrinus*, wanderer, might derive from the occasional sightings of the birds out at sea, using ships as rest stops during cross-sea wanderings.

Figure 2.4 *below*

Another illustration from *De Arte Venandi cum Avibus*, this one showing courtiers preparing for a day's hunting.

Pope, one of which led him to be excommunicated. Frederick was a gifted linguist, a lover of the arts, a rationalist at a time when superstition was the norm, and a scientist, though some of his experiments on people were both deeply unpleasant and cruel. He was also a gifted falconer who experimented with breeding to better understand the birds. His scholarly book mentions the Peregrine, but experts are undecided as to whether Frederick took his lead from Albertus or vice versa, or whether both were using a name common at

The Peregrine Falcon

the time. One illustration in Frederick's book (Fig. 2.3) also casts doubt on the suggestion that it was the trapping of falcons which was the basis of its name, while another shows a gathering of falconers preparing for a day's sport (Fig. 2.4).

But whichever source of the name peregrinus is correct, the name certainly pre-dates Linnaeus' Systema Naturæ by many centuries. Carl Linnaeus (1707-1778) was a Swedish botanist/zoologist famous for having formalised the binomial nomenclature system for animals, birds *etc*. The first edition of Systema Naturæ was published in 1735 with subsequent editions appearing through to 1761. But although Linnaeus included several falcons that bred or were seen in Sweden – Gyrfalcon, Hobby, Kestrel and Merlin – he did not include *Falco peregrinus*. That name was first seen in the *Ornithologia Britannica*, published in 1771 by the English ornithologist Marmaduke Tunstall (1743-1790). The Peregrine described by Tunstall in 1771 was British: the falcon is still held at the Natural History Museum in Tring, Hertfordshire, and is the type-specimen of the nominate (*F. p. peregrinus*) - photographs below and overleaf. The nominate is the most widespread of the numerous

Two views of Tunstall's Peregrine. The upper image has a glass eye as when the bird was exhibited in the Natural History Museum, London that side faced the public. That prolonged exhibition also explains the fading of the feathers. The lower image has no eye but less faded feathers. With permission of the Natural History Museum, Tring, England.

potential sub-species which have been mentioned briefly in Chapter 1, and which will be discussed in greater detail later in this chapter.

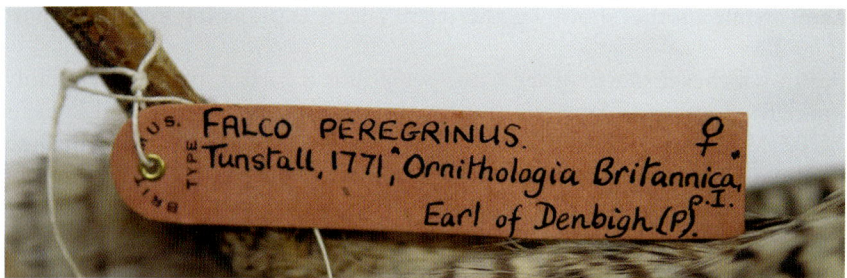

The type-specimen label on Tunstall's Peregrine. With permission of the Natural History Museum, Tring, England.

As with the Merlin, the collective noun for Peregrines, is 'a cadge', a name deriving from falconry, a cadge being an open square frame set on four legs on which a group of falcons can be carried. Since falconers in general use Peregrines, the term came to be applied to a group of this species. In earlier times, visiting falconers would often request of their host that their falcons be allowed to use the host's cadge to be carried into the field, hence the term 'cadge a lift' which is still in common use in English today. In that brief note on the collective name for the species, Peregrine has been used rather than Peregrine Falcon. While the latter is the formal name for the species, the former is the common vernacular name and will be used throughout the rest of this book. However, having mentioned 'Falcon' it is worth noting that the derivation of the word is not straightforward. Majority opinion favours an origin in the Latin *falx* (the genitive of *falcis*), a curved blade or sickle, probably from the shape of the bird's talons or bill, perhaps even the shape of the wing. That fits with the idea that falconry as a sport was known to the Romans who acquired the skills from the east. However, there is almost no evidence that the Romans practiced falconry, and such evidence as currently exists suggests falconry did not appear in Europe prior to the dissolution of the Roman Empire in 476AD. Proponents of Roman falconry most often quote the 1st century AD poet Marcus Valerius Martialis – commonly known as Martial – who wrote:

> *He used to prey upon birds, now he is the servant of the bird-catcher, and deceives birds, repining that they are not caught for himself.* [1]

During the Roman era, writers documented every aspect of life and there are apparently no other references to falconry, so while some see Martial's quote as an indication of Roman falconry, most authorities believe the poet was referring to the opportunistic use of raptors to scare birds into nets, a

---

[1] From Martial's *Epigrams* Book 14 Number 216. The translation is from Bohn's Classical Library of 1897, and includes the now rarely heard *repining*, meaning *upset* or *miserable*.

Nominate Peregrine, Isle of Man, UK. *Peter Christian.*

Two very wet *F.p. anatum* adults, San Pedro, Los Angeles, USA. *Gabriel Gruia.*

method known to have been practiced by the Egyptians and Greeks. As an example, Aristotle, writing in the 4th century BCE, notes the inhabitants of Thrace working with hawks in marshland. Birds would be flushed, hawks chasing them so that the birds panicked and came back to earth, allowing the hunters to easily gather them up. As a reward the hunters would throw some captured birds into the air for the hawks to take. Such commensal behaviour between human and animal is rare, but not unheard of, and might well have been the basis of Aristotle's observation. The behaviour may therefore not have required either ownership or training of a raptor, and so does not fit the true definition of falconry.

The origins of falconry are disputed, with no firm evidence prior to the third century BCE, but general opinion is that it may have evolved much earlier, perhaps as early as the second millennium BCE and probably in central Asia where the immense flatlands of the Asian steppes were well-suited to both watching the action and for ease of retrieval of both quarry and raptor, each of which offered better sport than the hills and woodlands of Europe. The sport

Figure 2.5 Drawings of images carved at Hittite sites in Anatolia. The *upper* image includes a falconer goddess being presented with a libation, while another servant is part-kneeling with a gift. The seated goddess has released a falcon, its jesses hanging below the bird. The remainder of the relief represents a stag hunt. The drawing is from a bronze seal, probably early 19th century BCE (Beran site). The *lower left* image is a falconer with a *lituus*. The word *lituus* derives from the Roman, meaning the stick carried by divinators, but the Hittite word was *kalmus*. It is believed the *kalmus* was thrown at prey (*e.g.* a hare), or to flush game for the falcon to chase. The shape suggests a primitive boomerang. The *lower right* image shows a falconer riding on a stag. In Hittite religion the stag was the steed of the Protector God. The two lower drawings are from the KÜltepe site, also known as Kamesh, in central Turkey. The images date from about the 19th century BCE. Images from Canby (2002) and Serjeantson (2009) with thanks to the University of Chicago.

probably spread east to China, Korea and Japan, moving westward sometime later, although evidence from the remains of the Hittite empire of Anatolia (modern Turkey) suggests the sport may have moved west much earlier than is often suggested (Fig. 2.5).

The Hittite Empire reached its height around the 13th century BCE, it declined significantly thereafter and had largely disappeared by the 8th

The Peregrine Falcon

Figure 2.6 A mosaic from the Villa of the Falconer, at Argos, Greek Peloponnese. The mosaic illustrates a falconer with a bird on his arm waiting for a companion to fix his shoe before the pair leave with their dogs.

century BCE. However, falconry was probably practiced in Mesopotamia (an area now covered by south-east Turkey, eastern Syria and Iraq) before the rise of the Greek and Roman civilisations. Falconry does not seem to have reached Europe earlier than 500AD, mosaics in the 'Villa of the Falconer' at Argos in Greece, which clearly show a form of falconry, dating from around that time (Fig. 2.6). While the mosaics certainly show a 'falconer' with his bird it is not clear exactly what sport was being practised. It is known that in medieval times the 'daring' of small birds involved one person having a falcon on his fist while others had nets or long poles with small nooses at the end. The group would go out into the country to find a flock of feeding small birds, preferably larks. The person with the falcon would then either fly it, or make it flap its wings. Seeing the falcon above them, or apparently getting ready to fly at them, the birds would freeze on the ground making it easy for those with nets or pole nooses to catch them.

The later arrival of any form of falconry to Europe, and also of a possibly more sophisticated form of the sport, is also supported by considerable archaeological evidence from Germany and Scandinavia dating from the late 6th century AD (Wallis, 2017). Wallis also notes that a recent study of excavation finds at Sutton Hoo, Suffolk, England presents persuasive evidence

Mouse, a male (tiercel) Peregrine and owner/falconer Oliver Tomlinson. The equipment attached to the bird is almost identical to that shown in the illustrations on p49 which date from the late 19th century. It is also true that the equipment has not changed radically since medieval times, although modern electronics means it is now easier for the falconer to stay in touch with his bird. The inset photograph shows RS's IMU clipped into *Mouse*'s harness. Data collected on flights with *Mouse* and other falconry birds is set down in Chapter 3.

that copper alloy bells discovered there may have been used in falconry, perhaps being attached to a neck loop rather than to the foot of a bird (Fig. 2.7).

 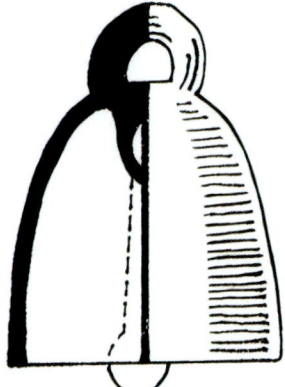

Figure 2.7 A drawing of the copper alloy bell (INV 212) from Sutton Hoo. The bell was *c*.28mm high and *c*.20mm diameter and is described in Bruce-Mitford (1975). The suggestion of use in falconry was made by Carver (1998). The drawing here was made by Robert Wallis from the Bruce-Mitford (1975) illustration. Reproduced by courtesy of Robert Wallis.

The Peregrine Falcon

Sutton Hoo has also provided evidence in the form of a magnificently decorated purse lid dated to the 7th century AD which illustrates a raptor attacking prey (probably a duck). Further evidence was unearthed at a site in Essex, England, to the south of Sutton Hoo. Dated stylistically to the late 6th/early 7th centuries this superb gold ring is decorated with both pagan and Christian imagery (see adjacent photographs and Fig. 2.8). While the motifs of the purse and ring do not of themselves confirm the existence of falconry in Anglo-Saxon England, other finds, for instance of raptorial remains, together with the remains of prey in quantities that suggest hunting rather than the mere presence of locally available prey, at settlement sites imply that falconry was practised in Britain in the late 6th/early 7th century.

A purse lid excavated at Sutton Hoo, Suffolk, England. The exquisite lid is of gold, cloisonné garnet and millefiori glasswork. At the lower corners a moustachioed man is attacked by large animals (wolves?) while at the lower centre mirrored birds of prey attack ducks. The work is cloisonné garnet with a blue/black millefiori chequer design on the hip of the raptor, and a blue ring around its eye. Millefiori is missing from parts of the duck on both mounts. Very small garnet cells are used in a chevron design to create a feather impression on the wings and tails of the raptors. The white background is modern material replacing the original purse material (which is speculated to have been wood, leather or perhaps bone or walrus ivory) which has not survived. For further details see Bruce-Mitford (1975). Courtesy of the British Museum.

An early Saxon gold ring excavated in north-west Essex, England. The 'crouched bird of prey' motif (Style II ornament) is typical of Anglo-Saxon art of the period (late 6th/early 7th centuries) and while it might be contended that it would be standard for a war-like society, falconry being a high-status sport would also make it a desirable decoration for such high-quality/high-class items.
© Saffron Walden Museum, Essex.

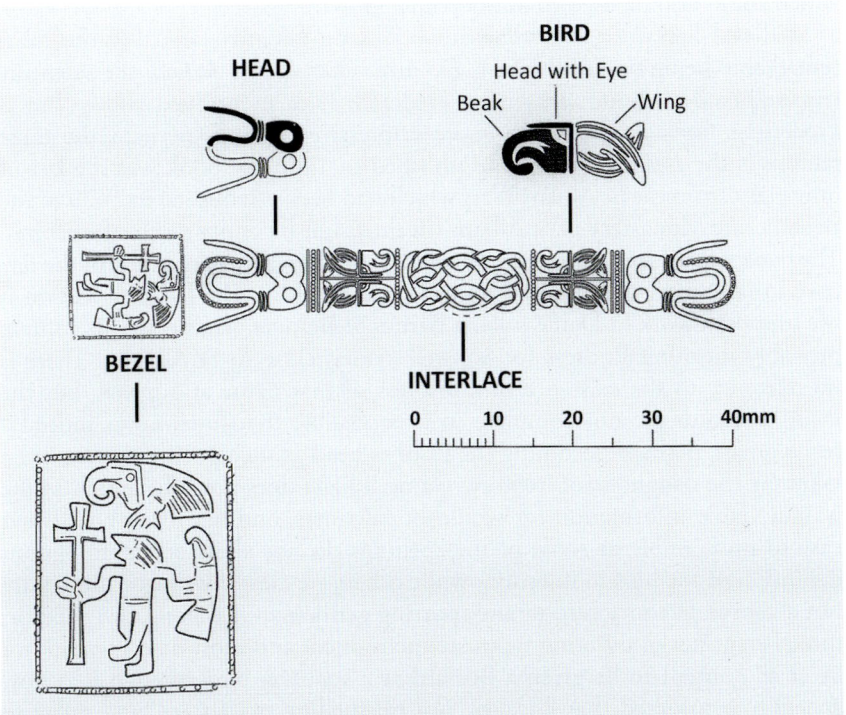

Figure 2.8 Drawings of the decoration of the NW Essex ring. Courtesy of Chris Fern.

But if the Saxons brought the sport to Britain, it was the Norman conquest which brought the language of falconry. The old French *faucon* was the word for a diurnal raptor and may well have derived from the Latin: raptor itself derives from the Latin *rapere*, to snatch or carry off. As French was the language of the English nobility after 1066, *faucon* became the old English form for the birds until about the 15th century when scholarly Englanders added an 'l' to conform to the assumed Latin origin. Other terms for falcons which also derive from falconry, and which remain in (relatively) common use, include tiercel (occasionally tercel) for a male falcon. This probably derives from the Latin *tertius*, meaning one-third, through the old French *terçuel* (occasionally *tiercelet*) because the male birds of the larger falcon species are, in general, about one-third lighter than the females. Another term is eyas (occasionally eyass) for a nest-bound chick. This is assumed to derive from the Latin *nidus*, nest, through the old French *niais*, with *un niais*, being modified over time. Chicks are also occasionally referred to as pulli, though this is often used to refer to the chicks of any bird rather than exclusively those of a falcon or raptor. Here the derivation is more certain, the Latin *pullus* meaning a young animal, though as it was also the Latin for chicken (the base of the French *poulet*, which became Middle English *polet* and hence pullet for a hen), there is room for negotiation, particularly as *pullulare* was the Latin verb 'to sprout' which could also be used to imply young growth.

Medieval literature abounds with books on falconry, arguably the most remarkable being one of the first, *De Arte Venandi cum Avibus*, the scientific treatise/practical guide written by Frederick II as mentioned above. But if Frederick's book is one of the more impressive treatises, perhaps the most famous is the *Book of St Albans*. Published in 1486 the book was the last of only eight from the St Albans Press which had been established in 1479 in the Benedictine Monastery of St Albans (in the English county of Hertfordshire). The book was a compilation of information of interest to gentlemen of the day covering the topics of falconry, hunting and heraldry. The falconry section is the supposed work of Dame Juliana Barnes or Berners, 'Dame' in this context probably meaning Prioress (of Sopwell Priory, close to St Albans). There is no reference to the existence of a prioress of that name at Sopwell, but the Priory's records are not complete so it is possible that Juliana was indeed a real woman, perhaps becoming the Priory's head as such positions were often taken by the daughters of nobility. Dame Juliana does not offer much in the way of viable information for would-be falconers, and her list is not always easy to unscramble as some of the spellings do not allow an unambiguous definition of the bird in question, while others are clearly wrong. Vultures are not a chosen falconry bird of any sporting gentleman, let alone an Emperor; three Peregrines of differing types are mentioned, and most dukes would not be at all content to be given a Bustard as a sporting bird (though it is now generally considered that 'bustard' is a misspelling of 'bastard' and refers to a hybrid falcon). Dame Juliana's list must also surely not be considered as an

The Peregrine Falcon

indication of which noble had, or wished to have, which bird. It was, rather, a list of avian raptor hierarchy set against the hierarchy of English nobility, a list which would likely have been seen as crucial in the class-steeped society of the time. And so we read that a King must have a Gyrfalcon, for a Prince it must be a Peregrine and a Lady must have a Merlin. For a Young Man it must be a Hobby.

Usually the list is extended to include 'a Kestrel for a Knave', but that association is not, in fact, from the *Book of St Albans*. Rather, it derives from the *Harleian Manuscripts* held at the British Museum. Collected by Robert and Edward Harley, the first two Earls of Oxford and Mortimer, in the late 17th/early 18th centuries, the relevant manuscript dates from the first half of the 15th century and replicates the list in the *Book of St Albans* (or was it in fact the original source?), but adds the famous line about the 'Kestrel for a Knave'. The latter became the title of a well-known book by Barry Hines, published in 1968, which itself became the subject of a film, *Kes*, directed by Ken Loach and released in 1969.

Falconry as a sport would not have arrived fully formed, but by the medieval period the furniture mounted on the bird (jesses, hood *etc.*) and the training and hunting techniques were largely defined. Modern materials and electronics have modified equipment, but the sport remains essentially as it has been for many centuries. The medieval period was in many ways when the sport was at its height in terms of its being a sport of kings and a

Images of falconry equipment from *Die Raubvögel Deutschlands unde des angrenzenden Mittel-Europas*. For further details of the book, see p4 of this book.

Gyrfalcon on the top of the *Falkehus* (Falcon House), Reykjavik, Iceland. During the medieval period gifts of falconry birds cemented friendships and alliances between the royal houses of Europe. Most prized were white Gyrfalcons. The resident population of Gyrs on Iceland were paler than those of Norway, but not white. However, regularly each winter white birds from Greenland would arrive on migration. The Danish Crown (which controlled modern day Norway and Iceland) employed trappers to capture these white falcons which were then carefully stored awaiting ships to Denmark. The Falcon House was superior to that of any of the resident human population. The carved Gyr was mounted on the new house when the original was demolished.

reflection of the behaviour of the highest rank of society, though it might be argued that modern Arabian interest and hard cash has provided a revival, though one limited in geographic scope. In medieval times the royal families of Europe vied with each other for the finest birds and falcons were presents between kings and emperors sent to foster good relationships. Nothing oiled the wheels of diplomacy as well as a white Gyrfalcon, these being treasured as possessions and gifts at least from the 13th century.

Falconry was also a reflection of the culture of the period, or at least of the highest echelons of society. The culture was based on chivalry and courtly behaviour and many writers have drawn a comparison between those attributes and falconry. The attitude of the falcon to its handler, the handler offering unconditional love, the falcon responding with nothing of the sort, combining minimal signs of interest (affection would be much too strong a word) with maximum periods of disdain. The responses echoed the approved chivalric behaviour of a knight to his lady, the knight offering undying love, the lady offering disinterest (in public at least). As a consequence, in many illustrations of the period a couple would often be shown with a falcon, the bird as a symbol of their togetherness (Fig. 2.9).

Plumage
The nominate Peregrine is described here, the variations on sub-species being detailed later in the Chapter.

The Peregrine Falcon

Figure 2.9 *Herr Wernher von Teufen* (Man and woman with hawk) from the Manesse Codex (also known as the *Heidelberg Book of Songs*) created in the first half of the 14th century. The Codex is an illustrated collection compiled by Rüdiger Manesse of Zurich. The Codex comprises 135 items, mainly Minnesang (from the German, best translated as 'love song') a traditional lyrical poem or song. Each *minnesang* was illustrated by a painted miniature several of which include falcons. The image reflects the culture of the time, as noted in the text.

Adult male

The upperparts are blue-grey, darker, occasionally black, on the crown and paler on the tail, faintly barred darker apart from the head, which has a solid colour. The cheeks are pale cream or white, apart from dark ear coverts and a broad malar stripe. While most falcons exhibit such a stripe, the Peregrine's 'moustache' is one of its defining characteristics. The throat is white. The underparts are cream or pale cream-buff with dark spotting, heavier towards the vent, which gives the appearance of barring. The underwing is greyer, again with dark spotting which suggests barring. The undertail is barred dark brown, the broadest bar being sub-terminal. The terminal bar is white. The iris is dark brown, the eye ring yellow, as is the cere. The bill base is also yellow, the remainder being slate blue, darker (even black) at the tip. The tarsi and feet are yellow, the talons black.

Adult female

As male, but often darker above and more buff below, the markings heavier. Differences in colour and marking overlap so that these slight plumage differences are not diagnostic in the field. However, when male and female are together the size difference is extremely obvious. In addition, the male often appears brighter, the feathering of the throat and ventral area being whiter.

Juvenile
The head is dark brown, the crown feathers with cream edges, the forehead entirely cream, as are cheeks and throat. Remaining upperparts are dark brown with chestnut feather edges. The underparts are cream or cream/buff with dark brown arrow-like markings, these larger and denser towards the vent. The undertail is barred with brown, again the broadest bar being subterminal. The terminal bar is white/cream. The underwing is cream/buff with heavy dark barring. The iris is dark brown, the eye ring and cere pale blue or pale blue-green. The bill is blue-grey, darker at the tip. The tarsi and toes are blue, blue-green, or a pale, washed-out yellow.

Nestling
The first down is white/pale cream. The second down (at 7-10 days) is pale grey. Feathers begin to emerge at about 10 days. The iris is black, the cere and feet pale grey. Nestlings are fully feathered at about 38 days, by which time down has virtually disappeared. The birds usually fledge between 39-46 days, males typically maturing slightly quicker.

Dimensions (*nominate*)
Overall length (bill tip to tail tip) 36–50cm – females are significantly larger than males.
Wing: male 300.8 ± 7.8mm; female 345.6 ± 10.4mm.
Tail: male 136.7 ± 3.9mm; female 164.4 ± 10.5mm.
Bill with cere: male 24.3 ± 1.0mm; female 27.5 ± 0.9mm.
Tarsus: male 47.3 ± 1.6mm; female 52.2 ± 3.1mm.
Toe: male 47.2 ± 1.3mm; female 52.2 ± 2.0mm.
Talon: male 19.6 ± 0.7mm; female 22.9 ± 0.9mm.

Juvenile wing: male 309.7 ± 13.0mm; female 354.0 ± 6.0mm.
Juvenile tail: male 149.1 ± 4.0mm; female 178.1 ± 5.0mm.

Weight (*nominate*)
Females are about 50% heavier than males.
Male 650 ± 80g; female $1,100 \pm 200$g. In both cases the weight is variable throughout the year and particularly during the breeding season.

Ocelli
Negro *et al.* (2004) noted that ocelli (false eyes) had been described on American Kestrels, which have two well-defined ovoids on the back of the head, each with a central dark spot, but had not previously been reported on Eurasian Hobbies. Most Hobbies have ocelli: very prominent in some cases, usually as brown ovoids with a white centre, but they are rarely seen on the large falcons. However, they have been seen occasionally on Peregrines, and were prominent on a nominate juvenile photographed (opposite) in England.

Juvenile nominate Peregrine with clear ocelli on the back of the head

Reverse Sexual Size Dimorphism

Female Peregrines are significantly larger than males, the species exhibiting Reverse Sexual Dimorphism (RSD), an attribute Peregrines share with other falcons (and raptors). While it is sometimes maintained that RSD is unique to raptors, this is not so, the trait also being seen in the Stercoraiidae (the skuas or jaegers), Scolopacidae (sandpipers) and Sulidae (boobies and gannets). It is occasionally maintained that RSD is associated with predatory behaviour, but although the skuas are predators, neither Scolopacidae nor the Sulidae are, and the Laniidae (shrikes), which predate birds, rodents and lizards, do not exhibit RSD. However, predatory behaviour is often chosen as a starting point for considering RSD because of the effect of body mass on flight and, therefore, hunting characteristics.

Andersson and Norberg (1981) suggested that RSD in raptors resulted from three factors. Firstly, role partitioning favoured one sex being larger to guard the nest while the other hunted for the pair and their offspring. This does not, of itself explain why females are larger, but accepting that one sex would be bigger, a second factor suggests it made more sense for the female to be the nest guard. The suggested reason was that females risked injury to the eggs they were carrying during prey attacks, and their role in egg laying predisposed them to stay close to the nest. Females exhibit pre-laying lethargy, this being followed by laying, incubating and chick brooding which tie her to the nest (for *c.*15% of the year), a larger size making covering eggs and chicks easier.

Once the female's role as nest guard and offspring 'nurse' was established, a larger female represented a better option. The third factor suggests that a smaller body mass favoured more agile flight and, therefore, better hunters so that males of species which hunted avian prey would tend to be smaller. In favour of this Andersson and Norberg pointed out that RSD was more pronounced in species which specialised in avian prey. This is certainly true of the British breeding falcons, the size differential of Peregrines and Merlins being greater than that of the rodent-hunting Common Kestrel, and of the Eurasian Hobby which preys on insects for much of the year.

However, while these reasons for the evolution of RSD are compelling, debate over the issue has resulted in a considerable number of papers over the

RSD is illustrated in this photograph of male (left) and female nominate Peregrines at a cliff nest site in Scotland.

years, both before and after the work of Andersson and Norberg. No consensus has emerged, either on the reasons for RSD or on whether it emerged in ancestral falcons because males became smaller or because females became larger (though data in support of either might be impossible to obtain). More than 20 theories have been put forward, with varying levels of success in terms of both explaining the phenomenon and in achieving a measured degree of scientific acceptance. The 'most successful' theories fall into three main groups.

The first is ecologically based, the idea being that the sexes operate in specific niches, in this case prey size, to avoid competition on a shared territory. Evidence in support of this theory is contradictory. Pande and Dahanukar (2012) noted that their work on Barn Owls in India indicated that the mean mass of prey brought to the nest by males was significantly lower than that brought by females, but that males brought a greater number of prey items. Pande and Dahanukar considered that this supported an ecological basis for RSD. However, the work prompted an interesting debate which highlighted a lack of consensus. Olsen (2013), replying to the Indian researchers from Australia, pointed out that a study across all raptors showed that while males did indeed carry lower mass prey than females in many species, in others the mean mass of the sexes was the same and some males carried heavier prey than their larger female partners. Pande and Dahanukar (2013) agreed these comments were valid and called for more research into the nature of the relationship between RSD and foraging in male and female raptors.

In Britain, separate studies on Peregrines, had also come to different conclusions. In Cornwall, Treleaven (1977) found no difference in the mean size of prey delivered to the nest by the two parent birds, while Parker (1979) in Pembrokeshire and Martin (1980) in the Lake District found that the prey delivered by males was significantly smaller than the prey of females. In his study of Peregrines in southern Sweden, Lindberg (1983) noted that heavier prey – gulls, ducks and grouse – tended to be a greater proportion of the prey during the later breeding season when the female Peregrines joined the hunt to provide food for the nestlings. Lindberg found that the average weight of the males' prey was 188g, that of females being 251g. But in a more recent study, Zuberogoitia *et al.* (2013), studying the falcons in northern Spain, found no difference in the size of prey delivered to the nest. These contradictory results suggest that the hunting prowess of individual Peregrines, of either sex, is more complicated than a straightforward relationship with the bird's weight implies. Provisioning of a brood clearly invokes an imperative which is not easily rendered into well-defined categories.

That said, other studies have noted that in single chick broods the male does much of the provisioning of the youngster, while in broods of three or four the female brings a greater fraction of the prey raising another possibility – as the female is larger, she is able to carry larger prey to the nest which is clearly advantageous. This adds a further argument to the ecological theory, positing that females are larger so that they can carry larger prey to the nest. Further evidence comes from studies which note that male Peregrines consistently bring small prey items to the nest even if they are killing larger prey, for instance grouse, which they feed on. Females may also intercept prey deliveries by her mate taking the heavier prey items to the chicks, though this might only apply to the later stages of chick growth, when the chicks are capable of dismantling prey themselves, as in general only the female feeds chicks bill-to-bill. The ecological theory has much to recommend it – but it does not explain why females are larger than males rather than vice versa and has failed, to date, to muster a consensus among researchers.

The second RSD theory is behavioural, suggesting one of three possibilities. The first suggests females are larger in order to dominate their male partners, maintaining the pair bond and ensuring food deliveries. The second suggests that larger females can outcompete conspecifics in competition for males. The third suggests that smaller males can make superior (more agile) display flights and so are more attractive to potential mates.

The third RSD hypothesis is physiological – either because smaller males, being more agile, are better hunters (an extension of the third behavioural hypothesis noted above, as flight agility indicates probable worth as hunters and, therefore, providers for their mates and broods), or that larger females can produce larger eggs and/or clutches and so enhance breeding success. One aspect of the latter idea is the 'starvation hypothesis' which posits that larger females are better able to withstand food shortages during the breeding season. A physiological basis for RSD is certainly supported by the observation of Newton (1979) that in female falcons, body condition is critical for reproduction and that females are at their heaviest during the laying phase of the breeding cycle and remain heavy (although weight is lost) during incubation and the early nestling phase. Although it is in the reproductive interests of the male to feed his mate and so increase her weight, the increased size of the female would also allow her to dominate her mate, 'bullying' him into hunting. In a study of Tengmalm's Owl (*Aegolius funereus* – now more usually called the Boreal Owl) in Finland, Korpimäki (1986) found that weight dimorphism in owl pairs was positively correlated with the laying date of the first egg, *i.e.* pairs in which the weight difference was larger bred earlier than pairs where it was smaller. As early breeding tends to produce larger clutches and more fledglings this supports the starvation hypothesis. However, Korpimäki found no evidence that heavier females were better nest guards. Korpimäki also found that male wing length was correlated with laying date, suggesting that females were selecting shorter-winged males as they were likely to be more agile and, therefore, better hunters.

Further evidence for a physiological aspect to RSD came from the work of a Norwegian group. Slagsvold and Sonerud (2007) studied seven species of owl, hawk and falcon (the latter being Merlin and Peregrine) and noted that the ingestion rate of prey was variable, and that this might influence dimorphism. Ingestion rates were higher for smaller prey and for mammalian rather than avian prey. Mammalian prey was also ingested faster by raptors which were primarily mammal feeders. The researchers argue that taking larger prey increases the feeding time for chicks, which both ties up the feeding parent and heightens the risk of the chicks chilling. Males, the main providers, should therefore be selected for taking smaller prey which would favour smaller size. Females, on the other hand, would benefit from being larger to better cover the eggs and chicks, and to survive a winter shortage of smaller prey. In a later study on Common Kestrels (Sonerud *et al.*, 2013, Sonerud *et al.*, 2014a), the Norwegian team noted that video evidence suggested that the assumption that females delivered larger prey to the nest was flawed. It seemed that females intercepted the larger prey items the male was carrying, part consuming the prey and then feeding the remainder to the chicks, while the male delivered smaller, more easily managed, prey items directly to the chicks once they could handle them.

The change from the male passing food to the female and directly provisioning the chicks occurred first for insect prey, then lizards, mammals and birds, and occurred earlier for smaller mammalian prey than for larger items. The female's ability to intercept prey deliveries depended on her size (allowing her to bully the male) while the male's ability to return quickly to hunting also depended on his size, lending support to the contention that ingestion rates may have been influential in the development of RSD. The Norwegian researchers followed up this study with another (Sonerud *et al.*, 2014b) covering nine raptor species, which showed similar results. From their study they noted that the size difference in the sexes should increase as the size of prey taken increased, from insects through reptiles to mammalian and then avian prey. Because of the range of weights seen in UK falcons it is difficult to assess with certainty whether the suggestion of Sonerud *et al.* (2014b) translates to the UK's breeding falcons. It is the case that female Peregrines are much heavier than males, but for the three smaller falcons (Merlin, Eurasian Hobby and Common Kestrel) the size differential is not as pronounced: Sonerud *et al.* (2014b) would suggest that the mammal-hunting Common Kestrels show a smaller disparity than the avian hunting Merlins, which they indeed do, and a similar, or slightly larger, difference to that of the primarily insect-feeding Eurasian Hobbies. Again, that is the case, but it must be remembered that Hobbies hunt avian prey during the breeding season. However, as the Hobby breeding season lasts only about 33% of the year, the finding would again appear to be in accord with the work of Sonerud *et al.* (2014b).

Fig. 2.10 is a pictorial representation of the three RSD hypotheses. The figure is from the work of Krüger (2005) who collected data on 237 species of Accipitridae (hawks), 61 Falconidae (falcons) and 212 Strigiformes (owls), and used comparative studies of 26 variables (both physiological – body mass, wing length *etc.*; behavioural – breeding system, displays, *etc.*; and ecological – habitat, range size *etc.*) to investigate potential correlations with RSD. Krüger found different levels of correlations between the three raptor forms, but in all three the dominant correlation supported the third of the hypotheses mentioned above, *i.e.* that small males evolved to efficiently hunt small agile prey. Krüger's findings suggest that the trend for smaller males occurred before falcons began to attack larger prey, which would explain why males are relatively smaller in larger falcon species. Fig. 2.11 (overleaf) shows the variation in RSD, defined as the cube of wing length ratios of the two sexes, across the studied species.

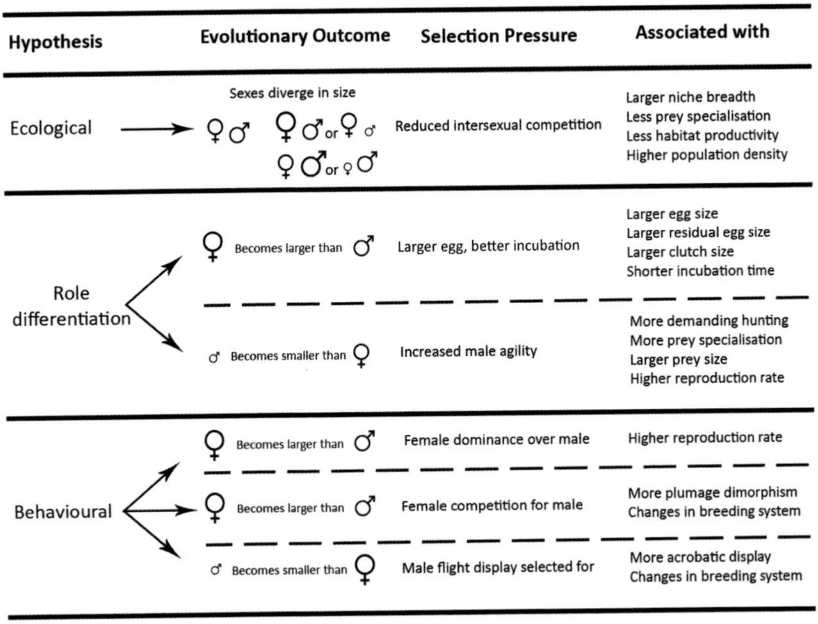

Hypothesis	Evolutionary Outcome	Selection Pressure	Associated with
Ecological	Sexes diverge in size ♀♂ ♀♂ or ♀♂ ♀♂ or ♀♂	Reduced intersexual competition	Larger niche breadth Less prey specialisation Less habitat productivity Higher population density
Role differentiation	♀ Becomes larger than ♂	Larger egg, better incubation	Larger egg size Larger residual egg size Larger clutch size Shorter incubation time
	♂ Becomes smaller than ♀	Increased male agility	More demanding hunting More prey specialisation Larger prey size Higher reproduction rate
Behavioural	♀ Becomes larger than ♂	Female dominance over male	Higher reproduction rate
	♀ Becomes larger than ♂	Female competition for male	More plumage dimorphism Changes in breeding system
	♂ Becomes smaller than ♀	Male flight display selected for	More acrobatic display Changes in breeding system

Figure 2.10 The three main hypotheses proposed to explain RSD, with their predicted outcome in terms of sex differential size, and the selective advantage of, and main correlations of, the resulting RSD. Redrawn from Krüger (2005).

Krüger's work was a step forward in that it explained why males had become smaller than females rather than for females to become larger than males. In more recent work Pérez-Camacho *et al.* (2018) collected data from 75 raptor species and scored four factors linked to RSD to create a mathematical model of their interaction. The factors were prey agility, the structural complexity of the hunting habitat, territorial behaviour and territory size. The analysis again suggested that the evolution of RSD strongly favoured a reduction in male size rather than an enlargement of females, and that as prey agility and the complexity of the hunting habitat increased (each of these factors influencing territorial behaviour and territory size) the degree of RSD would increase. A further mathematical analysis by Mills *et al.* (2019), who modelled the catch success of male and female Peregrines against a variety of prey – varying in size from the Eurasian Blue Tit (*Cyanistes caeruleus*) to the Mallard (*Anas platyrhynchos*), a differential of x100 in terms of weight – showed that in level flight males would be more successful hunters than females, though this advantage is significantly reduced for stoop attacks. However, Mills *et al.* note that, as expected, the carrying capacity of the female Peregrine is superior to that of the male. In combination, the work of Krüger (2005), Pérez-Camacho *et al.* (2018) and Mills *et al.* (2019) appeared to narrow the choices in terms of RSD theories, but within two years a further paper (Schoenjahn *et al.*, 2020) suggested that another physiological aspect could be at play, requiring females to be larger than males so they could better defend eggs and chicks against predators and so enhance breeding success.

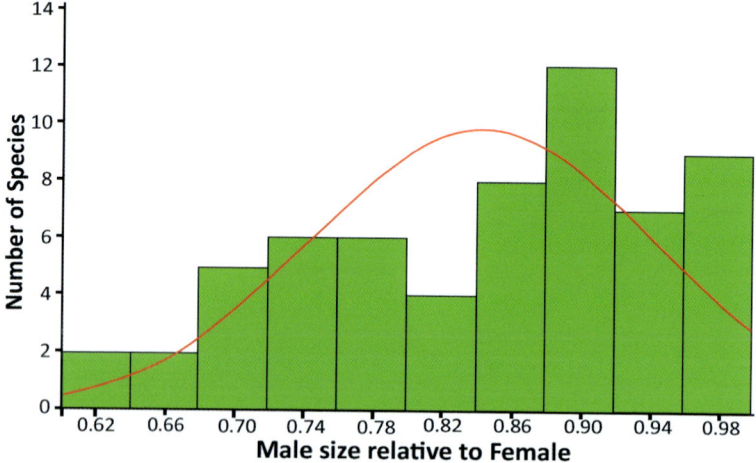

Figure 2.11 Distribution of RSD in the Falconidae. The histogram includes the 61 species recognised as forming the family at the time Krüger carried out his analysis. The lowest value (0.61) is for the Bat Falcon, the highest (0.99) for the Black Caracara (*Daptrius ater*). The red line is a normal distribution. The distribution is skewed from normal, indicating a shift to a relatively small male size across the falcons. Redrawn from Krüger (2005).

Schoenjahn *et al.* suggest their methodology overcomes one difficulty of all research on RSD – it distinguishes between the identified benefits of RSD being derived from the proposed causal process(es) or are incidental. Whether the academic community accepts the idea as definitive or just another option for an ongoing debate is itself a matter of debate.

Sub-species
The sub-species, which are considered in detail below, conform, in general, to both Gloger's Rule (that darker sub-species inhabit areas of higher relative humidity: *F. p. ernesti* of Malaysia and Indonesia is the darkest, though the African Peregrines are also dark) and Bergmann's rule (that larger sub-species occur at higher latitudes: *F. p. pealei* of the Aleutian Islands and southern Alaskan coast is arguably the largest, though the Nearctic and Palearctic Tundra Peregrines are almost as large). Gloger's Rule relates to the burden of feather-eating bacteria, the darker pigmentation of feathers making them more difficult to break down. However, darker feathers absorb radiation better than paler feathers so in areas of high temperature, but lower average humidity, paler plumage occurs more frequently (see Galván *et al.*, 2017, with a study in Spain).

Captive Peregrines are regularly cross-bred with other falcons, and this has caused some concern as falconry birds occasionally escape and are known to have bred with native, wild Peregrines. In the USA eggs have been collected where this is known to have occurred, for fear of diluting the wild stock, but given the numbers involved and the reduced fertility often associated with hybrids, this fear is probably unfounded. In a study of the effect of escapees in Spain (Rivas-Salvador *et al.*, 2021) noted that 64.3% of escaped raptors were not recovered. Of the escapees, most were Harris's Hawks (*Parabuteo unicinctus*) – 504 between 2006 and 2018 – with Peregrines and Peregrine hybrids the next most numerous, 405): the two species represented 45.6% of all escapes. In total over the 13 years 1995 raptors of 33 species and 27 hybrids escaped. The Peregrines and Peregrine hybrids totalled 1.04% of the estimated Spanish Peregrine population.

As noted in Chapter 1 it is also known that hybridisation between a wild male Peregrine and a female Prairie Falcon has occurred.

The currently agreed Peregrine sub-species are considered west-east and are listed in the distribution map on p83. In all cases, the excellent book of White *et al.* (2013a) gives more detail of plumage and size differences, and also includes data on the history and classification of the sub-species. Note that White *et al.* assume the Barbary Falcon is a Peregrine sub-species (*F. p. pelegrinoides*) and it is also included here, consistent with the view expressed in Chapter 1. The developing field of genetic analysis means that in future sub-species may be distinguished by genetics as well as by phenotype, with recent work by Price-Waldman and Stoddard (2021) suggesting that linkages between phenotype and genotype may also be a future tool for assessing sub-species.

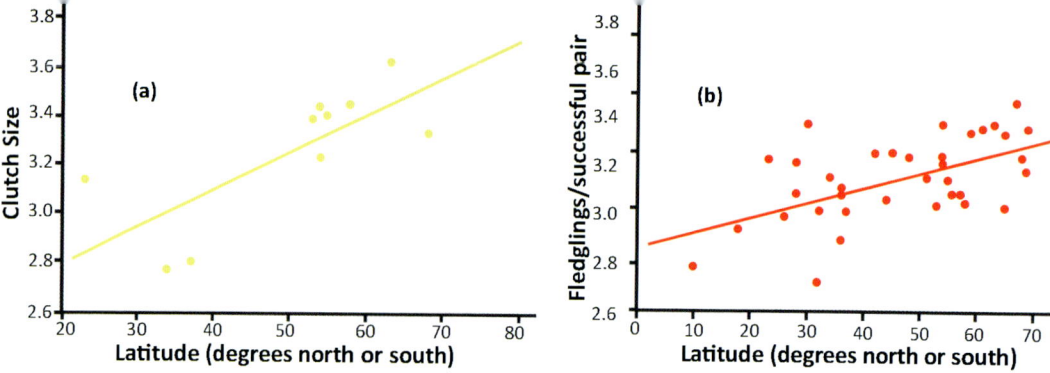

Figure 2.12 Peregrine breeding performance measured in terms of mean clutch size (a) and mean number of fledged chicks per successful Peregrine pair (b) in relation to latitude. The sub-species considered were nominate, F.p. anatum, F.p. cassini, F.p. macropus, F.p. minor, F.p. peregrinator and F.p. tundrius. In each case the line is the best correlation to the data. In the case of clutch size the significance of the line as a variation from a horizontal line is >95%. In the case of fledgling success the significance is >99.9%. Redrawn from Jenkins and Hockey (2001).

One obvious fact emerging from the study of the various Peregrine sub-species is the size of the populations of tropical falcons relative to those of northern and southern species. Jenkins and Hockey (2001) addressed this topic, noting how both the density and productivity (Figs. 2.12a and 12b overleaf) of Peregrines were correlated with distance from the Equator, while the specifics of habitat increased as the Equator was approached – but see, also, Hartley (1992) for a view on Peregrines in Africa. By studying the population of Lanner and Peregrine falcons in South Africa Jenkins and Hockey showed that the population of Peregrines was limited by both competition from Lanners and by resource deficiency. Morphologically Peregrines are adapted for high-speed stoops and chases of avian prey whereas congeners in the tropics have a wider prey spectrum which includes mammalian prey. In a separate paper (Jenkins, 1995) Jenkins noted that Peregrines were poorer at soaring and gliding than Lanner Falcons and so were outcompeted by them. Perhaps to compensate, Peregrine pairs preferentially chose higher nest sites on available cliffs, these being especially useful for the preferred hunting technique of perch hunting (though also being useful for a falcon intending to stoop hunt). Jenkins (2000) noted that those pairs with the highest nest sites had the greater hunting success rates. In addition, northern avian species typically breed at a specific (seasonal) time, breeding success being augmented by both the availability and relatively easy capture of juvenile birds, the juveniles appearing at precisely the time falcon chicks need maximum food intake. By contrast, as the Equator is approached the seasonal pattern differs and many potential prey species have aseasonal breeding patterns, disrupting the correlation between falcon breeding and prey availability. The combination of a morphology evolved for prey which is less available and congeners able to hunt local prey more successfully limits Peregrine populations in tropical areas.

One further issue must be borne in mind when considering sub-species and their plumages. In many areas local populations exist, these often accounting for the range of plumage colours seen, for instance, in *F.p. cassini* in its huge north-south range in South America and in *F.p. peregrinator* across its Asian range. In each case the colour range is striking. Subtler differences are also seen in more local populations, *e.g.* in *F.p. pealei*, with Aleutian birds being darker than those on the mainland.

Tundra Peregrine *(F. p. tundrius)*
Paler than the nominate, usually having a white forehead band and pale lores (Alaskan birds show this more prominently than Canadian falcons), a narrower malar stripe and minimal markings on a pale breast, the lower body being pinkish. A small number of birds are very pale. Comparable in size to the nominate. The wings are narrower and shorter than the other North American species. Migratory, some birds travelling very long distances to Chile and Argentina.

F.p. tundrius, Hood River, Nunavut, Canada. Per Michelsen.

This *F.p. tundrius*, photographed at its nest at Rankin Inlet, Nunavut, Canada, has head colouring very similar to some *F.p. cassini* in southern Argentina. *Alistair Franke.*

American Peregrine *(F. p. anatum)*

The former range of *F.p. anatum* in North America was extensive, with breeding at up to 3700m in the Rocky Mountains. However, the DDT crisis of the 1950-1960 period almost wiped out the Peregrine in the USA south of Canada and east of the Rocky Mountains, although birds in Alaska, and (to a lesser extent the western States), were less affected. Introduced birds in the US (from as far afield as the UK and Australia, as well as from Canada and Alaska) means they are no longer specifically anatum[2] and are marked with a hatched area in the distribution map (p83). One interesting aspect of the reintroduction has been the increase in urban populations. Caballero *et al.* (2016) note that 57.2% of Peregrines east of the Mississippi River are urban nesters, while in the midWest 91% of natal dispersion was to urban sites. In Illinois 90% of Peregrines are now urban breeders.

Peregrines have also been reintroduced to the eastern USA and this has not been without consequences. Between 1975 and 1985 over 300 captive-reared falcons were released and successfully produced a breeding population on the Virginia Barrier Islands off the eastern coast of the Delmarva Peninsula, which defines the eastern side of Chesapeake Bay: an area that had no recorded history of a Peregrine population. By 2007 a self-sustaining population of 55 Peregrine breeding pairs had been established, all on artificial nest substrates,

---

[2] DDT also took a huge toll on Peregrines in other areas of the world. In southern Scandinavia the species almost became extinct in the 1970s. Reintroductions, chiefly from Finland, Scotland and northern Sweden have resulted in a different genetic make-up from the past stock, but in this case all the reintroduced birds were *F.p. peregrinus* (see Jacobsen *et al.*, 2007). Similarly, reintroductions in Poland after the DDT crisis were from western European nominate stock (Puchała *et al.*, 2021). This issue will be considered again in Chapter 10 *The Pesticide Crisis*.

Above Male *F.p. anatum*, Three Arches Rocks, Oregon, USA. *Roy W. Lowe.*

Below *F.p. anatum* San Diego, California, USA. The falcon is carrying a Wilson's Phalarope (*Phalaropus tricolor*). *Will Sooter.*

including nesting towers. These Peregrine pairs had adjusted their breeding times to match the migratory timing of shorebird populations[3]. This has had a disastrous effect on a migrating sub-species of Red Knot (*Calidris canutus rufa*) and other shorebirds for which the peninsula represented a critical migratory staging post. Watts and Truitt (2021) found that in areas which lay within 6km of active falcon eyries the population of knots (measured as birds/km) was between 6.5 and 64 times lower than close to unused eyries, and that for active eyries there were always >6 times the density of knots at distances of >6km from the eyries than there was for distances <3km. The Red Knot population has declined from an estimated 100,000-150,000 to less than 30,000 over the period of Peregrine introduction and the knot sub-species is now listed as 'Threatened'. Habitat loss is considered the major reason for the reduction, but the impact on the population of predation by Peregrines is non-trivial.

Alaska birds are *F.p anatum*, but as they may breed with *F.p tundrius* (probably) to the north and *F.p. pealei* (definitely) to the south, may be similar to the nominate, or both paler and darker. In general they have a black forehead or a very narrow white band and a black crown and lores. Some birds have a rufous tinge to the breast. Alaskan *F.p. anatum* are sized between *F.p. tundrius* and *F.p. pealei*, though southern birds are closer to nominate coloration. Across the anatum range, some birds are resident, northern birds migrating to the southern states or Mexico.

Peale's Peregrine (*F. p. pealei*)
Peale's are the largest sub-species, though the size range of other sub-species means that individual birds of those races may be larger than the smaller Peale's. Males average around 42cm (bill tip to tail tip) and females 48cms, but larger specimens are seen. Most birds are heavily marked and very dark, though paler and less marked birds do occur. The darkest birds tend to breed on the Aleutian Islands. The black malar stripe is usually very wide. The Aleutian birds are resident, feeding on coastal birds, but mainland birds may migrate as far south as southern California.

The exact breeding areas of *F.p. tundrius*, *F.p. anatum* and *F.p. pealei* in Alaska, north-west USA and western Canada are subject to debate. For the most recent research see Talbot *et al.* (2017) who consider that *F. p. pealei* breeds along the coast of Alaska into Washington State, and along the Aleutian chain

---

[3] Ydenberg *et al.* (2004) noted changes in the behaviour of Western Sandpipers (*Calidris mauri*) following the increase in raptor numbers after restrictions in DDT usage. Throughout history there has been an arms race between migratory predators and migratory prey species. Lank *et al.* (2003) noted that some Arctic-breeding calidrids truncated parental care to their chicks in order to migrate earlier than the Peregrines that prey on them, arriving at their wintering grounds in time to moult their flight feathers before the falcons arrived. But, of course, that leaves juvenile calidrids at risk to the falcons, and there is always the chance that some Peregrines may also choose to migrate early as a response. It is not only prey species that alter their behaviour when Peregrine numbers increase. In a study of Merlins wintering in west Washington State, Buchanan (2012) noted that when the Peregrine population increased the small falcons hunted less frequently, reduced the time of hunting flights (which reduced the hunting success) and hunted closer to vegetation cover.

Juvenile dark *F.p. pealei*, North-west USA. Nick Dunlop.

to Russia's Commander Islands. Talbot *et al.* consider *F.p. anatum* breeds in a west to east corridor of Alaska and northern Canada, the corridor's northern limit being the southern limit of *F.p. tundrius* (Fig. 2.13). Environment and Climate Change Canada (2017) Part 2 suggests a more limited range of *F. p. pealei* in Canada, the sub-species breeding on the Queen Charlotte islands, along the coastal belt to about 50° north and along the northern coasts of Vancouver Island.

Figure 2.13 Distribution of the three North American Peregrine sub-species in Alaska and neighbouring north-west Canada. Redrawn from Talbot (2017). Note that the map is a flattened 3D projection resulting in the spacing of the lines of latitude appearing unequal.

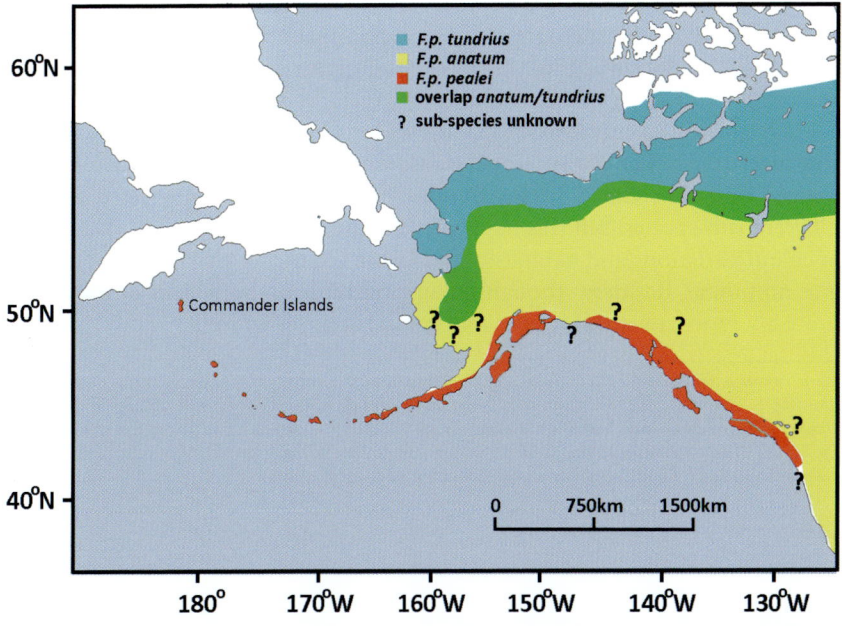

Paler *F.p. pealei*, seen in the southern section of the sub-species' range. *Nick Dunlop.*

South American Peregrine (*F. p. cassini*)

Occasionally called the Austral Peregrine which may cause confusion with the Peregrines of Australia and the islands off Indonesia. The significant range of this sub-species, extending northward perhaps as far as the Peru-Columbia border and south to Tierra del Fuego[4] and the Falkland Islands has resulted in a plumage coloration range. The predominant colour is similar to the nominate, though some birds are pallid and others are very much darker. The very dark birds have malar stripes merging into the nape. The pallid birds are very similar to tundrius in having white foreheads and very pale breasts and may have either darker or paler, more scalloped, backs. The palest birds were at one time called Kleinschmidt's Falcon and were suggested to be a different sub-species, but this has been dismissed. Ellis *et al.* (2020) has details of colour variations and habitats of the falcons in the southern continent. Size is as nominate. Resident, though the upland birds on both sides of the Andes probably move towards lowland and coastal areas in winter.

---

[4] *F. p. cassini* is the most southerly breeding Peregrine of all the sub-species, with breeding confirmed at about 56°S. It is not clear which sub-species can claim to be the most northerly breeding. *F.p. tundrius* breeds on Baffin, Banks and Victoria islands in Arctic Canada, and on the west coast of Greenland, while *F.p. calidus* breeds on the southern island of Novaya Zemlya and towards the northern edge of the Taimyr Peninsula in Russia. All of these sites could make a 'farthest north' claim.

Above The colour range of *F.p. cassini* seen across South America. *David Ellis.*

Below This *F.p. cassini*, photographed in south-western Argentina, has very similar colouring to the *F.p. tundrius* on p62. *Miguel D. Saggese.*

Female *F.p. calidus* with her clutch, northern Siberia, Russia.

Siberian Peregrine (*F. p. calidus*)

The northern Eurasian birds are very similar to *F.p. tundrius* in size and plumage. As with their American cousins they also migrate long distances, western birds heading to northern Africa, eastern birds to Indonesia and the Philippines.

Migrated *F.p. calidus*, photographed at the Little Rann of Kutch, Gujarat, India. *Dhairya Dixit*.

Mediterranean Peregrine (*F. p. brookei*)

Another sub-species with a significant range across which there are colour variations. In general *F.p. brookei* are similar to nominate birds, but tend to have a rufous blush on the breast and rather heavier marking in the form of horizontal lines, sometimes complete, rather than streaks. In north Africa *F.p. brookei* and Barbary Falcons may share breeding areas close to the Atlas Mountains, but data is scarce on the distribution of both. *F.p. brookei* are on average 15-25% less heavy than nominate falcons. Essentially resident.

F.p. brookei, Covadonga, Asturias, Spain. *Francisco Marzoa Alonso.*

F.p. brookei, Antikithira, Greece. Torsten Pröhl.

Cape Verde Peregrine (*F. p. madens*)

A very rare resident of the Cape Verde Islands in the Atlantic Ocean. The birds are classified as 'Endangered', with a population which may not exceed 10 breeding pairs (Anderson and White, 2000). The Latin name derives from the Latin *madeo*, usually translated as 'drenched', and refers to the rufous 'drenching' of the blue coloration of the back, this being most prominent on the head and in the edging of the malar stripe. In other respects the birds are similar to the nominate, though in size they are about 5% smaller. The coloration has suggested to some authorities that *F.p. madens* is an isolated population of Barbary Falcons. Resident.

Barbary Falcon (*F. p. pelegrinoides*)

The Barbary Falcon breeds on the Canary Islands and in remote spots across north Africa and the Middle East. Across the range the isolated nature of populations has resulted in variations from paler to darker birds, some with rufous-blushed breasts, but all with a rufous nape similar to *madens*, this varying from pale-brown to red-brown. Significantly smaller than nominate, with males weighing about 450g and females about 610g. Largely resident, though some eastern birds move short distances southward. In a worrying article, Rodríguez *et al.* (2019) noted that the number of lost falconry Peregrines (lost after being imported to the islands), were known to have interbred with the Barbary population and that this, together with the illegal harvesting of pure Barbary nestlings was threatening the Canarian Barbary stock. Rodríguez *et al.* noted that in a study across Tenerife only 54.1% of the 'Barbary' stock showed typical plumage, suggesting that regulations were required to conserve the pure stock.

Barbary Falcon, Gilgit, Pakistan. *Imran Shah*.

African Peregrine (*F. p. minor*)

The Latin name is a giveaway, some male birds weighing no more than 300g, though the average is about 500g, with females at about 700g. The species has a somewhat disjointed range across Africa south of the Sahara. Very dark back and malar stripe, but often with a buff-blush to the belly. Resident.

F.p. minor showing its dark back and malar stripe, Wondo Genet near Hawassa, in the Ethiopian Rift Valley. *Torsten Pröhl*.

Above   *F.p. minor*, Tsavo East National Park, Kenya. *James Kashangaki.*

Below   *F.p. minor*, South Africa. *Tomas Grim.*

Madagascan Peregrine (*F. p. radama*)

Very dark falcon: either this or *F.p. ernesti* is the darkest of all sub-species. Resident on Madagascar and probably also breeds, at least intermittently, on the Comoros. Few details of the species are known. From the extremely limited data of Benson (1960), Comores birds are slightly smaller than those on the larger island. The population is believed to be 150-300 pairs, based on the available habitat (Sutton *et al.*, 2021). Comparable in size to the African Peregrine from which it perhaps evolved. Red-listed Resident.

F.p. radama, Ankarafantsika, Madagascar. **Theofanopoulos Fanis.**

Red-naped Shaheen (*F. p. babylonicus*)

Shaheen is from the Persian, *shah* (king) and *een* (bird). This is the palest of the Peregrines (though darker birds are seen), as pale as Barbary Falcons. Most have a beautiful salmon pink blush on the front and on the nape where it may even be reddish. The back is grey, paler than on nominate. For an excellent description of the plumage range of this sub-species see Bhatt and Ganpule (2017) who report on the falcons which are occasional visitors to north-west India. Comparable in size to brookei, but looks more slender. Breeds in central Asia where the habitat includes steppe, desert and areas of forest, ranging from

The two photographs of *F.p. babylonicus* on this page were taken at the Little Rann of Kutch, Gujarat, India.
Above by *Nirav Bhatt* (this adult female falcon is eating a pigeon), *below* by *Dhairya Dixit*.

The Peregrine Falcon

sea level to 2700m in the mountains (Stepanyan, 1969a, 1969b). The exact range of the falcons is poorly known, but they do share areas with both *F.p. brookei* and Barbary Falcons to the west and are seen in the arid foothills of the Karakoram/Himalayas to the east. In the higher, central areas of the range, the Peregrines share a breeding range with Saker Falcons. Resident or altitudinal migrant over much of its range, but during winter some birds are seen in the southerly deserts of western India.

Black Shaheen (*F. p. peregrinator*)
Widely distributed across the Indian sub-continent, breeding at up to 2400m in the foothills of the Indian Himalayas (Pande *et al.*, 2017) and even higher, to 3000m, in Nepal (Grimmett *et al.*, 2000). Grimmett *et al.* state that the falcons have been seen as high as 4200m, though breeding was not confirmed. The sub-species also breeds eastwards to northern Myanmar, south-eastern China and the Andaman and Nicobar Islands (Samson *et al.*, 2017). Across this vast range the falcons vary in colour, but all are essentially a black-backed, black-headed bird with a rufous front, the extent of rufous coloration and black breast/belly thigh barring varying with area. In Pakistan and northern India,

F.p. peregrinator, south-west Gujarat, India. Nirav Bhatt.

Left An adult.

Right A juvenile.

The Peregrine Falcon

Nepal, Bhutan, northern Myanmar there is distinct barring, but on birds of southern India/Sri Lanka it is less pronounced. To the east the barring returns and the rufous coloration tends to fade. Resident or altitudinal migrant.

F.p. peregrinator, Bangladesh. *Mosharaf Hossain.*

Japanese Peregrine (*F. p. japonensis*)

The distribution of this sub-species is still debated. Peregrines on Japan are similar in size and appearance to nominate birds, while those of Kamchatka and eastern Siberian Russia, extending all the way north to the coast of the Arctic Ocean, are darker and significantly darker than the calidus birds to the west. Linking the two are the Peregrines of the Kuril Islands (Artukhin, 2009) which are known to breed on Kunashir, the island closest to Haikkado. Historically these eastern birds have been seen as a separate sub-species, *F.p. harterti*, though recent genomic studies have not supported this separation. Japanese birds are resident, the northern falcons definitely migrant, travelling south to Japan, eastern China and Korea.

Two photographs of *F.p. japonensis*, Chiba Prefecture, Japan. *Toshimitsu Nuka.*

Izu Peregrine (*F. p. furuitii*)

This Peregrine, occupying an island group lying south of Japan's main island (Honshu), formerly called the De Vries Archipelago, and another collection of islands (the Ogasawara Islands) further south, is the rarest and least studied of all the sub-species. How many falcons there are, and on which islands they may breed is not known with certainty. As dark as *F.p. pealei* but smaller. Red listed, but Data deficient. Resident.

Ernest's Peregrine (*F. p. ernesti*)

Named for Ernest Hose, a Victorian collector, *F. p. ernesti* is arguably the darkest of all sub-species, some birds being almost uniformly black, apart from a white chin, but most having a rufous wash to the breast and belly, overlaid with darker barring. Resident in Malaysia, the Philippines and Indonesia and nearby Pacific islands, though breeding in Malaysia was not confirmed until 1996: in 2007 the breeding population of Malaysia was 15 pairs, with six other pairs suspected, that small total breeding in an area of about 20,000km^2 (Molard *et al.* (2007) which implied a total population of 70-80 pairs in Peninsular Malaysia. However, this number has been considered an overestimate by later work (Ooi *et al.*, 2020). Baudat-Franceschi *et al.* (2013) found 80 potential territories on New Caledonia, with 40-60 breeding pairs. There is a debate as to whether the Solomon Island Peregrines are *ernesti* or *nesiotes*, the former being favoured as the Solomons are close to New Guinea.

Opposite and *above*   Adult *F.p ernesti*, Perak, Malaysia. *Amar-Singh HSS.*

Below   Juvenile *F.p ernesti*, Subic Bay Rainforest, Zambales, Philippines. *Vinz Pascua.*

Australian Peregrine (*F. p. macropus*)

Similar to, but a little smaller than, the nominate. This resident bird is found across Australia, but a debate remains about the birds of south-west Australia, centred on Perth, which have a distinct rufous blush on the lower breast and belly. This coloration has suggested a different sub-species (*F. p. submelanogenys* - see left-hand front endpaper), the name relating to the black cheeks of the birds, though in general the black cheeks are no different from those of falcons to the east.

Above   F.p macropus, Cobble Creek, south-east Queensland. *Aviceda*.

Below   F.p macropus. Ken Griffiths.

Melanesian Peregrine (*F. p. nesiotes*)

Very dark, rufous breasted and heavily barred falcon resident on Fiji and (probably) other island groups of the south Pacific including Vanuatu and New Caledonia. Dutson (2001) lists sightings of Peregrines on some of the smaller islands of Melanesia. White *et al.* (2000) suggest that the population on Fiji is declining, having been about 100 pairs in 1985-86, but with only a handful of breeding pairs seen in 1997. White *et al.* (2013a) consider that perhaps only 100-150 breeding pairs might occupy the 51,610km^2 of Fiji and Vanuatu, the reduction probably due to the increased human population, particularly on Fiji, coupled with the introduction of voracious predators such as the Small Indian Mongoose (*Urva auropunctata*) and the human appetite for eating the fruit bats on which the Peregrines prey.

Above   Adult male *(left)* and adult female *(right) F.p nesiotes*.

Below   Juvenile *F.p nesiotes*. All photos on New Caledonia. *Julien Baudet-Franceschi*.

Distribution

As noted in Chapter 1, Peregrines are found on all continents apart from Antarctica (map *opposite*). BirdLife International (www.birdlife.org) calculates that the combined range of the Peregrine and Barbary falcons (assuming the latter to be a Peregrine sub-species, though the addition of the Barbary's range has minimal impact on the numbers) constitutes 31% of the land surface of the Earth, an area of almost 46,500,000km^2 (*c*.18 million sq. miles). But the distribution is not uniform and includes curiosities. Peregrines are absent from some oceanic islands – not that curious since any island has to be both reachable and to hold enough prey to sustain a viable population. But there are no Peregrines on Iceland, which has been colonised by both the 'cold-weather' falcons – Gyrfalcon and Merlin. It cannot be the weather that influences the absence as Peregrines breed further north in America and Siberia than do Merlins (and almost as far north as Gyrfalcons) and it cannot be that they are outcompeted for prey by the larger falcon as RS has seen Peregrine and Gyr pairs breeding at opposite ends of the same short canyon in northern Canada. In that situation the two male falcons chose to avoid conflicts by leaving/entering the canyon from opposite ends: such skirmishes as did occur suggest neither was keen on full-on combat. Peregrines are also absent from New Zealand, despite having colonised Australia and islands of Melanesia/Polynesia. In their book on the Peregrine, White *et al*. (2013a) identify these and other islands not settled by Peregrines and ponder the reasons for the falcon's absence, suggesting that the fact that the species is philopatric – *i.e.* that it tends to return to the previous breeding places or, in the case of young birds, the neighbourhood of the birth site – militates against colonisation. Other islands, curious for the absence of Peregrines are the Faeroes (to the south of Iceland), the Bering Sea islands of St Lawrence, the Diomedes and the Pribilofs, the islands of the Caribbean, the Atlantic islands of Madeira and the Azores, the Galapagos and Juan Fernández islands in the Pacific Ocean, and the Seychelles in the Indian Ocean.

As White *et al*. also note, at some time in the past Peregrines did successfully colonise most of the world, and they conclude that it is enigmatic why some islands remain uncolonised when nearby islands or the continental mainland do have breeding Peregrines. That Peregrines can achieve trans-oceanic flights is clear from the fact that they have been seen on, for instance, South Georgia (Wood and Woods, 1997 and Clarke *et al*., 2012) and the Pitcairn Islands (Brooke, 1995). Colonisation requires a minimum of a breeding pair of birds to arrive (in reality several pairs to avoid in-breeding) and some small islands may be prey-limited, particularly in winter. While migration to and from large islands is straightforward, annual flights to and from mid-ocean specks of land is far from trivial. Such considerations mitigate against populations becoming sedentary, but the fact that Peregrines have managed this on islands such as Cape Verde makes the species' absence from large islands well-stocked with prey and relatively close to continental land masses (*e.g.* Iceland) surprising.

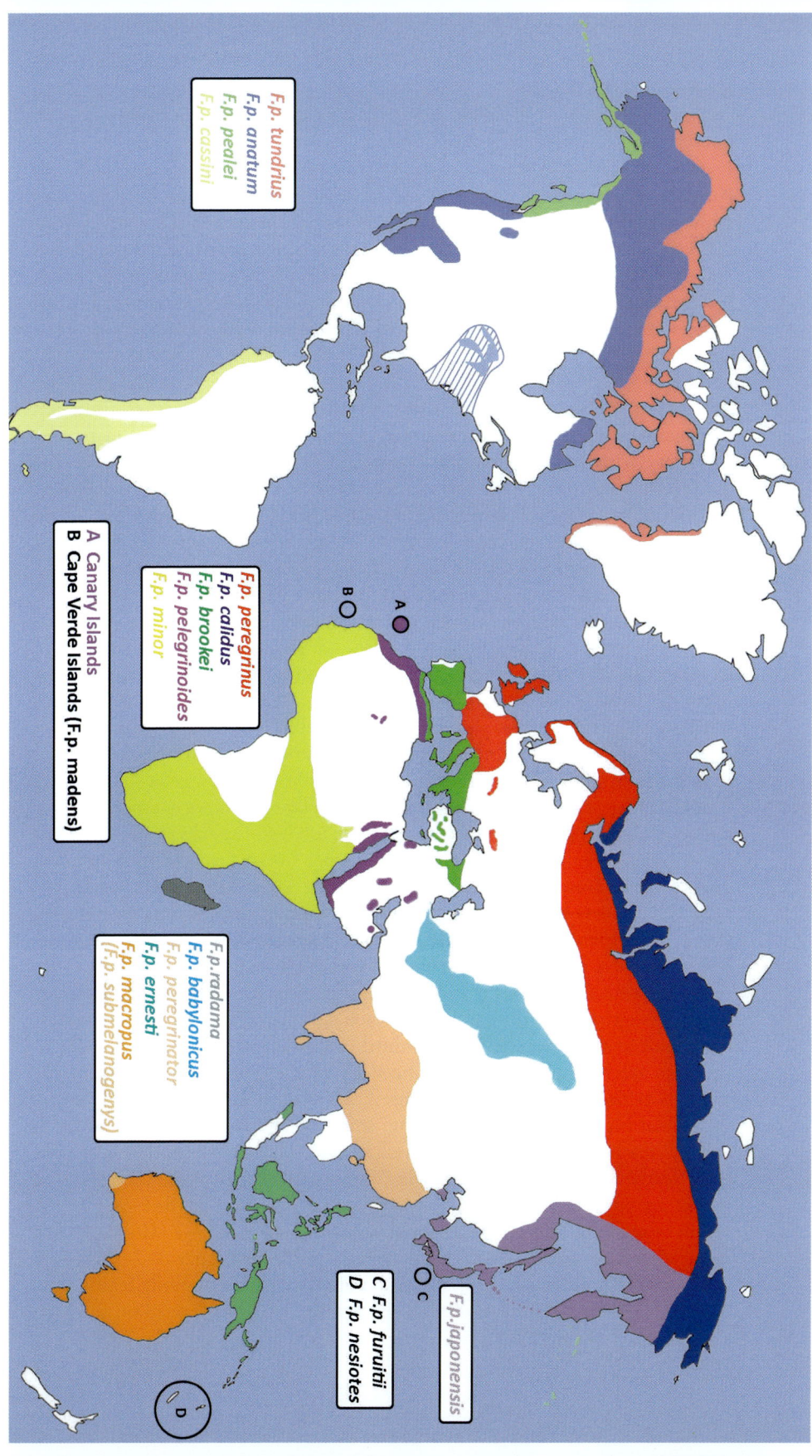

Even more baffling is the fact that Peregrines are not found in areas of South America that appear to have the correct habitat (though much of that area comprises the Amazon Basin and Peregrines are very definitely not forest-falcons), despite these being adjacent to areas which are home to the species and over which Peregrines migrate. It is possible that the absence of Peregrines is due to ecological factors not yet understood or competition from established species, but more work is required before a definite reason can be set down.

Moult
Feathers are delicate and under constant attack from the weather and from the parasites which thrive on them. All birds combat this wear and tear by preening, with some also using water and/or dust bathing to reduce the parasite burden. Peregrines are no exception, preening often (to stimulate and distribute preen oil from the uropygial gland at the base of the tail to waterproof the feathers) and having been recorded to bathe in both dust and water (see Schmidl, (1988) and references therein), though they are not frequent bathers. One very interesting example of 'bathing' (it having been accidental rather than intentional) was reported by Ford (2007) who observed a female Peregrine hitting the sea off Portland, Dorset, England after binding to a pigeon. The falcon spent 20 minutes making her way to shore, using a 'butterfly stroke'. The following day the female was seen standing in rain with outstretched wings which she shook periodically in what appeared to be an attempt to remove salt from her plumage. Another interesting example was that of Stirrup (2021) who observed a female nominate Peregrine 'bathing' by immersing herself in the leaves of a London Plane (*platanus* x *acerifolia*) tree after a spell of heavy rain had saturated the foliage. The falcon stretched out her wings, fanned her tail and fluffed out her breast feathers while moving among the leaves, apparently in a deliberate (and successful) attempt to soak her feathers before moving off, presumably to find shelter in order to preen. The observation was made at Ely, Cambridgeshire, England where the Peregrines were breeding on the famous cathedral.

But despite the preening and bathing, feathers wear and need replacing. Adult Peregrines moult completely each summer. The female may start her moult at the time she lays her third egg. The male may start as soon as the eggs are hatched, despite the need to ensure good flight ability in the arduous period of provisioning himself, his mate and the brood. However, individual birds may alter the timing, some not starting the moult of flight and tail feathers until breeding is well advanced, though even in these cases body feathers may be moulted earlier. The moult begins with the primaries which are moulted following the sequence 4–5–3–6–7–2–8–1–9–10. The moult of secondaries follows the sequence 5–7–4–8–3–6–9–2–10–11–1. The tail feathers are moulted sometime after P4 in the sequence 1–2–3–6–4–5, but full tail moult is often completed before the primaries. Body feathers are moulted at the same time, though a few feathers may be retained on the head and back. The full moult takes between 128 and 185 days (Mebs, 1960). Northern female Peregrines are known to suspend the moult sequence during breeding, which may explain, in part, the fact that some adults complete the moult much faster than others.

The timing of the moult is dependent on the latitudinal position of the population. Low and medium latitude sedentary (or partially migratory) populations exhibit a single seasonal moult, whereas the high latitude sub-species have two seasonal moults, arresting the moult in order to make their long migration flights, then completing the moult in winter quarters. Figs. 2.14 (a) and (b) show the moult intensity in a male Peregrine in two successive years, and a female Peregrine in a single year. In each case the falcons were housed in an outdoor aviary in Sweden (Lindberg, 1983).

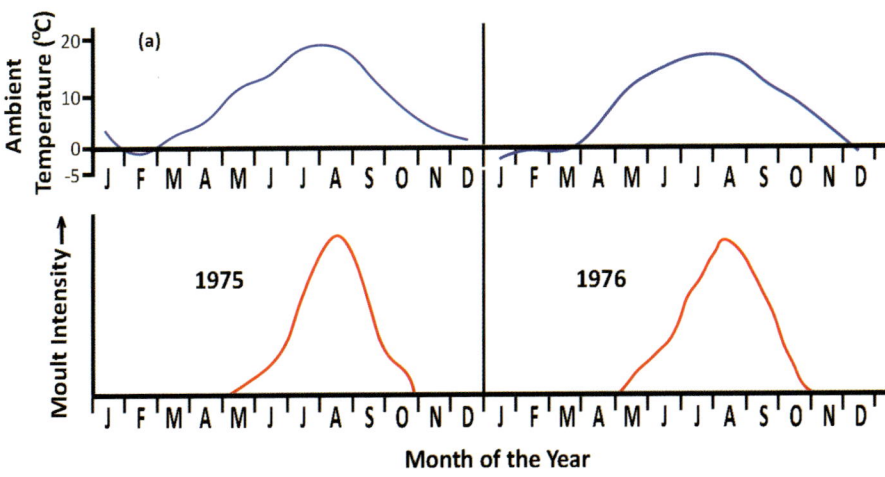

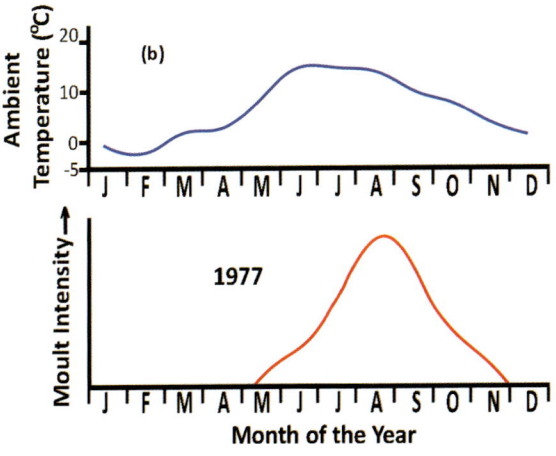

Figure 2.14 Moult intensity of a male Peregrine (a) *above* for two successive years and female Peregrine (b) *left*, each in an outdoor aviary, plotted against ambient temperature. Redrawn from Lindberg (1983).

In a study on Golden Eagles (*Aquila chrysaetos*) Ellis *et al.* (2016) noted that the necessity of maintaining flight during the moult, which can last much longer in the eagle than in the Peregrine, means that some flight feathers become very worn or severely damaged. By deliberately cutting some rectrices, Ellis *et al.* showed that the eagles could actually grow such feathers out of sequence, advancing their moult by up to a year. Ellis *et al.* suggest that this implies the existence of a neurophysiological mechanism which allows preferential moulting. While the Peregrine's moult is much shorter than that of the Golden Eagle, it is likely that a similar mechanism exists in the falcon as flight feathers are equally important for both species.

Juveniles start a complete moult in the early spring of their second calendar year, though there is evidence from North American Peregrines that juveniles may moult some body feathers during their first winter (Pyle, 2005). Pyle examined specimens of 20 American (both North and South) juvenile raptors from several collections and noted that 19% of first-year birds were moulting during September-November (with a maximum feather replacement of 10% on an individual bird), with 46% of birds having moulted by December-February (with a maximum feather replacement of 25%) and 85% of birds having moulted March-May (maximum replacement 20%). The juvenile moult sequence is the same as for the adults, though in general moulting of the tail starts sooner after the moult of P4 (as early as six days: Mebs, 1960). However, the time to completion of the full moult to adult plumage is very individual, with some birds completing by summer, others not until the winter.

Habitat

The Peregrine is primarily a predator of avian prey. The species therefore requires open country over which it can hunt, one irony of the Peregrine story in Britain being that it is likely that the felling of the forests which once cloaked the island and the agricultural landscape that was then created improved the country for the falcons, providing them with more open country, though the later use of pesticides to aid agricultural yields was to imperil the species. As well as hunting areas the Peregrine needs adequate populations of potential prey species (though the Peregrine's plasticity in terms of diet means that the list of suitable prey is a long one – see Chapter 4), and steep, inaccessible cliffs on which it can breed. The combination is best represented by sea cliffs and inland mountain areas, and these were traditionally the principal areas in

Opposite Peregrines occupy a variety of habitats across their vast range, from tropical mangrove forest to the edge of the Arctic, from the high Himalays to islands in the Pacific Ocean,

Top F.p. *peregrinator* in the mangroves of the Sundarbans National Park, West Bangal, India. *Swaroop Singha Ray.*

Centre F.p. *tundrius,* Hood River, Nunavut, Canada. *Per Michelsen.*

Lower Left Tadapani, Nepal. *Right* Central mountain chain, New Caledonia. *Julien Baudat-Franceschi.*

which the falcons bred in the British Isles until the end of the last century. As a consequence, there are few breeding Peregrines in the east of the country, the flat lands of East Anglia offering no suitable breeding sites from which to take advantage of the open land and supplies of prey. Humans have, however, aided the species by creating 'cliffs' on which it can breed, the immigration of Peregrines to towns and cities being one of the more intriguing aspects of the recent history of the British population, though urban living is not restricted to Britain, now being seen across the falcon's range. And in cities there is rarely any shortage of feral pigeons (*Columba livia (domest.)*).

The same basic requirements of open country and inaccessible nesting areas apply through the species' vast range, applying equally to Peregrines breeding at over 3000m in some areas, and on small oceanic islands in others.

In all cases the basic breeding requirements remain fixed, the Peregrine's ability to adapt, and particularly the flexibility in its diet, means that the vision of the falcon sweeping majestically across the wilderness now needs to be augmented by the vision of it using street lighting in cities to hunt bats or migrating birds, the vision of a fortress-like eyrie on some vast cliff being set beside the equally applicable one of a box on a cathedral or high-rise tower.

Voice

Peregrines are vocal birds during the breeding season, with calls beginning in winter close to traditional eyries, but are much more silent otherwise. The most common call, and one which, once heard, is never forgotten, is the alarm call, a typical falcon 'kek-kek-kek'. Monneret (1974) renders the call as a suitably Gallic kré in a fine paper which also sets down the display flights and stances of Peregrines. The call, which is usually the only one made outside the breeding season, is a staccato series of keks separated by fractions of a second (on sonograms the separation is 0.1–0.2 seconds, though on occasions it can be longer, perhaps 0.5 seconds), the series being as long as the bird chooses, with frequent repetitions after periods of silence, the persistence of which depend on the nature and duration of the perceived threat. In general, the female call has shorter gaps between the keks, so short that if the bird is being very aggressive the call can become a continuous scream. The male call is higher-pitched than the female's, and the call of individual birds is so regular that they can be recognised, with a high degree of accuracy, not only during a season but in subsequent years. The call is usually a response to disturbance or intrusion by any species other than conspecifics. Occasionally, however, it is used as a high intensity threat to conspecifics including a mate (Cade, 1982).

A second call is usually described as a wail or whine, and is plaintive, more frequently used in the breeding season. When heard, it clearly gives the impression that the wailing Peregrine wants something to change. It has been rendered as 'eeyaik' (Cade 1960) and 'yee-errk' by Ratcliffe (1993). One of the authors (SW) feels that Ratcliffe's rendition is the more accurate. This call is given by both adults and frequently by food-begging juveniles. It can become a continuous stream of sound and increases in intensity as desire for change increases, becoming more like a 'whine'. A version of this call is associated with the feeding of fledglings, both male and female Peregrines making the call when they are attracting the attention of the youngsters as they arrive with prey. In this case the call is less harsh. A slower, softer, higher-pitched and more drawn-out wail is also made. Both sexes use it as a contact call, by way of attracting the mate's attention when they are out of sight of each other during the early breeding season. Females also use it to request food from males, as an invitation call for copulation and to suggest an incubation changeover. Males use it when arriving with food. Wailing calls are also given when the falcons have lost eggs or young though this could, in these circumstances, be associated with courtship renewal. This has been witnessed by one of the authors (SW) and is referred to in Ratcliffe (1993).

A third call, once memorably described as resembling the creaking of an old gate (Rand and Gilliard (1967): this might be the source of the name, as it is sometimes referred to as the 'creaking call'), is now usually described as a two syllable 'ee-chip' or a three syllable 'ku-ee-chip', though since it can be either two or three syllables, the call is more generally referred to as chupping. Cade (1960) renders the male Peregrine creak as 'klee-chip' with the emphasis on the second sound, that of the female being 'klee-chuck', again the second sound being emphasised.

Various observers have described subtle differences between male and female versions. Chupping may vary in intensity, tone and pitch, but is a greeting and bonding conversational call, often used by the female to attract the male to the nesting ledge and occasionally as a copulation invitation (Fig. 2.15). (Many observers consider this softer, often shorter call is a separate call – 'chittering' – rather than a chupping variant.) Females also use it when feeding young chicks, when it can bring to mind the soft clucking of a mother hen or even a gentle cooing. Male Peregrines tend to chup when around the female: the size difference between the sexes means the female often bullies

Figure 2.15 Female and male Peregrines exchanging soft creaking calls early in the breeding season. Redrawn by Kerry Jane from Monneret (1974) a fine report on Peregrine calls and displays.

the male, particularly at the nest when she wishes to take over incubation and the male is reluctant to get off the eggs, or when demanding food – the male frequently chups in appeasement.

Carlier (1995), studying Peregrine calls in France, noted that the wail and chup (creaking) calls varied in frequency for males and females through the phases of the breeding cycle: chupping was most frequent during courtship and was more or less equally used by both sexes; males chupped more frequently during incubation and used the call when delivering food once the eggs had hatched (Fig. 2.16).

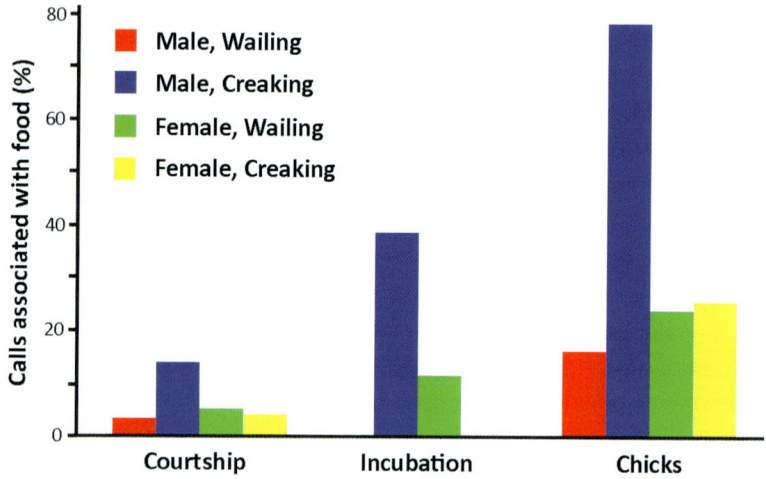

Figure 2.16 Percentage of creaking and wailing calls associated with food made by male and female Peregrine Falcons during the breeding season. The three periods are courtship, incubation and the time between the hatching and fledging of the chicks. Redrawn from Carlier.

A strident, loud and persistent variation of the creaking call is made when an intruding adult conspecific enters into the vicinity of the eyrie site, particularly in the breeding season. Cade (1960) suggests that whenever this more strident call is made, it is likely to be a precursor to conflict, a suggestion confirmed by one of the authors (SW) who has heard the call many times prior to conflict, mostly during the breeding season. In these circumstances the call is often made in conjunction with the territory-holding falcon showing signs of agitation – low horizontal body posturing, feathers tight whilst continuously watching the intruder – seemingly ready to launch an attack at any second. As a side note, one of the first indications for an observer that an intruding Peregrine is present is a chupping resident.

One of the authors (SW) has also heard a deep, loud, short monosyllabic grunt when a Peregrine has been surprised by the sudden arrival of a conspecific. Other calls have also been described, but these are all variations on those given above, most of the variation deriving from circumstance. For a comprehensive review of Peregrine vocalisations see Nelson (1970).

Young Peregrines begin to chirp in the egg up to 72 hours before hatching and continue to chirp for a day or so after hatching. The chirp then alters to a variation of the chupping call, but higher-pitched when the chicks are hungry. As they grow the nestlings develop the full range of Peregrine calls and from a young age will make the familiar 'kek-kek-kek' when disturbed, though when approached by, for instance, an observer from about two weeks of age, a chick will hiss loudly while lying on its back and extending its feet. However, although the calls develop as the chick ages, Peregrine chicks raised in isolation all demonstrate the complete range of calls once fledged, strongly suggesting that the vocalisations are inherent, not learned. Less well-known until recently was that unhatched siblings communicate with each other as well as with their parents. Working with eggs of the Yellow-legged Gull (*Larus michahellis*), Noguera and Velando (2019) found that information was also being passed by vibrations between embryos. To test this hypothesis, eggs were removed from clutches in the wild and placed in incubators within a laboratory. Some eggs of a clutch were then exposed to the calls of potential predators, while others were not. The unexposed eggs were then replaced in the clutch. A later examination of the eggs showed that embryonic changes and the levels of stress hormones in all eggs were similar: vibrations within individual eggs were being used to pass information between siblings. Noguera and Velando note that in the wild developing embryos are subject to the alarm calls of adults and that vibrations may assist in transferring this information between embryos at different stages of development and that this would explain why after just a few hours of hatching all chicks will crouch on hearing an alarm call.

In a study of alarm calls as a way of identifying individual birds, Telford (1996) noted that the difference between those of males and females were apparent, the female always being harsher and lower than the male. In each case, although the call might vary marginally day by day, in a way analogous to the way a human's voice might vary because of the person's emotional state, the calls of an individual were stable year-on-year. As a consequence, the calls could be used to identify particular birds at nest sites as an aid to understanding annual population turnover. Interestingly, Telford noted that the difference in calls between individuals was advantageous to a Peregrine pair, the male of a pair not having to waste time and energy responding to a female alarm call while he was hunting if the caller was not his mate[5].

For excellent sonograms of the calls of Peregrines, see Vasina and Stranck (1984) who depict the alarm call of a nesting female; the contact call of a male; the anti-aggression call of a perched male; the alarm call of a female with young; the alarm call of a male; and the call of nestlings in terms of frequency and duration.

---

[5] RS, working with Klaus Dietrich Fiuczynski on Eurasian Hobbies in Berlin, was amazed to discover that the Hobby expert was using the calls of individuals as a means of detecting whether the same birds were occupying a given territory or if one of the pair had not returned from its annual migration.

3   Flight Characteristics

Avian flight is complex, the mathematics used to explore it no less so. The mathematical problems are also compounded by the fact that flight is more than mere wing shape and beat frequency, flying being energetic and so involves biology as well as physics. For interested readers, the books of Pennycuick (2008) and Videler (2005) represent good starting points. Here we briefly discuss only the basic features of falcon flight.

In a classic paper on the flight of avian predators, Andersson and Norberg (1981) defined six flight characteristics which are important in the capture of prey – linear acceleration in flapping flight; maximum speed in horizontal flapping flight; terminal dive speed; maximum rate of climb; angular roll acceleration; and turning ability. The last two define the manoeuvrability of the bird[1]. What Andersson and Norberg found was that of the six, only terminal dive speed was improved by body mass, all the others varying with the inverse of mass, *i.e.* smaller birds fare better[2]. Higher body mass allows a

---

[1] In a classic paper on the influence of natural selection on the shape of birds' tails, Thomas and Balmford (1995) define two aerodynamic parameters which influence flight patterns. These are 'agility', defined as maximum rate of turn (usually associated with high-speed flight) and 'manoeuvrability', defined as minimum turning radius (which occurs at low speeds). The two parameters are essentially the same as those that define 'manoeuvrability' in Andersson and Norberg (1981).

[2] Manoeuvrability was also investigated by Howland (1974) who defined the relative importance of speed and turning circle radius for prey attempting to escape a pursuing predator. The ratio of speed and turning circle is particularly important for smaller, extremely, agile falcons such as the Eurasian Hobby (see Sale and Messenger, 2021).

Nominate Peregrine, Portland, England. *Verity Hill*.

stooping falcon to accelerate faster under gravity, the bird closing its wings to reduce the profile it presents to the air and so minimise drag. However, despite what is occasionally seen written, stooping falcons cannot fold their wings completely unless they dive vertically: in any high angle stoop the wings must be partially open to provide the lift necessary to maintain the stoop angle. But while a high body mass increases stoop speed, it reduces the possible rate of climb: acceleration and horizontal speed also decline with increasing body weight. It is, therefore, not surprising that stooping is most usually associated with the larger falcons, though smaller falcons also utilise it as a hunting technique on occasions.

While flight ability .v. body mass implies that a low mass would be an advantage to a falcon, body mass is important for practical reasons. An avian predator has to stop its prey, and then carry it away: while eating at the point of capture is acceptable to an adult bird, breeding requires prey to be returned to the nest. Both considerations require a body mass which is at least comparable to that of the prey. Secondly, as already noted, there is more than physics to falcon flight, and while body mass is undoubtedly important (and wing loading (see Box 3.1) perhaps more so) the power input to flapping flight, and into a quick take-off and fast acceleration are important in surprise hunting. Physiology must therefore be considered, initial power output favours short, fast attacks followed by longer resting periods.

Box 3.1 Wing Loading
Andersson and Norberg (1981) noted that in the two characteristics which affect the 'manoeuvrability' (see Footnote 1) of an avian predator, wing loading was important. Wing loading is defined as the ratio of body weight to wing area and is generally given in units of g/cm^2. The variation across the avian world is remarkable, being around 0.2 for hummingbirds and up to 2.3 for members of the auk family which are heavy, but have short, stubby wings used for both flight and underwater 'flight', water being a medium 1000 times denser than air. The wing loading of birds of prey is generally low, at about 0.3, reflecting the need to carry prey, but does vary with prey type. Wing loading also varies with hunting technique. Harriers and kites, which spend long periods in the air searching for prey which is, in general, small and slow-moving but (relatively) abundant, have lower wing loadings than those, such as hawks and falcons, which spend less time in the air and attack faster and heavier prey which is less abundant. Falcons feeding primarily on avian prey, such as Peregrines, have higher wing loadings which underpins fast flight. However, high wing loading decreases manoeuvrability and falcons which feed on insects (*e.g.* Eurasian Hobbies) have lower wing loadings: they are more agile in the definition of Balmford and Thomas (1995) – see Footnote 1 – and can make tighter turns, the radius of turn being proportional to the wing loading.

The calculation of wing loading is not trivial. The weight of adult predatory birds varies with the season, being lower during the breeding season, particularly for males whose workload increases substantially when feeding both the female and the growing brood: while wing size does not vary, wing loading must as the male loses weight. As we shall see in Chapter 5, it is possible for a male falcon to lose significant body weight during the breeding season: females also lose weight when her chicks are newly hatched and she prioritises their feeding over her own. Measuring wing loading must, therefore, not only be done carefully, but with an understanding of the timing of breeding for both males and females.

A further complication is that as well as the weight of a falcon varying with time of year, wing area also varies during each wing beat which means that assessment from photographs of flying birds is difficult. Assessing it from trapped birds is also difficult as live birds do not react well to having their wings stretched. Even falconry birds, which, when feeding, are willing to have their wings somewhat extended by their owners (with whom they are, of course, familiar) can be steadfastly intolerant of having the wing fully extended. The best solution is to work with carcasses, when both weight and wing area can be accurately measured. In his work on assessing the various physical parameters which influence the flight characteristics of birds Pennycuick (2008), having recognised the importance of wing loading in assessing fundamentals such as wing-beat frequency, noted a method for accurately assessing wing area on a carcass. Fig. 3.1.

Flight Characteristics

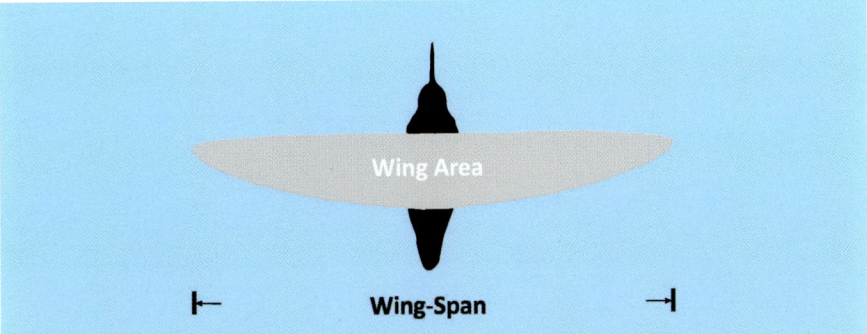

Figure 3.1 Wingspan and area as defined by Pennycuick (2008). The weight of a gliding bird is balanced by the pressure difference between its upper and lower surfaces multiplied by the wing area. While the wing provides most of the area, the pressure differential operates across the body as well and so the relevant portion of body area must be included. In practice, working with a carcass, the area of one wing, extended to include the body to the spine, is measured and then multiplied by two. The same procedure is used for a live bird, though, as noted in the text, producing full extension may be problematic and requires patience. Redrawn from Pennycuick (2008).

A dead female Peregrine killed when it collided with a window. The bird was used to calculate wing loading, one wing being laid out beneath clear polythene to maximise, as far as possible, the wing area. For further details see the caption on a separate image overleaf.

The falcon (above), an *F. p. brookei* with a beautiful pale red underside blush, was the sister of one born with a twisted spine. Although it was not an excellent falconry bird, the deformed bird was kept and looked after throughout its life. The sister grew into a superb falconry bird, and was known initially only as Twisted's Sister, the falconer not usually giving names to his birds. Over time, the name was shortened to *Twisted Sister* as this reflected the falcon's reaction to any attempts to attach gps units (such as RS's IMU - see text later in this Chapter) *etc.* to her, the response involving lunges with talons and bill, all intended to prevent any attachment. Flights with the IMU therefore involved RS carrying antiseptic wipes to apply to wounds (his own or those of the falconer) and becoming skilled at removing talons from his own, or the falconers', hands. Ultimately *Twisted Sister* (TS hereafter) became an affectionate name for this superb, but feisty, female. Most of the flying data presented below derives from flights with this bird. Her flying weight was 835g. TS's unfortunate death following a hunting flight is detailed in *The Hunting Flights of Falconry Birds* below. The bird's owner had plans for the bird after her death and did not want her wing damaged, so a second (nominate) female Peregrine, which had been killed in a collision was obtained: this bird also weighed 835g and was laid out ready for wing area calculation. A grid was overlaid on the photograph on the previous page to allow the calculation. It was not possible to reproduce the full size the live bird might have been able to sustain, but it is considered the wing area was probably within 5% of full spread. The calculated wing loading was 0.61g/cm^2. Further calculations based on measuring identical wing dimensions of both birds (*i.e.* the two carcasses) enables TS's wing loading to be calculated: the value was 0.6g/cm^2, again with a potential error of up to 5%.

Flight Characteristics

Nominate Peregrine, Isle of Man, UK. *Peter Christian.*

The Peregrine Falcon

In his classic book on falcons, Tom Cade (1982) collected wing loadings from various sources for both male and female birds of many Falco species. These data have been used to prepare Fig. 3.2 which identified falcons as a group which largely falls, when assessed by body size and wing loading, as sitting above the 'searchers', predatory birds which cover large areas of their habitat searching for prey, and below the line which defines 'chasers', predators which have the ability to pursue prey by one means or another.

The mathematics of flight show that Peregrines, being heavy birds, have a large turning circle in comparison to smaller, and therefore more agile, falcons. But the agility of Peregrines can often surprise - both prey and the human observer - as did this nominate adult turning towards its eyrie.

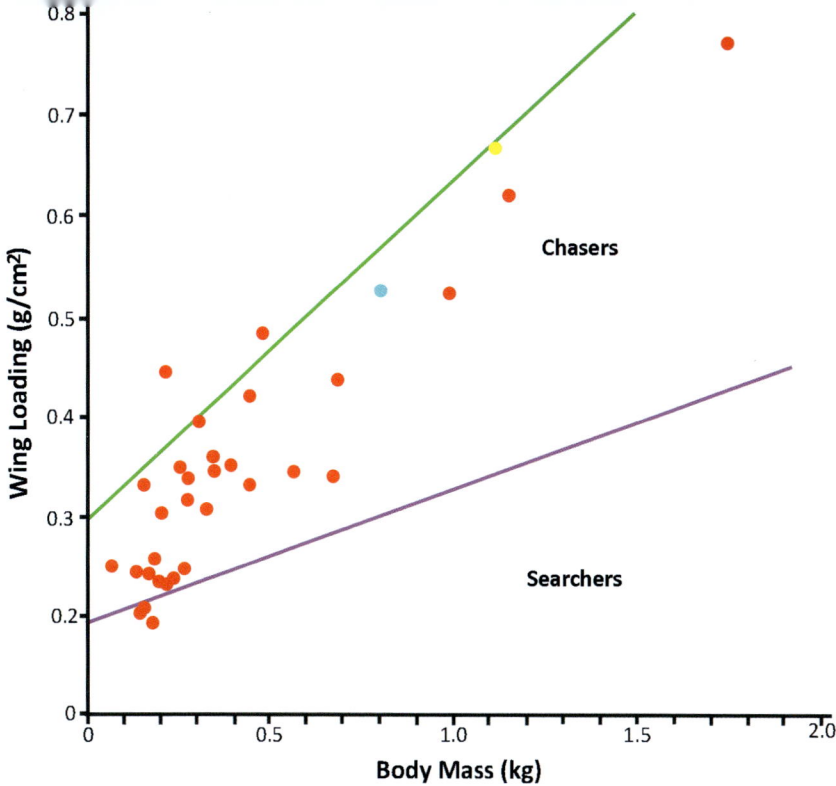

Figure 3.2 Wing loading and body mass for the majority of Falco species marked by red dots. In each case the red dot represents the mean value of body mass and wing loading for adult males and females. The male nominate Peregrine is represented by the blue dot, the female by the yellow dot (which is consistent with, though not overly so, the data on the opposite page). Below the green line would be situated Accipiter 'chasing' species, while below the purple line would lie 'searching' species such as harriers and kites. Cade (1980) positioned his Accipiter line on the basis of data collected from three North American hawks (Northern Goshawk (*Accipiter gentilis*), Sharp-shinned Hawk (*Accipiter striatus*) and Cooper's Hawk (*Accipiter cooperii*)). We have checked the line's position relative to three African hawks (Black Sparrowhawk (*Accipiter melanoleucos*), African Goshawk (*Accipiter tachiro*) and Shikra (*Accipiter badius*)), from data in Mendelsohn et al. (1989), and it is consistent with that data. The purple line is also consistent with data from African harriers and kites.

The heaviest Falco (the Gyrfalcon) is represented by the red dots (male, *left*, and female, *right*) furthest to the right. The other rightward red dot is the Saker Falcon. The cluster of red dots to the left of the figure are the Kestrels, Hobbies and other Falcos. The species below the purple line are, by order of increasing wing loading, the Western and Eastern Red-footed Falcons and the Lesser Kestrel, all of which are almost exclusively insect feeders and so have low wing loading for improved agility. The two species above the green line are the Orange-breasted Falcon, to the left, and the Mauritius Kestrel. The former hunts mainly over forests its hunting mode being chasing rather than stooping to avoid prey which is hit, but not captured, being lost in the forest: the species also has very large feet to increase the probability of successful capture. The Mauritius Kestrel is primarily a hunter of small avian prey by surprise attack or tail-chasing. In all cases the data for the figure have been taken from Table 4 of Cade (1982), the figure itself being based on Fig. 4 of the same reference. The data of Cade's Table 4 are based on a formulaic calculation of wing loading, not on the basis of imposing a grid on a dead bird as used in our study: the formula Cade used derives from Greenewalt (1962).

The figure shows that those Falcos which predominantly feed on avian prey have both high wing loading and body weight, the former to offer quick acceleration and rapid level flight speed, the latter to aid acceleration under gravity for stoop attacks.

Falcon Flight

While it is possible to define the characteristics which define raptor flight, analysing flight mathematically is, as noted above, difficult. The mathematics of fixed-wing aircraft, *i.e.* the interaction between the wing and the air, is now well-understood, particularly with respect to the vortices generated and their effect on preferred wing shape and performance. But the avian wing need not maintain the 'simple' geometry of a fixed-wing aircraft: birds can morph their wings by moving individual feathers in three dimensions allowing a vast number of potential geometries[3]. Combined with highly sensitive wind/air pressure sensing this wing morphing allows amazing active flight control even if the wing is not flapping. Flapping flight brings even more complexity as the bird's shoulder rotates during a complete up/down stroke and, as has recently been shown, the left and right wings, may move differently during the complete wing beat to allow for alterations in the wind across the body due to gusting[4].

The difficulty of mathematically analysing the flying bird led Colin Pennycuick, one of the world's leading experts on the modelling of avian flight whose death was sadly reported in 2020, to develop a theory based on defining the clearly important parameters – wingspan, wing area, mass *etc.* – and combining these using dimensional analysis rather than pure theory to produce a model which could evaluate the basic characteristics – wing-beat frequency, minimum power speed *etc.* – for any bird. That model is used below to define these characteristics for the Peregrine, in combination with details of an inertial measuring unit (IMU) designed and constructed for attachment to falconry birds in order to check the validity of modelling.

Flapping Flight

Peregrines usually hunt from a perch, particularly in winter when energy needs to be conserved, but will use flapping flight to make aerial searches for prey.

Wing-Beat Frequency

A fundamental parameter of bird flight is wing-beat frequency. While to an extent the frequency at which a bird beats its wings is controlled by the bird itself, all bird observers will have noted that smaller birds flap at a higher rate than larger ones, and that for birds of the same size, those with smaller wings

---

[3] Durston *et al.* (2019) note that birds can change wing geometry, altering camber, twist, sweep and dihedral, and offer insights on the effect of such changes based on the photography of a year-old Barn Owl and a three-year old male Peregrine.

[4] Ravi *et al.* (2020) implanted electrodes into the flight muscles of Ruby-throated Hummingbirds (*Archilochus colubris*) and exposed them to gusting wind in a wind tunnel. While hummingbirds are very small and so will very likely be more influenced by gusting than larger birds, there is no reason to doubt that all birds can morph their wings individually to counteract the effects of gusting.

flap faster than those with larger wings. From his work on the flight dynamics of birds Pennycuick (2008) derived an equation which relates the obviously important variables of a bird to derive what may be termed its 'natural wing-beat frequency':

$$F = M^{3/8} S^{-23/24} A^{-1/3} g^{1/2} \rho^{-3/8} \qquad (3.1)$$

where:
F is the wing-beat frequency,
M (kg), S (m) and A (m²) are the mass, wing span and wing area (see Fig. 3.1) of the bird,
g is the acceleration due to gravity,
and ρ is the density of air (kg/m³).

Over recent years the availability of electronic components for attachment to birds to track and analyse flights has improved to the point where IMUs can be constructed of sufficiently small size and weight to fly on relatively small species. With the assistance of interested colleagues RS constructed IMUs to investigate the flight characteristics of falconry birds. The availability of captive, but free-flying, Peregrines has allowed these IMUs to be flown to investigate some aspects of the flight of the species such as wing-beat frequency and cruising flight speed, as well as to make measurements during hunting flights.

The first IMU comprised a gps unit which ran at 1Hz, tri-axial accelerometer, gyro and magnetometer units which ran at 200Hz, and a barometer, running at 20Hz, which backed up altitude data from the gps satellites: while satellite altitude data are excellent if many satellites are in view, the data become less reliable if fewer are seen. Having learned from the behaviour of this unit, a second was constructed with tracking at 18Hz, barometer at 50Hz and tri-axial units running at speeds from 400Hz to 1.6kHz. Both IMUs were powered by small LiPo batteries which give flight times from about 25 minutes to an hour depending on battery size (*i.e.* mAh). In both IMUs data were stored by an onboard flash drive. The IMU was retrieved after each flight, data being downloaded for analysis. Each IMU measured about 33x18x8mm: the first weighed about 8g including battery, the second 3g plus the battery (0.7-1.2g, depending on power). Each requires a 1g metal spring which allows clipping to a standard harness worn by almost all falconry or demonstration birds primarily to allow the attachment of a location device in case the bird goes missing (Box 3.2 overleaf).

Box 3.2 Putting tags on birds

Size and mass were, of course, critical elements of the design. The mass of a unit has a direct effect on the bird as it represents an immediate increase in weight without the advantage of an increase in fitness which might otherwise compensate in the longer term. In studies by a Dutch team at the University of Groningen (Videler *et al.*, 1988a, 1988b), male and female Common Kestrels were loaded with 31g or 61g of lead (approximately the weight, and double the weight, of the voles which wild falcons hunt). Unloaded the kestrels flew at speeds close to the V_{mr} (maximum range speed – see below) predicted by theoretical models, but as the payload increased speeds declined, approaching the V_{mp} (minimum power speed – see below) predicted by the same models – these models are explored further in the text. As the payload mass increased, the decline in speed was accompanied by an increase in wing-beat frequency and an increase in tail inclination. The payloads carried by the falcons were large. The male weighed 160g, the female 190g, so 31g represented 19% and 16% of body weight respectively for male and female, while 61g represented 38% and 32%.

In the wild, smaller alterations in body mass are likely on a daily basis. Hambly *et al.* (2004), looking at the variation in flight energy costs and flight speeds in the Cockatiel (*Nymphicus hollandicus*) – an endemic Australian species also known as the Quarrion – noted that daily weight changes of 5-10% were normal. Hambly and co-workers noted declines in flight speed of the birds with payloads of 5%, 10% and 15%, but then a reversion to control flight speed (*i.e.* 0% payload) at 20%. The decreases in speed for lower payloads were initially small (of order 1% at 5% payload, 3% at 10% payload), but rose to 7% at 15% payload, a statistically significant difference). The reversion to normal flight speed at the higher payload was accompanied by a significant increase in wing-beat frequency: the Cockatiels were changing their behaviour rather than simply reducing their speed. The energy cost of flying increased with payload, but was not significantly higher than unloaded flight for any payload. While this is surprising, Hambly *et al.* note that Cockatiels regularly increase their body mass by 20% prior to migration, suggesting that the observed behavioural change at the higher payload was an inherent strategy.

While falcons differ from Cockatiels in many ways, the results of Hambly *et al.* suggest that provided the mass of a unit added to a falcon was minimised the effect on flight characteristics would be limited. This was confirmed by a study of Pennycuick *et al.* (2012) testing on Rose-coloured Starlings (*Sturnus roseus*). But a separate issue was noted in that work, namely interference with the air flowing over the back of the bird. In the study the payload (a dummy transmitter) projected 6mm above the back of the bird. This increased the drag coefficient of the bird (as measured in a wind tunnel) by 50%, but the addition of an angled aerial increased the drag coefficient by almost 200%. Pennycuick and co-workers then simulated the effect of the mass and drag

(*Branta leucopsis*) which had been fitted with a satellite transmitter to study its migration. The simulation suggested that the effect of increased mass due to the transmitter would have been small, but the effect of drag coefficient increase would have reduced range, and decreased energy reserves on arrival.

In a further test of the effect of tagging birds, Homberger *et al.* (2021) studied the survival of Grey Partridges (*Perdix perdix*) tagged as part of a reintroduction programme in the only remaining area of Switzerland where the species still bred. The results were not encouraging, the annual survival of untagged birds being five times that of those with tags. As the authors note, the risks and benefits of tagging need careful assessment in any reintroduction and, by implication, in any study.

A related concern was raised by Sodhi *et al.* (1991) who wondered if the radiotagging of urban-breeding Merlins in Saskatoon, Canada was affecting breeding success as well as survival. The tags used weighed 4g, representing 2.4% and 1.6% of the body weight of male and female Merlins respectively. In this study no difference was found in the breeding success of tagged and untagged males. In addition, survival rates of both males and females, as measured by the return rates of the birds the following year, were indistinguishable. By observation there was also no discernible behavioural difference between tagged and untagged birds, though those fitted with tags (either to the underside of tail feathers or legs) tended to peck at the tags for some time and to preen frequently immediately after release. In conclusion Sodhi *et al.* conclude that the small radiotags were having no behavioural effect on the Merlins. The difference in the results between the studies of Sodhi *et al.* and that of Holmberger *et al.* (2021) is striking, but the later study reinforces the overall view that putting tags on wild birds needs very careful assessment.

While considerations of reintroduction and breeding success were irrelevant in the studies carried out on falconry birds for this book, it is heartening to note that all the experiments reported above suggest that in the short term the use of tags on birds does not significantly affect flight performance.

The UK restricts attachments intended for wild birds to units weighing preferably less than 3% and definitively below 5% of body weight. For captive birds this limit is somewhat artificial as the IMU is removed after each flight: in the case of the Peregrines used to collect the data described below the unit represented no more than 0.5-1% of normal flying weight.

Using the IMU, the wing-beat frequency of the falconry Peregrine TS was measured in natural conditions, *i.e.* with the bird flying outdoors, wind speed being measured using a standard anemometer.

Allowing for the IMU mass, and assuming the wing area calculated above and an air density of $1.225 kg/m^3$, Pennycuick's equation (Equation 3.1 above) gives a wing-beat frequency of 5.0Hz for TS. Fig. 3.3 (overleaf) is the z-axis accelerometer data from level flight by TS. In this 6.5s flight section the bird made 32 wing beats, a wing-beat frequency of 4.9Hz. Given the probable

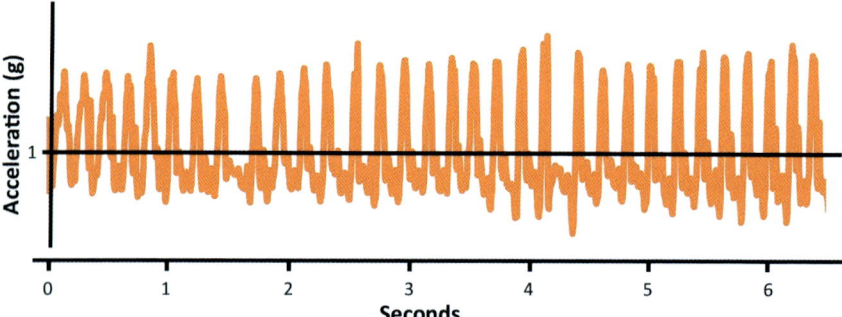

Figure 3.3 Wing-beat frequency for the falconry Peregrine TS during a period of cruising flight.

error in calculating wing span (and, therefore, wing loading) this frequency is in extremely good agreement with the theory. The calculated frequency is also in close agreement with those measured on migrating Peregrines above Switzerland, southern Germany, Mallorca, Malaga and southern Israel (Bruderer *et al.*, 2010) who measured a range of 4.9-5.1Hz, corrected from altitude to mean sea level, using high-speed photography.

From the work carried out on each of the UK's breeding falcons, what is noticeable is that an individual bird, having reached adulthood and, therefore, having a (more or less) fixed wing area and span will have a fixed (perhaps 'preferred' would be a better word) wing-beat frequency for a given weight and will use that frequency at all times, varying the power input per beat or utilising the wind to vary flying height and speed.

Cruising Speed
At any point on Earth, an object falling through any medium is subject to gravity, which accelerates it, converting potential energy into kinetic energy. The falling is also subject to drag due to the resistance to movement of the medium. This resistance clearly depends on the medium's density – for instance, things fall faster through air than through water. Drag is itself a complex aerodynamic force which comprises several differing forces dependent on the profile of the object (*i.e.* the shape offered to the medium – form drag), the density of the medium (friction drag), and the interference of the medium's flow pattern over the object (most obviously, the creation of vortices in the wake of the bird which create a drag on forward motion). The three forms of drag are collectively known as parasitic drag. For a bird, which evolution has formed to maximise lift and so minimise the energy input of flying, there is an induced drag consequent upon the inherent lift generated by the bird's shape in any non-vertical flight. From basic aerodynamics the power required to generate lift varies with speed, while the power to overcome drag, and therefore generate thrust, varies with the cube of speed.

F.p. ernesti, Perak, Malaysia, Amar-Singh HSS.

These theoretical considerations predict that the power required for a bird flying horizontally at a range of speeds will follow a U-shaped curve, *i.e.* the power output required to fly slowly, and to fly quickly, will be higher than for intermediate speeds. Proof of this theory is complex, but early experiments by Tucker (1968, 1972) suggested that the theory holds. Flying birds in a small wind tunnel, the bird fitted with a mask to measure respiratory rate, Tucker (1968) measured the breathing of two Budgerigars (weight 30-40g) at various speeds and produced a classic U-shaped curve (Fig. 3.4), as did Bundle *et al.* (2007) for two different bird species.

Over the three decades subsequent to Tucker's experiments, other researchers have measured the power .v. speed curve for differing species. Dial *et al.* (1997) used in vivo bone-strain measurements of the pectoralis muscle in three Black-billed Magpies (*Pica hudsonia*, weight about 175g) flying in a

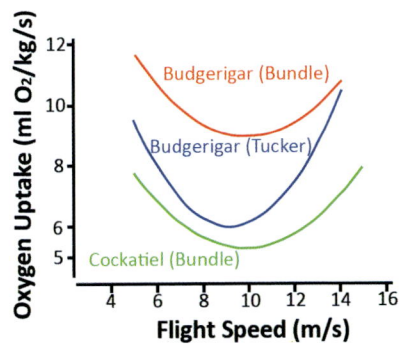

Figure 3.4 Variation of oxygen uptake with flight speed for Budgerigars and Cockatiels (Bundle *et al.*, 2007) and Budgerigars (Tucker, 1968). The Tucker data was extrapolated by Bundle *et al.* to fit their range of flight speeds. Bundle *et al.* consider the difference between the Budgerigar data in the two data sets is due to the differing experimental set-ups. Redrawn from Bundle *et al.* (2007).

wind tunnel with a range of flight speeds from 0m/s (*i.e.* 'hovering') to 14m/s (50.4kph). The resulting power curve resembled the theoretical U-shape, the birds requiring almost three times the power when hovering in comparison to minimum power flight speed, but was flattened at higher speeds. *i.e.* the range over which the power was minimal was greater than might have been expected. Tobalske *et al*. (2003), integrating *in vivo* measurements of the pectoralis

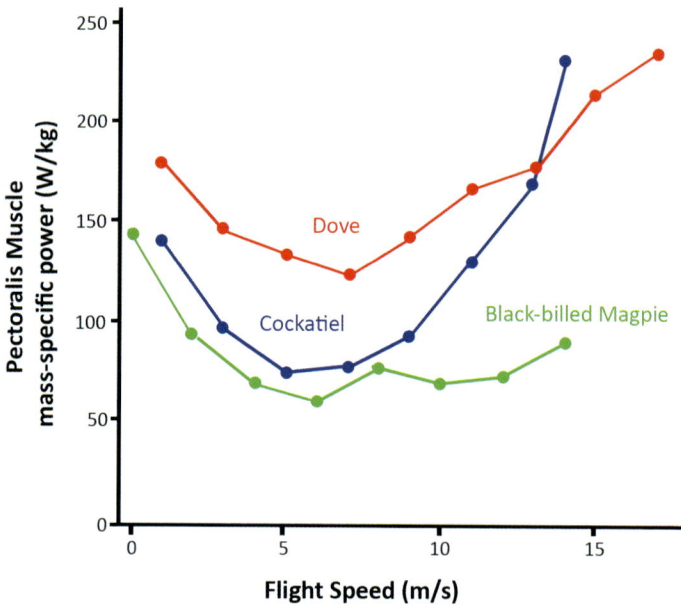

Figure 3.5 Power .v. speed curves measured for Cockatiels and doves by Tobalske *et al*. (2003), for Black-billed Magpies by Dial *et al*. (1997) and for Zebra Finch (Askew and Ellerby, 2007). Note that the dove used in the experiment of Tobalske *et al*. (2003) was identified as a Ringed Turtle-Dove (*Streptopelia risoria*), but in Tobalske (2007), which replicated the comparison figure, was identified as a *Zenaida macroura* (Mourning Dove). While the curve for Cockatiels in the work of Bundle *et al*. (2007 – Fig. 3.5) is very similar, there is a clear off-set in terms of the speed at the minimum of the curve – 6m/s Tobalske *et al*. (2003) against 10m/s Bundle *et al*. (2007). In the later paper by Tobalske *et al*. (2007) the researchers suggest the two distinctly different experimental methods (measuring mechanical power and respiration) results in differing flight efficiencies across speeds and that in future it is vital to understand the difference when making comparisons. Askew and Ellerby (2007) actually collected data from Budgerigars as well. These data lie close to, and occasionally overlap, those for the Zebra Finch, and also marginally overlap the Magpie data at lower speeds. The data has therefore been excluded to avoid confusion.

muscle with quasi-steady-state aerodynamic modelling, measured curves for Cockatiels (weight about 90g) and a dove species (weight 150-160g), while Askew and Ellerby (2007) used *in vivo* pectoralis muscle measurements to produce curves for Budgerigars and Zebra Finches (*Taeniopygia guttata*, weight 12g) which had been trained to fly in a variable-speed wind tunnel (Fig. 3.5).

Fig. 3.5 shows data from species differing in weight by a factor of 15, but despite that range some researchers suggested that true U-shaped flight curves were applicable only to lighter bird species. However, Tucker (1972) produced data on two 300g Laughing Gulls (*Larus atricilla*) and the curve in that case was similar in having a minimum point (at about 6m/s – 22kph) and an increasing power output at higher speeds. But the experiment did not include lower speeds. Hudson and Bernstein (1983) flew five Chihuahuan Ravens (*Corvus cryptoleucus*), mean weight 480g, fitted with masks in a wind tunnel, but they also obtained data over a limited speed range (8-11m/s – 29-40kph). The power requirement increased linearly throughout that speed range, a result which is consistent with a curve with a minimum of <8m/s (28.8kph), but hardly conclusive proof that U-shapes are 'real' curves across the range of avian body weights. That debate continues.

But assuming a true U-shaped curve applies to the mechanical power required for flight, the minimum point of the curve represents a bird's most

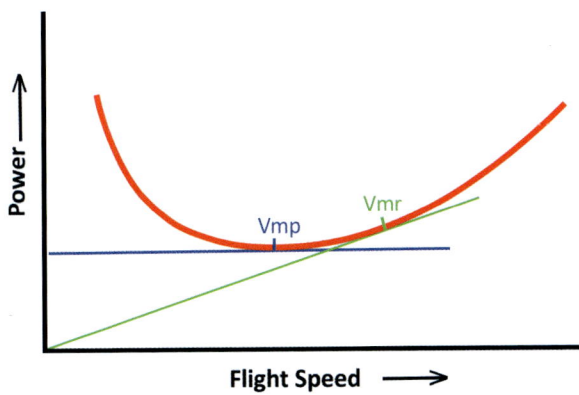

Figure 3.6 Theoretical mechanical power curve (power .v. flight speed) for a nominate Peregrine. The minimum of the curve reveals the minimum power speed, while the asymptotic line from zero to the curve reveals the maximum range speed.

economical speed, *i.e.* where the energy used per unit flight time is a minimum (the minimum power speed, V_{mp}). The curve also allows a second speed to be calculated, a speed at which the ratio of lift to drag is a maximum. This is the speed at which the energy required to travel a unit distance is a minimum (the maximum range speed, V_{mr}). V_{mr} is of critical importance to migrating birds (Fig. 3.6).

Pennycuick (2008) provides an equation for the calculation of V_{mp}, this again depending on the mass and wing dimensions of the bird, as well as the frontal area the bird presents to the air and the body's drag coefficient, together with air density, gravity and the induced power factor. To keep itself airborne the bird must power its wings. In an ideal world, all the downthrust power of wing movement would perform this task. In reality, the wing is performing two jobs simultaneously, both maintaining the bird airborne and creating forward thrust. The induced power factor is a multiplier on the ideal induced power to allow for this real world.

Pennycuick's equation is:

$$V_{mp} = \frac{0.807 k^{0.25} M^{0.5} g^{0.5}}{\rho^{0.5} S^{0.5} A_b^{0.25} C_d^{0.25}} \qquad (3.2)$$

where:
M, S, g and ρ are as in Equation (3.1) above, and:
k is the induced power factor,
A_b is the bird's frontal area (m²),
C_d is the bird's body drag coefficient.

The induced power factor usually differs from 1.0 by a surprisingly small amount, lying between 1.1 and 1.2. For flapping flight the value would be higher and Pennycuick suggests 1.2.

In an earlier paper, Pennycuick (1988) provided information for the calculation of A_b and C_d. Using data obtained from both museum specimens and living birds he produced an empirical equation for A_b:

$$A_b(m^2) = 8.13 \times 10^{-3} M^{0.666} \qquad (3.3)$$

from which $A_b = 7.21 \times 10^{-3}$ m² for TS.

Pennycuick (2008) suggests a value of 0.1 for C_d. Using these data Pennycuick's equation predicts a cruising speed of 13.2m/s (48kph).

Those who have watched Peregrines in flight, particularly breeding birds pursuing avian intruders into the nesting areas, will know that the falcons can readily exceed this speed at times, with speeds of up to 100kph having been stated for Peregrines in level flight pursuits.

Soaring and Gliding

Despite the adaptations for flying, it is energy intensive and birds utilise both spatial and temporal changes in wind velocity to gain height and, therefore, potential energy, by soaring. This energy can then be exchanged for kinetic energy as the bird loses height. Potential energy can be exchanged quickly or more gradually.

Soaring utilising spatial changes relates to the vertical gradient of wind velocity due to frictional losses in the air close to the Earth's surface, and the potential existence of updrafts created by local topography, while temporal changes arise from gusting as consistently homogenous winds are rare. In general, the use of the differential speeds of air layers is termed dynamic soaring as it is usually associated with seabirds (particularly albatrosses), while the use of spatial and temporal changes induced by land topography or temperature (the latter creating rising columns of warm air) is termed slope soaring. In each case the bird first travels across the wind, gathering energy in the way that sailing boats do while tacking. The bird then turns downwind, gathering further energy as it is pushed along by the wind. Now the bird turns across the wind again, before turning into the wind and converting kinetic energy to potential energy as it rises against the air flow. This is a necessarily simplistic explanation as in practice the conversion of kinetic to potential energy is complicated by the fact that kinetic energy varies as the square of velocity and velocity is varying, and drag must be included for a full understanding (as drag is also variable with time as the bird's attitude alters). However, the wing shape of birds reflects their use of either method. Albatrosses have long, narrow wings for dynamic soaring, the primaries showing little slotting (the discrete 'finger' structure of the primaries), while vultures and eagles have broad wings and significantly fingered primaries, the wing offering high lift, but relatively limited manoeuvrability. Only the outermost primaries of Peregrines (Nos. 9 and 10) are slotted at the wing tips, the slotting being much less pronounced than in the wings of species which habitually soar while searching for prey. Nevertheless, despite their wing shape and feather configuration, Peregrines are remarkably adept at soaring.

The exchange of potential to kinetic energy is most obviously seen in a stooping falcon, where the exchange is quick, a high-angled dive resulting in fast acceleration as prey is attacked. A more gradual loss results in gliding, the bird exchanging potential energy for position as it traverses the landscape in its search for food.

Although Peregrines undoubtedly soar and glide when they are able, they are not renowned for these flight patterns, and falconry birds very rarely utilise either flight mode. As a consequence no direct data was obtained with the IMU on either flight mode.

However, for completeness data on gliding has been calculated using the equations set down by Pennycuick (2008). There is a limit to how slow a bird may glide as ultimately the wing is no longer able to generate the lift required to stay aloft, and the bird stalls. Pennycuick's equations allow the minimum sink speed to be calculated, the speed at which the bird is just above stall speed. For a Peregrine the minimum sink speed is 7.6m/s (27.4kph). This compares with the speed for the best glide ratio (V_{bg}, the polar glide speed) of 11.8m/s (42.5kph). The difference between the two will be immediately apparent to a glider pilot, but is less obviously different for the layman. The minimum sink speed equates to the maximum time a bird can stay in the air during a glide (*i.e.* without wing flapping) for a given height loss (*i.e.* before reaching the ground), while the best glide speed defines the maximum distance that can be travelled for a given height loss. For a glider pilot the latter is obviously critical as it maximises the likelihood that a safe landing area can be reached. For a bird the difference is somewhat academic as it is difficult to imagine the circumstances in which the bird would be forced to choose between them. In a polar glide, a Peregrine travelling at V$_{bg}$ glides at an angle to horizontal of 3.9°.

Migration Flights
Pennycuick (2008) also sets down an equation for the calculation of V$_{mr}$, the maximum range speed, which a bird would be expected to use (in still air) on migration:

$$V_{mr} = \frac{k^{0.25} M^{0.5} g^{0.5}}{\rho^{0.5} P^{0.25} S_d^{0.25}} \qquad (3.4)$$

where k, M, g and ρ are as for Equation (3.2) above and:
P is the equivalent flat-plate area of the bird's body,
S$_d$ is disc area of the bird.

The equivalent flat-plate area of the body is defined as P = A$_b$C$_d$ with the two terms as in Equation (3.2) and S$_d$ is the disc area produced by the wings and = πS²/4, with S the wingspan (as in Equation (3.1)).

For TS V$_{mr}$ = 17.7m/s (64kph) which would be the expected migration speed for an adult female Peregrine, though many females are heavier than TS and so would have higher calculated speeds. Bruderer and Boldt (2001) and Bruderer *et al.* (2010) observed migrating falcons above several countries and measured speeds using radar and measured speeds which were comparable to the V$_{mr}$ calculated here. The studies covered many species, including all European falcons (Table 3.1).

F.p. anatum carrying prey, coastal Los Angeles, USA.
Gabriel Gruia.

Flight Characteristics

Falcon Species	Air Speed (kph)	Equivalent mean sea-level speed ± SD (kph)	Ground speed range allowing for wind (kph)	Mean wing-beat frequency (Hz)	Mean equivalent sea-level wing-beat frequency (Hz)
Eurasian Hobby	25.2–57.6	40.7 ± 7.9	21.6–64.8	5.2 5.8^3	5.0 5.6^3
Common Kestrel	21.6–68.4	44.3 ± 9.0	14.4–72.0	5.1 5.8^3	4.9 5.5^3
Merlin[1]	—	—	—	6.1	5.9
Peregrine	25.2–64.8	43.6 ± 12.6	25.2–75.6	5.1	4.9
Barbary[2]	50.4	51.8 ± 0.4	50.4–61.2	4.7 5.6^3	4.7 5.6^3
Eleonora's	43.2–50.4	46.1 ± 8.3	43.2	5.1	5.1
Lesser Kestrel	28.8–68.4	41.0 ± 10.1	28.8–64.8	5.1	4.9
Lanner	36.0–54.0	40.0 ± 9.7	39.6–61.2	4.4	4.2
Red-footed	43.2–50.4	46.1 ± 3.6	43.2–61.2	5.2	5.1
Sooty	50.4–64.8	54.7 ± 6.8	43.2–72.0	4.5	4.4

Table 3.1 Speed and wing-beat frequency of falcons during migratory flapping flight periods.

Notes:
1. Merlins were not observed in the study of Bruderer and Boldt (2001). Only two Merlins were observed in the study of Bruderer *et al.* (2010).
2. Only two observations in Bruderer and Boldt (2001).
3. These frequencies were observed from filming. The remaining frequencies were obtained using the variation in intensity of the radar echoes created by the wing beats. The wing-beat variations could be distinguished from phases of flight which involved no wing beats.

Cochrane and Applegate (1986) measured the migration speeds of Peregrines (and Merlins) in the USA using a combination of radiotelemetry and visual observation. The researchers noted that both species soared during migration and rarely flew in straight lines when hunting, but collected data from periods of flapping flight: these are set down in Table 3.2 overleaf.

Species	Duration[1] (mins)	Ground Speed (kph)	Air Speed (kph)
Merlin, immature male	20	48.5	48.5
Merlin, immature male	28.5	30	30.5
Merlin, adult female	30.5	74	44
Merlin, immature female	40.5	30	44
	Mean	45.6	39.3
Peregrine, adult female	55[2]	50	49
Peregrine, immature female	83	49.5	36.5
Peregrine, immature female	9	53	53.5
Peregrine, immature male	9	36.5	36.5
Peregrine, immature male	17	70	50.5
Peregrine, immature male	129.5[2]	35	36
	Mean	48.8	43.7

Table 3.2 Mean flight speeds of migrating Merlins and Peregrines measured on straight paths to the nearest 0.5km/h.

Notes:
1. Time over which speed was calculated.
2. The falcon was flying in a strong cross-wind.

The data for migratory Peregrines in Tables 3.1 and 3.2 show reasonable agreement, both with each other and with the value of V_{mr} calculated above, given the potential effects of wind, altitude and the imperative of individual birds on any given day of their migration.

Stooping
The Peregrine is usually quoted as the fastest creature on Earth and the (occasionally exaggerated) claims for its top speed make the falcon's stoop an obvious candidate for observation and research. For the human observer, stooping is also the most spectacular of all the falcon's hunting techniques, and it is often assumed that it is the sole means of hunting for Peregrines. In reality, while it is the most often used technique – stoops are seen in about 70-75% of Peregrine hunts – the falcons also use several other techniques (see *Hunting strategies*).

It is worth starting this discussion by noting that as stooping involves exchanging position above the Earth and, hence, gravity as the method of converting potential to kinetic energy, the velocity achieved depends on the mass of the bird: all other things being equal, a stooping Gyrfalcon would be faster than a stooping Peregrine. Stooping is, without doubt, visually sensational and allows a Peregrine to minimise the time taken to reach prey which heightens the likelihood of surprise. But pure speed would also seem to be disadvantageous. The obvious problem is that to hit prey at speed raises the risk of injury to the falcon, but it would also seem that dropping through the air at high speed means sacrificing control of flight path. But such is not the case, a theoretical study by Mills *et al.* (2018) noting that speed enables superior manoeuvring and higher roll agility, each of which raises the likelihood of intercepting the prey. At the last second before 'binding to' prey, the Peregrine reduces speed by opening of the wings and tail to enhance drag. Speed is also reduced if the Peregrine is hitting the prey, to a point where collision can be used to inflict serious injury on the prey without causing injury to the predator.

Both Cade (1982) and Ratcliffe (1993) quote claimed speeds for Peregrines from early observers, but not until the 1980s are claims from 'modern' equipment available for assessment. Alerstam (1987), while using tracking radar to monitor bird migration in southern Sweden, observed two stoops (and an aborted stoop) by a Peregrine apparently against Skylarks. The average stoop speeds for the falcon were 25m/s (90kph/56mph) over stoop periods of about 50s for the two full stoops, with a maximum speed of 39m/s (140kph/87.5mph) during one ten-second time interval. The Peregrine's stoop angle was 30° for the first attack, 64° for the second (in each case relative to the horizontal). Later Tucker *et al.* (1998) measured a speed of 58m/s (209kph/131mph) for a falconry Gyrfalcon, at which time the speed was the highest ever recorded for a bird. Peter and Kestenholz (1998) measured 51m/s (184kph/115mph) for a wild Peregrine. Tucker (1999) notes that in the case of the falconry Gyrfalcon (a bird called *Kumpan*) the speed of 209kph was reached in a dive starting at 500m during which the bird dropped 150m at an angle of 62°. The bird then altered its drag to maintain that speed: had it continued to accelerate it would have exceeded 256kph (160mph) by the time it reached the ground. In a study using IMUs on five falconry raptors including Lanner Falcons and Peregrine Falcons in South Africa, Hart *et al.* (2018) noted a maximum speed for the Peregrine of 196kph (123mph). At such speeds it is remarkable that the falcon's eyes are not damaged and its breathing not impaired by air pressure. The eyes are protected by a transparent nictitating membrane which guards against dust *etc.* during flight (and also against potential damage from an inadvertent claw by a hungry, food-grabbing youngster)[5]. The muscular structure of the eye, including the nictitating membrane, in woodpeckers is known to tighten and so prevent damage to the retina during high-speed bill movements (Wygnanski-Jaffe *et al.*, 2007) and it likely that the Peregrine is able to do something similar. The Peregrine also has a tubercle, a cone-shaped baffle, in

---

[5] In addition, see *Vision* below and the comments on the Harderian gland.

The Peregrine Falcon

the nostril, visible in the photograph of the head below, which deflects air from the nasal passages and protects them from shock, a feature copied in the cones of supersonic aircraft.

Head of an *F.p. anatum* showing the nasal tubercle. *Greg Hume.*

The bony tubercle protrudes slightly from the nares (nostrils), and is assumed to be an evolutionary development of the nasal conchae to aid the falcon's stooping flights by allowing sufficient oxygen to reach the bird's sensory systems despite the high back pressure developed by the high attained speeds.

Recent work has shown that the geometry of the bony tubercle in the falcon's nostril has indeed evolved to reduce dynamic pressure in the nostril. Using state-of-the art CT scanning Kesel *et al.* (2019) – see, also, Danter

(2018) have shown that the Peregrine's conchae and tubercle design is highly complex, involving several chambers and connecting passages which combine to reduce pressure.

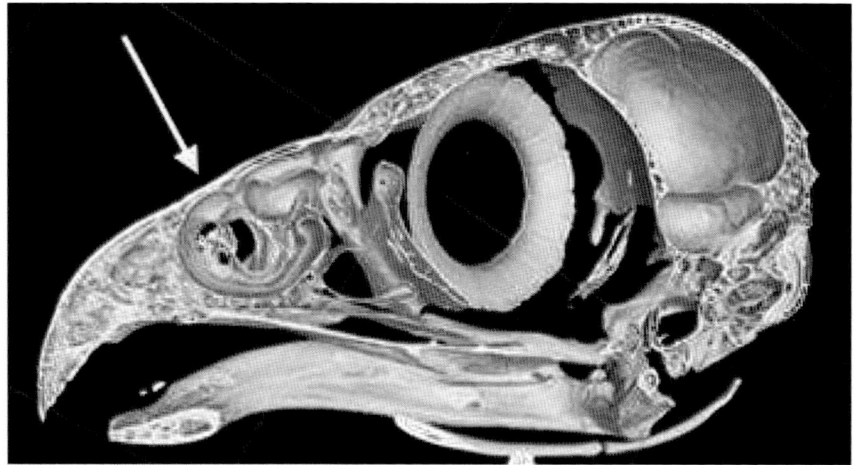

MicroCT image showing the complex structure of the Peregrine's nasal system. The arrow points to the tubercle. *With thanks to Prof. Antonia Kesel.*

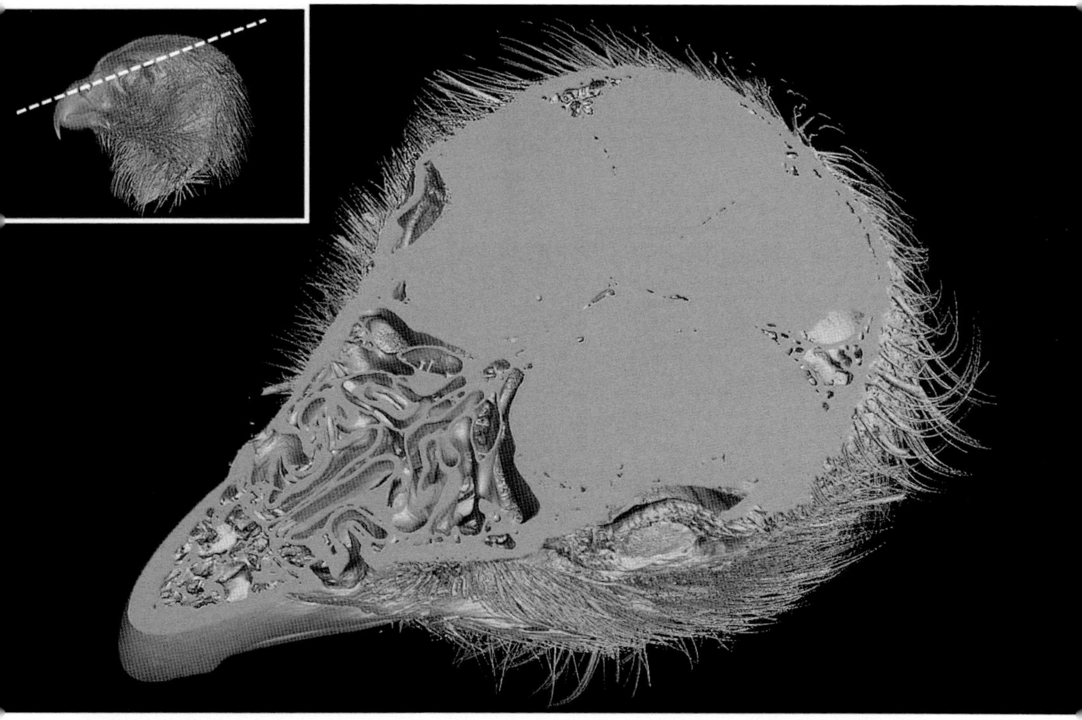

A slice cut through the head of the falcon under study reveals the complexity of the internal bill structure. The location and tilt of the cutting plane is indicated with a white, dashed line in the upper left corner. *With thanks to Prof. Antonia Kesel and Leon Danter.*

Using computational fluid dynamics and pressure measurements on simplified models in a wind tunnel, Kessel *et al.* (2019) were able to show that the tubercle design and position dampened both air pressure and velocity by reducing air volume flow by up to 25%. The tubercle also protects downstream nasal passages from shock loading. Interestingly, while similar bony structures are seen in other raptors, such as eagles, they do not form specific, distinct features, suggesting that the hunting technique of falcons – frequent short flights invariably involving high speeds, as opposed to long, slow soaring flights with occasional bouts of high-speed manoeuvring – has necessitated a specific evolutionary path[6]. Equally interesting is the fact that a similar design to Peregrine's conchae system has been copied in the inlet cones of supersonic aircraft.

The speeds quoted are high, but various, much higher, speeds have been claimed for Peregrines, usually as a result of stoops associated with chasing a skydiver. However, such claims are rarely published in peer-reviewed journals. In an article describing a series of dives and the various body shapes a female *F.p. anatum* (named *Frightful*) assumed in following a skydiver (Franklin, 1999), Franklin states that the falcon went through four identifiable shape phases as speed increased. The first was the 'classic falcon diamond', the wings folded back so that the wing tips and tail tip meet, but with the shoulders still discernible, the head forming the fourth point of the elongated diamond. In the second phase, the 'tight vertical tuck', the space between shoulders and body was reduced. In the 'mummy wrap' phase the interspace between shoulders and body at the base of the neck was zero, the shoulders being forced towards the crop to produce a 'mummy' shape. In the final shape the body appears elongated, the width reduced. Franklin refers to this phase as 'warpdrive vacuum pack'. In the view of the present authors, it is probable that this shape was imposed on the bird by air pressure rather than being chosen by the bird to increase its speed.

One flight by *Frightful* was observed by *National Geographic* magazine and a *YouTube* video was produced. That flight registered a top speed of 183mph (293kph). However, at the end of the video there is a statement that in a later flight *Frightful* achieved 242mph. This speed has not, apparently, ever been published in a peer-reviewed document. It was, however, the basis of the Guinness World Record speed claim of 389.46kph (an authentic looking record, given that it was quoted to an accuracy of 1 in 10^5, but merely a (very accurate) conversion of mph to kph). When RS approached Guinness for the peer-reviewed document which underpinned this speed, they hastily altered their record page to note a record of 'approximately 320km/h (200mph)', but

---

[6] While noting the flow reduction, Danter (2018) points out the work carried out on the modelling of the Peregrine's conchae was based on a single specimen which had been kept frozen for several years, so that further studies are required to confirm the results. Danter also notes that significant variation in conchae design across the avian world, referencing the study of Danner *et al.* (2016) on the sub-species of the Song Sparrow (Melospiza melodia) which showed differences in sub-species' conchae depending on whether the occupied habitat was dry, hot coastal dunes or inland moist areas.

F.p. anatum at start of a stoop. Will Sooter.

added a sentence about possible higher speeds by Peregrines up to '*389km/h (242mph)*'. This final sentence, and the associated speed was, again, set down without any formal justification.

Interestingly, the original Guinness claim stated that the 242mph flight had taken place in 2005. This was incorrect (the claimed flight had been in 1999), but may have been picked up from an article in the *Air and Space* journal of the Smithsonian Institution where Tom Harpole had published an article about *Frightful's* flight in the journal in 2005. The article has curiosities (not least claiming the flight was at '*nearly three miles up*', but later referring to '*17,000ft*', a very different height). RS approached the Smithsonian to ask the basis of the article. As the Smithsonian is an august organisation, to say the least, a studied response was expected. What was received was a three-line email stating that the *Air and Space* magazine had closed '*last December*' (*i.e.* December 2021), that the magazine staff had '*been disbanded*', and that Tom Harpole was '*no longer employed by the Smithsonian Institution*'.

As the speed of any falling object increases, so does the drag, but gravity remains constant so that ultimately gravity and drag are in equilibrium: the object has reached terminal velocity. Terminal velocity can be calculated from the equation:

$$V_t = \sqrt{(2Mg/\rho A_b C_d)} \qquad (3.5)$$

where:
ρ is the density of the medium through which the object is falling:
m is the mass of the object, allowing for the effect of buoyancy. To allow for this the true mass of the object must be reduced by ρV, with ρ as above and V is the object's volume,
g is the acceleration due to gravity (= 9.81m/s²),
A_b is the profile (cross-sectional area) of the object as seen by the medium;
C_d is the drag coefficient of the object.

The Peregrine Falcon

This equation can be used to calculate the fastest speed of a falcon diving vertically, although deriving values for A_b and C_d is no easy task. But if we assume at this stage that *Frightful*'s speed really was $c.390$kph at a height of around 5000m, then we can use Equation (3.5) to calculate speeds at different altitudes, assuming the falcon maintained the same shape and, therefore, profile and drag coefficient. The speeds at various heights can then be calculated (Fig. 3.7).

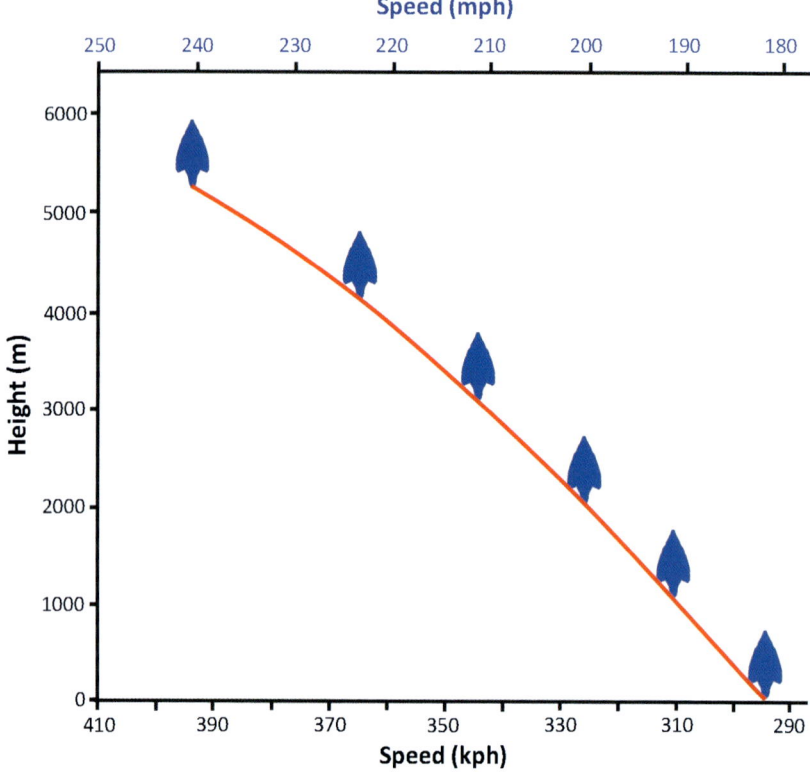

Figure 3.7 Variation of terminal velocity with altitude based on data with due allowance for 'standard' humidity and temperature at various heights. The curve gives the terminal velocity based on the equation set down in the text for each altitude. The value of 389.64kph (242mph) is as quoted for the female Peregrine *Frightful*. In principle the falcon would have taken a finite time to reach that speed from a starting vertical velocity of zero, but the bird exited a Cessna 172 aircraft which has a cruising speed of 226kph and a 'never exceed' speed of 302kph. It is not clear what the speed of the aircraft was when the bird was released. Converting horizontal speed to vertical speed would have given the bird a considerable head-start in accelerating to maximum speed.

Equation (3.5) shows that the velocity of a falling object varies with $1/(\sqrt{\rho})$ which results (Fig. 3.7) in a sea level speed of 294kph (184mph). While this is higher than the highest recorded speed for a falcon, it is not grossly so.

Stooping nominate Peregrine. *Tony Fallon*.

Interestingly it is also very close to the speed *National Geographic* magazine claimed for the bird (183mph) when they filmed Franklin and *Frightful*.

But there are several reasons for caution about the calculation, not least the fact that Peregrines may breed, and therefore hunt, at heights above 3000m in parts of their range and so may stoop in thinner air. More fundamental, the time taken to accelerate to high speeds has been ignored and wild falcons would never dive vertically as in such dives the bird would have no control over the flight[7]. To maintain control, in terms of both speed and direction, the Peregrine would have to abandon the 'shrink-wrapped' attitude, moving from a vertical stoop to an angled dive with the wings in a cupped position. This inevitably means an increase in the frontal area and an increase in drag coefficient, slowing the bird (stoops are normally at 30-45°, but can be both shallower and steeper) and the inclusion of a glide angle complicates the equation significantly, requiring solution in the form of a quartic (fourth-order quadratic) equation. The solution of such an equation allows the terminal velocity of a diving 'ideal' falcon to be calculated at various dive angles, and underpins the work presented by Tucker (1998).

Ideal falcons differ from real ones in not having to concern themselves with time and distance. While terminal velocity is a handy concept, the nature of it – the balance of drag and gravity – means it is approached asymptotically (and so, theoretically, is never reached). Tucker calculated the time to maximum

---

[7] SW has seen Peregrines appearing to dive vertically on occasions, extending a wing or tail to induce an angle away from vertical when required. Peregrine feathers are stiff and so can withstand the pressures required for manoeuvring at high speed. However, it must be remembered that such observations may only imply vertical movement as it is not always easy to see flights in three dimensions.

speed and the altitude loss during the dive, Figs. 3.8a and b. In each case the time is that required to reach 95% of terminal velocity. Tucker showed that the total distance travelled by the ideal falcon (the combination of horizontal distance and altitude loss) was almost independent of glide angle (within 6% of the mean in all cases) and was about 1,200m. As Tucker also notes, if the drag coefficient is assumed to be smaller (the parasitic drag coefficient was assumed to be 0.18 to produce Figs. 3.8a and 3.8b) the terminal speed of the falcon will be higher, but the distance travelled will be greater: if a coefficient of 0.07 is assumed, the value which yields a speed of 174m/s (626kph/389mph) then the total distance travelled must be almost 3,000m.

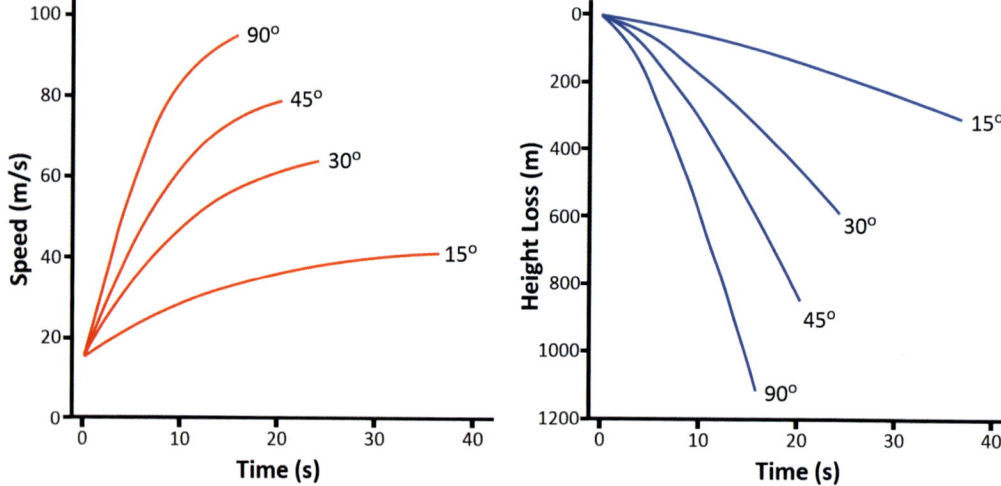

Figure 3.8a Variation of speed with time for an ideal falcon diving at different angles and reaching 95% of equilibrium speed at each angle. The ideal falcon weighs 1kg and begins the dive at a speed of 15m/s. Redrawn from Tucker (1999).

Figure 3.8b Variation of height loss with time for an ideal falcon diving at different angles and reaching 95% of equilibrium speed at each angle. The ideal falcon weighs 1kg and begins the dive at a speed of 15m/s. Redrawn from Tucker (1999).

Figs. 3.8a and b illustrate one of the problems with assessing the flight speed capabilities of Peregrines – there is, in all probability, a great difference between how fast a falcon can dive and how fast it does dive. While they may indeed be able to travel at or above 300kph (186mph), in the wild they are very unlikely to do so because to attain such speeds they need to be so far from their prey they may have difficulty in spotting it (despite their excellent eyesight – see below)[8]. Speed also has to be regulated in such a way as to reach the prey before it has time to make its escape, but at a strike velocity

---

[8] In an interesting, and very early, article regarding the speed of Peregrines, Orton (1975) noted that the argument of how fast a wild, hunting Peregrine could fly, as opposed to would fly, was reminiscent of medieval arguments about how many angels could dance on the head of a pin. Little seems to have changed over the 50 or so years since.

which allows the falcon to be uninjured and able to fly again tomorrow. This regulation of speed by the falcon was illustrated with a falconry bird (a Gyrfalcon weighing 1.02kg) in a study already noted above (Tucker *et al.*, 1998 when the falcon attained a speed of 58m/s (209kph - 131mph) during a dive at an angle of 62° (from the horizontal): the falcon dived from a height of 500m and travelled 250m horizontally to reach the falconer. In calculations of an ideal falcon following a similar dive path, the researchers suggest an acceleration phase followed by a flight at constant speed and a deceleration phase. To maintain a constant speed while continuing to dive, the falcon increases its drag coefficient, then increases it again more radically to decelerate. In each case the falcon has a number of options – it can change the angle of attack of the wing to vary induced drag, or lower its feet to increase parasitic drag. While, clearly, the bird was diving towards a lure rather than prey, it is probable that wild birds will follow the same general pattern, though observations of wild falcons accelerating into a dive by energetic wing flapping suggest that they may attain higher dive speeds, perhaps to 250kph (about 160mph, *cf.* the 'theoretical speed of 184mph above, and the speed of 183mph seen by *National Geographic* on *Frightful*).

During the rapid deceleration phase the falcon experiences a negative g-force, though the exact nature of this depends entirely on timescale. Tucker (1999), on the basis of accelerometers attached to falconry birds, suggests that in very rapid decelerations the bird may experience a 5-10g-force, which is extraordinarily high – high enough to tear the wings off most aircraft and to endanger even those built specifically for aerobatics, and to induce unconsciousness in pilots (Box 3.3).

Box 3.3 g-forces
While the decelerations imposed on falcons at the end of fast stoops seem extremely high, very short-duration imposed forces of order 4-5g are regularly experienced by the riders of high-speed roller coasters and by Formula One drivers. The loads imposed on the heads of American footballers in tackles have recently been shown, by sensors inserted in their helmets, to occasionally exceed 100g. While it might be assumed that such loads would inevitably lead to concussions, that is not the case, though the work has suggested that the footballers (and footballers (*i.e.* soccer players) in the UK and elsewhere) who routinely head a fast-traveling ball do sustain brain injuries in the longer term.

The problem experienced by aircraft pilots, who are usually cited in a discussion of imposed g-forces, is that in a seated position the pilot cannot prevent blood being drained away from the brain when the heart fails to overcome the effect of negative g, loss of consciousness following if the situation is prolonged. In much shorter duration events blood draining does not occur.

As mentioned earlier, Hart *et al.* (2018) noted a maximum speed for a falconry Peregrine of 196kph (123mph): same study registered a highest acceleration of 1.22g and a highest deceleration of 2.4g. In the work presented below on hunting falconry birds the highest instantaneous g-force registered by the IMU the birds were carrying was 6.5g.

That a falcon's wings remain intact during periods of high g-force is impressive, but noting that falcons can experience such high decelerations does not mean that wild birds routinely do so. Given that stooping is both visually spectacular and scientifically interesting as a hunting technique it is no surprise that several teams have investigated aspects of it. A combined team from Germany's Aachen University, Germany and the City University of London investigated the preferred shape of Peregrines during stooping, identifying two specific shapes, (Lagemann *et al.*, 2018: Fig. 3.9) and the loading on the wing in those two shapes (Gowree *et al.*, 2018) using both a live bird trained to fly down a reservoir dam wall in Germany (Fig. 3.10) and model birds in a wind tunnel: olive oil droplets were injected into the wind tunnel, with high-speed photography then used to watch vortex production downstream of the bird.

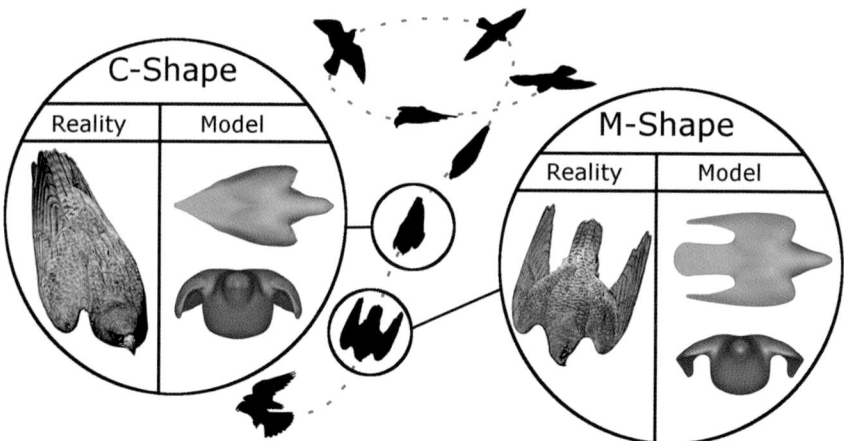

Figure 3.9 Montage of typical wing configurations during a stooping attack of a Peregrine Falcon. The Peregrine starts with a normal flight attitude, then folds its wings into a C-shape (see photograph p119) which can be modified to an M-shape (see photograph p117) several times during the stoop as speed and direction are modified. The montage ends with the Peregrine coming out of the stoop to grab its prey, though in some cases the falcon will strike the prey and then turn to capture it. Reproduced from Lagemann *et al.* (2018), with thanks to Erwin Gowree.

What the experimental work demonstrated was that the morphology and wing shape of the stooping falcon promoted vortices formed at the neck and dorsal region of the bird, with stronger vortices on both the wings and tail, reducing drag and so enabling the bird to travel faster. The vortices

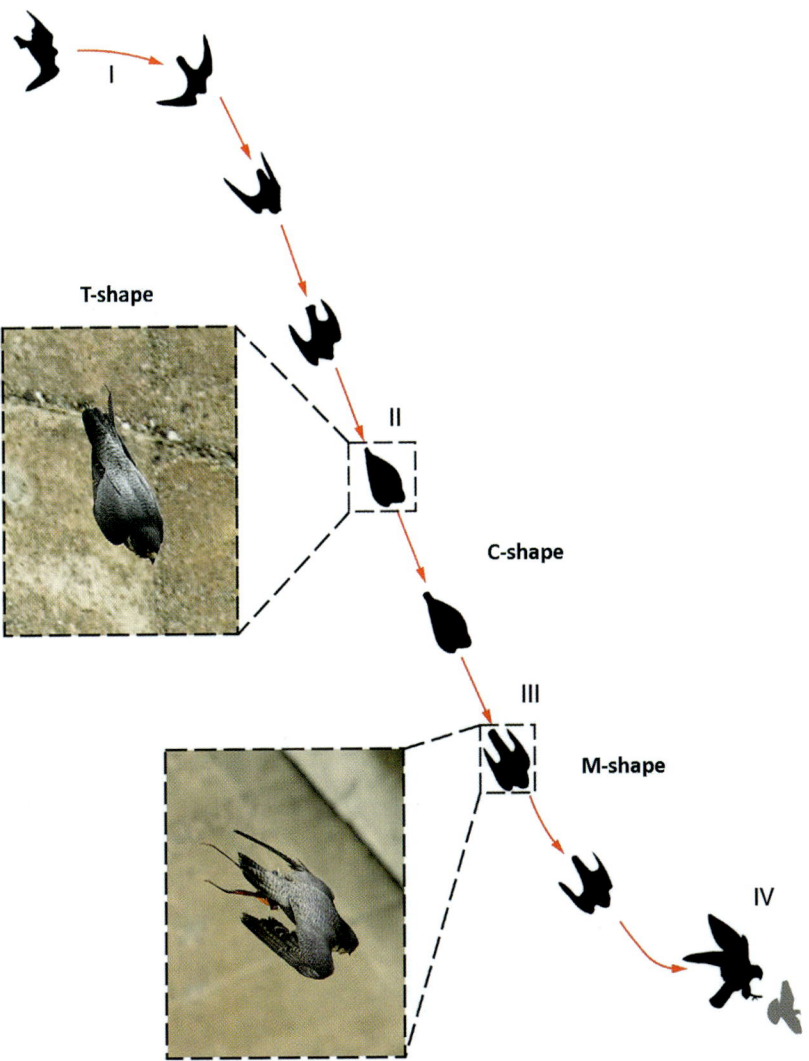

Figure 3.10 Shape variations in the stooping Peregrine. In the initial dive (I) the falcon flaps its wings to acquire speed before turning into the stoop. In phase (II) the falcon assumes a teardrop (T-shape, the equivalent of diamond shape of Franklin, 1999) to create minimum drag and so maximise acceleration due to gravity. Between phases (II) and (III) the bird cups its wings (C-shape) – see also Fig. 3.9 – to acquire control over direction and speed as each requires lift. As the Peregrine nears its prey it reduces its speed and maximises flight control by assuming the M-shape – see also Fig. 3.9 – where the reason for the name is revealed – prior to making a grab for the prey (IV). In instances where the Peregrine decides to strike the prey rather than grabbing it, choosing to circle and retrieve the prey either in the air or from the ground, the orientation of the bird may differ – see photographs in Chapter 9. Reproduced, with minor alterations from Gowree et al. (2018), with thanks to Erwin Gowree.

also enhanced manoeuvrability, providing both lift and control of pitch and roll, these being critical when the falcon comes out of the stoop in order to capture prey. Both morphology and the stiffness of the feathering of Peregrines have evolved to allow the bird to both achieve and control very high speeds (Fig. 3.11). The Peregrine C-shape appears unique to the species (though further research may show it also appears in the stooping of Gyrfalcons and Hobbies) and the specifics of it, in terms of the flow between the wing and body were investigated by German researchers (Ponitz *et al.*, 2014a and 2014b) who noted that in this configuration drag is reduced, while lift is increased, *i.e.* the falcon can go faster, but at the same time position control is enhanced (Fig. 3.12). Again, the important issue is vortex production beyond the tail. The high-speed photography used as part of the study of Ponitz *et al.* (2014b) showed that to prevent flow separation of the air travelling at speed across the Peregrine's back (which would distort

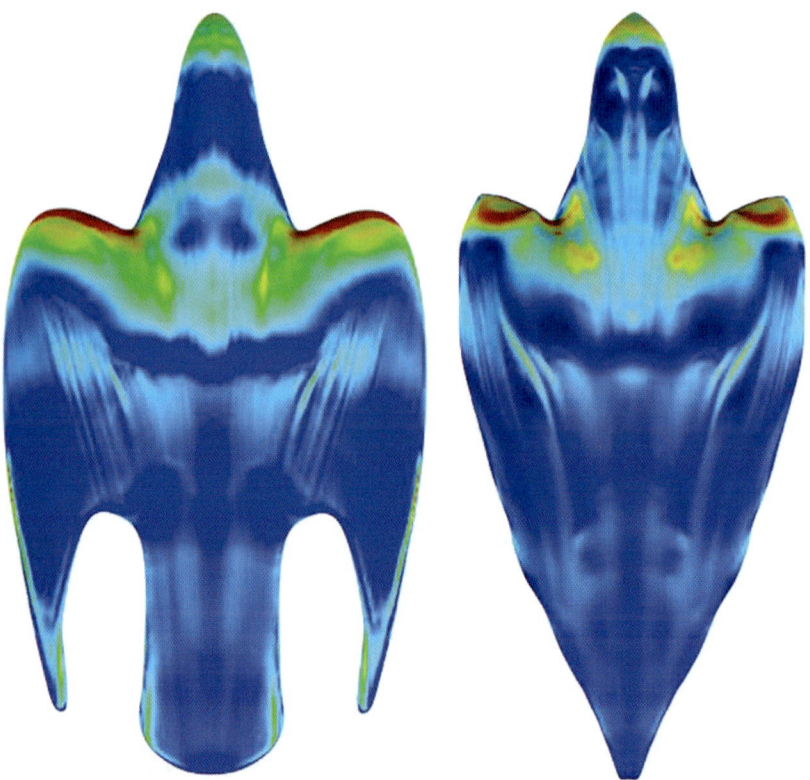

Figure 3.11 The modelled M- and C- shapes as photographed in a wind tunnel. The colours in the figure illustrate the variation of shear stress on the bird, stress rising from deep blue, through green and yellow and being a maximum in red. Reproduced from Lagemann *et al.* (2018), with thanks to Erwin Gowree.

Flight Characteristics

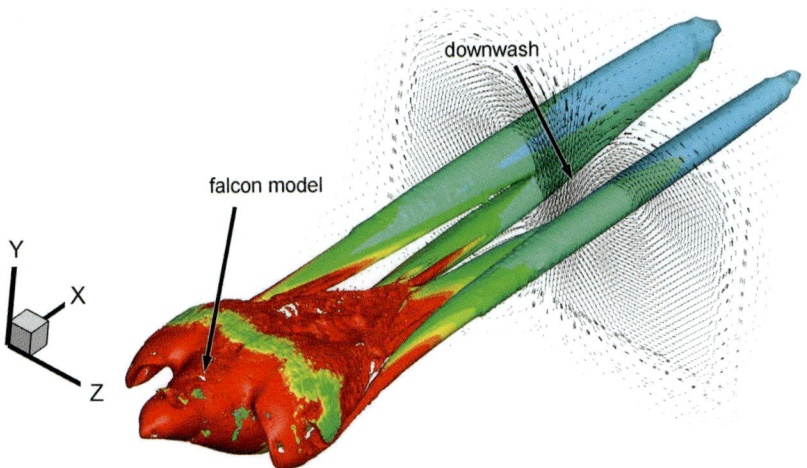

Figure 3.12 Visualisation of the spatial arrangement of wake vortices on a model C-shape Peregrine model in a wind tunnel. The colour variation from red through green to blue indicates vorticity, red being the highest. Vorticity is a measure of the rotation of a fluid (in this case air). In this figure the highest rotational speed of the vortices is shown in red, the lowest rotation being blue. Reproduced from Ponitz et al. (2014a), with thanks to Christopher Brücker.

the vortices at the tail and so act against the reduction in drag) individual feathers 'popped-up' at exactly the point at which theory suggested this should occur. In terms of flight, evolution has fashioned the Peregrine into a perfect high-speed flying machine.

As can clearly be seen from Figs. 3.11 and 3.12 the highest vorticity occurs at the front of the wing and while this aids high-speed stooping, it imposes the greatest stress on the humerus bone in the wing. Figs. 3.13a, b and c (overleaf) illustrate the cross-section of the humerus bone of a Turkey Vulture (*Cathartes aura*). Birds have hollow bones to reduce weight as an aid to flying. But the bones have to not only be light, but strong enough to resist bending and torsional loading. Calculating the values of I (planar) and J (polar) second moments of area it is possible to define the levels of resistance. CT scanning was used to measure the inner and outer radii of the mid-point of the humeri of a male Peregrine (Fig. 3.14 overleaf). For completeness, the analysis has been extended to cover the four UK breeding falcons (and will be published separately in the ornithological press). The humeri bone pairs of four male falcons are shown overleaf. In each case male bones were scanned as the bones of females undergo calcium loss while breeding (for the manufacture of eggshells) and using museum specimens it is not always possible to know when in the breeding cycle a female died.

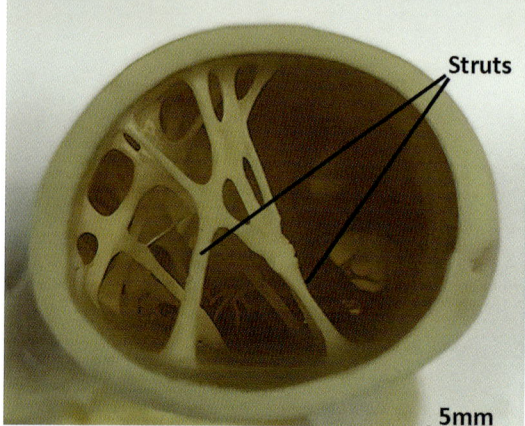
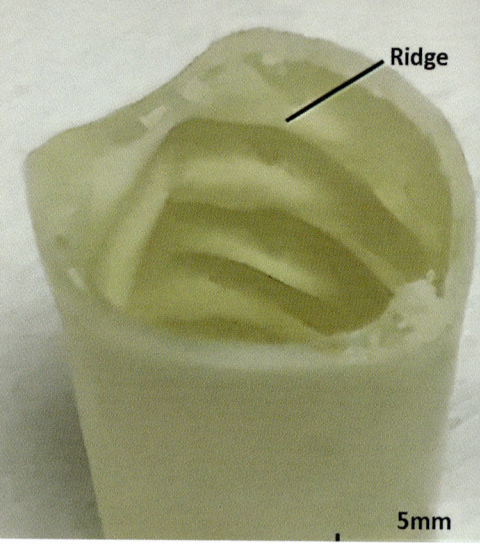

Figure 3.13a Cross-section of the humerus of a Turkey Vulture. The struts across the hollow bone tube strengthen the bone against the imposed loads of high-speed flight. Reproduced from Sullivan *et al.* (2017) with thanks to Marc Meyers.

Figure 3.13b Cross-section of the humerus of a Turkey Vulture. The ridges running along the internal wall of the bone resist torsion and local buckling. A similar design is seen in the inner cardboard tubes of kitchen rolls (Fig. 3.12c below) for exactly the same purpose. Fig. 3.12b is reproduced from Sullivan *et al.* (2017) with thanks to Marc Meyers.

Figure 3.13c A cardboard kitchen roll with the position of spiral strengthening ridges highlighted. The tube is manufactured from parallelogram (rhombus) sections glued together so as to produce the spiral ridges. The ridges resist both torsion and local buckling.

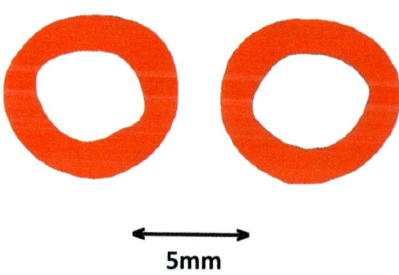

Figure 3.14 CT scans (artificially coloured) of the midpoint of the left and right wing humeri of a male Peregrine. Birds have essentially hollow bones to reduce weight and, therefore, aid flying. A hollow structure is also mechanically advantageous as higher diameter and thinner cortical bone resists both bending and torsion. For a stooping bird a circular cross-section is also an advantage as this geometry resists bending in all directions whereas an elliptical cross-section has the highest resistance to bending along the major axis. In our CT scanning of the humeri of all four UK breeding falcons, the humeri were essentially circular in cross-section, though the main axis was always along the bone's length rather than vertical to it. With thanks to the Natural History Museum for loan of the bones and Tom Dutton for the scans.

Landing *F.p. anatum*. Gabriel Gruia.

From the data the humeri of falcons are not truly circular (varying from circularity by a maximum of 16%), but for simplicity it has been assumed they are. On that assumption we can calculate the resistance to bending and torsion for each humeri, as given by I and J, from the formulae $I = \pi R3t$ and $J = 2\pi R3t$, where R is the radius of circle defining the bone and t is the cortical thickness. By using the maximum and minimum radii for both left and right humeri the calculated ranges for the male Peregrine are:

$$I = 134.6\text{-}242.3 mm^4 \quad J = 269.1\text{-}484.6 mm^4$$

Schmitz *et al.* (2020) claim that a Peregrine travelling at 288kph might experience a loading on the wing bones of up to 360g. This value is stated, with no underlying calculation, but given the cross-sectional area of the humerus it is not unreasonable. Not surprisingly, Schmitz *et al.* noted that the wing bones have been strengthened to combat the torsional and bending moments imposed directly by high speed and during the manoeuvring required to maintain control and direction during the stoop. Schmitz *et al.* used state-of-the-art laboratory equipment and computational fluid dynamics to study the humeri strength of Peregrine, Common Kestrel and a domestic pigeon. The calculated values for the Peregrine were $I=133.0\pm35.5 mm^4$ and $J=240.3\pm52.8 mm^4$. Given the range of calculated values by the present authors, and the standard deviations quoted by Schmitz *et al.* the agreement between the two measurements is reasonable.

We also measured I and J for the other three UK breeding falcons. The data for all four species are set down in Table. 3.3.

Species	Measured Radii (mm)	Cortical thickness (mm)	I (mm^4)	J (mm^4)
Peregrine	3.33–3.91	1.16–1.29	134.6–242.3	269.1–484.6
Eurasian Hobby	2.30–2.65	0.84–0.97	32.1–56.7	64.2–113.4
Common Kestrel	2.2–2.46	0.71–0.84	24.1–39.3	48.2–78.6
Merlin	2.30–2.59	0.78–0.90	29.8–49.1	59.6–98.2

Table 3.3 Calculated I and J values for the four UK breeding falcons. In all cases the data are from the two humeri of the same skeleton. With thanks to the Natural History Museum for the loan of the bones and Tom Dutton for the CT scanning.

Notes:
The data for the Eurasian Hobby is higher than that calculated by Sale and Messenger (2021) for a single bone removed from a wild Hobby. Since that bird was found dead after the date at which it should have migrated, it is possible the bird was emaciated. It is also the case that the calculation had used minimum radii to allow for lack of circularity. With that in mind, the difference in the two measurements is not unreasonable.

The Common Kestrel data differs from that of Schmitz *et al.* (2020) who quote 11.8±2.3mm^4 (max) for I and 22.0±3.6mm^4 for J. However, the Schmitz *et al.* photo of the Kestrel humeri cross-section shows a cortical thickness which is uniformly thin, whereas the mid-point of our CT scan shows a cortical which is not uniform. This difference in uniformity of cross-section explains the difference in parameters. In each case, the values for Kestrels are substantially different from those of the Hobby which is of comparable size: Hobbies often stoop which would account for bone strengthening, whereas Kestrels chiefly flight-hunt (*i.e.* 'hover'), a technique which imposes much less load on the wing bones.

The data for the Merlin – bones being strengthened in comparison to the Kestrel which is significantly larger – reflects the main hunting technique, pursuit flying.

Despite the bone strengthening, experienced wild birds are almost certainly capable of moderating both speed and deceleration in ways which minimise stress on their bodies while allowing success rates at reaching prey that are high enough to avoid starvation. Exactly how this is achieved has been the subject of an increasing number of experiments in recent years. From this work it is apparent that the stoop of a Peregrine goes through several phases. These are illustrated in Fig. 3.10, but for a more comprehensive analysis see data set down in Tables 1 and 2 of White and Nelson (1991) who followed two hunting sorties by a male Peregrine in Alaska, the first lasting 80mins., the second 23mins.

Flight Characteristics

Take-off and Landing

Dolnik (1995) calculated the energy requirement for ground take-off as 12xBMR (BMR = Basal Metabolic Rate: see Chapter 5 for further details). In calculations on the Eurasian Hobby, Sale and Messenger (2021) calculated *c*.14xBMR, while for the Common Kestrel, Sale (2020) calculated *c*.16xBMR. Whichever value is taken, taking off from the ground is energy intensive and Peregrines (and most other birds) seek to take-off from a high point to minimise energy expenditure.

Above   Juvenile nominate Peregrine taking flight by essentially falling forward off a high perch.

Below   Nominate Peregrine taking-off. The Evesham Bell Tower where this pair were breeding (see Chapter 7 for more details) had limited space between a stone wall and a steep roof, so the falcons were forced to expend significant energy when taking off. The male had delivered prey to his mate. *RS/SW*.

Above   Nominate Peregrine landing sequence. *Chris Skipper.*

Below   Nominate Peregrine landing sequence, Scotland.

Landing is less energy expensive as the bird is shedding kinetic energy so as to land with minimal forward velocity. But the bird also has to avoid stalling which could be disastrous. To do this the alula is extended, creating a slot in the wing's leading edge which allows a higher angle of attack than normal and so prevents a stall. In the upper image opposite the bird is seen sideways on and the alula is difficult to see, but in the image immediately below, the alula is seen clearly as the ground is approached, preventing stall. As soon as the ground is touched, the alula returns to its usual position of lying flat along the wing edge. For an excellent article on the physics of the alula see Álvarez *et al.* (2001).

The Hunting Flights of Falconry Birds

The use of falconry birds to study the hunting of wild Peregrines has the obvious drawbacks. The work of Feenders and Bateson (2013) on groups of age-matched Starlings, one group hand-reared, the other not, suggests that basic cognitive performance was identical in the two groups, though some emotional responses differed (for example, the hand-reared birds were less neophobic). While this suggests that well-trained, well-looked-after falconry birds do not significantly alter their flight behaviour there is still a major problem that cannot be overcome: the inability to replicate the hunting imperative of a hungry, potentially chick-rearing, wild falcon. Therefore, in the studies presented below on the hunting activity of falconry birds, while aspects of the behaviour are believed to be appropriate for wild birds – such as the strafing attack on a large Pheasant (*Phasianus colchicus*) – the attack and strike speeds and the g-force resulting from them may reflect, but not exactly mirror, the behaviour of wild birds.

Fig. 3.15 is the stoop of the male Peregrine *Mouse* (a bird which will be mentioned again later in the Chapter) towards an unseen (but missed) target. The flight lasted over 15mins, with *Mouse* reaching a speed of 162.5kph (101mph) in a stoop of only 140m.

Figure 3.15 Flight of the male Peregrine *Mouse* on 20 January 2022.

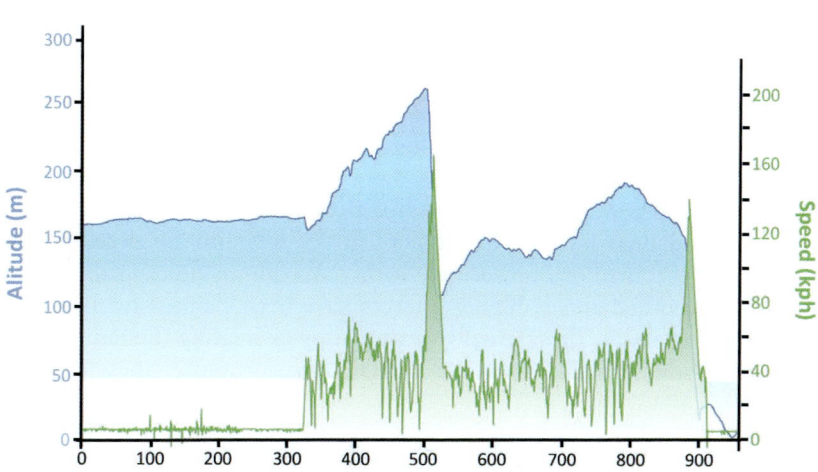

The falconry Peregrine TS, which has already been mentioned previously in this Chapter. Fig. 3.16 is a 3D image of a waiting-on and attack flight against Red Grouse (*Lagopus lagopus scotica*) in Scotland. The attack was successful, though the fastest speed seen on the final stoop was 108kph (67.5mph).

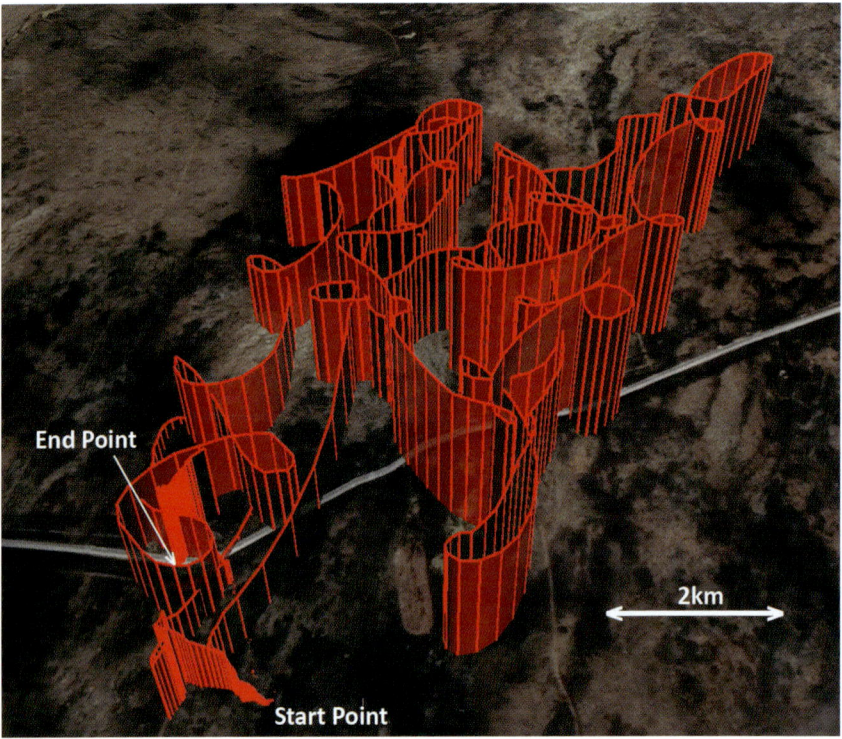

Figure 3.16 Flight of the falcon TS which resulted in the capture of a Red Grouse. The 3D imagery is from the processed IMU data.

TS had a habit of stooping very close to the ground. Figs. 3.17 and 3.18 show the falcon's attack on a Grey Partridge. It was a very steep stoop during which the bird reached a speed of 133kph. When it was getting close to the ground the falcon must have realised it might impact and needed to come out of the stoop very quickly. The satellite and barometer data on the IMU differed during this phase of the flight, though not grossly: the satellite data claimed TS came within about 3m of the ground. The barometer is more conservative, suggesting about 4m. By eye it seemed that impact was imminent and unavoidable. To accomplish the rebound TS decelerated at 4-5g over the course of one second, with a maximum deceleration of just over 6.5g. As noted above, higher g-forces have been claimed for Peregrines pulling out of stoops, these usually measured by estimating the radius of the circle carved through the air by the bird. RS is suspicious of these estimates

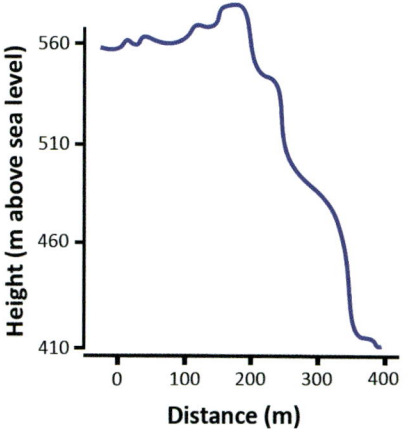

Figure 3.17 Profile of the flight of falcon TS which resulted in a successful attack on a Grey Partridge.

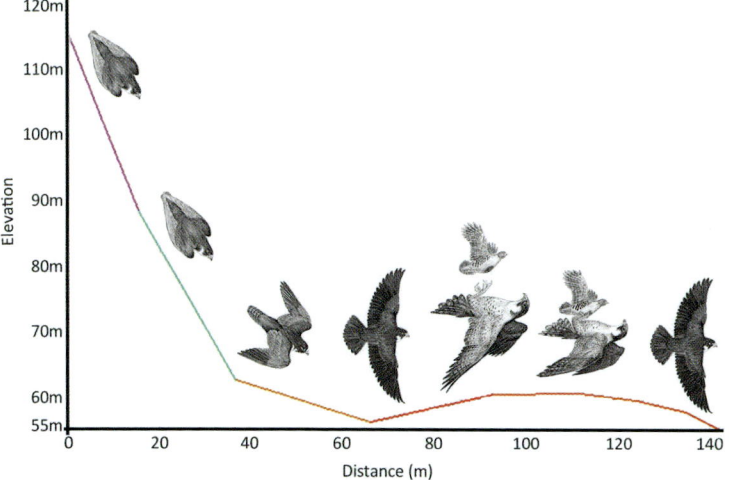

Figure 3.18 Final stage of the stoop and capture of a Grey Partridge by the falcon TS. *With thanks to Kerry Jane (Peregrines) and Martin Buckley (Partridges) for the drawings.*

and is not aware that 'true' higher g-figure has been measured to date on a Peregrine. Once it had avoided hitting the ground, TS was still travelling fast, initially going into level flight at about 100kph, but decelerating. The falcon turned through almost 180°, bound on to the Partridge at about 65kph and brought it quickly to the ground. The stoop, turn and grab of the bird shows its single-minded nature: within about 1.5s of having escaped annihilation it was grabbing at the prey.

All of the falconers RS has worked with have had, or know someone who has had, a falcon hit the ground when stooping – not always fatally, though some crashes are indeed fatal. Other observers have seen Peregrines hit the ground when hunting, and hitting the water when stooping at waders over the sea in winter (*e.g.* Ford, 2007). Sadly, TS repeated the trick of taking prey very close to the ground on a Red Grouse in Scotland some weeks after the Partridge catch. This time the grouse and falcon hit the ground. The grouse was killed, but TS seemed initially to have escaped serious injury. However, it is presumed she had sustained internal injuries as her condition deteriorated rapidly and she died a few hours later.

In a different hunting flight (Fig. 3.19) a tiercel (male) Peregrine was cruising in circles as it waited on the falconer and dog. When a Grey Partridge lifted, the tiercel was facing the wrong way so there was a pre-stoop phase in which the bird banked and dropped, increasing its speed as it went. On the stoop it accelerated at 1.6g to a speed of 132kph (82mph), but lost speed to about 80kph (50mph) as it hit the Partridge. The Partridge was hit at 'X' in Fig. 3.19. After the strike the Peregrine decelerated at -2g to travel at 29-36kph as it circled over the Partridge. As it did so the Partridge took off and flew in the opposite direction. The Peregrine therefore had to perform a circle prior to attempting to chase it down, decelerating, then accelerating as it went from about 10m/s south-west to 18m/s (65kph - 41mph) north-east. The deceleration was about 1.5g, but the acceleration was higher at almost 2g, although this time without the help of gravity. For comparison, if the Peregrine was accelerating at close to 2g, then it compares reasonably favourably with a Formula 1 racing car. The Peregrine easily overtook and caught the injured partridge.

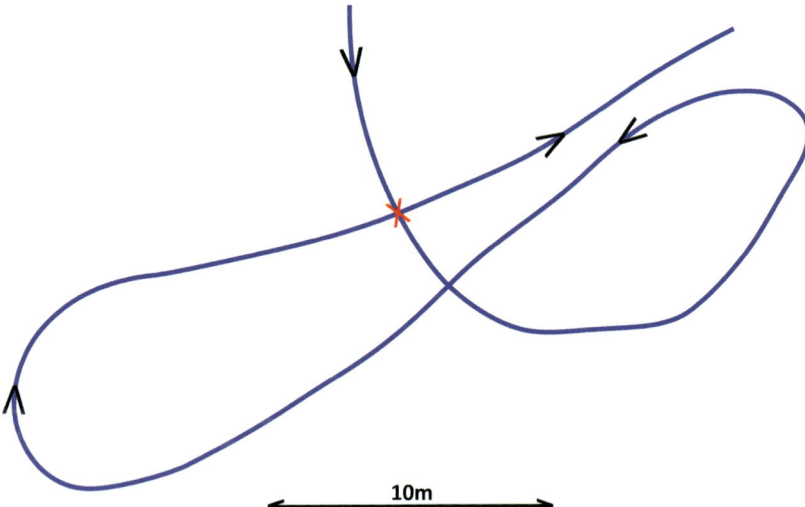

Figure 3.19 Final stage of the stoop, strike and then capture of a Grey Partridge by a male Peregrine.

A different form of behaviour was shown in a tiercel Peregrine attack on a female Pheasant (Fig. 3.20). The Peregrine attacked the Pheasant from a height of only 34m, having gently drifted down from about 50m. Despite

Flight Characteristics

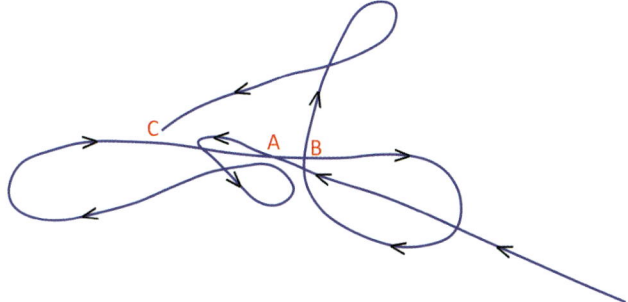

Figure 3.20 Strafing attack on a female Pheasant by a male Peregrine.

being at a low altitude the falcon accelerated rapidly by turning into a short, steep stoop and hit the Pheasant at a speed of approximately 29m/s (104kph) (point A). The Pheasant, stunned but alive, fell towards the ground. The Peregrine, still travelling quickly, though slowing rapidly, performed two tight left turns and approached the prey again. Though stunned, the Pheasant was presumably not badly injured, and the falcon decided against locking on and taking the prey to the ground where a struggle and, perhaps, a lucky strike from the Pheasant's bill or claw could cause its injury. It therefore struck the Pheasant again at more or less the same position though closer to the ground. Once on the ground the Pheasant was still active, and the Peregrine, circling right this time, passed over it, not striking it but apparently gauging its condition. The Peregrine then circled right again and this time did make a further strike (point B) which sent the quarry off to the left. The Peregrine then made a lazier turn right, by which time the Pheasant had moved a little further (to point C), but was now badly injured and essentially helpless, allowing the falcon to grab it on its final attack and to dispatch it soon after. This was the only strafing attack RS observed, but apparently such a technique is often used to subdue or immobilise potentially dangerous quarry.

Vision

As well as being capable of speeds which raise the Peregrine to the level of myth and legend, falcons are also renowned for having excellent eyesight. Their eyes are large in both absolute and relative terms, and are globose (globe-like – spherical). The large avian eye offers a wide field of view, but lacks the ability to move within its socket. Pursuit raptors such as Peregrines can move the eye 10-12° within the eye socket, allowing them to shift (at least slightly) the focal point without moving the head, which would in itself create drag. For comparison, the human eye can move about 44° left and right, and about 30-40° up and down[9]. The size and position of the eye, coupled with the limited ability to shift the focal position means that binocular vision is limited to

---

[9] For comparison with the Peregrine, O'Rourke *et al.* (2010) note that Red-tailed Hawks have about 5° of movement, while Cooper's Hawks (*Accipiter cooperii*) have about 8° of movement. However, the movement for the American Kestrel is less than 1° reflecting its standard hunting method.

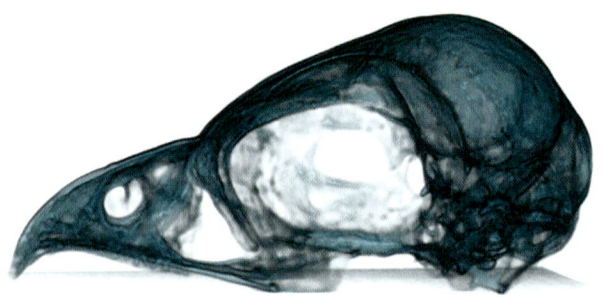

CT scan of a Peregrine's skull showing the size of the eye socket in relation to the rest of the skull.

the forward direction, the vast majority of the field of view being monocular. Fig. 3.21 is a representation of the vision of the Saker Falcon in 3D and 2D. The intention had been to replicate these images for a Peregrine (with the help of French scientist Simon Potier. Unfortunately, Covid, followed by a devastating avian flu epidemic in 2022 prevented the work. However, it is believed that the vision of the Saker and the Peregrine are essentially similar.

In general, Peregrines bind on to their prey, moving their talons into the binocular field of view. They will also extend the talons in similar fashion when seeing off other raptors encroaching on their territory. However, there are instances when a stooping Peregrine will strike prey with the talons held behind the binocular field of view, using the hallux (rear talon) to inflict serious, potentially fatal damage. In this situation the falcon is not 'seeing' the talon, judging its position by a knowledge of the prey's position and its body size and shape.

This wide monocular field limits the blind spot at the back of the head. Compared to those of humans, if falcon eyes are measured as a percentage of the weight of the head, the figure is ≥10%, in comparison to about 1% for a human. If measured by cranial volume, the falcon's eyes take up 50% in comparison to ≤5% for humans. The size means the eye has a long focal length, for better magnification, a large retinal area and a greater depth of field. The eye's structure is also different from that of humans, birds (along with most other vertebrates, but notably not primates) having both the two eyelids which humans have and, additionally, a 'third' eyelid. This nictitan or nictitating membrane lies beneath the external eyelids and sweeps the cornea (originating from the nasal side) to remove debris and maintain lubrication in a stoop. The nictitan is lubricated by the lacrimal gland comparable to a similar gland in humans, but in the case of falcons (and a few other avian species) there is a second lubricator, the Harderian gland, which produces a viscous fluid that also moistens the cornea: this fluid is especially useful during

Flight Characteristics

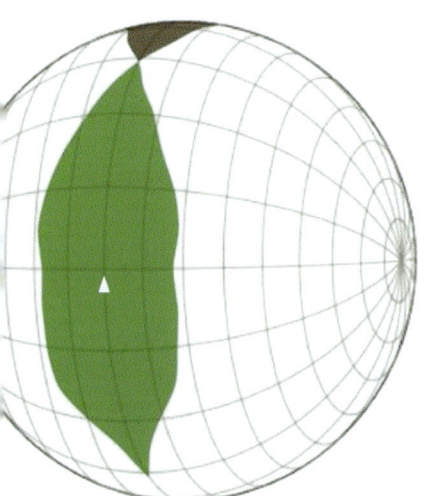
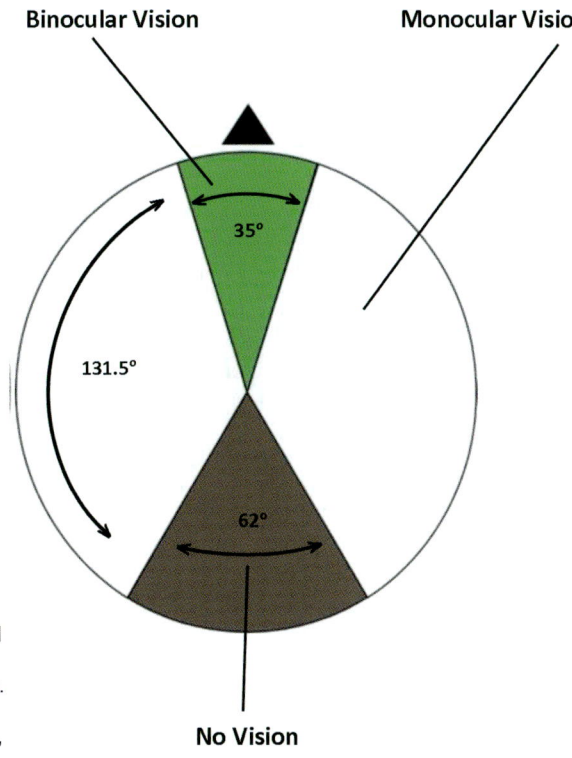

Figure 3.21 3D (*above*) and 2D (*right*) of the vision field of a Saker Falcon. The 3D image is an orthographic projection based on a global system with the equator aligned vertically in the median sagittal plane, with 20° latitudinal and 10° longitudinal intervals. The green areas represent binocular sectors, the white areas monocular sectors, the brown areas blind sectors. The 2D image indicates the size of these vision fields.

Nominate peregrine showing the nictitan half across the eye. See text overleaf for more details. *Peter Christian.*

high-speed falcon stoops when the less viscous 'lacrimal fluid' would dry out quickly. The avian nictitan is transparent (or translucent) and rapidly sweeps across the cornea. In the Peregrine the nictitan also has what amounts to a feather duster which aids the cleansing of the cornea on each sweep (Schwab and Maggs, 2004). Another adaptation of the avian eye is the pecten oculi (Fig. 3. 22) which is set in the rear chamber of the eye (the posterior segment), close to where the optic nerve leaves the eyeball. In the human eye the blood vessels course across the retina, transporting oxygen and nutrients to sustain the retina, retinal function and the eye's vitreous humour (which fills the eye's posterior segment between the retina and the lens), these retinal blood vessels partially obscure the retina, creating multiple shadows across the field of view. Birds require greater visual acuity and so to avoid this shadowing they possess an additional structure, the pecten. The pecten, which is present in all avian eyes (and in the eyes of some reptiles), pumps nutrition directly into the vitreous humour (the fluid of the posterior segment), obviating the need for retinal blood vessels. Korbel *et al.* (2000) used video fluorescein angiography to study the way in which nutrients reach the vitreous humour of the eyes of Common Buzzards and Tawny Owls, using several mammal species as controls. The process in the two avian species was very different, nutrition being 'squirted' into the humour in short bursts of a few seconds at regular, short (a few seconds) intervals (Fig. 3.23). There is no reason to suppose that the situation is any different in Peregrines (or other falcons/raptors). The exact shape of the pecten varies between families of birds, but its presence always produces a much-reduced obstruction of the retina. However, the retinal area shaded by the pecten is devoid of photoreceptors and may contribute in part to the blind spot behind the bird's head.

The size of the avian eye, coupled with the fact that Falcon retinae are thicker than in other species (up to 500µm, compared to 200-350µm in other birds and 230µm in humans) means it requires very high rates of metabolic energy and, therefore, oxygen. The latter nutrition is supplied in part by the pecten, but in addition Blank *et al.* (2011), studying the eye of domestic chickens, found a new neuroglobin type (GbE) which also provides oxygen supply to the retina: so birds have evolved this more efficient system for supplying nutrition to the retina, than exists in humans. The falcon retina is thicker because of information pathways. While in the human eye information is passed from the eye to the brain for interpretation, in falcons information is passed in both directions. The function of this dual pathway is not understood, but what is clear is that the falcon brain is influencing the information the bird's visual system is collecting in ways which may not be shared by humans and, therefore, making direct comparisons between falcon and human eyesight is more difficult than simple tests (as detailed below) suggest.

While the structure of the avian eye differs from that of humans it was assumed for many years that although the vision of falcons was probably much more acute than that of humans, they viewed the world in much the same way

Flight Characteristics

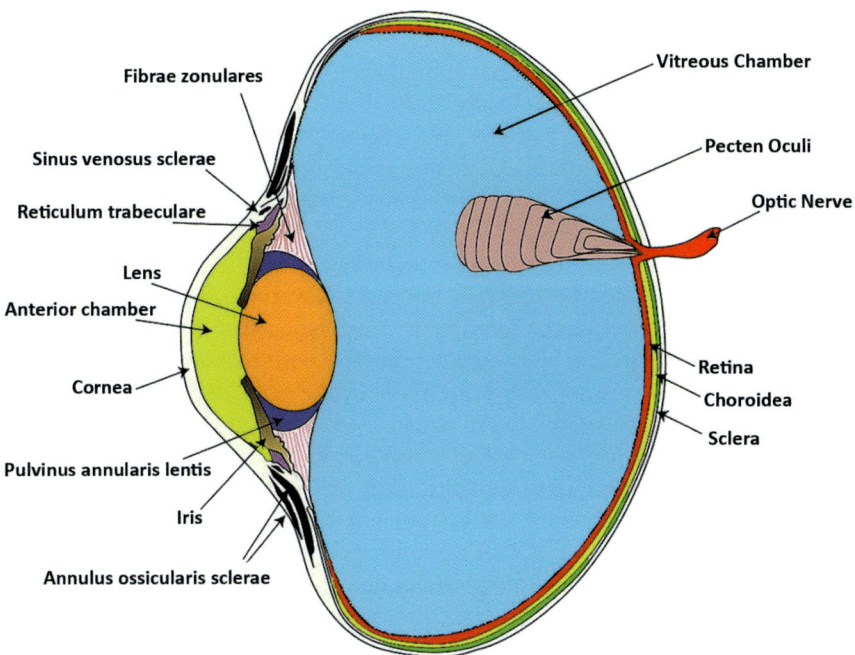

Figure 3.22 *Above* Schematic of a 'standard' avian globular eye. Other avian eyes can be flat or tubular, but the Peregrine eye is globular, as illustrated.

Figure 3.23 *Below* Schematic pattern of the ejection of fluorescein dye into the vitreous humour arising from the dorso-temporal aspect of the pecten (right eye) of a Tawny Owl, as seen through the opthalmoscope in the non-anaesthetised bird. Redrawn from Korbel *et al.* (2000).

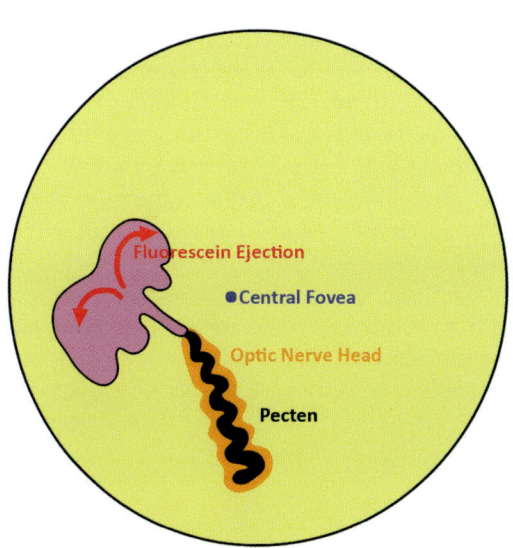

as we do, *i.e.* that their vision encompassed the same range of wavelengths as ours does (the visible spectrum). Visible light is a very small range (in terms of wavelength) of the electromagnetic spectrum (Box 3.4).

Over time, this suggestion that humans and birds see in much the same way has needed to be revised. In the 1970s it was discovered that some birds could see UV light (see, for instance, the seminal paper of Bennet and Cuthill, 1994). It is still unclear whether Peregrines can see UV, although it must be assumed they can as recently published work by Hart *et al.* (2016) on the Emu (*Dromaius novaehollandiae*) suggested that UV sensitivity is likely to be ancestral in avian evolution and therefore present in all birds unless it has been lost. The use of UV light in the flight-hunting ('hovering') mode of Common Kestrels, was suggested some years ago, the idea being that since rodent urine and faeces fluoresce in UV, and the mammals use both to mark trails, a Kestrel needed only to observe moving urine trails in the grass to find a meal. This idea enlivened many a falcon display when the handler explained the hunting technique, particularly if the audience was made up chiefly of children and their potentially embarrassed parents, but the suggestion has now been (mostly) discarded as it seems unlikely that Kestrels can distinguish the UV from a urine trail from the UV of the underlying ground. While UV vision by falcons seems to allow plumage to be seen in a way that humans cannot replicate, whether it is also useful as a hunting technique is still debated.

But while the ability to see in UV meant that birds had better vision than the standard three colour (blue, green, red) vision of humans, it was soon apparent that the avian retina was more complex than had been assumed. As

Box 3.4 The Electromagnetic Spectrum
This is not the correct forum for a discussion on whether electromagnetic radiation is particle-like, wave-like or both at the same time: it is sufficient to note that the spectrum is most conveniently considered by wavelength, and covers a range from radio, where the wavelength can be measured in metres to X-rays, where it is measured in nanometres ($1nm=10^{-9}m$, *i.e.* one thousand-millionth of a metre) and beyond to γ-rays (gamma rays) with wavelengths measured in picometres ($10^{-12}m$, *i.e.* one million-millionth of a metre). Visible light is usually defined as covering the range 400-700nm with red light at the upper end and blue light at the lower, though it is worth noting that many people can see colours up to 1000nm and both children and young adults may see down to about 310nm. Radiation with wavelengths below 400nm is known as ultraviolet (the last colour of the 'rainbow' spectrum of visible light being violet) while at wavelengths above 700nm it is known as infra-red. Infra-red wavelengths are used in night-vision cameras *etc.* as the heat produced by warm objects (including warm-blooded animals) is in the infra-red part of the spectrum. Ultra-violet (UV) is beneficial to humans as it allows the synthesis of Vitamin D, but over-exposure is also harmful, causing sunburn and, potentially, skin cancers.

in humans, the avian retina has two types of photoreceptors, rods and cones. Rods are of a single form and aid vision in low light, being sensitive to shades of grey. Cones are two forms, single and double. But in the avian eye there are four single cones rather than three. Light incident on any of these must pass through an oil droplet before reaching the cone's photopigments.

There are six types of oil droplet, labelled T, C, Y, R, P and A, each varying in its carotenoid content and transmittance, and which therefore correspond to specific areas of the spectrum. The four single cones are sensitive to light of maximum wavelength – T (412nm), C (474nm), Y (541nm) and R (605nm). The four cones of the Peregrine's eye mean that the spectral output of the receptors differs from the three colours associated with the human eye, as well as adding a fourth, 'UV'. The double cone, associated with oil droplets P and A are not seen in the central (deep) fovea of Peregrines, but are seen in other raptors (*e.g.* the Eurasian Sparrowhawk). It is believed that double cones provide achromatic (colour-free) vision and so eliminate chromatic aberrations, which would be of assistance to the motion detection of a fast pursuit raptor, particularly one such as the sparrowhawk which hunts in woodland. Fig. 3.24 is a schematic of the single cone photoreceptor of the avian eye, while Fig. 3.25 (overleaf) is a representation of the four cones of the Peregrine's eye.

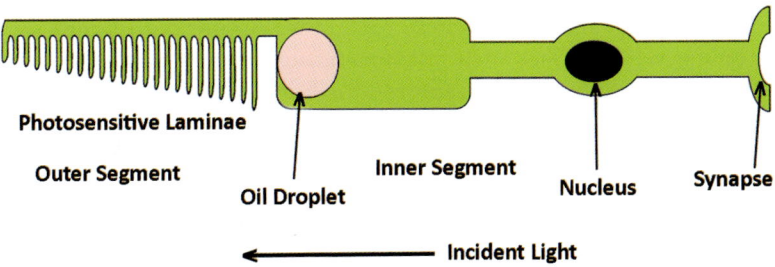

Figure 3.24 Schematic of an avian cone photoreceptor. Redrawn from Jones *et al.* (2007).

Stoddard *et al.* (2020) studying the Broad-tailed Hummingbirds (*Selasphorus platycercus*) have suggested that the extra cone means that the birds almost certainly see more natural colours than humans. Humans see the non-spectral colour purple when red and blue cones are stimulated simultaneously, but it is likely that birds see other non-spectral colours such as UV+red, UV+green, UV+yellow and perhaps other combinations, suggesting that birds live in much richer colour environment than we humans do. As birds may also see in the UV part of the spectrum, their world is also much more visually rich overall[10].

---

[10] Humans and, it would seem, most other mammals are trichromatic, *i.e.* they see three colours. Most birds, together with amphibians, reptiles, fish, most insects and some mammals are tetrachromatic (*i.e.* they see in four colours). Pigeons, some other birds and butterflies are pentachromatic – they see the world in five colours.

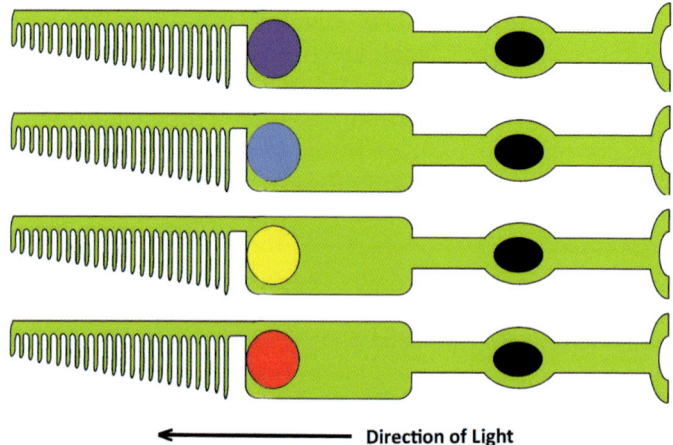

Figure 3.25 Schematic of the four cone receptors in a Peregrine's eye. The oil droplets are coloured to aid identification but are not a true representation of the four colours seen by the bird. The violet droplet corresponds to the near UV/violet part of the spectrum while the other three colours are better representations of the spectral regions.

While the actual chromacity of all avian eyes, including the Peregrine, is unclear, it is known (Palacios and Varela, 1992) that Pigeons can visualise four colours, seeing yellow rather than 'manufacturing' from the three primary colours of human vision, as well as seeing a further 'colour' in the UV part of the spectrum (Fig. 3.26).

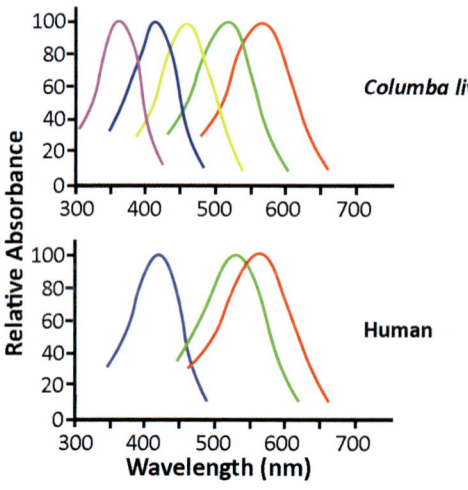

Figure 3.26 Relative absorbance of pigeon and human eyes relative to the wavelength of incident light. Humans see three colours whereas the pigeon sees four colours and a fifth UV 'colour'.

Interestingly, raptors usually have their pupils wide open, even in strong sunlight, (most species constrict the iris, reducing their pupil size in bright light). In camera terms this means they have a low F-stop which restricts the depth of field but maximises acuity. Barlow and Ostwald (1973), researching pigeons, suggest that the pecten may act as an inter-ocular shade, reducing the direct impact of sunlight on the retina, something which would clearly aid a raptor with its eyes wide open. As the foveae have only cone receptors Peregrines have poor night vision, though since the falcons have taken to urban living, they have begun to hunt at night using the copious street lights erected by their human neighbours (who have also, of course, provided suitable nest sites by erecting high-rise buildings or pseudo-cliffs as the falcons see them).

The way that falcons see the world, both temporally and spatially, differs from that of humans. Temporally, birds see the world faster as they have a higher flicker-fusion frequency (FFF) than humans. FFF is a measure of the ability to resolve rapid changes of light (or rapid movements). Films are shown at a frame rate of about 30Hz as at rates below this, humans can resolve the change from one frame to the next and so see the flicker as frames change. Some humans can resolve higher frequencies, though few are able to detect the 50Hz flicker of a fluorescent light. By contrast birds have an FFF which is at least 100Hz, with some species having even higher rates (and dragonflies having an FFF of around 300Hz). Such high FFF rates in the avian world are required by both hunters and hunted: flying through woodland at speed, humans would collide with branches which birds easily avoid as their eyes, and, of course, brains, see and process visual information more quickly. While this ability is at a premium for woodland prey species and such predators as the Goshawk, falcons also need to be able to react quickly to changes in flight direction by fleeing prey or to avoid lunges by captured prey. Jones *et al.* (2007) noted that many birds had an FFF greater than 100Hz, with that of pigeons being 116-146Hz. More recently, Potier *et al.* (2020a) have accurately measured FFF for Peregrines (129Hz), Saker Falcons (102Hz) and Harris Hawks (81Hz). Potier *et al.* suggest a link between fast vision and hunting strategy, so that raptors chasing fast moving prey would have a higher rate than those which hunt slower prey[11].

In addition to a falcon's ability to see the world faster than humans it appears their eyesight is also spatially superior to that of humans. Spatial discrimination means that some birds, including many raptors, can distinguish smaller objects, *i.e.* they have greater visual acuity, although the degree to which this is the case is a matter of debate, in large part because of the difficulty of comparing like-for-like. In humans it is an easy task to measure a person's visual acuity as the individual being studied can answer straightforward questions (as anyone who has had their eyes tested by an optician will know). For a falcon more subtle (and, therefore, more difficult) tests are required. In an early study Fox

---

[11] In a separate report (Potier *et al.*, 2020b) noted that the FFF of Peregrines fell off significantly as light intensity decreased: Peregrines are well equipped visually as diurnal predators, much less so in poorer light.

et al. (1976) the subject was an American Kestrel, (gender not stated). The experiment was a standard two-choice discrimination, the bird flying 1.8m to a window showing a vertical grating of differing spacings, the spacing being altered to identify the minimum spacing that the bird could differentiate. When stimulated by the window the bird would fly to it to receive a food reward. In the study the falcon flew 1.8m to the window which offered a subtended angle of 1° to the bird. Fox *et al.* reported an acuity of 160cpd for the Kestrel (see Fig. 3.27 for an explanation of cpd) and suggested the falcon's eyesight was 2.6x that of a human. Hirsch (1982) used an essentially similar technique in a study of a male American Kestrel (named Argus) – the bird was offered two screens, one of which showed the same basic grating: the distance from the screen to the falcon was 2m, the angle subtended by the screen being 3°. Hirsch found an acuity significantly lower than that of Fox *et al.* (1976) and calculated a factor of 1.33x for the improvement of falcon over human. However, Hirsch's result was queried by Dvorak *et al.* (1983) who wondered whether the lighting conditions in the experiment were sufficient to allow a real comparison. He noted that American Kestrels normally hunt in bright sunshine and that it is known that their visual acuity falls as luminance decreases, something which had also been noted by Fox *et al.* (1976). In response, Hirsch (1983) stood by her factor of 1.33x. While the degree to which luminance can affect visual acuity is a reasonable basis for discussion, the cpd claimed by Fox *et al.* (1976), and other researchers for different raptors, has continued to be debated as some claims seem at odds with the anatomical resolution calculated from the spacing of photoreceptors and the density of ganglion cells (the final outputs) of the retina. The measured acuity in humans is known to vary from 30-70cpd, while the theoretical value for the American Kestrel is 46cpd (Dvorak *et al.* (1983)[12]. The latter figure certainly suggests that the falcon's acuity should be closer to that of humans rather than more than double, despite the eye of the former being only half the size of a human eye, the difference being explained by superior optics and closer-packed photoreceptors (about double the number). Bringmann (2019) lists a value of 59cpd for the Common Kestrel based on data presented by Oehme (1964), while Reymond (1987) notes 73cpd for the Brown Falcon, which is comparable in size to the Peregrine, though mainly hunts mammals. Reymond noted that in the case of this strongly diurnal raptor, the maximum acuity was not constrained by changes in luminance.

Reymond (1985) noted a value of 132-143cpd for the Wedge-tailed Eagle (*Aquila audax*)[13] and suggested the eagle's vision was probably twice that of a human, but did note that, as with the American Kestrel, the eagle's vision was more sensitive to light levels than that of a human, performance deteriorating more rapidly as luminance declined. Building on Reymond's

---

[12] In a study of nine American Kestrels, Gaffney and Hodos (2003) measured acuities varying from 39.7-71.4cpd, with a median of 46cpd.

[13] González-Martín-Moro *et al.* (2017) quote a value of 140cpd for the Peregrine, but do not give a precise reference for the claim.

Flight Characteristics

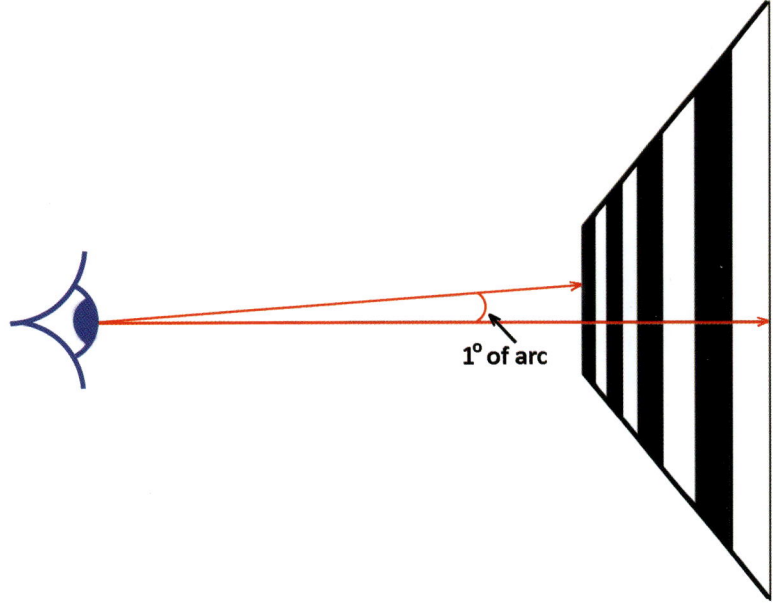

Figure 3.27 The use of a vertical grating to measure visual acuity. In the figure there are 4 grating cycles, *i.e.* 4 sets of black and white vertical bars) within the 1° arc subtended on the subject's eye. If that were the subject's limit, vision would be described as 4 cycles/degree (4cpd). Redrawn from Jones *et al*. (2007).

work, Martin (1986) was able to suggest that for the eyes of large falcons the visual acuity might be twice that of a human. Jones *et al*. (2007) came to the same conclusion. If a factor-of-two vision improvement does not sound enough, it has to be remembered that human vision is itself remarkably acute, so that falcon vision is extraordinarily good.

However, there are further complications which make direct (*i.e.* like-for-like) comparisons difficult. The first of these involves the mechanics of the eye. Fig. 3.28 is a cross-section of an avian eye showing the position of two important muscles. Brücke's Muscle is common to both humans and birds and changes the shape of the lens of the eye. Crampton's Muscle is seen only in

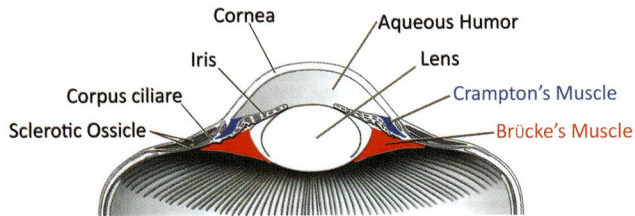

Figure 3.28 Schematic of the avian eye showing the position of the Brücke's and Crampton's muscles.

birds and pulls the cornea, adjusting its curvature and, therefore, its refractive power. The effect in each case is to alter the accommodation (*i.e.* altering the focal length of the lens and hence the point of focus) of the eye. Alterations in accommodation are usually measured in dioptres (the reciprocal of the focal length in metres), Crampton's Muscles having been shown to offer an accommodation of up to 28 dioptres in Red-tailed Hawks and 16 dioptres in American Kestrels (Glasser *et al.*, 1997). The effect of these alterations of the lens and cornea is to provide the Peregrine with the ability to manipulate the eyes in ways which are not possible for a human, even to the extent of allowing the bird to turn its eyes into a telephoto lenses, which makes a straightforward comparison of Peregrine .v. human acuity difficult.

The fact that the Peregrine also has a much higher FFF means that not only may it have greater visual acuity but also that it can see faster, and so may be able to perceive motion at significantly greater distances than a human.

One of us (SW), watching breeding Peregrines at Symonds Yat, England, has several times noted a female preening, apparently oblivious to the world around her. But then the bird's feathers tightened as she stared into the distance. SW records one such instance – '*Suddenly animated, her head bobbing fast from side to side, apparently using parallax to estimate distance. Her calls and body language means she has seen an intruder, (another Peregrine) and it won't be her mate. The female leaves, heading for the peak of Coppett Hill, 1.7 miles away. Despite my use of 10x magnification binoculars she disappears from my view. Several minutes later a Peregrine, flapping hard on a straight line in determined flight, comes directly towards us from the north. It is not 'our' female, but another, smaller female. The territorial female is chasing the intruder down. She closes in, then hits her: feathers fly and the intruder, chastised, continues on deep wingbeats, fast on her way south and away*'.

The observation recalls a section from Dick Treleaven's book (Treleaven, 1977) where he notes '*Interceptions made a long way out to sea are baffling. I have many times watched a peregrine fly straight out to sea until it has disappeared from view in my binoculars (10 x 40), obviously in pursuit of quarry beyond the range of my vision. Several minutes later I have picked it up again carrying a kill. This type of flight is more common in fine weather when the visibility is exceptionally good, and pigeons are taking short cuts across the bays on their way up the coast*'.

These observations do not, of course, amount to science. But they are not uncommon among Peregrine watchers and suggest that the visual abilities of the falcons far exceed those of humans.

Further observations also imply that the vision of Peregrines is significantly better than that of the pigeons that usually form a significant fraction of the falcon's diet, both in terms of visual acuity and spatial resolution (*i.e.* flicker rate). Fig. 3.29 indicates the field of vision of a Pigeon. Hahmann and Güntürkün (1993) measured the spatial (*i.e.* visual) acuity of the frontal field of the Rock Dove by subjecting 10 pigeons 140 times each to a grating

Flight Characteristics

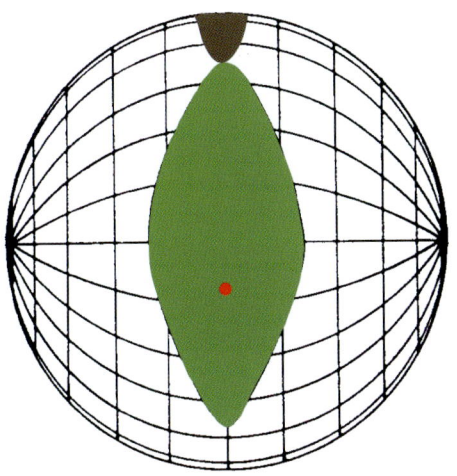

Figure 3.29 The visual fields of the Rock Dove (*Columba livia*)/Feral Pigeon in 3D. The pigeon's *area dorsalis* (the brain area associated with feeding) is indicated by the red dot and lies 10-15° below the eye-beak axis. Redrawn from data in McFadden and Reymond (1985).

similar to that of Fig. 3.27 at a distance of 83cms and concluded that the mean acuity was 12.6cpd (the maximum observed in an individual bird being 20cpd. In later work, Rounsley and McFadden (2005) agreed with the mean acuity (their value was 12.8cpd: calculated from five birds, each tested 96 times) but considered that this acuity was seen only when the birds were 10cm from the grating, with the acuity falling rapidly at greater distances (to only 1cpd at 1m). This surprising difference in the spatial acuity of Peregrines and Pigeons (a factor of at least eight) explains why, when sighting a flock of pigeons within their territory, from some distance away, Peregrines will not hesitate to fly towards the flock before climbing above them to gain position for a stoop – the falcons are clearly aware that the pigeons cannot see them if they maintain a reasonable distance during their manoeuvring.

Avian eyes are adapted to meet the physiological needs of the species, *e.g.* vultures who fly great distances in search of carrion have greater acuity than motion hunters (including Peregrines). But the differences between carrion and motion hunters is not merely visual acuity but behavioural. Vultures hunt for prey by both scanning the ground and scanning the sky – they are both looking for prey and watching other vultures, this explaining why large numbers of birds quickly arrive at prey. Similar behaviour was noted in observations of Snowy Owls (*Bubo scandiacus*) by Potapov and Sale (2012) who developed the 'boid theory' to explain why Snowy Owls all breed in the same area when there is a lemming year: the birds are moving across the landscape in loose aggregations (boids). Each bird of the boid watches both the ground and its boid companions and when a single bird sees a lemming mass all the owls of the boid close in on it and begin to breed. Similarly, other species may have vision and behavioural traits which are adapted to colour, functionality in dim light *etc.*, offering greater visual abilities in certain situations.

But there is a further complication in falcons, as they do not have a single fovea in their retina. The foveae are pit-like areas where the photoreceptor cells are most closely packed and so provide the most acute vision. In humans the eye's single fovea is set centrally giving us stereo vision and, therefore, an appreciation of depth. Falcons have a similar fovea (the shallow or temporal fovea) allowing the same visual signals, but in common with some other avian species, they also have a second fovea which is more visually acute. This deep (or central) fovea is set at an angle of about 40° to the main axis of the bird, *i.e.* the centre of the tail, the spinal column and the tip of bill (Fig. 3.30)[14]. Fig. 3.31a shows the cross-sections of the deep and shallow foveae of the Peregrine, while Fig. 3.31b are similar cross-sections for other avian species which illustrate certain aspects of the foveal biology which enhance the visual acuity. The scale bar in Fig, 3.31a is an indicator of how tiny the optical system of the Peregrine (and the human) is considering its fundamental position in the falcon's (and a human's) senses. Further indications of the scale are provided in Table 3.4 (overleaf).

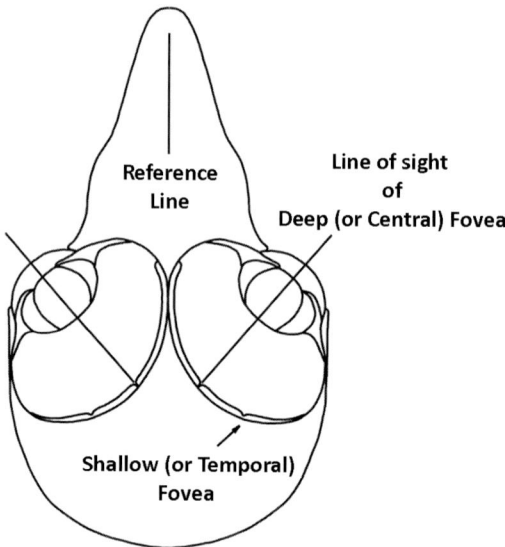

Figure 3.30 Schematic of a falcon's skull, showing the position and line of sight of the shallow (or temporal) and deep (or central, sometimes nasal) foveae.

---

[14] In a study of the eye of the Yellow-billed Gull, Victory *et al.* (2021) noted that in addition to the bird being bifoveal there was a band of higher acuity which extended across the field of view between the two foveae. It is likely that the Peregrine's eye has a similar band. In addition, Victory *et al.* noted that the gull's vision allowed the formation of three distinct fields of view, two wide monocular fields covered by both foveae, and a single binocular field defined by the central fovea. Again, this accords with vision in the Peregrine.

Flight Characteristics

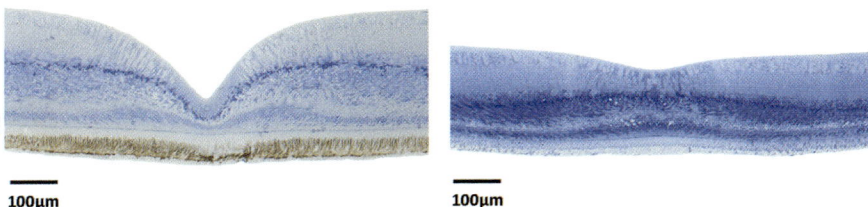

Figure 3.31a Cross-sections of the deep (left) and shallow (right) foveae of the Peregrine Falcon. Reproduced from Mitkus *et al.* (2017) with thanks to Mindaugas Mitkus.

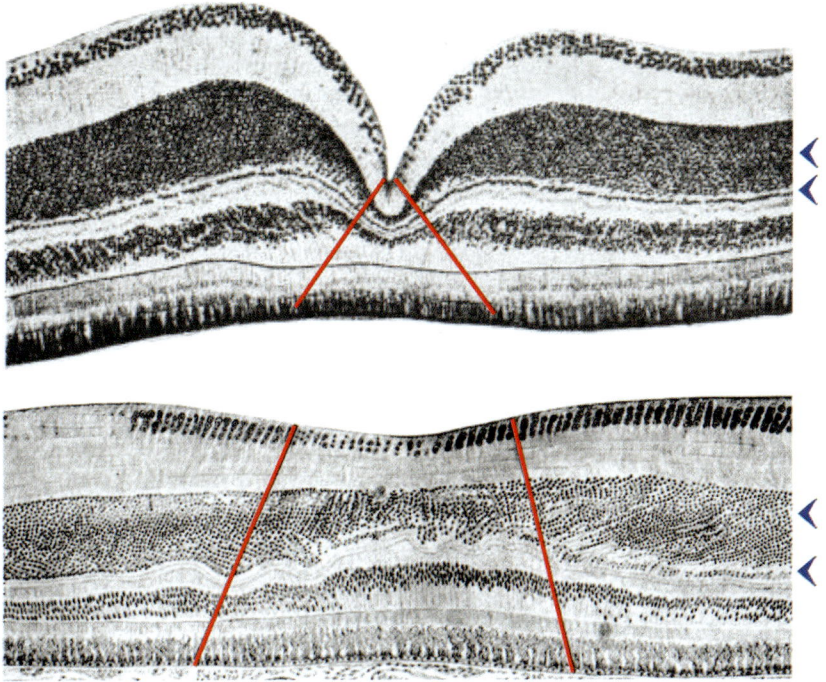

Figure 3.31b Cross-sections of (*above*) the deep fovea of a Barn Swallow, and (*below*) the shallow fovea of a Golden Eagle. The red lines indicate the magnification of the resulting image from the different anatomy of the two foveae: as noted by Snyder and Miller (1978) the spherical portion of the deep fovea acts in a similar way to a telephoto lens enhancing magnification. The blue arrows indicate the levels of the Müller cell layers at the inner and outer boundaries of the bipolar cell layer. See text for further information. Reconstructed after Bringmann (2019).

As Bringmann (2019) notes, Walls (1942), who prepared the base layer upper image, suggested that the construction of the deep fovea produced a magnification of the incident image which did not happen elsewhere in the

	Fovea Pit depth	Thickness of Fovea Centre	Wall Thickness	Pit Width	Reference
Central (Deep) Fovea	198	94	292		Schulze (2016)
	166	75	241	833	Bringmann (2019) from information in Mitkus et al. (2017)
Temporal (Shallow) Fovea	105	171	276		Schulze (2016)
	49	136	185	517	Bringmann (2019) from information in Mitkus et al. (2017)

Table 3.4 Measurements of the two foveae in the Peregrine Falcon. The dimensions are listed in μm (micrometres), 1μm being one-hundredth of a millimetre.

retina. The magnification of the shallow image, as indicated in the lower image, is much reduced. Bringmann noted that the Müller glial cells, non-neuronal cells which provide mechanical stability for neurons aided both magnification and light focusing by forming a refractive layer in the fovea walls. The obvious question which arises from falcons having twin foveae is what is the purpose of the deeper fovea? As its position offers monocular rather than stereo vision what benefit does it offer a falcon in the wild? From our personal observations we believe the answer is an aid to detecting far-off objects, prior to hunting a mobile prey item or to police its territory, activities which are clearly related.

While observing Gyrfalcons in the wild in Arctic Canada in locations where they were chiefly hunting Ptarmigan and Arctic Ground Squirrel (*Urocitellus parryii*), but also waders and gulls, RS was surprised to see that when stooping after prey the falcons were not travelling in straight lines. Discovering that the falcon eye was twin-fovea, and that the more visually acute fovea was set at an angle of about 40° to the main linear axis of the bird (*i.e.* the axis running through the bill, the centre of the head, back and tail), and knowing that falcon eyes do not move in their sockets, RS calculated the path a falcon would travel if it maintained the deep fovea on the prey. The flight path is shown in Fig. 3.32.

Flight Characteristics

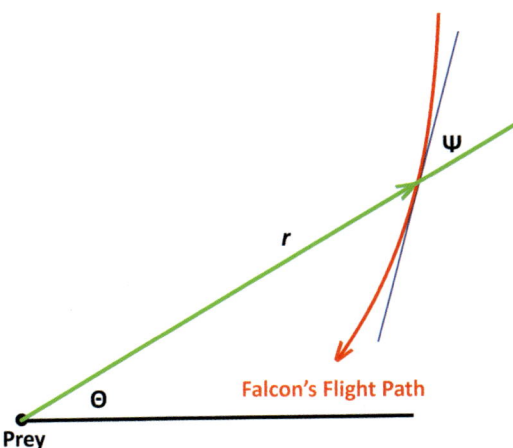

Figure 3.32 Flight path of a falcon relative to its prey. Assuming the prey remains stationary, the distance, r, to the prey and the angle, θ, between the line of sight of the falcon and the ground plane changes as the falcon moves, but the angle Ψ remains constant. Ψ is the angle between the tangent to the falcon's path and the line of sight at any instance. After RS had calculated the falcon path he discovered that Tucker (2000a) had made the same calculations some years earlier. In deference to that original work the same symbols are used in this figure.

At any instant the angle (Ψ) subtended by the tangent to the falcon's flight path and the line of sight to the prey is constant (assuming the prey remains stationary), and in polar coordinates the equation of the flight path is:

$$r = ae^{b\theta} \quad (3.6)$$

where:
r and θ as defined in Fig. 3.32,
e the base of natural logarithms,
a and b are constants which define the tightness of curvature of the resulting curved path. b is dependent on Ψ, the relationship between them being:

$$b = 1/\tan \Psi \quad (3.7)$$

If Ψ = 90°, then b = 0 and the curve is a circle of radius a. As Ψ reduces towards zero, b increases and the curvature becomes less tight: ultimately, Ψ = 0°, b approaches ∞, and the spiral becomes a straight line.

In mathematics the curve is well-known and named a logarithmic spiral: shows a typical spiral. It is equally well-known in nature – fossil ammonites and Nautilus species illustrate the spiral, which in their cases are formed by the production of new camerae (chambers) into which the animal moves as it grows. Romanesco, also known as Romanesque cauliflower or Romanesque broccoli, a vegetable which has become popular in recent times, also grows in a logarithmic spiral (see photos overleaf). Insects will also fly in logarithmic spirals, though in this case the reason is not the eye, but the fact that they fly in straight lines by keeping a light source at a constant angle to the direction of

Above left An ammonite.

Above right Section through a nautilus shell.

Left Romanesque vegetable.

travel. When the light source is the sun or moon, it is effectively at infinity and so, as noted above, the 'curved' flight path is a straight line. But when the light source is a light bulb or a candle flame the insect spirals into it along a curve of form as defined by Equation 3.6.

Once the form of flight path of the hunting Gyrfalcons on Canada's Bylot Island had been defined, the obvious question was, why were the falcons choosing to deviate from the idea that the shortest distance between two points is a straight line? If the falcon were to follow a straight-line path to its prey, then to maintain observation with the deep fovea the bird would need to move its head sideways by 40°, a turn which would inevitably increase drag. While this seemingly makes the reason why the falcon chooses the spiral path clear, this obvious reasoning works only if the decrease in speed (caused by increased drag) along the straight path is such that the falcon will always reach the prey quicker along a spiral path. Importantly, Tucker (2000a) considered this issue in his work. He noted that there was a value of the ratio of drag coefficients in the two cases (falcon flies with head bent off axis (C_{db}) or

Flight Characteristics

falcon flies with head straight (C_{ds})) at which the increased speed on the spiral path and the increased drag on the straight path compensated and the falcon reached the prey at the same time whichever path was chosen. One extremely useful property of logarithmic spirals is that the ratio of the curved and straight path lengths is essentially constant at all times. If the ratio of drag coefficients was lower than this 'break-even' value then the straight path is still the best route. If not, the spiral path wins. Tucker calculated that the break-even value of C_{db}/C_{ds} was 1.8. In a separate paper Tucker (2000b) noted that the ratio was in the range 2–6 as speed increased, and so was always above the break-even value: the falcon would follow the spiral path.

So, having calculated theoretically for an ideal falcon that a spiral path was better than a straight path with the head bent to one side, the next objective was to see if real falcons were following spiral paths. This was tested by observing a pair of Peregrines hunting from an eyrie in central Colorado, using the naked eye, binoculars, tripod-mounted telescopes and an optical tracking device fitted to a telescope (Tucker *et al.*, 2000). To eliminate the possibility that what they were observing were in fact straight lines, the research team calculated the 95% confidence limits on the tracking device used to define paths. In all cases the observed curved paths lay significantly outside the confidence limits applied to a straight path. Over three weeks of observation Tucker and co-workers were certain that curved flight paths were being observed (Fig. 3.33), though the researchers had some reservations. Principal of these was the difficulty of being precise about the path being followed by a fast-travelling falcon viewed at a distance, though the tracking device employed eliminated most of this concern. The Peregrines were also noted as invariably curving to the right, a result consistent with RS's observations of wild Gyrfalcons and Peregrines which also saw all curved flights bending to the right.

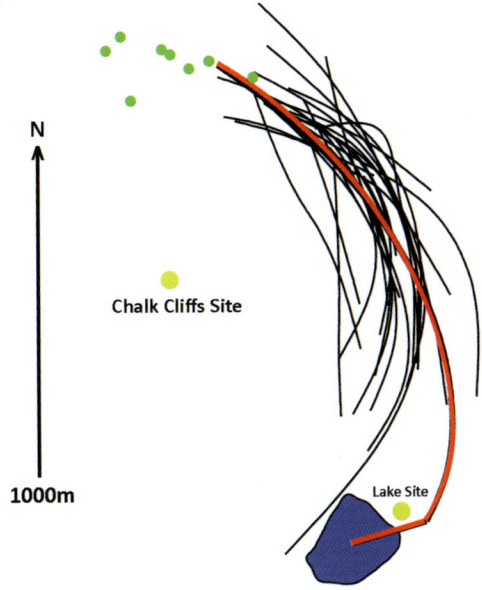

Figure 3.33 Observation of a pair of wild Peregrines breeding at the Chalk Cliffs (at an altitude of up to 3050m) of central Colorado, USA. The observations were carried out at two sites, one close to the Cliffs, the other close to a prominent feature, known locally as Dead Horse Lake (each marked with a yellow dot). The green dots are perching places used by the male falcon prior to embarking on hunting flights. The thin black lines are the flight paths of the Peregrines, the thicker red line is the ideal path of a falcon using deep fovea vision to approach prey. Redrawn from Tucker *et al.* (2000).

The dogleg in the ideal path is not explained in Tucker *et al.*, but accords, in length, with RS's view based on his personal observations that Peregrines (and Gyrfalcons) switch from deep fovea to stereo vision at about 250-300m from the prey.

The observations of Tucker *et al.* were almost exclusively in the morning when such a path would bring the falcon out of the sun as it approached its prey. Since Peregrines are known to dive out of the sun, this might, therefore, explain the observations. However, the preponderance of right-curving paths did suggest that the falcons were favouring their right eye. RS's observations of Gyrfalcons during the perpetual day of the Arctic summer did not suggest the curved flights were always made to bring the attacker out of the sun. Tucker *et al.* also considered the possibility that a curved path might mislead prey into believing they were not under attack. However, the number of curved paths observed were believed to justify the idea that use of the deep fovea for prey monitoring was the explanation for the behaviour. One further concern is that the topography of the area would tend to lead the Peregrines over a well-defined point in the cliff line and from that point the lake 'target' area would be to the right of the falcon which might, therefore, induce a rightward curve.

More recently, the use of video cameras on falconry birds – attached either to the falcon's back or mounted on a modified hood – have suggested that hunting falcons may occasionally use saccadic (*i.e.* very fast) head movements so as to scan prey with the deep fovea, with minimal impact on drag and that use of such movements may negate the need for a spiral flight path (Kane and Zamani, 2014; a subsequent paper, Kane *et al.*, 2015, extended the study of hood cameras to the Goshawk while a third paper, Ochs *et al.* (2017) extended it to three hawk species). However, the falcons involved in the study were hunting birds in flight and, it would seem, from relatively short distances, which would favour the use of the binocular vision afforded by the shallow fovea as this offers excellent depth of field which monocular vision does not: even if the deep fovea and a spiral flight path are used in pursuit of ground, or close-to-ground, prey, when acute vision is required to observe small prey from a significant distance and to isolate it from a confusing background, it is assumed that for the final metres before interception a falcon would change from deep-fovea to shallow-fovea vision, as binocular vision would be required for attack. However, Kane and Zamani do also suggest that for longer falcon-prey distances their data are still consistent with use of the shallow fovea, though it is not clear what definition of longer distance is referred to. In a later study, Brighton *et al.* (2017), falconry birds, in this case Peregrines, were again fitted with head cameras and chased lures which were either thrown in the air or towed behind a drone. Incidentally the Peregrine also attacked live prey several times, unsuccessfully. What Brighton *et al.* discovered was that in attacks on moving targets the Peregrines' movements mirrored those of guided weapons. In other words, the mathematics which underscore missiles aimed at flying targets (*i.e.* hostile aircraft) also readily described the movements of the falcon as it closed in on prey. The essence of the programming of a guided missile is known as proportional navigation (PN) in which the rate of change of the attacker's attack angle is related to, but not exactly equal to, the rate of change of the target's angle of movement. The effect is that the attacker is attempting to head off the target at the same time as shortening the distance between the two by outflying it. In PN, the basic equation of pursuit is:

$$\gamma(t) = N\lambda(t) \qquad (3.8)$$

The PN which flows from Equation (8) is shown schematically in Fig. 3.34. The value of N depends on the relative speeds of attacker and target. For missiles attacking high-speed aircraft, N is usually about 5. For Peregrines which, though fast in avian terms, are much slower than missiles, N is about 2.

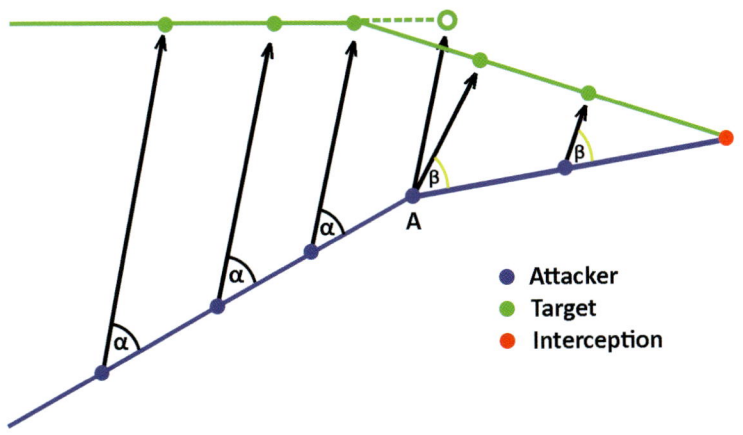

Figure 3.34 Attack on a target. The relative speed of the two are not illustrated by the distances between points on the figure, the diagram being schematic only. To the left, the attacker is flying at a line-of-sight angle α° to the flight path of the target, and this angle remains constant until the target makes an attempt to escape by changing direction. The attacker then modifies its line-of-sight angle (of attack) and the catch is made. In practice with a falcon and its prey, the prey can move in three dimensions and as many times as it may, but despite these movements, as long as the prey does not reach cover, and the falcon's speed is greater than the prey's, an interception will be made.

Brighton *et al.* concluded from their work that deep fovea vision was not the basis for Peregrine attacks. However, again the attack distance studied was significantly less than that identified by Tucker *et al.* (2000) or that observed by RS. The obvious question which arises from the studies of Kane and Zamani (2015) and Brighton *et al.* (2017) is: if the shallow fovea is used at both shorter and longer distances, what was the evolutionary purpose of the deep fovea? In the wild, falcons will sometimes fly at heights above 300–400m. Spotting small prey from such heights, and perhaps at significant horizontal distances as well, would certainly seem to require use of the deep fovea for tracking, at least in the early stages of a stoop. In order to test the deep fovea hypothesis RS's IMU has been flown on falconry birds. The number of flights has been limited as there are few falconers who are willing to train their birds to fly long distances to swung lures. The use of drones to ease the burden on the falconer was dismissed as that would have given the falcon an audible clue as the falconer's position which would have defeated the object of the flight. It was also necessary to change the topography over which the flights took place so that the bird did not produce an internal map which would aid subsequent flights. When all these factors were included in conversations, the number of interested falconers dwindled sharply.

RS observed curved flight paths on two occasions with the female Peregrine TS, but on each occasion, while the flights were longer than the 300-350m assumed transfer to stereo vision, they were not long enough for a definitive result. In each case the falcon curved right. On a tiercel Peregrine called *Mouse*, three significantly longer flights were also curved, in each case to the left. While left-curving flights differ from those of Tucker (and RS) there is no reason to suppose that an individual falcon would favour one eye over the other for deep fovea vision so that 'left-eyed' birds would be seen (though 'right-eyed' birds might predominate). Fig. 3.35 shows the overlay of two flight-paths of *Mouse* in the Brecon Beacons National Park in Wales. The two flight days were distinctly different, the blue track being a day with a westerly wind with a steady speed of around 10m/s (22.5mph) gusting to 16m/s (58kph): the temperature was 3.5°C. The severe wind and high wind-chill factor meant a shorter flight, the later part only being shown. The red flight was on a quieter day, again with a westerly wind but gusting only to 1m/s (3.6kph), though it was still cold (around 3°C). The red track shows that Mouse initially circled above the release team (the authors) before spotting the falconer 2km away. The bird then flew in a marked curve before transferring to stereo vision at about 300m from the lure. The blue flight, though much shorter followed the same curve despite the fact that the bird was battling a strong wind from the left and again changed to stereo vision at about the same point position (*i.e.* c.300m).

These two flights do not confirm curved flight based on deep fovea vision until a transfer to shallow fovea, stereo vision. For such confirmation 10 falcons would need to show curved flight on each of 10 flights over various distances with varying winds from varying angles. But Mouse's flights do suggest that Peregrines use deep fovea vision to locate targets at a distance and may also take flight paths to those targets which are dictated by deep fovea vision until the requirement for depth of field, *i.e.* stereo vision, takes precedence.

Figure 3.35 Two flights of the tiercel Peregrine *Mouse*. The release point of the bird for the red track is shown. North is at the top of the map. The two tracks are correct in terms of latitude and longitude, but have been altered in terms of elevation by adding 50m to all height positions to provide a clearer 3D image.

The Peregrine Falcon

Before leaving vision, it is worth noting a remarkable observation of a Peregrine in Pau, France (so probably *F.p. brookei*). There, the male of a breeding pair had only one eye, the right eye (Mouze, 2022). The male had been hunting and breeding successfully since 2012. Mouze suggests the male had been without the left eye since birth, which perhaps had aided the overcoming of the handicap, but given the enormous input of vision in hunting behaviour, that a male could not only hunt to feed itself, but could also successful breed year on year is extraordinary.

Above The one-eyed male Peregrine with a recent kill.

Below A close-up of the male.

With thanks to Jean-Philippe Crabbe, Michel Mouze and Marc Duquet.

Nominate juvenile failing to catch something from a pond. *Colin J. Lea.*

Hunting Success

The hunting success of Peregrines has been the subject of debate for many years, with falconers claiming remarkable statistics for their birds, these often at odds with the rates seen by observers of wild birds. A full examination of the data is set down in Chapter 4 *Hunting Success*, but a quick overview is included in Roalkvam (1985) who produced insightful data based on observations in the wild at various times of the year and during migration, as well as data from falconers. Rates on migration and for juvenile birds are as low as 7-8%. For breeding birds the figures improve, with rates of 15-20%. While these success rates seem low particularly as high-speed stooping appears to a human observer to be the ultimate attack method with guaranteed success, as Hustler (1983) points out, a small adjustment in the flight of the attacked bird is sufficient to avoid capture.

Cooperative hunting produces higher rates, with figures around 30-40% being quoted. Higher rates are seen by observers who differentiate between high and low intensity hunting. The former is when the falcon is really hunting rather than just cruising the sky and making the occasional attack, in part as a way of practice. For high intensity hunting observers have claimed up to 69% *e.g.* Treleaven, 1998). That figure is occasionally dwarfed for falconry birds with rates of over 80% being claimed, though such rates are, of course, unrealistic as the Peregrine is then usually assisted by dogs and beaters, and is sometimes bring flown in areas where the prey density is artificially high.

Olfaction

For many years it was assumed that birds had no, or a limited, sense of smell, but recent research has suggested that in many species it is more important than had been imagined – see, for instance, the work of Driver and Balakrishnan (2021) for the developing field of genomics and avian olfactory receptors. Vultures are known to smell carrion and Potier (2020) noted that all the raptors which have been studied are known to have well-developed olfactory bulbs (neural structures responsible for the sense of smell). Mäntylä *et al.* (2020) showed that insectivorous birds were attracted by the smell of pine trees that had been damaged by foraging herbivores, as the damage attracted insects which arrive to lay eggs and the insect larva which feed on resin. Wikelski *et al.* (2021) found that European (White) Storks (*Ciconia ciconia*) were attracted to the volatile organic compounds emitted when grass was freshly cut, offering evidence that it was olfaction that was responsible for the well-known, but not understood, appearance of birds of various species when fields were mown or ploughed. While these later reports do not offer any suggestion that avian predators may use olfaction to seek out prey, the work of Mahr and Hoi (2018) suggests that prey species may be able to use smell as a means of early detection of a predator. In the study Red-legged Partridge (*Alectoris rufa*) were not only able to detect the smell of mammalian predators, but to detect the scent of an alarm secretion released by European Hoopoes (*Upupa epops*). If other passerines emit an alarm secretion as well as having an alarm call, this might offer two methods by which the partridges might be alerted to the presence of an avian predator.

Unusual Peregrine hunting activites.

Above In California, an adult *F.p. anatum* knocked this Willet into the ocean 100m off-shore. There was a large swell and the Willet kept diving until the exhausted falcon gave up and retreated to a cliff top perch. The incoming waves then brought the Willet almost to shore. Now the Peregrine could easily walk over to catch the Willet in water too shallow for it to dive for cover. *Will Sooter.*

Left Peregrine hunting bats, north-west USA. *Nick Dunlop.*

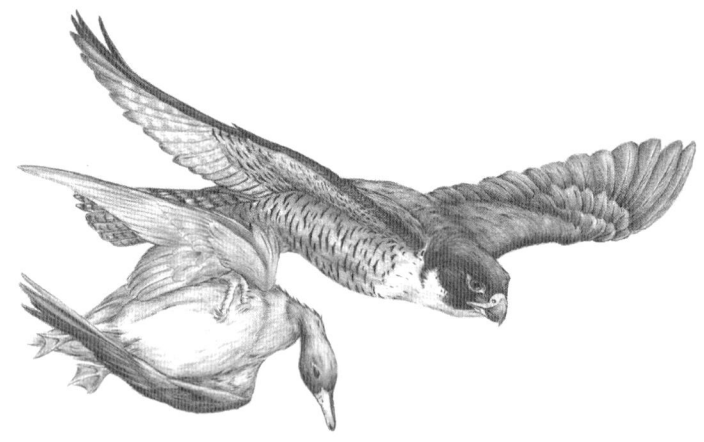

4 DIET

Not surprisingly for a species which has occupied such a vast area of Earth, even allowing for the fact that the chosen habitat across the range – open country over which to hunt – is similar, the Peregrine is flexible in terms of prey. Uttendörfer (1952) notes a total of 210 species which formed the diet of Peregrines in central Europe, a list which suggests that the falcon will take on any flying prey which it considers it has a reasonable chance of taking. Preferred prey lies in the weight range 50-500g, a very wide band, covering a significant number of species. Feral pigeons weigh about 300g which represents *c.*50% the weight of an average nominate male Peregrine (650g) and *c.*30% that of a female (1100g). Pigeons, either feral or their wild counterparts form a significant fraction of the Peregrine diet across its range. However, in general, Peregrines take prey averaging about 200g[1]. Dawson *et al.* (2011), studying *F.p anatum* breeding along the Yukon River of north-west Canada, found that while the 50-500g weight range applied in general, heavier prey – shorebirds, jays and thrushes – were taken more often than would be expected by abundance, while smaller prey – sparrows *etc.* – were taken less often than abundance implied: the Peregrines were maximising prey deliver to their chicks. Heavy prey, classified as >1000g, was least selected except where the female Peregrine was making an increased contribution to offspring feeding. In those, few, cases, presumably where the female was large enough to successfully hunt large prey (and also capable of laying a larger

---

[1] In a study of nominate Peregrines in Exeter, England across 20 years, Drewitt and Dixon (2018) collected 5426 prey items with a total weight of 1044.64kg an average prey weight of 192.5g.

clutch) the increased biomass delivered to the nest meant that more fledglings were raised. There is some evidence to suggest that Peregrines are attracted by species with conspicuous colour patterns, and to individuals with odd flight or movement patterns which might suggest they are sick or injured.

The preferred method of attack – stooping – allows the Peregrine to take species much larger than itself. There are several records of Peregrines attacking geese as large as Greylags (*Anser anser*), males of which may weigh up to 3.5kg (almost three times the weight of a female Peregrine), while historically in Europe falcons were frequently flown against Grey Herons (*Ardea cinerea*), which, though of comparable weight to a female Peregrine, are very large and formidably armed. Equally formidable is the Great Black-backed Gull (*Larus marinus*), which may weigh 1.8kg and has a reputation as a fierce predator, but which has also fallen victim to Peregrine attack[2]. Since there are also accounts of Peregrines taking Goldcrests (*Regulus regulus*) weighing 5.5g, it is clear that the falcons are opportunistic hunters of any bird that comes within their territory and appears vulnerable to attack. There is, of course, a limit to the weight which a hunting Peregrine can carry back to its eyrie if it is hunting to feed chicks, this appearing to be about the weight of the falcon itself. This raises the question of why the falcon will attack prey it cannot carry away. Avian predators taking prey to be carried will consume the prey at the killing site, then return later to either continue eating or to carry off the now, reduced weight, kill. Taking heavier prey may also be a winter survival strategy if other prey is not available. It may also, of course, be that the presence of prey within the Peregrine's territory and the favourable positions of hunter and target trigger an automatic attack response in the falcon.

While the majority of prey is avian, other prey is also taken. Insect feeding is known, primarily in recently fledged birds when their hunting skills are developing. However, there seem to be exceptions. Pruett-Jones *et al.* (1981) noted the copious remains of cicada and grasshopper species at the eyrie of *F.p. macropus* in Victoria, Australia, while White and Brimm (1990) watched an immature *F.p. nesiotes* take large (6cm) locusts fleeing from a grass fire on Viti Levu, Fiji. In Arizona, USA, Ellis *et al.* (2007) found that cicadas formed 1% of prey remains in Peregrine eyries, but noted that the insects formed 69% of the falcon attacks they witnessed and were particularly important to females on guard duty at nests as they represented prey which could be taken quickly from the guard perch. Perhaps the most remarkable insect feeding was that seen by Davis (2008) who notes two independent observers watching adult Peregrines catching salmonflies (*Pteronnarcys californica*, a very large stonefly). In one episode in 2004 both male and female Peregrines of a breeding pair took the insects: the female took 23 insects in three minutes catching each in her talons and transferring them to her mouth; the male took 14 and then

---

[2] Nethersole-Thompson (1931) records an adult pair of Peregrines repeatedly stooping on a Great Black-backed Gull one after the other until the male hit it hard, allowing the female to grip onto it. The male then joined her, the three birds fluttering to the ground. After a struggle the gull was finally killed.

returned to the nest, apparently delivering them to the brood. In 2007 another observer watched a male take 22 insects, averaging 22s between catches: the female of the pair also flew from the nest a dozen times to catch an insect, returning to the nest each time. Both observations were in Montana, USA.

Mammals are also taken, records including Rabbits (*Oryctolagus cuniculus*), hare leverets, shrews and voles. Court (1986) found that the population of *F.p. tundrius* in northern Canada increased by 30% during a year when the rodent population reached a peak[3], and in a study carried out subsequent to the rodent peak Bradley and Oliphant (1991) noted that mammalian prey – Northern Collared Lemming (*Dicrostonyx groenlandicus*) and young Arctic Ground Squirrels (*Spermophilus parryi*) – constituted a third of the Peregrine diet by weight even though the rodent population had declined to a more normal level. Bradley and Oliphant examined pellets rather than remains at plucking posts, and suggest that this gives a better idea of the degree to which the falcons take mammalian prey, implying that it is possible that throughout the Peregrine's range, while avian prey is dominant in the diet, the mammalian fraction may be underestimated. Elsewhere in the USA, Pagel and Schmitt (2013) found the leg of an American Marten (*Martes americana*) in a Peregrine nest in the Yellowstone National Park, hypothesising that it may have been killed by an adult falcon when it attempted to predate fledglings. Pagel and Schmitt also note that others have found the remains of Mountain Beavers (*Aplodontia rufa*) and Snowshoe Hares (*Lepus americanus*) in Peregrine nests. Ratcliffe (1993) notes rabbits, hare leverets, voles and shrews as Peregrine prey. Swann (1998) records finding an uneaten young Stoat (*Mustela erminea*) in a Peregrine nest in Scotland, while Tyler and Ormerod (1990) record the cub of a Red Fox (*Vulpes vulpes*) in prey remains in Wales. While mammals represent a small fraction of the Peregrine diet, in a fascinating study at two sites in Arctic Russia (the Nenetsky Nature Reserve and in the southern Yamal Peninsula), Pokrovsky *et al.* (2020) found that this lack of enthusiasm for mammalian prey was benefitting another raptor. Peregrines are highly aggressive when breeding, and this aggression was reducing the number of mammal feeders close to their nests, allowing rodent numbers to increase. This increase encouraged Rough-legged Buzzards (*Buteo lagopus*) to move closer to Peregrine nest sites than was normal to feed on the rodents. In general, the buzzards take a spectrum of prey, including larger mammals (rabbits, hares) and avian prey such as grouse. But it seems that in some cases, the buzzards were reducing their prey spectrum, taking advantage of increased rodent numbers to benefit from Peregrine aggression towards other competitors. This curious association between falcon and hawk is mentioned in more detail in Chapter 9 *Friends*.

---

[3] Such peaks are a characteristic of the rodent populations of the Arctic: the 'lemming years' in Scandinavia are famous for population increases that have historically meant the ground almost heaving with rodents. Such events became famous for the idea of the rodents committing mass suicide by throwing themselves into water when the reality was that they could not distinguish a small pond from a large lake or an ocean. While local Peregrines remained primarily bird feeders, they were sufficiently flexible in their hunting abilities and diet to take advantage of the rodent surplus.

This *F.p.anatum* adult in California, USA has just captured a Californian Ground Squirrel (*Otospermophilus beecheyi*). Will Sooter.

Bats are also taken: Sprunt (1951) in Texas, USA, records Peregrines stooping into the bats as they streamed out of their roost, or flying alongside the prey 'river' and then swerving sideways to grasp a nearby victim. Stager (1941) also watched six Peregrines who daily waited for bats to emerge from a different cave in Texas. The falcons hunted the bats for over an hour, flying into the bats as they streamed out from the cave. Lee and Kuo (2001) also observed Peregrines and Red-tailed Hawks hunting at a cave in Texas: the hawks were better at catching the bats in the evening, the falcons better at dawn: no reason was given for this difference and none is obvious to the present authors. Lee and Kuo estimated that the two raptor species (an average of five of each species on the 23 days of observation) took 2,153 bats, a large number, but only 0.02% of the estimated population of 10 million Mexican Free-tailed Bats (*Tadarida brasiliensis*) at the site. On Fiji, Clunie (1976a) noted that Flying Fox fruit bats (*Pteropus tonabus*), which weigh about 700g, were the staple food of the local Peregrines (*F.p. nesiotes*) at an eyrie on southern Viti Levu, with the much smaller Long-tailed Fruit Bat (*Notopteris macdonaldi*) also being taken. In late August/early September in Illinois, USA, Byre (1990) watched a group of five, recently hacked juvenile Peregrines kill 29 bats from three migrating species (with a further 15 probable kills) during three days of early morning hunts. Byre notes that the falcons had made their first kills only two weeks earlier, implying that the falcons were using bats for both food and to hone hunting skills. Byre noted occasions when several falcons surrounded a single bat, with one eventually taking it, but thought this was evidence of 'close-focused' hit-and-miss hunting rather than a truly cooperative activity. Twice bats were forced into a nearby lake and then taken from the water surface. In a study in France, Duquet and Nadal (2012) listed a total of 11 bat species which had been taken by avian predators, with Peregrines responsible for 29% of the total number. Bats have also been identified in the prey remains of urban Peregrines in Britain, while Rollie and Christie (2006) record the capture of a bat in the middle of the day (12.20) in Kirkcudbrightshire. The bat – probably a Noctule (*Nyctalus noctula*) – took evasive action when approached, but the Peregrine captured it on the third attempt: the Peregrine bit the bat in flight, apparently to kill rather than consume it.

Cade (1982) records seeing a Peregrine catch an Arctic Grayling (*Thymallus arcticus*) as it broke surface in Alaska, and other attempts to catch fish. Feeding on carrion has also been observed, though information on this is fragmentary and contradictory. Anecdotal information from falcon watchers suggests the behaviour is rare, and literature searches bring up few incidents, which suggest the same. One striking example of the behaviour was reported by Lockhart *et al.* (2020) who had mounted a camera to record American Herring Gull (*Larus smithsonianus*) behaviour on the Canadian shore of Lake Superior. At one nest the female gull became increasingly lethargic while incubating and eventually died. Four days later an adult (probably female) Peregrine (presumably *F.p. anatum*) arrived at the nest and scavenged the carcass, decapitating and partially plucking it, and feeding on the pectoral area for 86mins (see images opposite). The Peregrine did not return to feed on the carcass. Subsequently the male Gull returned to the nest and removed the carcass from the eggs, revealing that both pectoral muscle and some internal organs had been eaten by the falcon. The male gull did not attempt to incubate the eggs, which were later predated by an American Crow (*Corvus brachyrhynchos*). Lockhart *et al.* surmise that the falcon may not have realised the gull was dead when it landed on it, but believe this unlikely given the falcon's acute eyesight: that the Peregrine landed 'quietly' on the gull also suggests it knew it was dead. Either way, the falcon continued to feed.

Carrion feeding seems to occur only when hunting is poor and being usually seen in juvenile birds whose hunting skills are still developing: *e.g.* Holland (1989) observed a juvenile Peregrine one summer in California that perched near a roadkill California Ground Squirrel (*Otospermophilus beecheyi*), then dragged it to the verge, mantled it and fed on it for 25 minutes, while Buchanan (1991), in west Washington State, USA, noted juvenile Peregrines feeding on carrion during winter – a dead Guillemot (*Uria aalge*) in one case, a White-winged Scoter (*Melanitta deglandi*) in another. However, Varland *et al.* (2009), made 1109 surveys of Peregrines at three coastal sites in Washington State, USA, and of 172 observations of falcons with prey 29% were carrying

Opposite   F.p.anatum(?) female scavenging a dead Herring Gull. With thanks to Craig Hebert.

carrion, chiefly seabirds and shore birds. While these observations were mainly of juvenile falcons, they were not exclusively so, adults also being seen. There was also no apparent change in scavenging behaviour with season. However, as Peregrines often cache prey, particularly during the breeding season, some observations will most certainly have been of birds retrieving cached remains.

From the various data on the diet of the Peregrine it is clear that the falcon is indeed an opportunistic hunter with an eclectic menu, a behaviour which has led to occasional dismay among birdwatchers when rarities such as Wryneck (*Jynx torquilla*) are taken: Ratcliffe (1993) reports that in 1980 a juvenile Peregrine took 36 Roseate Terns (*Sterna dougallii*) at a site on Anglesey, UK, hunting which was not likely to have endeared the youngster to local birdwatchers. Elsewhere there have been reports of Peregrines causing potentially species-threatening damage to rare species. One such is the burrowing endemic Macgillivray's Prion (*Pachyptila macgillivrayi*) on St. Paul Island in the southern Indian Ocean. The island lies roughly midway between Africa and Australia (and is equally close to Antarctica, to the south). It was visited by fur seal hunters from the late 17th century onwards who introduced several mammal species including Black Rats (*Rattus rattus*). Although the rats were exterminated by 1997, by then the only surviving colony of the rare Prion was on an off-shore 1ha islet (Roche Quille) where 150 breeding pairs had their burrows: the population of adults and juveniles was believed to be 540 birds. In 1999 a Peregrine (unknown sub-species) arrived on the island, probably as a consequence of a tropical storm. The Peregrine was seen for the first time on 27 January and for the last time on 15 February. Such visits by falcons are extremely rare, but of the five falcons known to have reached the island in the 55 years to 2007, four have arrived between 1999 and 2007 (one Hobby, one Peregrine, one Sooty Falcon and one unidentified falcon). Global warming is bringing an increase in ocean storms and, consequently, an increase of falcons to St. Paul. During its visit in 1999 the Peregrine killed 27 breeding prions (from a total 54 remains found: 19 of the others had been killed by skuas, a permanent local threat, with 8 deaths from unknown causes. Jiguet *et al.* (2007) calculated that if falcon visits occurred regularly at 5 five-year intervals and reeked the same havoc each time, Macgillivray's Prion would become extinct in 200 years as the intervening periods between falcon visits was insufficient for the population to recover. Reynolds *et al.* (2015) also report a vagrant Peregrine killing around 4%, in 2006, and around 2%, in 2008, of the endangered Laysan Teal (*Anas laysanensis*) which had been translocated to Midway Atoll in the Pacific Ocean in order to boost a population endemic to Laysan Island. A vagrant Peregrine on Laysan Island itself also took more than 70 endangered Laysan Finches (*Telespiza cantans*) in 2008/9. Reynolds *et al.* note that while the observed predation did not, of itself, result in significant population declines of either species, such attacks by vagrants, if they were to increase, when combined with weather related catastrophes which might underpin the vagrancy, could be a threat.

F.p. calidus, wintering at the Little Rann of Kutch, Gujarat, India, eating a feral pigeon. The curious pink coloration suggests the pigeon might have been taken from a local village. *Dhairya Dixit*.

Composition of diet

While the comments above emphasise the catholic nature of the Peregrine's diet, the diet of individual falcons is usually more restricted, the range of prey often being a good match to the composition of local species. Coastal species will take a larger fraction of seabirds, moorland Peregrines may take more game birds, urban birds will take pigeons, and birds wintering at estuaries will take waders. One generalisation is that if they are available, pigeons, wild or feral, form a significant fraction of the diets of all Peregrines, as will be seen later in this Chapter.

Feral pigeons in towns and cities can often be found with racing bands and wing stamps indicating that the birds are racing stock which have failed to return to their lofts and joined the feral population which has no doubt been swelled in this way for at least a century. Occasionally in adverse weather conditions whole races may be wrecked with hundreds of birds being lost, some of which then manage to make a living for themselves in the wild but increasingly in urban environments where they breed throughout the year. Although feral pigeons breed mostly between March and July, they can breed throughout the year. Pigeons, on average, produce 2-5 broods and typically produce a clutch of two eggs. UK population estimates in 2009 were 450,000-650,000 pairs or around 1,000,000 birds.

At 230-370g pigeons make an ideal prey for Peregrines and occur in traditional coastal and inland territories as well as in urban territories. Lindberg (1983) noted that the average energy value of a Common Starling (*Sturnus vulgaris*) was 160 kcal, while that of a pigeon is 600kcal: with 100 attacks on Starlings the 'captured energy' is 8,800kcal, assuming a capture rate of 55%, while for the pigeons it is 15,600kcal, with a capture rate of only 26%. Ignoring the energy requirements of hunting and transport, Lindberg calculated that the success rate against pigeons would need to drop below 14% before they would be an unattractive target in energy terms. In an equally interesting study, Tornberg *et al.* (2016) who studied the impact of the returning Peregrines (after the decline in falcon numbers due to agrochemical poisoning) on avian populations in Simo (south-west Lapland), Finland, close to the Gulf of Bothnia. The 'protection hypothesis' suggests that although apex predators may negatively affect primary prey populations, they may benefit other species by suppressing mesopredators and so limit egg and chick predation. The study showed that the hypothesis was supported for ducks, but was less clearly supported for other species as it was difficult to separate out the negative effect of habitat loss.

The fact that pigeons are a mainstay of British Peregrines has also led to conflict with humans, as we shall see when considering the UK Peregrine population in Chapter 10. In general cities have significant populations of pigeons, and this obviously contributes to their occurrence in the Peregrine diet. Pigeons also need water regularly, which takes them to drinking places, but they can be fast, agile fliers and so are not a trivial target though as noted in Chapter 3 *Vision* their size and limited vision are likely contributory factors to the Peregrine's enthusiasm for taking them.

In a fascinating study in California, Palleroni *et al.* (2005) sought to understand the vulnerability of pigeons to attack, assuming that this must contribute, in some way, to the high proportion of pigeons in Peregrine diets wherever they are found. Palleroni and co-workers noted that wild pigeons (Rock Doves) had a conspicuous white rump, a feature which is often absent in Feral Pigeons as a result of negative assortative mating (*i.e* the feral birds do not preferentially choose mates with plumage similar to their own). The result is that over time the white patch has been lost in much of the feral stock whose plumage patterns are now a general mix of grey, blue, and white. Palleroni *et al.* noticed that pigeons with the white rump patch were killed much less often than those with other colour patterns. To test if this was a real effect, the feathers of 756 trapped pigeons were switched, so that those without a white patch acquired one, and those with one lost theirs (Fig. 4.1). The effect was dramatic, with the mortality rate of birds which had acquired a white patch dropping, and that of the birds that had lost their white patches increasing in tandem. Palleroni *et al.* conjecture that the evasive action of attacked pigeons, a roll at close approach of the falcon, causes a flash of the white patch against the more cryptic coloration of the remaining plumage,

Diet

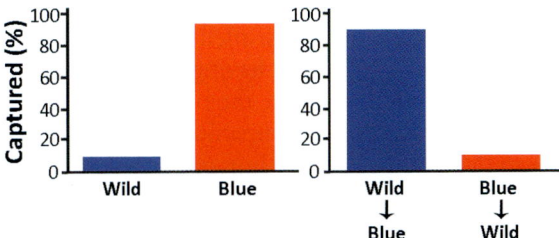

Figure 4.1 Percentage of successful captures of feral pigeons by Peregrines in Davis, California. The pigeons were defined by coloration as 'wild' (blue-grey birds with a white rump patch, resembling the urban stock's Rock Dove ancestry) and 'blue' (birds having a blue rump). The left-hand capture data note that 'wild' birds were almost ten-times less vulnerable to successful attack. But when the rumps of the birds were artificially coloured the capture data was reversed. Redrawn from Palleroni *et al.* (2005).

Two feral pigeons photographed in a town in England. The upper bird shows the ancestral white rump of its wild forefathers, the other is without a white patch.

and that this acts to confuse the falcon, which misses the roll manoeuvre and so misses the prey. In pigeons without the patch, there is no white flash and the roll can be observed by the falcon, which can compensate for it and continue its attack. Palleroni *et al.* comment that a similar anti-predator tactic is employed by many fish species, which roll to alternate dark dorsal and white ventral surfaces. It is also speculated that the loose feathers of pigeons have evolved as a defence mechanism, the stooping Peregrine which attempts to bind to its prey occasionally catching only a foot-full of feathers while the partially disrobed pigeon makes good its escape.

Interestingly, for a species which tends to form flocks, pigeons do not have an alarm call. In a neat study of their behaviour, Stephan and Bugnyar (2013) noted that individual pigeons increase their vigilance and scanning of their local environment if they either hear or see a raptor (in the case of the study, Common Buzzard calls or the sight of a stuffed Buzzard). Pigeons, it seems, take responsibility for their own safety, not having evolved a system for alerting members of their flock – though of course the prompt flight of one bird will usually initiate similar behaviour in its companions. In a separate study, Kano *et al.* (2018) fitted an IMU to the heads of homing pigeons to investigate head movements during flights when released a few kilometres from their lofts. They found that pigeons move their heads more frequently than appeared necessary for manoeuvring in flight, the movements being saccadic and most pronounced in yaw (left and right) and roll (side to side) rather than pitch (up and down). The pigeons were clearly constantly vigilant against potential attack, though the relative lack of interest in pitch suggests the birds were less concerned about being attacked from above. At first glance this seems at odds with the most spectacular of the Peregrines attack modes, the stooping attack which rakes the hallux across the victim. However, while such attacks do occur, they are outnumbered by stoops which finish behind the prey, converting potential energy into kinetic energy for a horizontal attack or even one which takes the falcon upwards to the prey, attacking its most vulnerable region. Pigeons, it seems, are more concerned about attacks from pursuit predators such as hawks, perhaps reflecting the learned likelihood of such attacks.

As with other falcons, Peregrines prepare their avian food by plucking, removing as many feathers as possible. But no matter how thorough the falcon, inevitably all meals will involve the ingestion of some feathers as well as other indigestible parts – bones, claws and bills, and the fur and bones from mammalian prey – which cannot easily pass through the digestive tract. This detritus is formed into a mass within the bird's gizzard and regurgitated at regular intervals, regurgitation having the added advantage of purging the upper section of the digestive tract. Known as casts by falconers, but as pellets by others, the masses are cylindrical, about 30-50mm long, 10-18mm in diameter, usually rounded at one end and pointed at the other. The production of pellets is probably assisted by gastroliths (known as rangle by falconers), a collection of ingested grit/small stones. Though widely known to be ingested by herbivorous birds to aid the breaking down of their forage, the ingestion by falcons is inferred from falconry birds, though it was seen by Albuquerque (1982) who saw an adult female *F.p. tundrius*, wintering in Brazil, picking up grit from a sand store.

When cast, pellets are slimy with mucus which sometimes means they stick to branches close to nest sites. Pellets are normally regurgitated at dawn. As this is usually when the bird is at its roost, or at the nest in the breeding season, this aids discovery, but Peregrines' choice of nest site – often on tall, near vertical cliffs, or at the tops of buildings with minimal access possibilities – means they can be difficult to obtain. Analysis of collected feathers at plucking posts, or of pellets has allowed the diet of Peregrines to be established in certain areas. Ratcliffe (1993) collated data from several studies over a

Diet

Typical nominate Peregrine pellets.

combined period covering 1904-1975 (though not continuously during that time) on the falcon's diet during the breeding season in six areas of mainland Britain (Fig. 4.2 overleaf).

Nominate Peregrine regurgitating a pellet on Lincoln Cathedral, England. *Peter Day.*

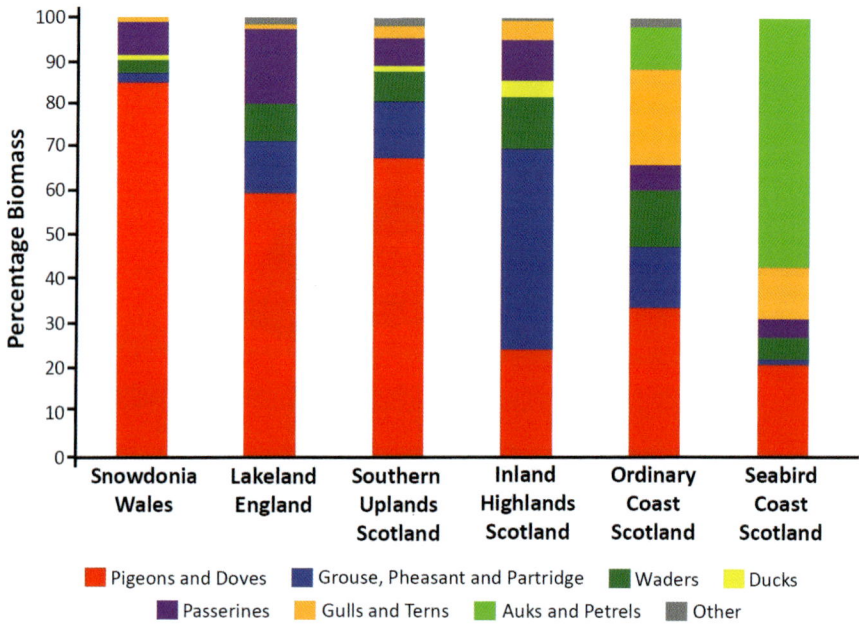

Figure 4.2 Frequency of prey by biomass taken by Peregrines in six areas of mainland Great Britain. 'Ordinary' coastline is not a specific seabird breeding coast, whereas 'seabird coast' includes seabird breeding colonies. The dates over which data on prey was collected were:

Snowdonia: 1950-1957 and 1978-1979.
Lakelands: 1904-1962, except for the years 1928-1934.
Scottish Southern Uplands: 1923-1974.
Scottish Highlands: 1961-1962, 1964-1975.
Scottish 'ordinary' coast: 1961-1962, 1971.

In the actual prey tables in Ratcliffe (1993) which have been condensed to produce Fig. 4.2, a total of 108 species are mentioned. Of these, there are two instances of Great Black-backed Gull, mentioned above as a formidable opponent. Common Buzzard, Eurasian Sparrowhawk[4], Merlin, Common Kestrel and Peregrine as raptorial prey: the single Peregrine mentioned by Ratcliffe may have been taken in a territorial dispute – such disputes will be mentioned again in Chapter 9 *Foes*. Five owl species are mentioned as prey (Long-eared, Short-eared, Tawny, Barn and Little (*Athene noctua*)). Ratcliffe's prey list also includes five mammal species (Rabbit, Brown Hare (*Lepus europaeus*), Mountain Hare (*Lepus timidus*), Water Vole (*Arvicola amphibius*) and Field Vole (*Microtus agrestis*)).

While pigeons are clearly important to the Peregrine diet in Britain (Fig. 4.2) and also in much of the rest of Europe – see, for instance, Mlíkovský

---

[4] Rudebeck (1951), studying the diet of Swedish nominate Peregrines observed one that specialised in hunting Eurasian Sparrowhawks, Rudebeck noting '*the day before when there had been a dense fog, I observed a male Peregrine flying by with a screaming Sparrowhawk in its claws*'.

Male Red Grouse photographed in the Scottish Highlands. *Stan Maund*.

and Hruška (2000) in the Czech Republic – in areas where feral, domestic or wild pigeons are few or absent, such as northern Fennoscandia, ducks or other species dominate the diet (see, for instance, Sulkava, 1968, whose study in Finland showed that in terms of numbers taken the diet was dominated by Lapwing (*Vanellus vanellus*), Eurasian Teal (*Anas crecca*), Mallard (*Anas platyrhynchos*) and Black-headed Gull (*Chroicocephalus ridibundus*).

In winter many Peregrines stay close to their breeding areas, and even those that move tend to occupy similar habitats, the requirement for open country over which to hunt being paramount when the necessity for nesting cliffs has diminished. In his study in southern Scotland, Mearns (1982) noted that 86% of the inland breeding territories he visited were occupied, as were 88% of the coastal territories. Mearns also studied the winter diet of Peregrines at coastal and inland sites in southern Scotland. The coastal sites were backed by farm and moorland and included estuaries within 1km. The inland sites were for sheep grazing or were moorland, and ranged from 250m to 550m in height. Prey remains were identified at plucking sites. At the inland sites passerines accounted for 47% of the diet (by number) across a four-year period (but were 71.4% of the diet in one winter: in that winter Redwings (*Turdus iliacus*) and Fieldfares (*Turdus pilaris*) constituted 44.6% of the total diet). The bulk of the remaining prey (from across the four-winter period) were pigeons (47%), waders (13%) and grouse (5%). At the coastal sites pigeons were 58% of the diet (again across four winters), with passerines constituting 32% and waders 7%. Interestingly, in the same winter in which passerines constituted a greater fraction of the diet at inland sites, they also made up a higher fraction (47.6%) at coastal sites, though in this case Redwings and Fieldfares were only 16.5% of the total diet. Mearns also collected 91 Peregrine pellets, all but four from inland sites. The prey fractions in these were similar to those from the inland plucking sites, but included 7.1% of duck remains: ducks had been absent from inland plucking sites and only formed 2% of the remains from the coastal sites. Two of the pellets also included the remains of voles. The species identified in the study of Mearns did not include any not already identified in the data underlying Fig. 4.2. In another study in Scotland, this time at an estuary near Dunbar, to the east of Edinburgh, Cresswell and Whitfield (1994) noted that the prey spectrum was almost entirely shorebirds with Oystercatchers (*Haematopus ostralegus*), Redshank (*Tringa totanus*) and Curlew (*Numenius arquata*) topping the list by number caught.

Ratcliffe (1993) noted that in the Lake District Peregrines were roosting close to their breeding sites during the winter, but while some were hunting locally, taking a greater proportion of corvids and seabirds, others were journeying to coastal sites to hunt, returning to their roosts each evening. Interestingly, despite its reputation as a solitary bird, Kelly and Thorpe (1993) reported a communal roost on the Isle of Man, with up to nine Peregrines sharing a stand of mature conifers. Even more interesting was their observation that sometimes the roost was also shared by Common Kestrels, Eurasian Sparrowhawks, Ravens (*Corvus corax*), and two Merlins on one occasion. The roost was used regularly from late August until early December and then more sporadically. The reasons for communal roosting are not obvious, and are even less so when the congregations of other raptors and Ravens are also considered.

Weir (1977) noted that in Scotland's Spey valley wintering Peregrines took advantage of migrating flocks of thrushes in autumn and spring, and hunted Red Grouse (*Lagopus lagopus scotica*) and Rock Ptarmigan (*Lagopus muta*) during the winter months. Other studies in Britain and North America have shown the importance of shorebirds to wintering Peregrines (*e.g.* Dekker, 1988; Cresswell and Whitfield 1994; Cresswell, 1996; Buchanan, 1996: these studies are considered in the following section), while on Vancouver Island, British Columbia, Dekker (1995) noted that 88% of Peregrine prey (by number, 46 of 52 prey items) were ducks (of five species) which were taken in flight or from land in equal numbers.

Moving east from North America, the prey of the various Peregrine sub-species are set down below.

F.p. tundrius (Figs. 4.3a and 4.3b)

At Rankin Inlet in the Canadian Arctic, Bradley and Oliphant (1991) were able to identify 14 avian prey species (though many of the remains could not be identified). The major remains were of local passerines, chiefly Horned (or Shore) Lark (*Eremophila alpestris*), Lapland Longspur (or Lapland Bunting) (*Calcarius lapponicus*) and Snow Bunting (*Plectrophenax nivalis*), followed by shorebirds, including Dunlin (*Calidris alpina*) and Semi-palmated Plover (*Charadrius semipalmatus*), and other species including Long-tailed Duck (*Clangula hyemalis*) and Rock Ptarmigan. Interestingly, as noted above, the Peregrines were also taking significant numbers of the local mammals: Northern Collared Lemming and Arctic Ground Squirrel. In west Greenland, Rosenfield *et al.* (1995) noted the main prey to be passerines, chiefly Lapland Longspur, Northern Wheatear (*Oenanthe oenanthe*), redpolls and Snow Bunting, but including Rock Ptarmigan, Mallard and a single White-fronted Goose (*Anser albifrons*). The same prey spectrum was found by Robinson *et al.* (2018), also in Nunavut, Canada, but the mammalian prey included not only the species mentioned by Bradley and Oliphant (1991) above, but Arctic Fox (*Vulpes lagopus*), Arctic Hare (*Lepus arcticus*) – probably juveniles in each case – and all three Arctic mustelids, Short-tailed Weasel (or Stoat), Least

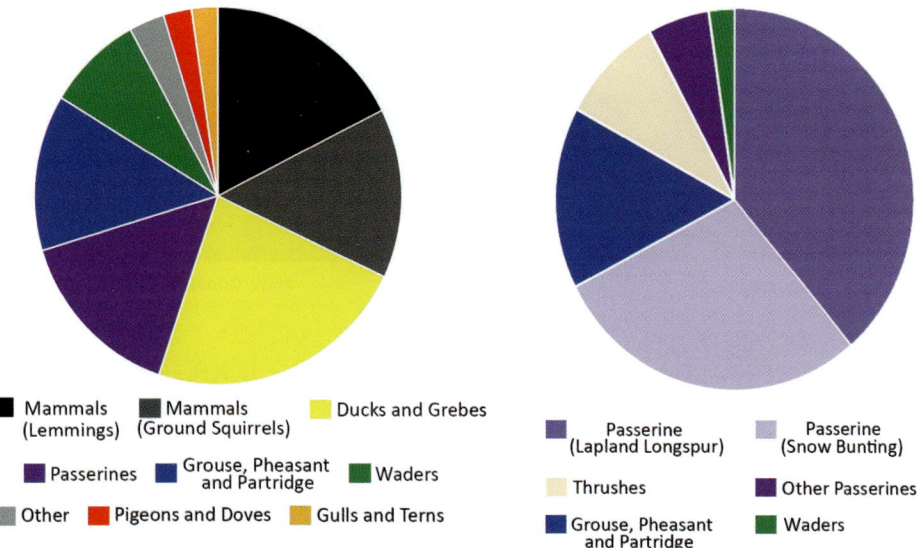

Figure 4.3a *Left* Prey spectrum by biomass percentage of *F.p. tundrius* at Rankin Inlet, Canada during the breeding seasons of 1986 and 1987. Constructed from data in Bradley and Oliphant (1991). The study showed an unusually high fraction of mammalian prey.

Figure 4.3b *Right* Prey spectrum by biomass percentage of *F.p. tundrius* near Søndre Strømfjord, west Greenland during breeding seasons from 1972-1990. Constructed from data in Rosenfield *et al.* (1995).

Weasel (or Weasel) (*Mustela nivalis*) and American Mink (*Neogale vison*). The study also compared the data on diet from analysing blood samples taken from falcon chicks with that from trailcams: the latter were found to provide better information.

In another study in west Greenland, where the local Peregrines were hunting the same passerines, Meese and Fuller (1987) found that while most prey species were choosing to nest further from Peregrine eyries, the Snow Buntings were nesting close to the foot of the eyrie cliffs as their preferred nest sites were amongst cliff-foot rock piles which created clefts and caves inaccessible to Arctic Foxes. However, the buntings altered their behaviour, choosing to fly very close to the ground when foraging to discourage Peregrine attacks, trading increased risk of falcon attack for decreased chick losses.

In wintering quarters the falcons take advantage of the local avian population, as well as feeding on bats.

Willow Grouse (*Lagopus lagopus*).

Arctic Ground Squirrel.

F.p. pealei (Fig. 4.4)

The limited coastal area inhabited by this large falcon is reflected in the limited prey spectrum, chiefly comprising seabirds. Beebe (1960) refers to the sub-species as the 'Marine Peregrine'. Studying the falcons in coastal Washington State and nearby British Columbia Beebe noted prey included a number of Pacific auks and petrels, together with sea ducks such as scoters, Long-tailed Ducks and Buffleheads (*Bucephala albeola*). This prey spectrum was confirmed by Nelson and Myres (1976) who noted the Ancient Murrelet (*Synthliboramphus antiquus*) as the main target for falcons breeding on Langara Island, British Columbia. On Amchitka Island in the Aleutians, White *et al.* (1973) listed a total of 31 avian prey species and the capture of several Brown Rats (*Rattus norvegicus*) which had been unfortunately introduced to the island. The avian species included Pacific auks (Crested Auklet (*Aethia cristatella*) and Least Auklet (*Aethia pusilla*)) which were the main prey items,

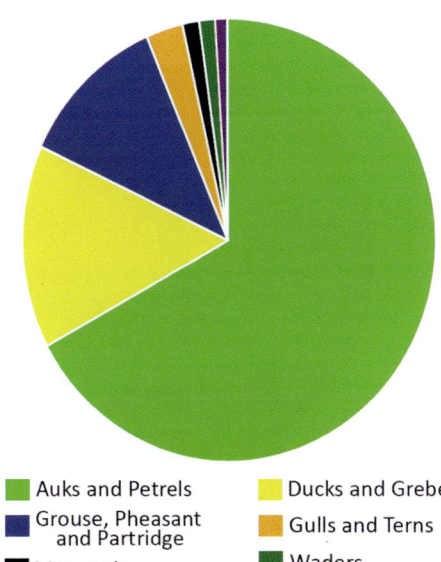

- ■ Auks and Petrels
- ■ Grouse, Pheasant and Partridge
- ■ Mammals
- ■ Ducks and Grebes
- ■ Gulls and Terns
- ■ Waders

Figure 4.4 Prey spectrum by biomass percentage of *F.p. pealei* on Amchitka Island, Aleutians, during breeding seasons from 1968-1970. Constructed from data in White *et al.* (1973).

Above Bufflehead. *Below* Crested Auklet.

together with terns, petrels, gulls, waders and a small number of passerines and resident ducks. Interestingly, despite the size of *F.p. pealei* there is no mention of them taking the Aleutian Canada Goose (*Branta canadensis leucopareia*) which weighs about 2kg.

F.p. anatum (Figs. 4.5a and 4.5b)

Because of the reintroduction of varied sub-species in the eastern USA (see Chapter 2), data from sites in western North America would have been preferred, though as there is no reason to believe that the introduced falcons will not share the same plasticity as 'unadulterated' populations in terms of prey spectrum, data from rural Kentucky (Carter *et al.*, 2003) has been included. The data cover both breeding and wintering periods and so differs from that of Castellanos *et al.* (2006). The differences illustrate the plasticity of the Peregrine's diet in the two very different habitats. Castellanos *et al.* (2006) collected prey remains from both breeding eyries and wintering sites on Baja California, Mexico in the late 1990s. The bulk of the prey during the breeding season comprised seabirds (gulls and terns), though the Black-crowned Night-heron (*Nycticorax nycticorax*) provided the highest count for an individual species. During the winter, migrating shorebirds formed the bulk of the falcon diet, the main species taken being Long-billed Curlew (*Numenius americanus*), Willet (*Tringa semipalmata*) and Short-billed Dowitcher (*Limnodromus griseus*).

The Peregrine Falcon

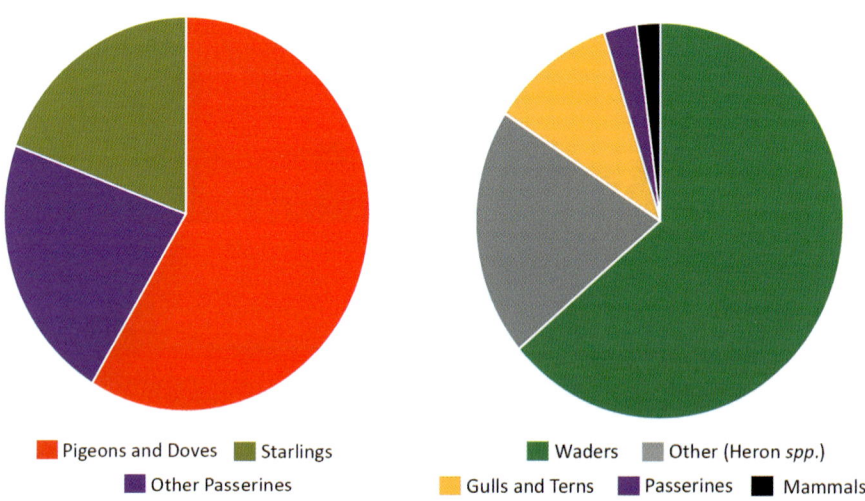

Figure 4.5a *Left* Prey spectrum by biomass percentage of *F.p. anatum* in rural Kentucky (Ohio River Valley) during the period March 1999 to December 2001. Constructed from data in Carter *et al.* (1995).

Figure 4.5b *Right* Prey spectrum by biomass percentage of *F.p. anatum* in Baja California, Mexico during the breeding seasons 1980 and 1981. Constructed from data in Castellanos *et al.* (2006).

Sandhill Crane.   Willet.

The hunting of seabirds by Peregrines was confirmed by Velarde (1993) who noted Heermann's Gull (*Larus heermanni*) and other gulls, together with alcids and terns in the diet of Peregrines breeding on Rasa Island in the Gulf of California. In the USA, the prey spectrum of the falcons is astonishing,

Diet

with White *et al.* (2013a) noting that 309 avian species, ten bats and a further nine mammals having been recorded as prey (190 avian species having been recorded in California alone). Not only the sheer number of species, but the size differential is remarkable, with a recorded killing of a Sandhill Crane (*Grus canadensis*), potential weight 3.1kg, at one end of the spectrum, and stoneflies, length 3.8-7.6mm in length, at the other.

Arguably the most interesting of the various papers which refer to the diet of *F.p. anatum* is that of Ellis (2006) who noted that in Arizona, 22% of the falcon's prey was White-throated Swifts (*Aeronautes saxatalis*). While the bulk of the kills were juveniles, this percentage of an extremely fast, agile prey is remarkable.

F.p. cassini (Figs. 4.6a and 4.6b)

Ellis *et al.* (2002) studying the diet of Peregrines in southern Argentina (Patagonia) – but excluding Tierra del Fuego – and southern Chile, identified 55 prey species as well as several other unidentifiable remains. The largest number of prey was the Eared Dove (*Zenaida auriculata*), though significant numbers of Tawny-throated Dotterel (*Oreopholus ruficollis*), Southern Lapwing (*Vanellus chilensis*) and Least Seedsnipe (*Thinocorus rumicivorus*) were also found. Other species ranged from a juvenile Lesser Rhea (*Rhea pennata*), through gulls, prions, petrels and terns for falcons nesting in coastal areas, further shorebirds, ducks and passerines, a total of 26 avian families

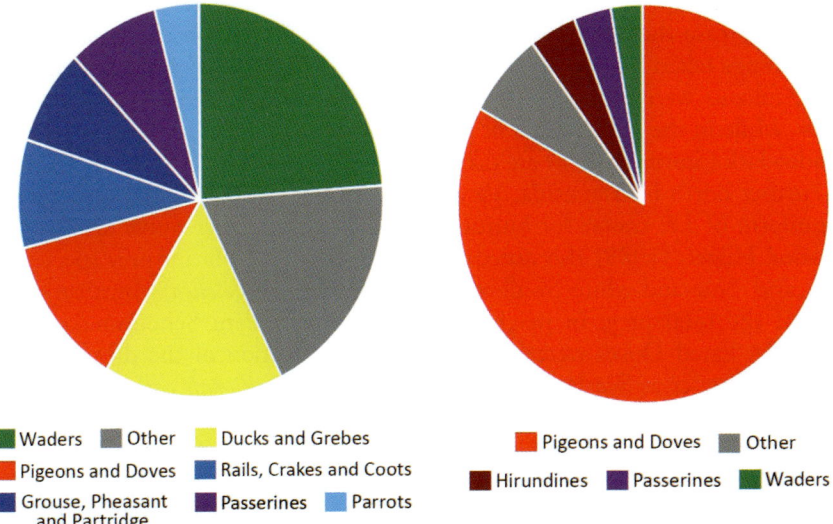

Figure 4.6a Left Prey spectrum by biomass percentage of *F.p. cassini* in southern Argentina and Chile during the breeding seasons 1993 and 1994. Constructed from data in Ellis *et al.* (2002).

Figure 4.6b Right Prey spectrum by biomass percentage of *F.p. cassini* in Ecuador during the breeding seasons 1982 and 1983. Constructed from data in Hilgert (1988).

Pintado Petrel.

Long-tailed Meadowlark.

in all. One interesting prey species was the Buff-necked Ibis (*Theristicus caudatus*), with an attack noted on a full-grown adult (adults weigh on average about 1.7kg). In a further study (Ellis *et al.*, 2019), again in Patagonia, the researchers noted Peregrines searching through Ibis breeding colonies for unguarded nestlings. Recently-hatched nestlings weigh about 100g, but the falcons were also taking much older chicks, weighing 500g and even on two occasions 1000g. Ellis *et al.* conclude that at some Peregrine eyries Ibis chicks are an important source of biomass for falcon chicks. The hunting of Ibis was confirmed in another report from Coastal Patagonia, (Saggese *et al.*, 2019), but in this case the species was the Black-faced Ibis (*Theristicus melanopis*) the nestlings taken having an average weight of 500g. Other primary targets of the falcons were South American Terns (*Sterna hirundinacea*) and local ducks, while in the wilder parts of Patagonia such species as the Long-Tailed Meadowlark (*Sturnella loyca*) are taken.

Further north in Argentina, Vasina and Straneck (1984) noted the same enthusiasm of *F.p. cassini* for hunting Eared Doves. Further north again, in Ecuador, Hilgert (1988) also noted Eared Doves as a significant prey species, these and Blue-and-white Swallows (*Notiochelidon cyanoleuca*) constituting 13% of the total prey. In neighbouring north-west Peru Schoonmaker *et al.* (1985) Eared Doves were again a significant fraction of the diet, along with such exotics as Scarlet-backed Woodpeckers (*Veniliornis callonatis*) and Pacific Parrotlets (*Forpus coelestis*). Interestingly, Schoonmaker *et al.* also saw a male falcon chase an Andean Condor (*Vultur gryphus*) away from its cliff-face breeding eyrie.

F.p. cassini will also hunt seabirds at sea, Tamini *et al.* (2016) noting a single bird using a fishing vessel, midway between the mainland and the Falkland Islands, as a base to capture Wilson's Storm-petrel (*Oceanites oceanicus*) and a Pintado Petrel (*Daption capense*). On the Falklands themselves, the breeding population of *F.p. cassini* prey mostly on seabirds such as prions and shearwaters (Woods and Woods, 1997).

F.p. peregrinus

The prey spectrum of the nominate in Britain in non-urban areas has been explored in Fig. 4.2. The dietary spectrum in an urban British Peregrine is set down in Figs. 4.14 and 4.15. In London, what was notable was the way in which the growth in population of Ring-necked Parakeets (*Psittacula krameri*) resulted in the urban Peregrines expanding their prey spectrum to include this exotic newcomer (*e.g.* Hancock and Martin, 2015 – see also Chapter 7 *Charing Cross Hospital, London*). In Sweden, Lindberg (1983) collected 2810 prey remains at 18 Peregrine nest sites. 99 species were identified, the weight average being 242g in northern species, where ducks and waders formed a significant fraction of the prey, and 232g in southern Sweden where gulls, pigeons and passerines predominated. The prey diversity was larger in northern Sweden. Mammals (lemmings, voles) were taken, but their contribution to the diet was minimal (*c.*1%). Interestingly, the prey remains also included the Northern Pike (*Esox lucius*), European Perch (*Perca fluviatilis*) and frogs. In Russia, at a riverine Nature Park close to Yekaterinburg, Khlopotova (2013) noted the diet of breeding nominate falcons in 2008 and 2009. In 2008 pigeons formed the bulk of the diet, with crows and waders each providing a significant fraction, but the following year waders formed the bulk of the diet, with both thrushes and plovers representing a greater fraction than pigeons. Not surprisingly, rails and ducks also formed part of the prey spectrum in this riverine park.

While the spectrum of prey will vary across the huge range of the nominate falcons, which extends well into Asian Russia, the overall make-up will be largely similar to the diets in the UK and Sweden.

F.p. brookei (Figs. 4.7a and 4.7b)

Diaz del Campo (1974), studying Peregrines across Spain noted that the bulk of the prey taken comprised various doves – Rock Doves/Feral pigeons, Stock Doves (*Columba oenas*), Wood Pigeons (*Columba palumbus*) and European Turtle Doves (*Streptopelia turtur*) – a diet which it supplemented with woodpeckers, various larks, including the Calandra Lark (*Melanocorypha calandra*) and Thekla Lark (*Galerida theklae*), Starlings and various finches, Goldfinches (*Carduelis carduelis*) seemingly a particular favourite. Diaz del Campo states that the falcons ignored Quails (*Coturnix coturnix*), though these are taken in small numbers elsewhere, but took Red-legged Partridges (*Alectoris rufa*), once seeing a falcon flying in at great speed to snatch one which had escaped a tussle between a Bonelli's Eagle (*Aquila fasciata*) and a Golden Eagle (*Aquila chrysaetos*), escaping by outflying the pursuing Golden Eagle. Diaz del Campo also reported a female Peregrine attacking a hare. Zuberogoitia *et al.* (2013) studied the diet of the Peregrines in northern Spain (Biscay Province, close the Cantabrian Sea) – Fig. 4.8a – collecting the prey remains of 2832 birds of 128 species from 185 nests. The highest percentage of prey was Rock Doves (25.0% by biomass), the next highest

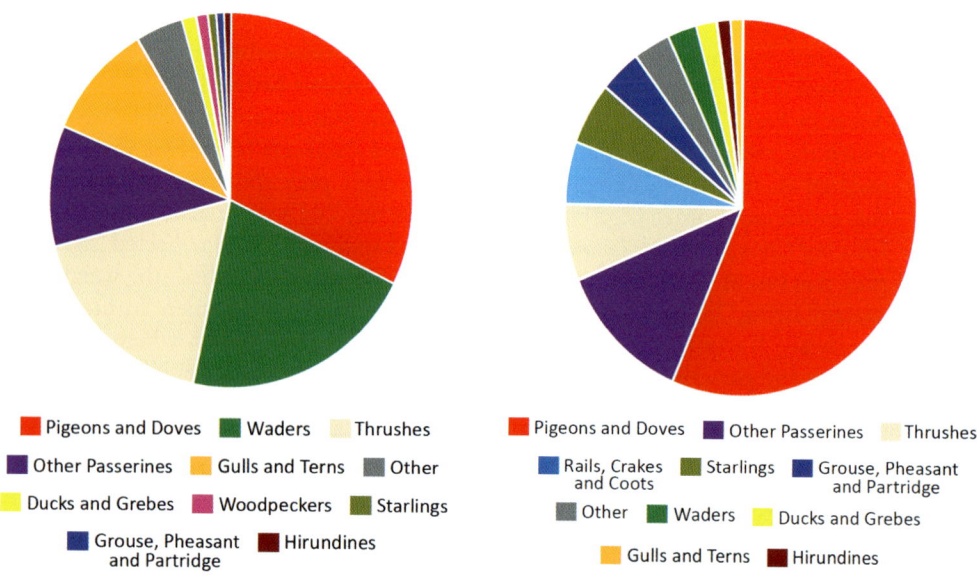

Figure 4.7a Left Prey spectrum by biomass percentage of *F.p. brookei* in Biscay, northern Spain during the breeding season across the period 1998-2010. Constructed from data in Zuberogoitia *et al.* (2013).

Figure 4.7b Right Prey spectrum by biomass percentage of *F.p. brookei* in Sicily, Italy during the breeding season across the period 1978-2015. Constructed from data in Bondi *et al.* (2016).

being Whimbrel (*Numenius phaeopus* – 9.1% by biomass) and Yellow-legged Gulls (*Larus michahellis* – 7.3% by biomass). Surprising additions to the diet were Montagu's Harriers (*Circus pygargus* – 10 birds, 0.9% of the prey by biomass) and Hobbies (19 birds, 1.1% biomass), and a single Northern Gannet (*Morus bassanus*, average mass 2.8kg) though it was not clear if the bird had been hunted and killed, or scavenged.

On the island of Malta, Fenech and Sammut (2017) noted that feral pigeons formed the bulk of the diet of the island's Peregrines (53.2%, by number), with Yellow-legged Gulls making up 6.8% (by number). The falcons also took Scopoli's Shearwaters (*Calonectris diomedea*), both Great Crested and Black-necked Grebes (*Podiceps cristatus* and *P. nigricollis* respectively) and the occasional escaped Budgerigar (*Melopsittacus undulatus*).

In Morroco, Barreau and Bergier (2001) record Peregrines taking Little Egrets (*Egretta garzetta*) and Choughs (*Pyrrhocorax pyrrhocorax*), as well as thrushes and swifts. Studying the Peregrines breeding on Zembra Island (a National Park off the coast of Tunisia) Thiollay (1988) noted that the majority of prey taken during the breeding season comprised migratory species, principally swallows and swifts, attacked over the sea rather than on the island itself. The resident Rock Doves were taken, but to a lesser extent, while larger

Feral Pigeon, Siena, Italy.

Thekla Lark, Andalucia, Spain.

seabirds on the island were attacked rarely. Overall, Thiollay noted that the breeding success of the island's falcons was limited by the lack of food, the clear evidence being the lack of food caching, small clutch sizes, the incubating female hunting, the lack of prey remains at nest sites and chick starvation. While there is no reason to doubt Thiollay's conclusions, this is curious behaviour for Peregrines, a species renowned for its eclectic prey selection.

Further east, the diet spectrum is similar to that of *F.p. peregrinus*. In Italy, Bondí *et al.* (2016) collected data from across the country (the Alps to Sicily), and from 1974-2015, to accumulate the diet of the falcons. The data amounted to 1550 items from 110 species (Fig. 4.8b). Serra *et al.* (2001) studied the diet of Peregrines in the city centre of Florence. Eighteen species of prey were discovered, with, not surprisingly, pigeons forming the bulk of the prey (30.4% by number). Two species of bats (Savi's Pipistrelle (*Hypsugo savii*) and Kuhl's Pipistrelle (*Pipistrellus kuhlii*)) formed 15.1% by number of the captured prey, with, surprisingly, Swifts forming a further 8.7%. Black-headed Gulls, which winter in the city, also formed 8.7% by capture, but were 13.1% by biomass. The weight spectrum of the prey varied from the small Savi's bat (7g) to a Woodcock (*Scolopax rusticola*) at 305g, but 53% of the prey (by number) weighed less than 150g.

Eastward again on the Eurasian mainland, Vitovich *et al.* (2000), working in the Caucasus, a mountain range which separates Russia from Georgia and Azerbaijan, and the Black Sea from the Caspian Sea, found that the *F.p. brookei* diet comprised 23% bats by number, with the remaining diet comprising more than 30 avian species. Of these the largest proportion by far were Stock Doves and feral pigeons (15.5%) followed by the Eurasian Bee-eater (*Merops apiaster*) which contributed 10.9%. The remaining diet comprised ducks, larks, thrushes and finches, as would be expected, but also included a single

Eurasian Hobby and Common Kestrel, two Common Buzzards, two Northern Goshawks, two Eurasian Sparrowhawks, and two unidentified harriers. The raptor fraction of the diet was, therefore, 9.1%, a surprisingly high fraction.

F.p. pelegrinoides (Fig. 4.8)

For completeness the Barbary Falcon is included, briefly, here. On the Canary Islands the falcons take Rock Doves and feral pigeons. Siverio *et al.* (2011) in a long-term study of the falcons on Tenerife found that 93% of the prey delivered to the eyrie during chick rearing were these Columbids.

Across the rest of their range the falcons take bats, available small mammals and, probably, reptiles given the desert nature of much of the range. Shafaeipour (2014) studied the diet in south-west Iran and noted 14 species of bird and two insect species in the diet of a breeding pair. Birds were 95.3% by frequency, 99.9% by biomass (Fig. 4.8). The primary prey by frequency was the Eurasian Bee-eater, but Wood Pigeons and Middle-spotted Woodpeckers (*Dendrocopos media*) provided the greatest biomass. Medium-large open habitat species such as Syrian Woodpeckers (*Dendrocopos syriacus*), Cuckoos, Common Magpies (*Pica pica*) and Chukars (*Alectoris chukar*) were taken more often than their availability implied. Other prey included local passerines. Elsewhere in the Middle East, the main prey appears to be sandgrouse, coursers, pratincoles and plovers, as well as available passerines.

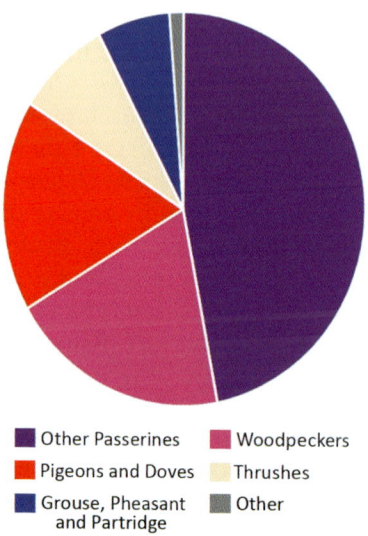

The White-spectacled Bulbul (*Pycnonotus xanthopygos*) is a potential prey of Barbary Falcons across the eastern part of their range.

Figure 4.8 Prey spectrum by biomass percentage of the Barbary Falcon *Falco pelegrinoides* during the breeding seasons of 2011-2013 in south-west Iran across the period 1978-2015. Constructed from data in Shafaeipour (2014).

Above   Juvenile *F.p. peregrinator* with Ring-necked Parakeet (*Psittacula krameri*), Hyderabad, India. Hari Patibanda.

Below   Adult *F.p. minor* bringing a Yellow Wagtail (*Motacilla flava feldeggi*) to its chicks at Wondo Genet, near Hawassa, in the Ethiopian Rift Valley. Torsten Pröhl.

The Peregrine Falcon

F.p. madens
There is no information in the literature for this very rare sub-species, but the knowledge on the position of eyries suggests the falcons take seabirds (gulls and petrels), shorebirds and land birds such as pigeons.

F.p. minor (4.9a, b, c and d)
Despite the footprint of the Peregrines in Africa, data on diet are surprisingly sparse. Jenkins (1998) and Jenkins and Avery (1999) studied the diet in three areas of South Africa (Cape Peninsula, the lower Orange River and

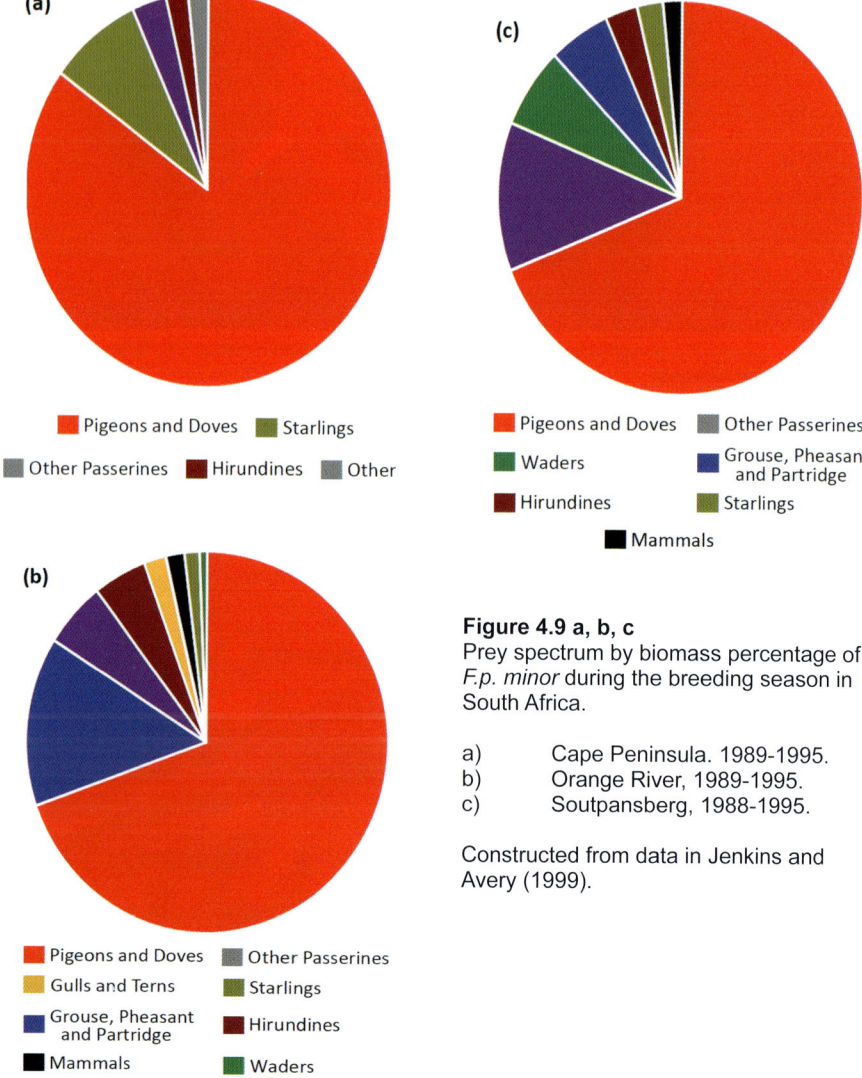

Figure 4.9 a, b, c
Prey spectrum by biomass percentage of *F.p. minor* during the breeding season in South Africa.

a)   Cape Peninsula. 1989-1995.
b)   Orange River, 1989-1995.
c)   Soutpansberg, 1988-1995.

Constructed from data in Jenkins and Avery (1999).

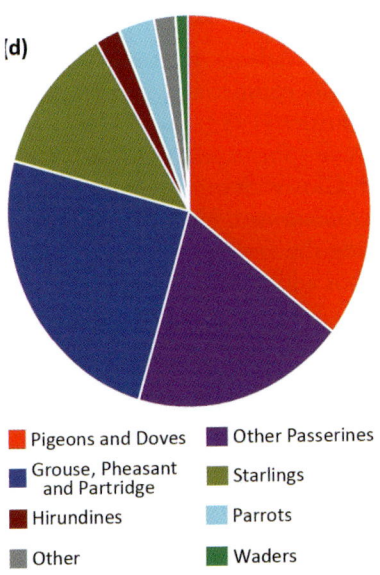

- Pigeons and Doves
- Grouse, Pheasant and Partridge
- Hirundines
- Other
- Other Passerines
- Starlings
- Parrots
- Waders

Figure 4.9d Prey spectrum by biomass percentage of *F.p. minor* during the breeding season of 1981 in the Sengwa Wildlife Research Area of Zimbabwe. Constructed from data in Hustler (1983).

Images to the right

Above Lilac-breasted Roller (*Coracius caudatus*), Namibia.

Below White-backed Mousebird (*Colius colius*), South Africa. Each of these species may form part of the prey spectrum of *F.p. minor* in southern Africa.

the Soutpansberg mountains in the north-east of the country: the areas are both well-separated and offer diverse habitats). The researchers found that Columbids formed the bulk of the prey at all three sites, varying (in biomass) from 68.4% at the northern sites to 85.1% on Cape Peninsula. The widest prey spectrum being seen in the Orange River area where swifts, sandgrouse, larks and free-tailed bats comprised 34% of the diet by number (but only 20% by biomass). Also in north-eastern South Africa, Tarboton (1984) noted a Peregrine diet rich in pigeons/doves during the incubation stage of breeding, but including other species such as the African Hoopoe (*Upupa africana*) and Black-headed Oriole (*Oriolus larvatus*).

In Zimbabwe Hartley (2000) noted that 48.7%, by weight, of the Peregrine diet was pigeons/doves, with another 12.9% of Phasianidae *spp*. The next highest fraction were hornbills. Hartley's work revealed an interesting dietary difference between Peregrines, and the Taita and Lanner falcons that shared its habitats: the smaller Taitas were mostly taking starlings, weavers and other small passerines, while the Lanners were primarily competing with Peregrines for pigeons and doves. Also in Zimbabwe, Hustler (1983) noted doves as a main target, particularly the Cape Turtle Dove (*Streptopelia capicola*), with other main targets being the Long-tailed Glossy Starling (*Lamprotornis mevesii*) and the Natal Francolin (*Francolinus natalensis*). Hustler (1983) also studied the Peregrine diet in Zimbabwe noting the same preference for dove species and the Long-tailed Glossy Starling, but adding Natal Francolin (and the Bearded Woodpecker (*Thripias namaquus*)) to the list.

F.p. radama

Benson (1960) examined the stomach contents of a Peregrine found on Grand Comoro and discovered the remains of two Foudia *spp.*, presumably the Madagascar Red Fody (*Foudia madagascariensis*) and the Forest Fody (*Foudia omissa*) each of which are still relatively common. The remains of a Streptopelia *spp.* were also found, probably the Malagasy Turtle-dove which has since been renamed *Nesoenas picturatus* (formerly *Streptopelia picturata*). Thorstrom *et al.* (2003) added the Madagascar Bulbul (*Hypsipetes madagascariensis*) and Streptopelia and Treron pigeons to the falcon's dietary list.

Over the last 60 or so years the Madagascar environment has been extremely degraded, a fact exemplified by the observation of Razafimanjato *et al.* (2007), who studied the diet of three nesting pairs of Peregrines at two sites on the island, that at one of the two studied sites (close to a lake on the island's central plateau), 99% of the diet of a single breeding pair comprised domestic chickens (the 1% was Rock Dove). At the other site, in the Tsimanampetsotsa Natural Reserve, the falcons took 19 different species, the bulk of the prey comprising island endemics, principally the Madagascar Bulbul and the Madagascar Red Fody.

Madagascar Red Fody, Mahajanga, Madagascar. *David Dennis.*

F.p. calidus (Figs. 4.10a and 4.10b)

In 1994 Henny *et al.* (2000) took blood samples from Peregrines breeding near the Ponoy River on Russia's Kola Peninsula (the results of which are mentioned in Chapter 10 *The Pesticide Crisis*), attached satellite tags to four birds (see Chapter 8) and also studied the falcons' diet. The Peregrines were mainly taking Ruff (*Philomachus pugnax* – comprising 52% of the diet), but were also taking Common Snipe (*Gallinago gallinago* – 8.6%), Jack Snipe (*Lymnocryptes minimus* – 4.8%) and Wood Sandpiper (*Tringa glareola* – 4.5%). In an earlier study (Osmolovskaya, 1948) noted that the falcons arrived very early in the Yamal, at least two weeks before the bulk of migratory species. This limited the available prey spectrum to resident Ptarmigan, a few avian early arrivals (Eurasian Teal, Long-tailed Duck, Red-necked Phalarope (*Phalaropus lobatus*) and other waders) and rodents, chiefly the Arctic Collared Lemming (*Dicrostonyx torquatus*) which formed a significant part of the diet – Fig. 4.10a. Later in the season when the avian migrators arrive the diet changes, the mammalian percentage declining (but not disappearing) and species such as Arctic Tern (*Sterna paradisaea*), waders and passerines being added – Fig. 4.10b.

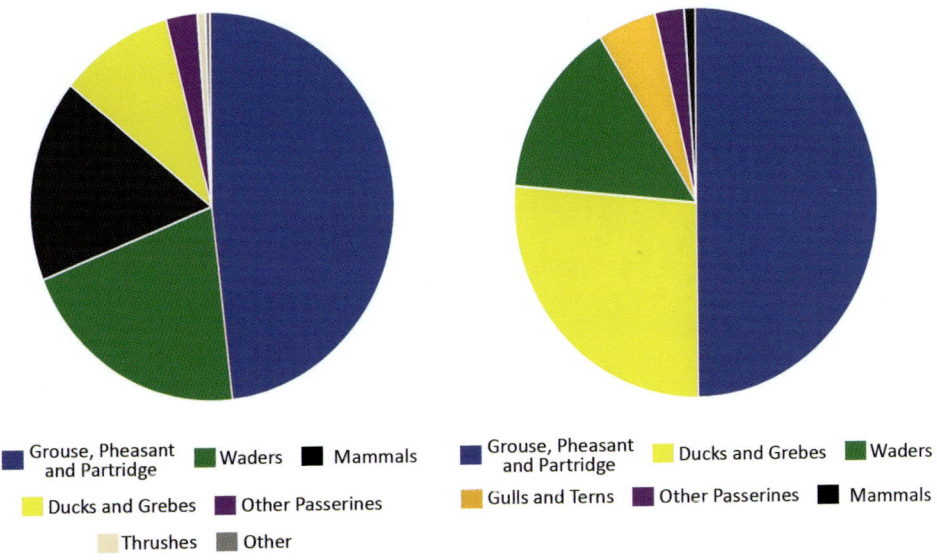

Figure 4.10a Left Prey spectrum by biomass percentage of *F.p. calidus* during the early spring at the watershed Lake Yarro-to, Yamal, Russia. The mammalian prey was chiefly lemming, though there was a vole contribution, this being less than 5% (of the mammal contribution by both number and biomass. Constructed from data in Osmolovskaya (1948).

Figure 4.10b Right Prey spectrum by biomass percentage of *F.p. calidus* during the breeding season at the Shchuchya river valley, Yamal, Russia. Constructed from data in Osmolovskaya (1948).

Studying the breeding of Siberian Peregrines in the Yamal/Ob River area Paskhalny and Golovatin (2009) noted a similar spectrum of prey to that of the Tundra Peregrines of North America. This is hardly a surprise as many of the potential avian and mammalian prey species are circumpolar, or have similar sub-species, in the two Arctic areas. Prey would therefore include waders, seabirds and passerines. Paskhalny and Golovatin note specifically Ptarmigan and a White-fronted Goose gosling and reporting information from other observers note Long-tailed Skua (*Stercorarius longicaudus*) and a Red-breasted Goose gosling (*Branta ruficollis*). Most surprisingly they note the capture of a Greater-spotted Woodpecker (*Dendrocopos major*) which had strayed far from its usual breeding grounds.

Wintering in the India sub-continent and also in north-eastern Africa, the falcons will again take a range of local avian species and also large numbers of bats.

Lekking male Ruff.

Long-tailed Skua.

F.p. babylonicus

Dementiev (1957) provided early data on the prey of the Red-naped Shaheen in Turkestan, noting the falcons feeding on House Sparrows (*Passer domesticus*), but probably also the Indian Sparrow (*Passer simplex*)), Alpine Swifts (*Tachymarptis melba*), Rock Doves, Bee-eaters (Merops *spp.*), Pin-tailed Sandgrouse (*Pterocles alchata*) and Chukar. Dementiev also mentions the falcons hunting ducks (Gadwall (*Mareca strepera*) and Teal – not specified, but probably the Eurasian (Green-winged Teal)). The hunting of Pipistrelle bats is also mentioned. In an earlier work (Dementiev, 1940) Dementiev also mentions the falcon's enthusiasm for Jackdaws (*Corvus monedula*) in what was then Kirghizia where the crows nested in large colonies, something which was later endorsed by Yanuschevich *et al.* (1959), who studied the diet in the same area, noting that the falcons specifically bred close to large Jackdaw colonies, the crows forming the bulk of the falcons' diet.

Stepanyan (1969) studying breeding falcons in Uzbekistan notes a diet which also included Alpine Swifts, as well as Desert Lark (*Ammomanes deserti*), several species of Wheatear, Eastern Rock Nuthatch (*Sitta tephronota*) and larger prey such as the See-See Partridge (*Ammoperdix griseogularis*). Stepanyan also mentions the falcon's enthusiasm for hunting Rosy Starlings (*Pastor roseus*) as do other observers.

White *et al.* (2013a) mention further prey species based on other researchers, but the overall prey spectrum remains the same, with the addition of small gulls for falcons breeding close to the Caspian Sea.

Above   Gadwall.

Right   Rosy Starling. *Koshy Koshy.*

F.p. anatum adult feeding its chick, San Pedro, California, USA. *Sam Lin.*

F.p. peregrinator

It is believed this sub-species evolved close to forests, explaining the fact that it very rarely stoops on prey, invariably binding on to its prey (*i.e* clutching the prey with its talons while in flight) as anything struck and falling under gravity might disappear into the forest before the falcon has a chance to circle and grab. That means that large species, which are knocked to the floor by other sub-species, are rarely attacked, prey size being limited by what can be successfully carried away.

Studying the falcons across its sub-continent range, Naoroji (2006) noted that Rose-ringed Parakeets (*Psittacula krameri*) were a favourite prey species, particularly on Sri Lanka, but that ducks, chukars, quails and doves were also popular, as were local hirundines and swifts. Bats were also taken when available.

Left Alexandrine Parakeet (*Psittacula eupatria*). **Right** Pied Bushchat (*Saxicola caprata*). Both species may form part of *F.p. peregrinator*'s diet.

Diet

F.p. japonensis (Figs. 4.11a and 4.11b)

The debate over whether this sub-species covers a vast (north to south) area of the eastern Asian landmasses, including a portion of the Russian mainland, has been mentioned in Chapter 2. The diet of the northern birds (breeding in Kamchatka and Chukotka) was investigated by Probst *et al.* (2007) who studied eyries on the inland Anadyr River and the coasts of Anadyr Bay and nearby Cape Navarin (Fig. 4.11a). The falcons at the latter site caught Crested Auklets almost exclusively, the only other prey species seen being a single Little Auk (*Alle alle*) and a single Tundra Vole (*Microtus oeconomus*). The other breeding pairs were more eclectic in their menu, the river pair taking Common Terns (*Sterna hirundo*), local waders and passerines, including a single House Martin (*Delichon urbica*). The Anadyr Bay pair took Arctic Tern and Long-tailed Skua, but also several Pacific Golden Plovers (*Pluvialis fulva*) as well as local passerines. Further south, near the Kolyma River, Dorogoi (1988) noted many different waders in the diet as well as auks and other seabirds where the river emptied into the Sea of Okhotsk.

Further south on mainland Japan, Brazil and Hanawa (1991) noted that doves, particularly Rock Doves/feral pigeons made up the bulk of the diet in all seasons with ducks, shorebirds and seabirds contributing to the prey spectrum dependent on habitat. The winter stomach contents of all raptorial

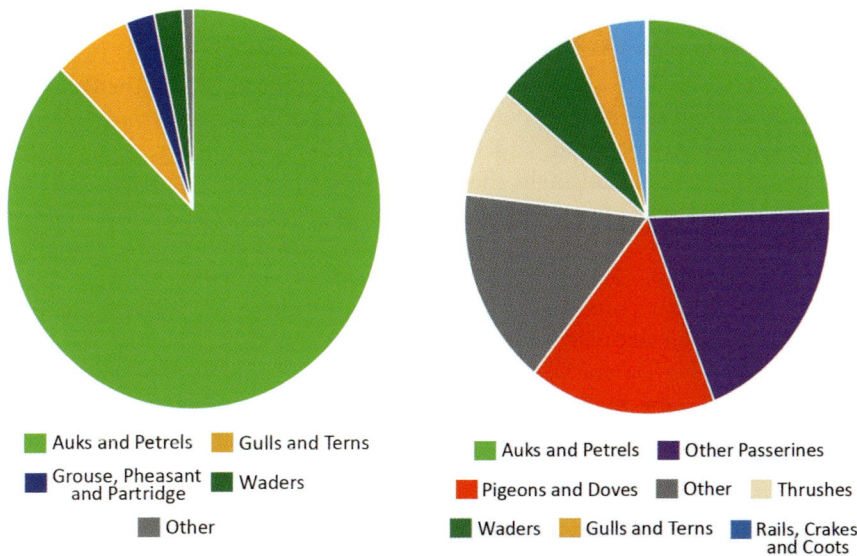

Figure 4.11a Left Prey spectrum by biomass percentage of *F.p. japonensis* during the breeding seasons 2001 and 2002 near Anadyr, Chukotka, Russia. Constructed from data in Probst *et al.* (2007).

Figure 4.11b Right Prey spectrum by biomass percentage of *F.p. japonensis* at sites across South Korea during the springs of 2001-2013. Constructed from data in Choi *et al.* (2015).

Above Spotted Nutcracker (*Nucifraga caryocatactes*).

Left Hokkaido Jay (*Garrulus glandarius brandtii*).

Both species may form part of *F.p. japonensis*'s diet.

birds on Japan were examined by Ishizawa and Chiba (1967). The Peregrines examined had fed on Copper Pheasant (*Syrmaticus soemmerringii*), various thrushes and unidentified larger birds. There were also the remains of a rat and a hare, and, surprisingly, the Miyama (or Japanese) Stag Beetle (*Lucanus maculifemoratus*).

Across the Sea of Japan, Choi and Nam (2015) studied the diet of the falcons in all seasons for 13 years in a variety of habitats. A total of 77 avian species were identified in total throughout the year. Of these 77, 56 were identified in spring, 15 in summer, 29 in autumn and 22 in winter (when *F.p. japonensis* is joined by *F.p. calidus* from the Russian Arctic). Two insect species were also identified, Globe Skimmers (*Pantala flavescens*) and Lesser Emperor (*Anax parthenope*)[5]. Excluding migratory species which can only be taken in spring and summer, the most obvious variation was that ducks were mainly taken in winter. Overall, the most often seen species in the diet were doves (Rock Dove and the Oriental Turtle-dove (*Streptopelia orientalis*)) and the Japanese Murrelet (*Synthliboramphus wumizusume*). Perhaps the most surprising finding was the heavy toll of Oriental Scops-Owl (*Otus sunia*). Overall, the bulk of the prey taken weighed less than 240g (roughly three equal totals of <80g, 80-160g and 160-240g). However, prey weighing over 1kg was very occasionally taken, this including the Black-faced Spoonbill (*Platalea minor*)

---

[5] For further information on the hunting of dragonflies by Korean Peregrines see Choi and Nam (2012).

and European Herring Gulls (*Larus argentatus*). Fig. 4.11b (p195) shows the spring (*i.e.* breeding) spectrum of prey of the Korean Peregrines. Fig. 4.17 (p205) shows the seasonal variation of the diet.

F.p. furuitii
This rarest of the sub-species is sparsely studied and understood, and apart from the general suggestion that the bulk of the diet will be seabirds and shorebirds, little is known.

F.p. ernesti
Again the data for this widespread sub-species (breeding on the Malay Peninsula, across Indonesia, the Philippines, Papua New Guinea and the Solomon Islands) is sparse. In his book on the Dutch naturalist Henri Jacob Victor Sody, Becking (1989) notes Sody's list of prey species of the Peregrines on Java, which included doves and pigeons, the White-faced Partridge (*Arborophila orientalis*) and the Blood-breasted Flower-pecker (*Dicaeum sanguinolentum*), swifts and swallows, and domestic chickens. Sody also identified the Olive Bee-eater (*Merops superciliosus*) as Peregrine prey, but this species has a range confined to central and southern Africa, and Madagascar. It is possible that Sody misidentified the Blue-tailed Bee-eater (*Merops philippinus*) which looks very similar when viewed from the front, its blue tail then being obscured. The falcons also took bats, together with local lizards and beetles. The Peregrines' enthusiasm for bats has not waned since the work of Sody, Wells (1998) noted one Peregrine walking along the ledge of a cliff on which bats were roosting in crevices, pulling bats from those crevices which were not too deep. Wells also notes a Brown Booby (*Sula leucogaster*) as prey of a Peregrine on Perak Island.

Blue-tailed Bee-eater. *Charles J. Sharp.* Blood-breasted Flower-pecker.

Early in the last century Wolfe (1938) a Captain in the US Army, stationed in Luzon, northern Philippines, found an eyrie littered with bat wings and unidentified feathers.

Later data suggests that across the region, doves and pigeons form the bulk of the diet. In Malaya, Molard *et al.* (2007) noted that one pair of Peregrines breeding on a small cliff close to a suburban factory were primarily hunting House Swifts (*Apus nipalensis*) which were nesting on the factory buildings in thousands. In a later report, (Molard, 2009) notes that the male Peregrine would wait on a ledge for the swifts to emerge through slits in the building wall and flying vertically downwards: they were then taken by the Peregrine dropping off the ledge. In New Guinea, Rand and Gilliard (1967) found that parrots and pigeons were among the prey. On Borneo, Thiollay (1983) recorded Peregrines entering caves at twilight to hunt for swifts and bats. The species of swift is not mentioned, but it was likely to have been the Cave Swiftlet (*Collocalia linchi*).

None of these records of *F.p. ernesti* prey allow the construction of a prey spectrum for the Peregrine at any point on its extensive range.

F.p. macropus (Figs. 4.12a-d)
Studies across Australia suggest that away from the coast Peregrines feed mainly on doves and pigeons, on parrots and introduced European Starlings. In Victoria, Pruett-Jones *et al.* (1981) noted that 51% of prey was feral pigeon, Galah (*Eolophus roseicapilla*) and Common Starling, though a further 86 species of small-medium birds were identified. The predation of Galahs during the breeding season places a burden on male Peregrines as they are about half the tiercel's weight.

In a study near Canberra, Olsen *et al.* (2008) studied the prey remains and pellets at 16 Peregrine eyries over four breeding seasons, 1991-1992 and 2002-2003. In the earlier periods Starlings were the main prey, though the fraction had reduced in the later years because of a decline in the Starling population. What was particularly interesting was that that in addition to drought taking a toll of the Starlings, competition by Common Mynas (*Acridotheres tristis*) had also been significant – but the falcons had not compensated for reduced Starling abundance by hunting the increased Mynah numbers, probably because they were both agile and tended to forage close to cover. The falcons took 37 identified avian species, the larger birds being Rock Doves, Galahs and Gang-gang Cockatoos (*Callocephalon fimbriatum*). On the coast, the falcons also took Silver Gulls (*Larus novaehollandiae*). Two species of mammal were recorded, Gould's Wattled Bat (*Chalinolobus gouldii*) and a juvenile European Rabbit. Various insects were also taken, chiefly beetles and cicadas.

Olsen *et al.* (1993a) studied the diet of Peregrines in three areas of New South Wales (two inland, one coastal), noting pigeons and parrots as the main prey at the inland sites, but with significant numbers of seabirds, chiefly shearwaters and cormorants, at the coastal site. In another study (Olsen *et al.*,

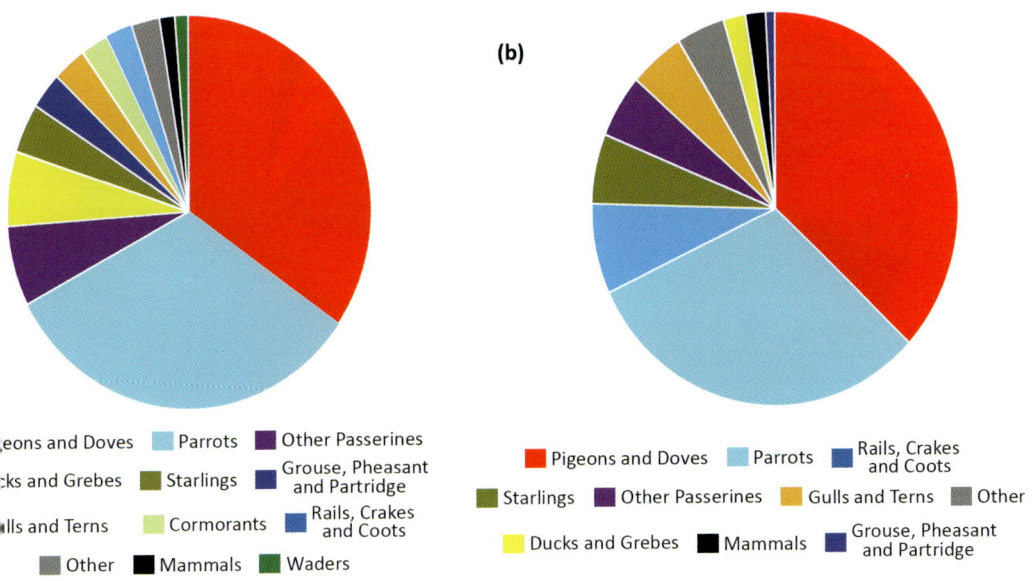

Figure 4.12a *Left* and **Figure 4.12b *Right***
Prey spectrum by biomass percentage of breeding *F.p. macropus* in Victoria, Australia in 1976 (a) and 1977 (b). Constructed from data in Pruett-Jones *et al.* (1980).

1993b) in sites at South Australia the Peregrine diet included the Australian (Nankeen) Kestrel, as well as significant numbers of pigeons and parrots. Olsen *et al.* (1993c) studied the variation of the Peregrines' diet during the year. The diet was surprisingly constant, with feral pigeons forming the bulk at all times. Galah and Rosellas (Platycercus *spp.*) were also popular throughout the year. Interestingly, in winter the falcons took Black Swans (*Cygnus atratus*), though few in number: the falcons hunted Noisy Friarbirds (*Philemon corniculatus*) during all seasons except summer. The study also looked at the influence of diet on reproductive success during the breeding season – this will be considered again in Chapter 7 *Chick Growth*.

Further north, in south-east Queensland Czechura (1984) presents a long list of prey species, including not only doves, but waders, quails, lorikeets and ducks, prions at the coast and Australian White Ibis (*Threskiornis molucca*).

In all studies across Australia White *et al.* (2013a) note that 142 species of bird and seven mammalian species (chiefly bats) have been found in the diet of Australian Peregrines, this long list being extended by reptiles and insects, and the addition of one (unnamed) fish.

Figs. 4.12a and b show a remarkable similarity of prey in consecutive seasons in Victoria, the loss of waders as prey being the only significant change from 1976 to 1977. By contrast, over the decade the prey spectra in nearby Canberra between 1991/2 and 2002/3 Figs. 4.12c and 4.12d (overleaf) show very distinct changes as weather, habitat loss *etc.* alter the avian landscape.

The Peregrine Falcon

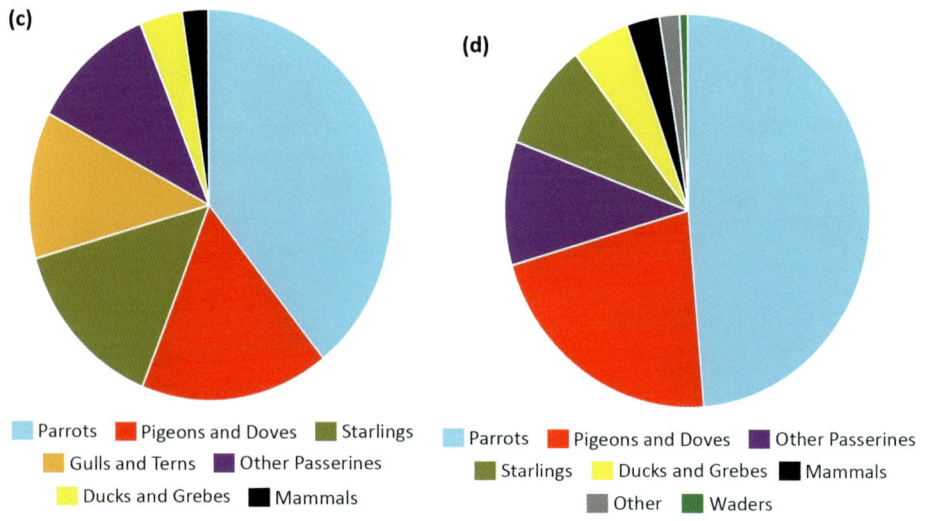

Figure 4.12c *Left* and **Figure 4.12d** *Right*
Prey spectrum by biomass percentage of breeding *F.p. macropus* in New South Wales, Australia in 1991/2 (c) and 2002/3 (d). Constructed from data in Olsen *et al.* (2008).

Above   Common Myna.

Left   Galah. *David Cooley.*

F.p. nesiotes (Figs. 4.13a and 4.13b)

This sub-species has been well-researched on Fiji. Clunie (1972), studying the falcons on Viti Levu, found at least a dozen species in the falcon diet. The list included local dove species and an array of birds vulnerable to attack. The nature of the habitat, with extensive dense forests, means that many of the local species, which number over 80, are not available to an aerial hunter. The prey list included several honeyeaters and the White-collared Kingfisher (*Halcyon chloris*). As noted earlier in the Chapter (Clunie, 1976a) also stated the Falcons taking Flying Fox bats: in one area these were the staple food of the falcons.

In a later study, in the urban/marine environment of Suva, Fiji's capital, Clunie (1976b) was able to add several more prey species to the list in his earlier work (Fig. 4.14 relating to urban prey). Within the city, doves and feral pigeons were taken, these forming the bulk of the prey biomass. Closer to the ocean the prey included several northern migratory shorebirds (American Golden Plover (*Pluvialis dominica*), Ruddy Turnstone (*Arenaria interpres*), Bar-tailed Godwit (*Limosa lapponicus*) and Wandering Tattler (*Tringa incanus*)), together with several southern seabirds including Red-footed Booby (*Sula sula*), Lesser Frigate-bird (*Fregata ariel*), Black Noddy (*Anous minutus*), Crested Tern (*Sterna bergii*) and Black-naped Tern (*Sterna sumatrana*) some of these being driven to coastal waters by bad weather. Most interesting is Clunie's suggestion that one Peregrine was preferentially hunting after sunset, the falcon seeking birds at a harbour, its success rate equalling that during bouts of daytime hunting.

White *et al.* (1988a) extended the study of prey species taken (Fig. 4.13a). Overall, Fiji offered 94 species (56 land, 31 marine/aquatic and 7 mammal),

Figure 4.13a Prey spectrum by biomass percentage of breeding *F.p. nesiotes* in three habitat types in Fiji. Constructed from data in White *et al.* (1988a).

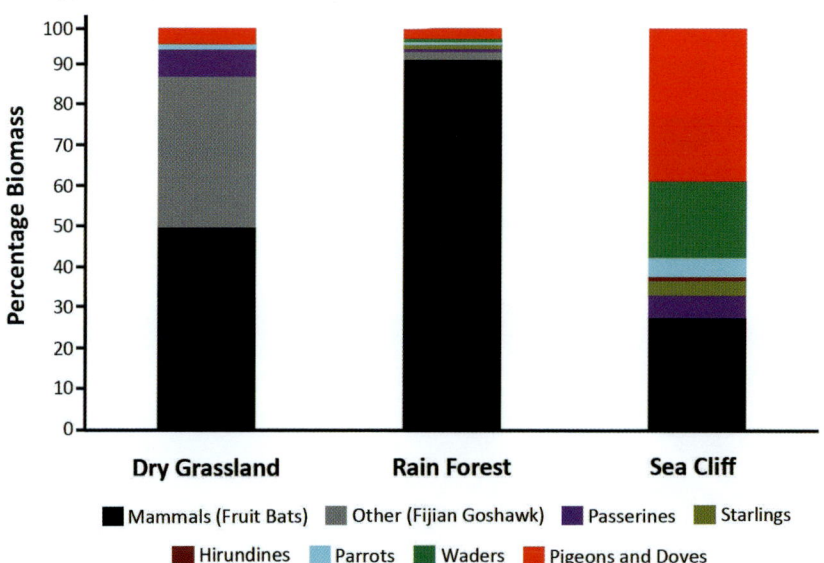

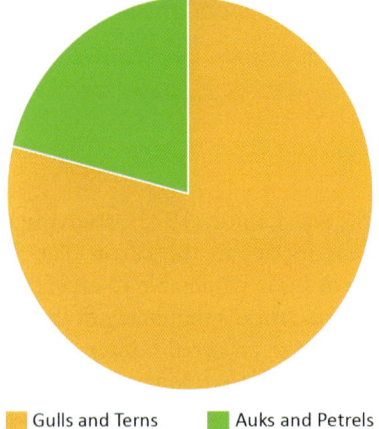

Figure 4.13b Prey spectrum by biomass percentage of breeding *F.p. nesiotes* at Grand Lagoon, New Caledonia. Constructed from data in Baudat-Franceschi *et al.* (2013)

■ Gulls and Terns ■ Auks and Petrels

but as some were forest dwellers, not all were available to the falcons. Of those that were (65 avian + 4 mammal), 45 were found in prey remains (41 avian and 4 mammal). The majority of mammalian prey was bats (chiefly Pteropus fruit bats, but including a rat). Of the avian prey, 14 were marine/aquatic, six were doves/pigeons, three were parrots, the remainder were passerines. One Voracious Gecko (*Gehyra vorax*) was also found, though this had probably been captured fortuitously rather than hunted.

In another study on Vitu Levi, White and Brimm (1990) saw a juvenile Peregrine catching locusts that were flying ahead of a spreading grass fire, catching 11 insects in an eight-minute period.

On Vanuatu Bergulla (1992) again notes that doves and pigeons are a primary prey source, though other small-medium sized birds such as the Rainbow Lorikeet (*Trichoglossus moluccanus*) are taken. The falcons also feed on bats and sometimes raid local domestic chicken flocks.

In New Caledonia Baudat-Franceschi *et al.* (2013) noted 23 prey species, the majority of which were marine-based, these including the Tahiti Petrel (*Pseudobulweria rostrata*) – Fig. 4.13b. However, the falcons also took inland species such as the Goliath Imperial Pigeon (*Ducula goliath*), which weighs up to 750g, and passerines such as the Grey-eared Honeyeater (*Lichmera incana*).

Below left   The remains of a Tahiti Petrel in an *F.p. nesiotes* nest. *Julien Baudat-Franceschi.*

Below right   Tahiti Petrel.

The above dietary snapshots were taken during the breeding season and in all cases were from studies of 'wild country' Peregrines. In recent years Peregrines across their range have been moving into urban environments, a habitat where the available prey spectrum may differ substantially from that in wild, or even rural, areas. Fig. 4.14 sets down the prey spectrum of differing urban-dwelling sub-species from across the world. As can be seen, the combination of pseudo-cliffs for nesting and significant numbers of pigeons (and parrots in urban Australia) has allowed Peregrines to increasingly become an urban species.

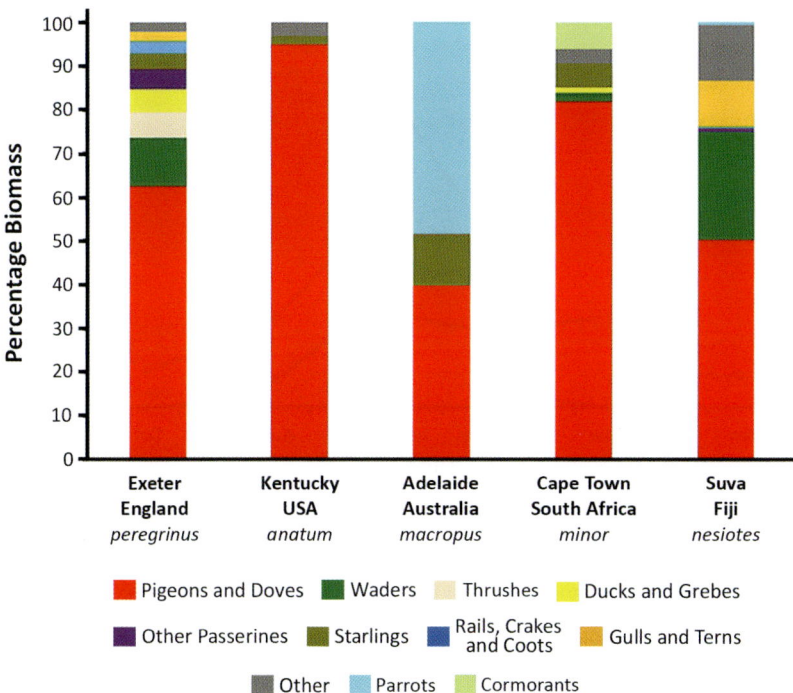

Figure 4.14 Prey spectrum by biomass percentage of breeding Peregrines in urban sites across the falcon's range.

Constructed from data in:
Exeter, England Dixon and Drewitt (2018), a study across 20 years (1997-2017), including 5426 identified prey items of 102 avian species and three mammals.

Kentucky, USA Carter et al. (2003), a study across 2 years, including 61 identified prey items.

Adelaide, Australia Olsen et al. (1993b), a study across 2 years, including 97 identified prey items.

Cape Town, South Africa Walker (2019), a study across 2 years, including 439 identified prey items.

Suva, Fiji Clunie (1976b), a study across 3 years, including 425 identified prey items.

As the snapshots of the sub-species diet set down in Fig. 4.14 were taken during the breeding season, the data can also mask the extent to which the dietary spectrum may change with the season. Fig. 4.15 illustrates the variation of urban, nominate Peregrines in Exeter, south-west England. Carter *et al.* (2003) also logged the variation of the diet of *F.p. anatum* in Kentucky, USA, during the year (Fig. 4.16).

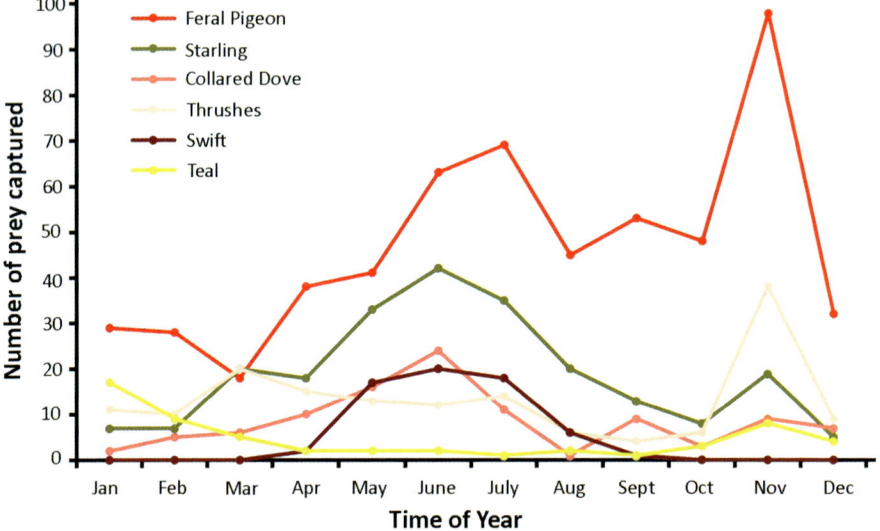

Figure 4.15 Variation of nominate Peregrine diet by biomass throughout the year (during the period 1998 – 2007) in Exeter England. Redrawn from Drewitt and Dixon (2018). RS is sceptical of the sudden rise in prey taken during November, relative to October and December, and notes that Drewitt and Dixon refer to an annual cleansing of gulleys, gutters and drainpipes in Exeter which takes place in November and which may, therefore, have skewed the data for that month.

Fig. 4.15 is based on over a thousand prey items. Fig. 4.16 is based on more limited data for *F.p. anatum* – identified prey numbered 112 in spring, 319 in summer, but only 14 in autumn (fall) and 20 in winter – which suggests that when the Peregrines were not hunting to feed chicks, and so were spending less time hunting, Starlings replaced pigeons as the prey of choice. The surprising addition to the diet in winter was the American Coot (*Fulica americana*). As with the Starlings, this may indicate a change to migratory species with the Peregrines hunting incoming birds which were not used to living with such a formidable predator. Fig. 4.17 is a further seasonal study of the Peregrine diet, in this case *F.p. japonensis* in Korea. Again, the data were limited (197 items in spring, 33 in summer, 51 in autumn and 81 in winter), but the shift of prey spectrum is equally notable. It would seem that in winter Korean Peregrines move to heavier prey – ducks and seabirds – which may be a better return for the hunting investment.

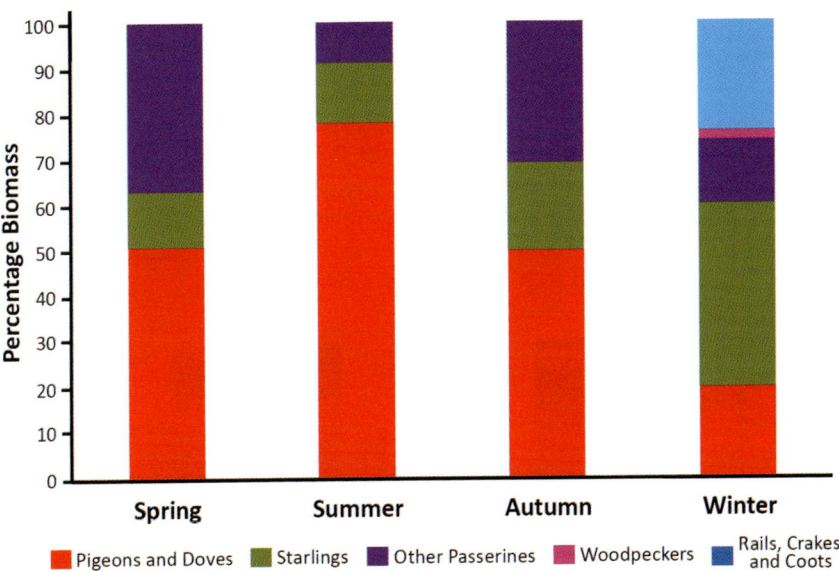

Figure 4.16 Variation of *F.p. anatum* prey spectrum by biomass percentage throughout the year (during the period 1999-2001) in Kentucky, USA. 112 prey items were identified in Spring, 319 in Summer, 14 in Autumn (Fall) and 20 in winter. Constructed from data in Carter *et al*. (2003).

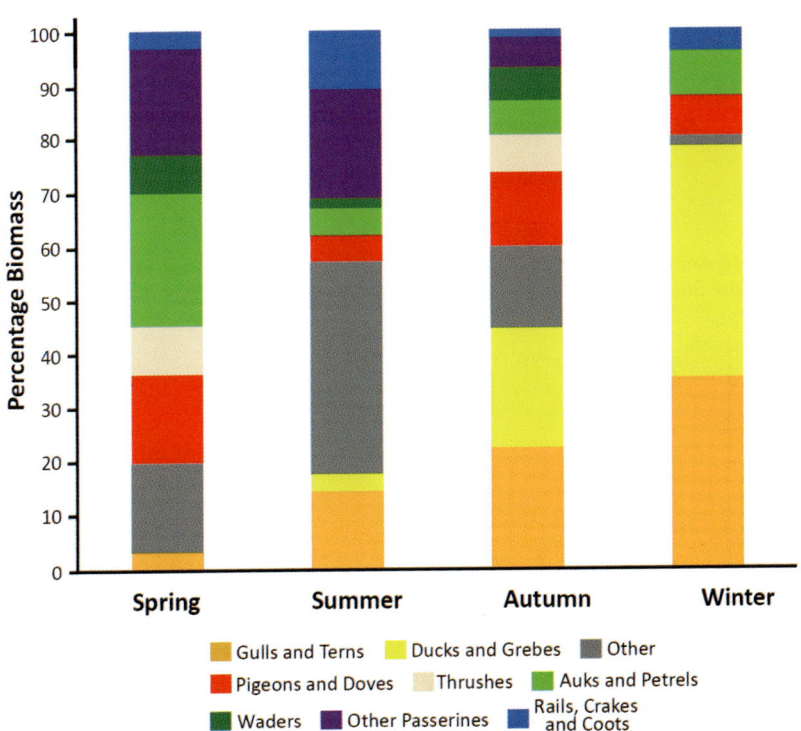

Figure 4.17 Variation of prey spectrum by biomass throughout the year (during the period 1999-2001) of *F.p. japonensis* in Korea. 197 prey items were identified in Spring, 33 in Summer, 51 in Autumn (Fall) and 81 in winter. Constructed from data in Choi *et al*. (2013).

Data collected across the Peregrines' range shows that in addition to seasonal changes in diet there may be changes from year to year. This is well illustrated by data collected in New South Wales, Australia (Olsen *et al.*, 1993c) for 'wild country' *F.p. macropus* (Fig. 4.18) and for urban nominate falcons in Warsaw, Poland (Rejt, 2001) and Rejt and Sielicki, 2007) – Fig. 4.19.

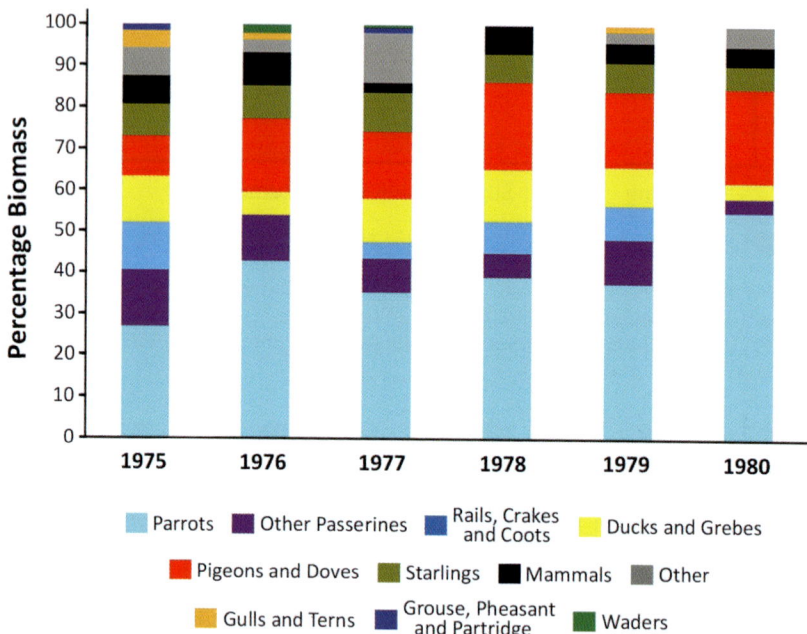

Figure 4.18 Annual variation of *F.p. macropus* diet by biomass during the period 1975-1980 in New South Wales, Australia. Constructed from data in Olsen *et al.* (1993).

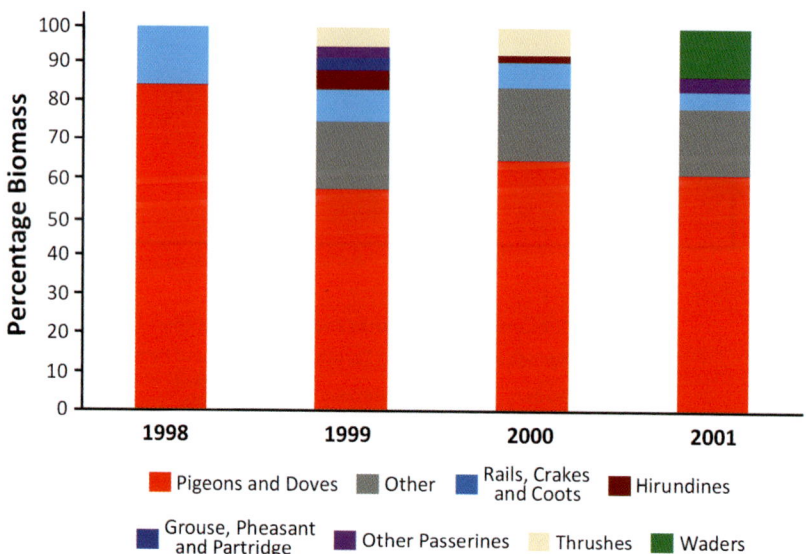

Figure 4.19 Variation of diet by biomass in Warsaw, Poland across four years. (2009), but see also Rejt (2001). of urban breeding nominate Peregrines Drawn from data in Rejt and Sielicki

In Australia the diet was surprisingly constant, with feral pigeons forming the bulk at all times. Galah and Rosellas were also popular throughout the year. Interestingly, in winter the falcons took Black Swans, though few in number: the falcons also hunted Noisy Friarbirds (*Philemon corniculatus*) during all seasons except summer. The study also looked at the influence of diet on reproductive success during the breeding season – this will be considered again in Chapter 7 *Chick Growth*.

In Warsaw (Fig. 4.19), while the fraction of pigeons remained more or less constant across the years 1998-2001, thrushes were not represented in 1998, but in later years formed a significant item in the prey spectrum. Equally interesting is the occurrence of waders in 2001, a prey source which had not been utilised in previous years. These changes beg the question of the underlying causes. Is it that there has been a change of breeding pair, either the male or female, or perhaps both, the newcomers having different specialities in hunting, or has there been a change in the prey species themselves, with some showing a decline in population while others show an increase?

One final issue is worth considering before leaving the topic of the dietary preferences of the various sub-species of Peregrine – in what way does the physical size (weight, wing dimensions) of adult Peregrines influence their choice of prey? Serendipity meant that the largest prey spectra, in terms of number of prey items caught, were for the three ideal sub-species: nominate Peregrines in England (which are among the heaviest falcons, only some *F.p. pealei* being heavier) – Dixon and Drewitt (2018); *F.p. minor* in southern Africa, the lightest sub-species (Hartley (2000) in Zimbabwe and Jenkins (1995) in South Africa; and *F.p. brookei* in northern Spain (Zuberogoitia *et al.*, 2013) which are a mid-weight. The relevant data for these three sub-species are set down in Table 4.1.

Subspecies	no.	Wing-length (mm) mean	low	high	no.	Wing-width (mm) mean	low	high	no.	Tail (mm) mean	low	high	Weight (gm) mean	low	high	Ref
Peregrinus																
F	27	346	324	363	28	166	138	190	29	164	138	184	1125	925	1300	1
M	26	301	280	312	25	139	129	147	26	137	128	143	680	580	770	1
Brookei																
F	11	323	309	333	12	154	147	162	12	149	138	159	875	775	935	2
M	20	286	267	319	18	129	118	145	21	126	114	141	576	492	697	2
Minor																
F	9	313	302	330	8	148	138	155	8	147	138	155	700	594	806	3
M	10	275	264	291	10	123	116	131	10	121	113	130	495	459	531	3

Table 4.1 Relevant data for the three Peregrine sub-species.

Data on dimensions are from White *et al.* (2013). Data on weights are from (1) Ratcliffe (1993), (2) Iñigo Zuberogoitia, pers. comm. and (3) Hartley (2000).

Data on prey weights in the Figures overleaf are from:
Fig. 4.20 Drewitt and Dixon (2018), Exeter, UK.
Fig. 4.21 Zuberogoitia *et al.* (2013), northern Spain.
Fig. 4.22 Jenkins (1995), South Africa, and Hartley (2000), Zimbabwe.

The prey spectra for the three sub-species were converted from number to biomass as necessary and is set down in Figs. 4.20-4.22.

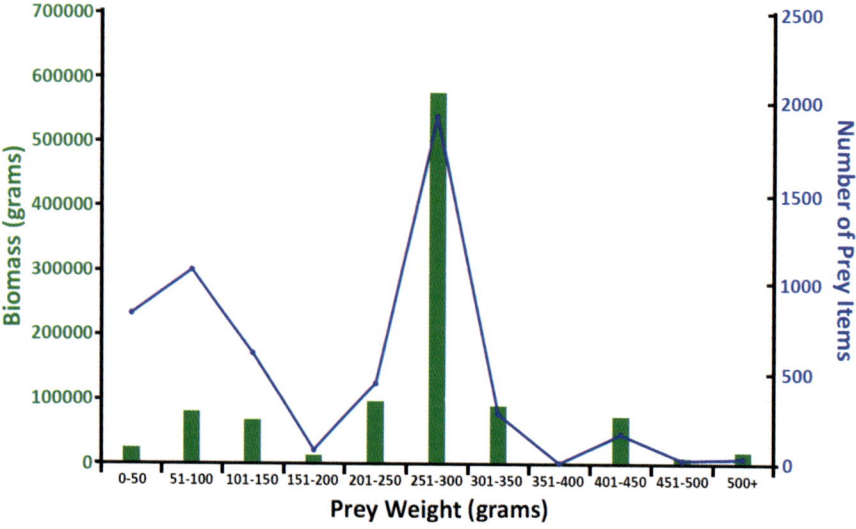

Figure 4.20 Prey spectrum of nominate Peregrine. Data from Dixon and Drewitt (2018). Number of prey items 5426, total prey weight 1,044,693g, mean prey weight 192.5g.

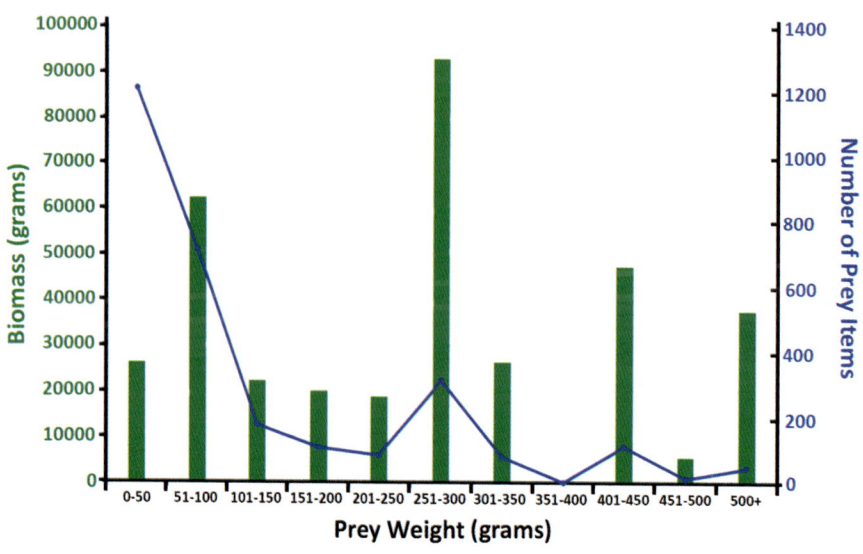

Figure 4.21 Prey spectrum of F.p. brookei. Data from Zuberogoitia et al. (2018). Number of prey items 2832, total prey weight 359,171g, mean prey weight 126.8g.

Diet

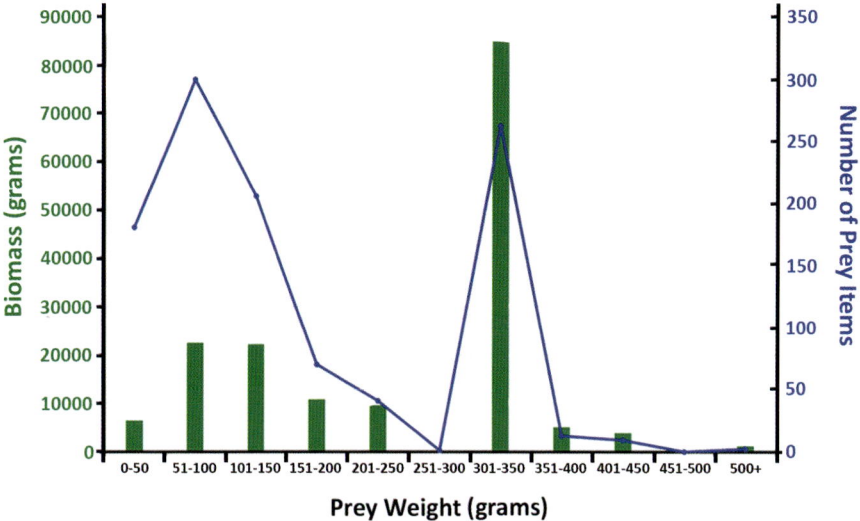

Figure 4.22 Prey spectrum of F.p. minor. Data from Hartley (2000) in Zimbabwe and Jenkins (1995) in South Africa. Number of prey items 1082, total prey weight 166750g, mean prey weight 154.1g.

The results were unexpected as it had been assumed that the mean prey weight would vary linearly with mean falcon weight. Once again Peregrines had proved to be surprising. In correspondence with the authors Iñigo Zuberogoitia noted that in his study the Peregrines were hunting small passerines more frequently than had been expected, probably due to the fact that they were simply more numerous. Iñigo believed that the size of the falcon does not limit the prey species size, noting that in his study the Peregrines also took Yellow-legged Gulls (816g) and perhaps even a Northern Gannet (2800g): the next largest prey were a single 1000g Pheasant and two Mallards (976g). The smallest prey was a Firecrest (*Regulus ignicapilla*), 6g. At other times Dr. Zuberogoitia had seen both Grey Heron (1500-2000g) and Eurasian Spoonbill (*Platalea leucorodia* – mean weight 1500g) captured. In correspondence with the authors, Arjun Amar of the University of Cape Town, South Africa noted that mean prey weight was more likely to reflect the available prey spectrum and so, again, mean prey weight would be dependent on that and environmental conditions rather than being defined by falcon weight. Given that Peregrine attacks on prey may be opportunistic and are likely to be devastating, the criteria would appear to be more what the falcon can transport back to the nest during the breeding season rather than what it can overcome.

Hunting strategies

The mathematics of the hunting techniques of Peregrines has been dealt with in Chapter 3. What follows here is a more observational analysis. Herbert and Herbert (1965) memorably suggested that the popular idea of the Peregrine was of a *'swift "engine of death" clashing and brawling about the skies with unabated fury, striking into oblivion every hapless bird of passage'*. Hyperbolic admittedly, but perhaps not so very far from the truth. Rudebeck (1951) was more constrained in his description, noting that the stoop of the Peregrine gave 'an overwhelming impression of strength, vigour, and swiftness'. But although the stoop is the most spectacular of the Peregrine's hunting techniques it is not the only one. Rudebeck notes that 'sometimes birds are pursued violently in the horizontal direction, or the falcon may try to catch a bird *en passant*, even quite low over the ground.' Peregrines, like other predators, work with a defined programme of attacks, choosing the most appropriate for the particular occasion and executing it to the best of their abilities, not always successfully.

Peregrines will search for prey from a perch, this usually being a high position, close to the eyrie during the breeding season or within the bird's winter territory, the latter because perch-hunting reduces energy loss, prey then being pursued in fast horizontal flight. The falcons will also search for prey while crossing their territory by flapping flight, or by soaring in wide circles above their territory, the equivalent of the falconer's 'waiting-on'. If prey is scarce such hunting flights can extend over large areas: Chapter 6 Fig. 6.2 notes Peregrines travelling more than 40km from their eyries in search of food during the breeding season. The direct pursuit of prey is occasionally observed following a failed stoop attack, but this rarely results in prey capture, the element of surprise having been lost. Treleaven (1980) watched a female Peregrine in south-west England which crouched in a crevice on a steep cliff and darted out to intercept passing Razorbills (*Alca torda*). Treleaven states that this was the commonest form of hunting he witnessed, but also noted Peregrines patrolling the cliffs in poor weather in the hope of surprising unsuspecting prey. If prey was spotted it would be pursued by fast, flapping flight. On occasions Peregrines will actually ring-up in pursuit of prey, in the manner of Merlins. Merlin ringing flights after Skylarks (*Alauda arvensis*) were the aim of female falconers in the royal courts of medieval Britain. In the classic ringing attack the falcon pursued a lark which was positioned too far from cover and so had to attempt escape by climbing, hoping to be able to outclimb its pursuer. It is popularly assumed that in such flights the two birds corkscrewed upwards, one below the other, but in reality the two rarely shared an airspace. Such flights are rarer for Peregrines as their turning circle is much wider than that of the smaller Merlin and so offer small prey good opportunities to escape. One of the authors (SW) has observed this on four occasions at a Gloucestershire (England) long-term study site, these all occurring after a failed stoop had scattered a feral pigeon flock. The ringing flight was performed by continuous high energy flapping flight in tight circles immediately underneath the intended prey. Each observation culminated in a successful capture, either by outflying or outlasting the prey. Ringing flights have been seen against heavier prey, and were favoured by medieval French falconers who referred to them as *haut vol* (great flight) with Peregrines ringing up after herons, cranes and kites (see, for instance, Macdonald, 2006).

Peregrines will also quarter the ground in the way of hawks and harriers, hugging the contours or using field hedges to conceal approaches in the hope of surprising prey. In an interesting article on Peregrine (*F. p. tundrius*) hunting on Alaska's North Slope, White and Nelson (1991) note that the contour hugging they observed (from a helicopter) looked similar to that seen by a Peregrine hunting alcids and storm-petrels off the coast, the falcon hugging swell tops to surprise its prey. Similar behaviour has been seen with *F. p. pealei* hunting alcids off the Aleutians (Nelson in Sherrod, 1988). Peregrines pursuing flocking waders, particularly during winter is seen wherever the falcons are found in coastal areas. The Peregrine will harass a flock, attempting to create panic so that a single bird zigs when the rest of the flock zags and so presents a single target for the falcon – see, for instance, Dekker (1998) who suggests that the falcons are reluctant to dive into the midst of a flock for fear of sustaining damage in collisions and so choose harassment in the hope of isolating an individual[6].

The falcons will also search for insects, mammals and even nestlings by walking (Dekker, 1980, Rosenfield *et al.*, 1995), though this seems to be a specialised behaviour for certain habitats: Rosenfield *et al.* studied Peregrines in west Greenland noting adult falcons pursuing the nestlings of ground-nesting passerines (passerines formed the bulk of the falcon diet), and record similar observations by others. Peregrines will also follow humans, dogs, agricultural vehicles *etc.* in the hope they flush prey. Peregrines may also 'ring' prey which they find themselves initially below and surprise prey by flying low and fast as they contour the landscape in an effort to flush prey from the ground. There are also records of Peregrines following cars or farm vehicles in the hope they startle prey into the open. Rudebeck (1951), who watched 252 Peregrine hunts in Sweden during the 1930s lists 10 different hunting techniques he had seen, though seven of these are different forms of stooping. Stooping is the most spectacular, most discussed and, occasionally, most overstated of all Peregrine hunting modes.

While often assumed to be of a standard form, stooping has many variations, the angle of attack varying from shallow, perhaps at only 30° to the horizontal, to the near-vertical. Shallower angles are also seen, these hunts amounting almost to the Peregrine chasing down its prey. At the end of the stoop the falcon either strikes the prey or binds on to it. The former may render the prey unconscious (or dead) or semi-conscious (and injured), resulting in it plummeting towards the ground. The Peregrine may quickly circle and take the prey before it lands or, as was noted in Chapter 3, may circle the prey on

---

[6] Rudebeck (1951) gives an alarming, though perhaps not entirely surprising, description of the outcome of a Peregrine attack on a Common Starling flock flying above Lake Krankesjön, southern Sweden, when '*The starlings were extremely agitated and zig-zagged continually. When they passed over a stretch of open water, a loud splash was heard – it was the excrements of the birds falling into the water like a shower of rain.*'

the ground to assess its capability to put up a fight, attacking in a series of loops to strike further times until the prey is no longer capable of resistance. By contrast, the Peregrine may bind on to the prey while in flight. Once secured, the prey will be dispatched in flight by a bite to the neck with the tomial tooth, the bite disarticulating the cervical vertebrae. This allows for a controlled flight back to an elevated position where the prey will be cached or consumed. Alternatively, and more often outside the breeding season, a large prey item may be 'ridden down' to the ground where it is then dispatched. Stoops may also end with the Peregrine diving beneath the prey and turning up to catch it from below.

Such hunts usually end with the falcon binding on to the prey. Dekker (1980), watching Peregrines migrating through central Alberta, Canada, tabulated the results of almost 700 prey attacks and the success rate of each – see Table 4.2. Dekker's falcons were primarily hunting ducks and waders (94% of all the prey taken: the remainder were passerines and gulls). Dekker also noted that the majority of captures were on targets which failed to employ the evasion tactics used by others of their species. The latter would swerve in flight and/or drop to the ground or onto water at the last moment before the attack, whereas those which were captured tended to maintain a level course and were overhauled.

Attack Method	Percentage of Attacks	Percentage Success Rate
Stoop[1]	21.5	11.0
Long distance flapping flight[2]	18.3	8.1
Low level surprise[3]	29.1	4.6
Short range[4]	23.3	12.0
Other[5]	7.8	14.0

Table 4.2 Hunting techniques of Peregrines in central Alberta, Canada. Data from Dekker (1980).

Notes:
1. Includes stoops towards both ground level prey and higher-flying prey.
2. Involves a flap-flying search over the ground followed by an attack at ground level prey.
3. Involves a flap-flying search followed by short-range attack at flying prey.
4. Surprise attack from a perch.
5. Includes tenacious pursuit, co-operative attacks by two Peregrines or an attack by a falcon on the ground.

Note also that the success data are for all Peregrines observed and so includes both adult and immature birds. The success rates quoted by Dekker (1980) are considered again below.

Nominate Peregrine with prey, Lincoln Cathedral, England. *Colin J. Lea*.

Cresswell (1996) studied wintering Peregrines in East Lothian, Scotland, which were also taking waders (Dunlin and Redshank) and noted that a significant fraction of attacks were initiated from perches, which implies surprise and pursuit hunting, though Cresswell also saw stoops and 'open' attacks, in which the falcon was visible to the prey before the attack. Cresswell's attack breakdown was 36.3% stoops, 36.1% surprise attacks and 25.1% open attacks. Such open attacks conform to what other researchers would term pursuits, though, of course, a failed stoop or failed surprise attack might also lead to a chase. Cresswell also noted 2.5% of ringing attacks. Cresswell's data can be compared to Dekker's (1980) in Table 4.2 with the proviso that individual observers may use different terms to describe what are identical hunting methodologies.

Most attacks seen by Cresswell (1996) were launched from the air (82.6%) rather than a perch and were directed at ground-based prey on 55.6% of occasions (airborne prey 44.4%). Prey was less than 100m away from the falcon when an attack was launched on 39.8% of occasions, at 100–500m on 36.7% of occasions and more than 500m away on 23.5%. On average Peregrines made two attacks (range 1.8±0.4) on each hunting flight: Cresswell saw one flight which involved 11 attacks. During his observations Cresswell found that the Peregrines' winter time budget was 72% perching, 21% flying, either moving position or hunting, and 2% feeding: the remaining 5% were not credited. Cresswell noted overall success rates of 9.5% against Redshanks and 7.6% against Dunlin, which are very similar to the overall attack successes in Table 4.2. Cresswell also noted that chases were mainly single stoops of short duration (49.7% lasted less than one second), but some were longer (36.6% were 1-10 seconds, 8.9% 11-30s, 2.4% 31-60s and 2.4% were >60s). These timings can be compared to those of Treleaven (1980) studying Peregrines in Cornwall, England who noted that hunts that ended with a kill lasted less than four minutes, some less than one minute: for one Peregrine all observed hunts lasted less than one minute – at the end of that time the hunt was either successful or abandoned.

Baker (1967) describes the kill at the end of a long stoop based on his observation. The feet are thrust forward, the long hind toe extended so that it rips into the body of the prey at speed, causing sufficient damage to maim or kill. Saxby (1874), in his observations on Shetland, concurs, noting that on examining Peregrines shot while they were hunting, blood and prey feathers were frequently seen on the hind talon. These observations are in accord with high-speed photography (Goslow, 1971). Goslow filmed strikes, but the situation was somewhat artificial – the captive falcon being freed and usually circling to a height of about 16m, at which point a short-tethered, blindfolded pigeon was realised in front of the filming screen. Filming was at 800-1000 frames/second. While the strike speed of the falcon was clearly limited, two distinct methods of attack were seen (Fig. 4.23). In one the feet were extended forward, the toes stretched out (Fig. 4.23a), as Baker suggested.

Goslow noted that the time required to deploy the talons was 60-100ms. More recently high-speed cameras have been used to analyse binding strikes, and these confirm Goslow's findings. Kane and Zamani (2014) note the capture of a crow by a Gyr-Saker hybrid in which the falcon spread its wings and tail less than 230ms prior to impact and extended its talons, the time to deploy the latter being 67ms, consistent with the data of Goslow (1971). This position – tail and wings spread, talons forced forward – seemed, in Goslow's study, to be taken when there was an attempt to bind on the prey. In the second method identified by Goslow, the toes are still open, but the body position alters (Fig. 4.23b). Some observers have claimed that the strike is with the toes closed, and that the hind toe extends due only to its length, but Goslow (1971) suggests this is not the case. The altered body position in the strike posture would allow the falcon to better absorb the shock of contact by allowing the feet to be pushed backward, while still allowing both the impact and the talons, particularly the hallux (rear talon), to inflict damage. The high-speed photography suggests the toes are bunched immediately after a strike, so it may be that some birds do actually strike with a bunched foot. It is possible that the falcon's decision whether to bind or just to strike is made only immediately before the prey is reached, when the Peregrine assesses any evasive manoeuvre by the prey and the speed differential and attack of angle this might induce, and alters its body position to compensate. Striking, rather than binding, is clearly advantageous, as it allows a greater contact speed. If the falcon intends to bind, its speed and that of the prey need to be approximately the same. The Peregrine may also need to see its feet to coordinate the bind, which would require a significant slowing as moving the feet into the area of binocular vision would massively increase drag: binding without seeing the

Figure 4.23 Drawings of prey captures based on Goslow (1971). *Martin and Hazel Buckley.*

position of the feet would require great dexterity, but as the Peregrine has this in abundance perhaps visual spotting is not necessary.

Striking at speed and not retrieving before the prey hits the water is a problem for coastal Peregrines hunting seabirds. Buchanan (1996) observed three occasions on which a Peregrine hit the water before pulling out of a stoop: he does not say whether the falcons survived, though the implication of there being no mention of the outcome is that they did. This has also been seen in Cornwall, England where the Peregrines survived a dip in the Atlantic Ocean (Peter Welsh pers. comm to SW). Ratcliffe (1993) also notes an attack on a Mallard on a hill loch which resulted in the Peregrine hitting the water at speed and throwing up a plume of water 3.5m high: the duck escaped by diving, the falcon presumably escaping unhurt as, again, there is no contrary record. Ruthven (2013) records a pair of Peregrines retrieving a Wood Pigeon from water in Scotland which, it was assumed, had been struck. Each bird took it in turns to manoeuvre the pigeon to the shore by dragging it, the male Peregrine completing the task by swimming with its wings as it gripped the prey. The two Peregrines, presumably a bonded pair – the observation was in February – then plucked and ate the pigeon.

There are several observations of Peregrines collecting prey from the water surface. Fisher (1978) saw a Peregrine forced to drop prey (a pigeon) into the sea by pursuing European Herring Gulls which then settled on the water around the prey, interested, but not attempting to feed. The Peregrine circled and made repeated dives, touching the prey twice before finally retrieving it and, again pursued by the gulls, flying off.

One of the most extraordinary over-sea hunts was observed by Rogers and Leatherwood (1981), who were aboard one of two ships working for the US National Oceanic and Atmospheric Administration in the Pacific Ocean about 65km from Clipperton Island, a small atoll 2600km west of Costa Rica and 2400km (1490 miles) north-west of the Galapagos Islands. For 2½days there had been a storm with easterly winds, then conditions calmed and an Osprey and Peregrine Falcon appeared, landing on the ships. Rogers and Leatherwood noted that in competition for the highest mast perch (19m on one ship, 25m on the other), the Peregrine dominated the Osprey. From its high perch the Peregrine surveyed the ocean and periodically flew off to attack Leach's Storm-petrels (*Hydrobates leucorhoa*). Attacks were direct, the falcon not choosing to gain height before attacking: if an attack failed the Peregrine returned to its perch rather than making a second attempt at capture in the same flight. Rogers and Leatherwood saw 11 attacks, eight of which were successful, the petrels being taken as they pattered across the water (their normal feeding behaviour). The Osprey, which had been feeding on flying fish, stayed with the ships for four days, the Peregrine for five. Walker (1988) also observed Peregrines catching Leach's Storm-petrels on the Isle of Man: the species was also found in the diet of a West Country urban Peregrine as well (Fig. 4.14, listed under 'other').

A nominate Peregrine retrieves a pigeon knocked into the sea off the Isle of Man, UK. *Peter Christian*.

Voous (1961) also records a Peregrine feeding on European Storm Petrels (*Hydrobates pelagicus*) caught from a ship in mid-Atlantic. Equally interesting was the behaviour of a Peregrine observed by Matsyna *et al.* (2010) in Russia's Kuril Islands. The Peregrine had harried a flock of migrating shorebirds, one of which had dropped into the sea to escape. Unable to take off from the water, the bird had paddled its way to the shore, where it was plucked from the surf by the waiting Peregrine.

These attacks suggest surprise or (limited) pursuit hunting, and Buchanan (1996) suggests that the danger of hitting the water means that the falcons are reluctant to stoop at prey (which in Buchanan's observed cases were Dunlin) over water, particularly if the sea is rough. Buchanan did see stoop attacks that were successful when the sea was calm (so that retrieval from the water's surface would have been easier). The Peregrines also used over-water horizontal pursuit of a Dunlin isolated from flocks, this technique allowing the falcon to outfly the prey and to grab it in mid-air. Ratcliffe (1993) also suggests that coastal Peregrines might wait for prey to cross from sea to land on occasions before initiating a stoop. However, while these observations show a reluctance to risk hitting the water, with the inherent danger of not being able to successfully take off again (with or without the prey), the flexibility of Peregrines to local conditions is indicated by their hunting of Ancient Murrelets on Langara Island and one of the Queen Charlotte Islands of British Columbia, Canada. There, Peregrines hunted at dawn and dusk when the auks left for, and returned from, their feeding grounds. However, Dekker and Bogaert (1997) noted that during the day Peregrines would occasionally fly out over the ocean, travelling beyond telescope range. Although 13 Peregrines returned without prey, on two occasions returning falcons were pursued by Bald Eagles (*Haliaeetus leucocephalus*): as the eagles are known to pirate prey from Peregrines it was assumed that some of the other 11 falcons had made successful attacks far out over the ocean. No information was available on the hunting technique employed by the Peregrines.

Peregrines are also known to fly low and fast across open country areas in the manner of harriers so that prey on the ground might be flushed into the air, a technique also used if prey has been driven into ground cover, in the hope of flushing it back into the air. There are also reports of Peregrines hunting in twilight and even at night. Beebe (1960) records observations during a Peregrine study on Canada's Cox Island, one of the Scott Islands off Vancouver Island's north-west coast. The falcons were hunting Ancient Murrelets and Cassin's Auklets (*Ptychoramphus aleuticus*), both of which leave their breeding grounds before dawn (usually an hour before) to feed, and arrive back after dusk (again about an hour after). Beebe notes the Peregrines hanging in the air ('as though hung on strings') at dusk when the auks were leaving the island and stooping on them when they were 'in silhouette against the sea'. Beebe does not mention whether his observation was on moonlit nights, but from personal experience (RS, with Ancient Murrelets), observation is best

F.p. macropus flying low and fast, Cape Baily, Kamay, Botany Bay, Australia. *Kytabu.*

when moonlight is reflected from the water as then the birds are silhouetted against the sea. Beebe thought the birds were catching auks, then caching them before hunting again. It is likely that the Peregrines were also hunting in similar fashion in the hour or so before dawn. Beebe's observation is certainly of crepuscular hunting, perhaps even nocturnal, but the latter has certainly become familiar in urban settings when the Peregrines use street lighting to hunt. Many species migrate preferentially at night and are often attracted by the lights of cities, Peregrines taking advantage of the fact that the nominally cryptic plumage of shore and water birds – dark dorsally, pale ventrally – is an aid to hunting when the prey is lit from below.

Such behaviour is clear from the prey composition of British urban Peregrines, which includes Quail, Water Rail (*Rallus aquaticus*) and Woodcock (*Scolopax rusticola*)[7], and has also been seen elsewhere (for a general review of nocturnal hunting behaviour of urban Peregrines, see Mebs, 2009)[8]. An interesting aspect of nocturnal hunting arises from a study in south-west China by Zhao *et al.* (2020) who showed that night-flying birds were strongly attracted to lights at the blue end of the visible spectrum, with the least attraction being seen with red light. Town planners could therefore reduce the take of nocturnally hunting Peregrines by preferentially using lights with outputs at the red end of the spectrum. Equally interesting are the observations of Rejt (2004) in Warsaw, Poland and Kettel *et al.* (2016) in Nottingham, England, who each report the nocturnal feeding of chicks. Rejt (2004) reports most feeding between midnight and 4am, with a mean of 1.5 feeds/night. Kettel *et al.* (2016) report four events in April 2015, timed between midnight and 02.18.

Very rarely, Peregrines will also 'hover' (a technique well-known in, for instance, the Common Kestrel, the technique then being called flight-hunting as true hovering is only really seen in hummingbirds). Roberts (1946) noted a Peregrine quartering marshland near Fort William, Scotland which hovered for 45s over three, week-old Redshank chicks before dropping on to one and flying off (pursued initially by the frantic mother Redshank). However, the present authors consider such a long period of hovering unlikely and wonder whether the observer mistook the species.

Peregrine pairs will also hunt cooperatively in the breeding season, both before egg laying and when the young are approaching fledge or recently fledged, with some observations suggesting that one falcon will drive prey into the path of the other, while others note one falcon scattering prey while the second waits to pick up a bird which has moved into a vulnerable position as a result of panic (Fig. 4.24). One of the authors (SW) has observed this behaviour on multiple occasions but, in particular, one pair observed between 2008-2012 at Avon Gorge, Bristol, UK would pair-hunt regularly and to great effect, raising 22 young in the space of five years, during which period they raised three broods of five, all of which fledged. Studying pairs of *F.p.*

---

[7] Interestingly, the same three unusual species were also seen in the prey spectrum of nocturnally hunting nominate Peregrines in Belfort, eastern France where the medieval castle is illuminated each night (Marconot, 2003). Belfort sits on the migration route between the Vosges and Jura mountains.

[8] Perhaps the most interesting nocturnal urban hunting was that studied by DeCandido and Allen (2006) from the outside observation deck of the Empire State Building in New York (325m above ground level). Watching from sunset to about midnight in the autumn and sunset to 11pm in spring the researchers saw both male and female falcons hunting (with up to seven falcons some evenings) and capture success rates of 28-53% against migrants. Identifying individual prey species was difficult, but the Peregrines definitely took Baltimore Oriole (*Icterus galbula*) and Yellow-billed Cuckoo (*Coccyzus americanus*).

Diet

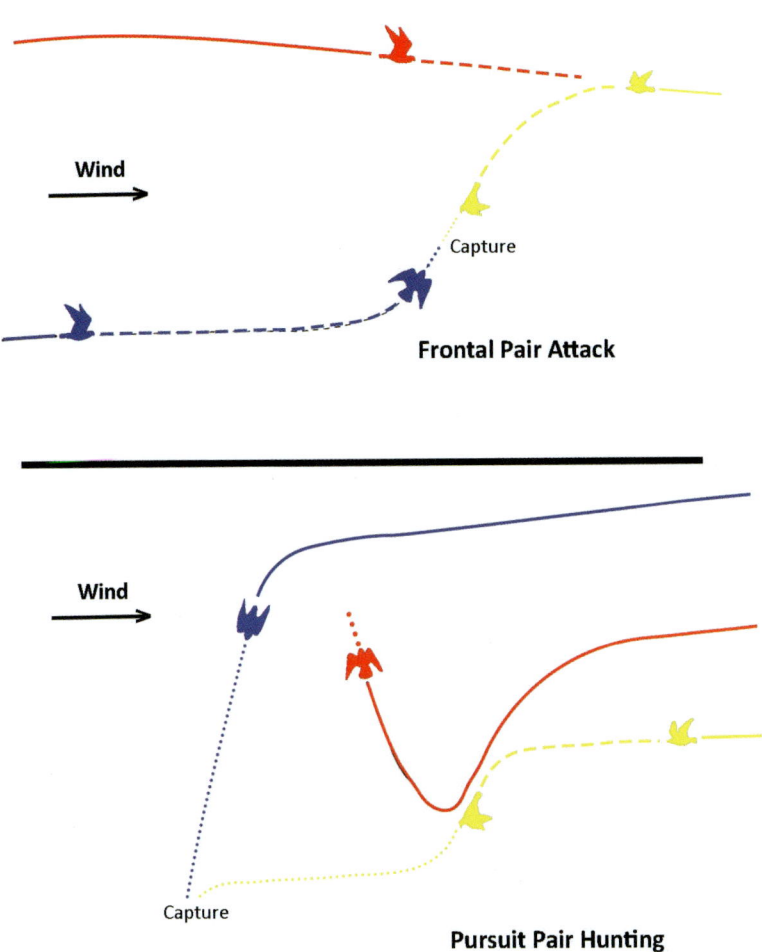

Figure 4.24 Cooperative hunting modes. Redrawn from Vasina and Straneck (1984).

anatum in Arizona, USA, Ellis (2006) noted that solo attacks on White-throated Swifts occasionally required many stoop attacks (Ellis states 'dozens', with some from 1000m) for a capture, but that falcon pairs occasionally hunted single swifts together, one from above, the other from below, timing their attacks so that they arrived at the Swift more or less together. In his study in southern Sweden, Lindberg (1983) found such cooperation increased the success rate of hunts from 34% to 45%. However, other studies suggest cooperation is not always so successful.

Defence techniques of prey

When an individual bird is under attack it has four main possibilities for escape. The bird can dive in an attempt to reach cover before being overtaken; it can attempt to outfly the predator in a horizontal chase, a tactic often employed if there is no time to reach ground cover, but tree/shrub cover is available at flying height, or if distance can be gained which can be converted into time for a controlled ground landing, or which persuades the predator to abandon the chase; it can out-manoeuvre the predator by making one or more tight turns in order to gain the time needed to reach cover or to persuade the predator to look elsewhere; or it can fly upwards in an attempt to outfly the predator vertically, again hoping to persuade the predator to abandon the pursuit.

Diving for cover is a tactic which requires sufficient time to be available for a controlled landing on the ground, and so is dependent on where the prey bird is situated when it comes under attack. In general, therefore, it does not require specific flight abilities by the prey: rather it requires sensible foraging and vigilance. The remaining three escape strategies are dependent on the flight capability of the prey in comparison to the flight capability of the Peregrine. In general prey species will be able to out-turn a Peregrine in level flight since the falcon's body mass mitigates against tight turns, and are likely to out-climb the falcon, at least in the short term, for similar reasons. Outflying a Peregrine is a different matter, particularly if the falcon has acquired significant speed in a stoop. While hunting from a perch, with a subsequent short, fast attack, will allow a Peregrine to outfly prey it is no surprise that stooping is the favoured attack mode of Peregrines in most circumstances. Assuming a failed stoop attack, the response of a grouse (for instance) to an attack by a Peregrine is to attempt outflying by fast horizontal flight, the aim being to reach safety before being captured. In a study by Pennycuick *et al.* (1994) on the contest between the Peregrine and the Sage Grouse (*Centrocercus urophasianus*), a North American grouse species, calculations of maximum flight speed indicated that the grouse was faster than the falcon in level flight, but that the morphology of the grouse was not suited for aerobic flight and that the power input for acceleration and fast flight would rapidly result in oxygen debt. The grouse would therefore need to reach safety quickly and to have created a sufficient time gap to allow a safe landing if a short flight to cover was needed. Similar reasoning would apply to the smaller British grouse species: Red Grouse have been timed at speeds to 100kph, but again they have a limited range at such speeds. As grouse have very high wing loadings – being heavy, in part because of large flight muscles, and having short wings which are excellent for acceleration and fast flight – they cannot make fast turns and, when pursued, make fast flights in straight lines, using their knowledge of their territory to head for the nearest position of good cover.

Other defence strategies involve more than an individual bird. One obvious strategy is the use of an alarm call to alert conspecifics to an approaching predator. At first glance this appears to be an altruistic behaviour but, of course, if you warn your neighbours of approaching danger, they may warn you the next time. Flocking is another defensive technique, one based on the idea of safety in numbers – if you are one of hundreds, the probability you will be attacked is much reduced. The defensive use of the white rump on feral pigeons when attacked by a Peregrine has been mentioned above. As pigeons also flock, flocking represents a second line of defence (one also seen in some other species). Buchanan *et al.* (1988) studying hunting Merlins in Washington State, USA, noted three flocking defence techniques of Dunlin. As with pigeons, in flashing flight the Dunlin changed direction or tilted their bodies synchronously so that the dark upper and pale lower side were flashed successively. In rippling flight, movements were not synchronous, a wave of movement passing through the flock. Finally, in columnar flight the flock rose in a 'tornado-like vertical column'. Combinations of the techniques also occurred: in all cases both flashing and rippling might be observed.

The various techniques presumably confuse the predator and each seems successful: flashing, which was used in 64% of attacks, had an 85% success rate at avoiding a Merlin attack; flashing and rippling occurred in 28% of attacks and had a 91% success rate; a combination of columnar flight, flashing and rippling occurred in 8% of attacks and had a success rate of 71%. Since Peregrines also hunt Dunlin, there is no reason to suppose the same techniques are not employed against the larger falcon. In a later study Hemelrijk *et al.* (2015) used computer simulation of observed movements in Eurasian Starling flocks which showed that under predator attack 'agitation waves', the choreographed movements of the birds, were set up by the rolling motion of individuals. An extension of this work (Storms *et al.*, 2019) identified six collective escapes which defined observable wave patterns in the flock, one of these waves being the 'flashing flight' identified by Buchanan *et al.* (1988). A different flocking defence mechanism was seen by Johnstone and Earl (2018) who witnessed the remarkable formation of a 'baitball' of Starlings when they were attacked by a nominate Peregrine on the Isle of Man. Johnstone and Earl's use of the term 'baitball', one used often to describe the way fish huddle together when attacked by a predator, seems entirely appropriate in this instance.

All of these anti-predator flock manoeuvres require that the birds stay in synch, and so must be carried out in such a way that the birds do not collide and at a speed which suits every bird in the flock. The former has been much debated, no true consensus having been reached. In all suggestions the essence is that each individual bird watches its nearest neighbours, though the number of neighbours is debated: seven neighbours has garnered many advocates, but also has its critics. From a point of view of speed, a delightful piece of work by Sankey *et al.* (2019) suggests that the 'Goldilocks Principle' applies, *i.e* that the speed is not too slow or too fast, but just right for the members of the flock so that no one is left behind or races on ahead. The basis of the Principle is set down in Fig. 4.25.

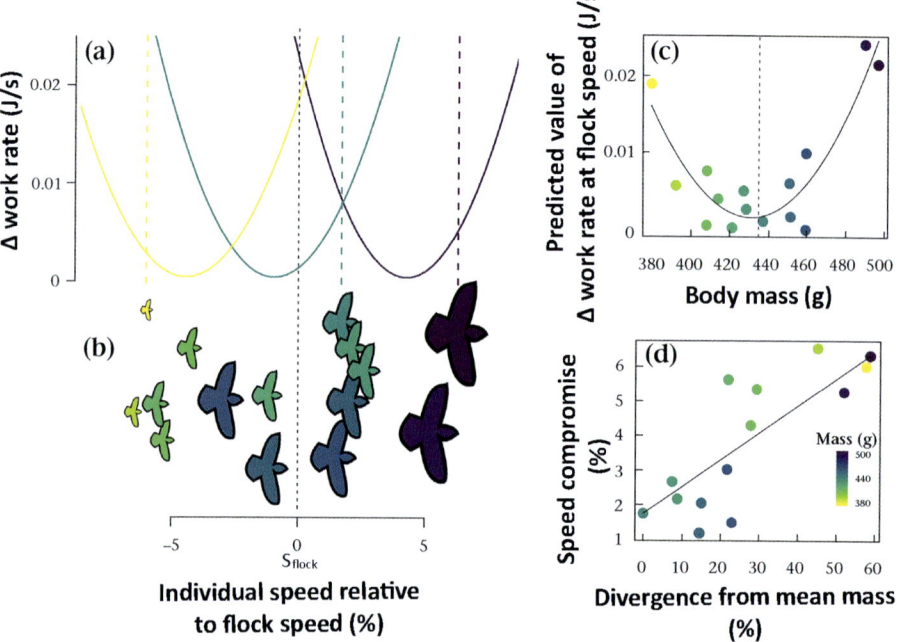

Figure 4.25

a) Predicted value of the difference (Δ) of energy output (J/s) of birds of different weights.

b) Variation of speed (as a percentage) and Δ work rate of an individual bird relative to the mean speed of the flock as a whole in comparison to its weight for the weight spectrum of birds within the flock.

c) Predicted work rate at flock speed of birds of differing weights.

d) Percentage speed compromise of birds of differing weights.

Percentage speed compromise of birds of differing weight. The 'Goldilocks Principle' for deciding the mean speed of a flock of pigeons with weights across a realistic weight spectrum. The birds in (a) and (b) are coloured according to mass. (a) are the work rates (J/s) of birds of a given weight, calculated from Heerenbrink et al. (2015). The curves are fitted polynomial regressions. (b) shows the speed variation of birds of differing rate relative to the speed of the flock. (c) shows the variation of work rate at flock speed of birds of differing weights. (d) shows the speed compromise that each bird in the flock has to make/tolerate in order for the flock to remain cohesive. The straight line is the best linear fit to the data. Reproduced, with permission, from Sankey et al. (2019).

Diet

Boyce (1985), studying Merlins hunting Dunlin at Humboldt Bay, California noted that often a Merlin would knock a Dunlin into the water rather than capturing it, or occasionally the Dunlin would deliberately dive into the water to escape capture. Assuming the Dunlin survived, it would resurface, and as the Merlin slowed in order to return to seize it, the flock wheeled around and positioned itself over the water-bound bird, shielding it from attack and giving it a chance to return to the flock. Often this tactic succeeded, though if the downed bird was injured it obviously did not.

Other flock-based defensive tactics by potential prey have also been observed. Dekker (1998) noted that during the period when their intertidal feeding grounds were under water, the Dunlins flocked over the ocean about 2km from shore, apparently as a defence against predation. Buchanan (1996) noted similar behaviour and that when at sea the Dunlin engaged in synchronised flights in the troughs between waves. Buchanan also noted that flocking over the ocean created a hazard for the predator, observing three occasions on which Peregrines hit the water before pulling out of a stoop (and also three occasions when Merlin did the same). Buchanan does not say whether the falcons survived, though the implication of there being no mention of the outcome is that all six birds did.

Prey also seem to be least vulnerable when they are feeding at low tides – many eyes to watch for predators – and most vulnerable when roosting – fewer eyes watching – and when the tide begins to ebb. In a study in North

Starlings forming a 'baitball' when attacked by a nominate Peregrine on the Isle of Man, UK. Baitballs are more often associated with the response of fish shoals when attacked by bubble-netting whales. In this case it seems the general panic caused by the attacking falcon created a tightly packed ball of Starlings. *Stephen Johnstone.*

America, Dekker and Ydenburg (2004) noted that the kill rate of Dunlins was highest at this time, the waders returning to feed then being closest to shoreline vegetation and so most vulnerable to surprise attack.

If flocks are such a successful anti-predation strategy, then it would be anticipated that predator attacks on flocking birds would be limited, but in his Scottish studies Cresswell (1996) found that Merlins, nominate Peregrines and Eurasian Sparrowhawks all attacked larger flocks of Dunlin more frequently than would have been expected by chance Fig. 4.26. This suggests that although a large flock enhances the chance of an individual bird surviving a raptor attack, it also improves the raptor's chance of a kill. But Cresswell's data did not support the idea that the raptor's chances of a kill were improved: for both Merlins and Eurasian Sparrowhawks the attack success rate was better against small flocks (defined as 1-10 birds) than against larger ones (11-200 birds), while for Peregrines the success rate was similar. However, the difference in capture rates for the two smaller raptors between small and large flocks was not significant, so the result is not counterintuitive (*i.e* it does not necessarily imply that raptors appear to consider that having more targets improves the likelihood of a kill even though defence strategies by flocks are successful, an idea which would imply that raptor attack strategies are not influenced by experience). The more likely explanations are that raptors attack larger flocks because this increases their chances of spotting a weak bird, and that flocking as a defence has its limitations, particularly for species such as Dunlin which are less successful at maintaining close separation and the cooperation needed for complete safety. Invariably some birds at the periphery of the flock detach from it and become vulnerable to attack.

Figure 4.26 Relative percentages of the availability of Dunlin flock of different sizes at Tyninghame estuary, Scotland, and the percentages of observed attacks by Peregrines. Redrawn from Cresswell (1994).

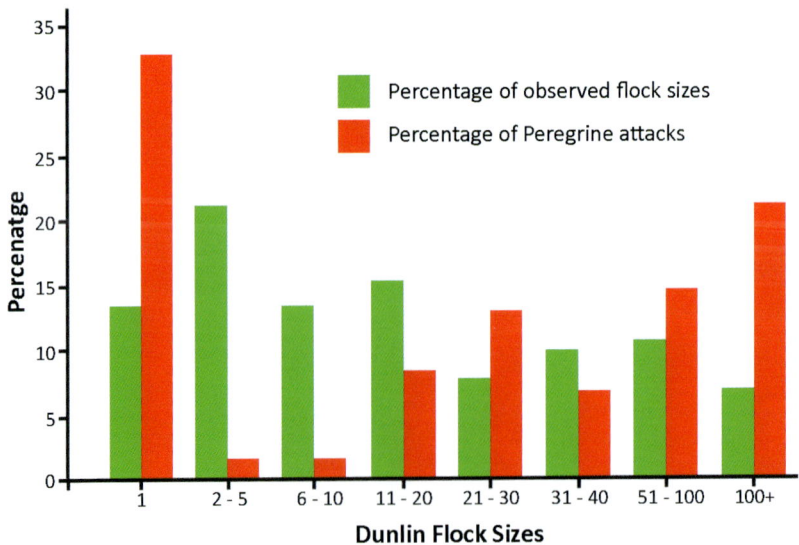

A murmuration, the name given to the beautiful wave formations created by Starling flocks when they are attacked by a raptor, or as they arrive at dusk to a preferred roosting site. In this photo, taken at Lith, North Brabant, The Netherlands. *Henk Bogaard.*

One very interesting aspect of flocking was discovered by Carere *et al.* (2009) studying Starling flocks in Rome, Italy. The researchers showed that not only were the choreographed movements of the flocks a defence mechanism, but that there was a social behavioural side to flocking. Each winter as many as 10 million Starlings congregate in Rome, arriving from all over northern Europe to enjoy the warmth of central Italy augmented by the heat of the city itself. The Starlings created wonderfully choreographed patterns in the evening sky above the city (but rather less wonderful excretions which damaged vehicle paintwork and made pavements treacherous). Of the vast number of birds, Carere and co-workers chose two flock areas to study, a city-centre square with rows of trees surrounded by buildings where about 20,000 birds collected, and a more open parkland area south of the city where 50,000 birds gathered. The predation risk from Peregrines was lower in the city-centre area than in the parkland. The researchers catalogued a series of flock forms, and found that larger, more compact forms were found in the high-predation-risk area, whereas smaller, looser flocks and singleton birds were found in the low-risk area. Peregrine success rates in the low-risk area were higher. While these findings are intuitively unsurprising, what was interesting was that the researchers found that the behaviour of flocks at a distance from the roosts influenced local behaviour, *i.e.* if a distant flock showed anti-predator flock forms, the local birds adopted similar forms, implying that social information was being passed between flocks.

Flocking is one form of defence against predation, but prey species have other alternatives. One obvious one is that they can seek cover, either by crouching on the ground or, for waders, by diving. Cresswell (1993) noted these strategies, as well as outflying the predator, by observing the escape responses of Redshanks under attack by Merlins, Peregrines and Sparrowhawks in eastern Scotland. Cresswell noted that if Redshanks crouched they were captured by Merlins in 22% of attacks, while all waders which flew or dived escaped capture. For the other two predators the figures were 8% capture if flying, 20% if diving and 91% if crouching (Sparrowhawk), and 14% flying, 2% crouching or diving (Peregrine). However, it should be noted that very many more Sparrowhawk attacks than either Peregrine or Merlin attacks were observed – for instance, over 500 Redshank responses to Sparrowhawk attacks were observed, compared to about 150 for Peregrines and only about 50 for Merlins. What was clear, however, despite the difference in numbers of observed attacks, was that the Redshanks were altering their escape behaviour dependent on which predator attacked: the risk of capture in crouching when under attack by Sparrowhawks meant that 86% of attacked Redshanks flew, only 4% crouching. For Peregrine attacks the higher risk of aerial capture meant that over 60% of Redshanks chose to crouch or dive.

In a separate study, Cresswell *et al.* (2010) looked at the influence of available cover on the feeding strategy of prey species. In the study the prey were Redshanks feeding on an estuary in Scotland. The predator was the Eurasian Sparrowhawk, but aspects of the results seem applicable to Peregrines which are also known to hunt Redshanks at the estuary. Cover in the context of the study was defined not as might be expected – that it allowed the prey to hide from the predator – but the reverse, that it allowed the predator to hide, and so more easily surprise the prey if it came close enough. The Sparrowhawks' chances of success fell exponentially with distance from the prey at the start of an attack, and the researchers found that there was a limit, about 30m, beyond which the Redshanks had adequate time to detect the presence of the predator and to make good their escape if attacked. Within the 30m band the Sparrowhawk's chances of success increased and the researchers found that only on days when the ambient temperature was below 5°C were the Redshanks feeding within the 30m limit, as the risk of starvation then increased to the point where the predation risk was in balance: on colder days the prey fed closer to the cover. In a related study, Cresswell and Quinn (2013) looked at the behaviour of Redshanks when they were faced with two predators, again at an estuary in Scotland. To avoid attack by Sparrowhawks, as noted above, the Redshanks moved further away from the cover that allowed the hawks to make surprise (ambush) attacks. But as the Redshanks moved further from the shore, the risk of capture by a Peregrine increased: the Peregrine's speed meant they were better suited to the open country of tidal flats. The effect of the two predators was to create a narrow 'safe zone', as illustrated in Fig. 4.27.

Diet

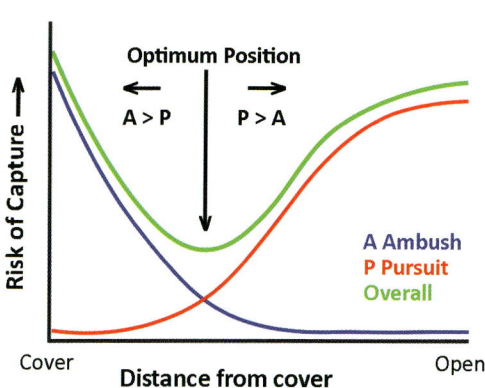

Figure 4.27 Risk of capture as a function of distance from cover for Redshanks under potential attack from an ambush (Sparrowhawk) or pursuit (Peregrine) predator. Redrawn from Cresswell and Quinn (2013).

One additional anti-predation strategy that potential prey may use has been identified by Oro (1996) who studied Peregrines (*F.p. brookei*) attacking Audouin's Gulls (*Larus audouinii*) at a colony on the Ebro Delta, on the Mediterranean coast of Spain. Oro found that predation rates were lower for gulls nesting in small, dense sub-colonies rather than in larger, looser colonies. Hunt (2012) also noted an unconventional defence mechanism from a Greenshank (*Tringa nebularia*) when it was attacked by a Peregrine, the shorebird landing close to a Mute Swan (*Cygnus olor*). The swan took no active part in the wader's defence, but its bulk limited the Peregrine's attacking options and eventually the falcon gave up.

Another fascinating aspect of Peregrine hunting strategy and prey defence was investigated by Beauchamp (2016a) who wondered whether predators can manipulate fear in their prey to their advantage. Beauchamp studied Peregrine attacks on roosting colonies of migrating Semi-palmated Sandpipers in the Bay of Fundy, between Nova Scotia and New Brunswick in western Canada. Theory suggests that the falcons should not attack roosts too early as this might lead the sandpipers to abandon the roost and seek an alternative. But the falcons cannot wait too long to attack in case the sandpipers have already left. For the sandpipers maximising responsiveness at all times reduces rest, so it is better to be responsive early and then relax so as to gain sufficient rest. Beauchamp found that the falcons did indeed wait initially, and then hunted across a spectrum of times to maximise uncertainty in the sandpipers, ensuring that either unrested birds or relaxed birds were available to attack, and so maximising hunt success. For the sandpipers, being very vigilant early before the light faded was the best option. As Beauchamp notes, '*falcons and sandpipers are engaged in a war of attrition.*'

In a separate paper Beauchamp (2016b) also investigated the effect of glare from the setting sun on sandpiper behaviour and Peregrine hunting success. Not surprisingly, as the glare increased, the sandpipers became sparser in comparison to numbers on cloudy days, those that remained pecked less often and so were less successful at foraging: these birds became increasingly skittish.

The Peregrines, of course, took full advantage of the conditions, attacking out of the setting sun.

Hunting Success
There are also numerous reports of Peregrines chasing or stooping at prey with little or no real intention of catching it, apparently either as a way of sharpening skills or in the manner of cats toying with mice. While these observations are well-attested and made by serious researchers (see below), they are viewed with scepticism by some who consider such behaviour the preserve of 'higher' animals and note that there are also well-attested instances of Peregrines killing numerous prey and caching those they cannot immediately eat.

Treleaven (1980) studying Peregrines in Cornwall, England, claimed a success rate of 52% – 30 kills from 58 attacks. However, Treleaven differentiated between what he termed High and Low intensity attacks, the former being true attacks, the latter 'to keep the Peregrine in 'good practice' for High Intensity hunting'. Of 55 attacks in which Treleaven could identify either High or Low intensity, 45 were considered High, these having a success rate of 69%. The success rate for Low Intensity hunts was 30%. A seasoned Peregrine observer can differentiate between the two intensities: essentially it is a question of the perceived urgency of a Peregrine's flight – flapping intensity and cadence – coupled with level of persistence.

Dekker (1980) also noted the differences in hunting and defined a 'half-hearted' or 'warming-up period ' hunting phase which lasted an extended period, followed by a short burst of serious hunting. Dekker concluded that while it was difficult to define the end of one and the start of the other, the success rate of serious hunting would be higher than the values quoted in Table 4.2 above though Dekker does not set down a value.

It seems, therefore, that taking the various studies as a whole some Peregrines at least do indeed have an initial period where attacks are not 'real' but allow the falcon to prepare itself for true attacks, and that if only the latter are considered then high (say 30-40%) success rates are achieved.

Another, related, problem in judging hunting success is that a single flight might include several attacks. As an example, Buchanan *et al.* (1986), studying Peregrines hunting Dunlin during winter in western Washington State, USA, noted that the success of hunting flights was 47%, but that of individual attacks was 14.6%. Of the attacks, 83% were directed against flocks (70% stoops plus 13% horizontal attacks), with 17% chases of individual Dunlin.

The problem of 'dummy' attacks, whether real or not, makes assessing the hunting efficiency of Peregrines difficult: as Ratcliffe (1993) notes, assumed figures vary from 100% success (an assumption of those who view the Peregrine as the ultimate hunter rather than a number based on observation) to much lower figures – Ratcliffe quotes the renowned Swedish ornithologist Gustaf Rudebeck as a source. Rudebeck (1951) noted 19 successes from 252 hunts, a success rate of 7.5%, comparable to the data of Table 4.2. But higher

Dunlin, a favourtite prey of northern Peregrines.

percentages have been quoted, seemingly from well-observed situations. Cade (1982) quoted values for a particular Peregrine (a falconry bird named Red Baron) whose hunts could be very well observed: in 1978 the falcon was successful in 73% of 81 hunts, in 1979 in 93% of 102 hunts, and in one sequence of 68 hunts in 44 days of that year achieved 100%. However, hunts by falconry birds are artificial and so will overestimate the success rate. Falconry birds are trained to 'wait on' above the falconer and, probably, his dog, watching for the prey to be flushed, so they start with a considerable advantage over wild birds.

Watching wild falcons in southern Quebec over two breeding seasons, Bird and Aubry (1982) saw 197 hunting attempts by a male, of which 69 were successful, a success rate of 35%. Lindberg (1983) observed the same success rate (35%) in southern Sweden, and noted that the rate varied with prey species, being 55% against Starlings and thrushes, but lower against heavier prey – 32% on gulls, 26% on pigeons: as noted above, early in *Composition of Diet*, Lindberg (1983) considers the lower rate is compensated by the greater food resource obtained from larger prey.

The success rates quoted above seem very high in comparison to those of other seasoned researchers, but similarly high rates were found for Peregrines hunting various species in western Japan (Yamada, 2001). In over 800 observations, divided into classes according to terrain, Yamada found success rates of around 45-52%, contradicting many of those quoted above. Interestingly, Yamada noted that the Peregrines were most successful when hunting over open water, which again contradicts some of the observations noted above. One possibility, in terms of success rate, is that Yamada was not differentiating between 'true' and 'dummy' attacks.

The success rate of 14.6% observed by Buchanan *et al.* (1986), and mentioned above, is higher than, but comparable to, that seen for wintering Peregrines on the Firth of Forth in Scotland's East Lothian by Cresswell and Whitfield (1994), though the sample size of the former was smaller. Cresswell and Whitfield noted a Peregrine overall success rate of 6.8%, which was lower than the rates of Eurasian Sparrowhawks (11.6%) and Merlins (8.8%)[9]. In a

---

[9] Cresswell and Whitfield (1994) noted the success rate against individual species. It was 0% against Curlew (22 attacks), Grey Plover (*Pluvialis squatarola*) (7 attacks), Lapwing and Oystercatcher (single attacks on each), 12.0% against Dunlin, 12.1% against Redshank, and 50% against Turnstone (one success in two attacks).

later winter study in the same area, Cresswell (1996) noted success rates of 9.5% against Redshank and 7.6% against Dunlin: there were also no successes in 21 attacks against Skylark. Cresswell noted that the Eurasian Sparrowhawk success rate was higher against both Redshank and Dunlin, while the Merlin success rate was also higher for Dunlin.

In his study of migrating falcons in Alberta, Canada, mentioned above, Dekker (1980) states that the overall average success rate for adult Peregrines was 9.8%, that for autumn immatures (*i.e.* birds hatched that breeding season) it was 2.4%, and that for spring immatures (*i.e.* birds hatched the previous breeding season that had survived the winter) was 7.1%. In a later study, Dekker (1998) in British Columbia, Canada found that the success rate of Merlins was higher than that of Peregrines, particularly for small passerines (Merlin 12.2%, Peregrine 3.8%). For waders the Peregrine's success rate improved in comparison to the smaller falcon (Merlin 12.6%, Peregrine 8.8%). In a later study in the same area, Dekker (2003) observed higher success rates, 9.1% when the falcon made open attacks on large flocks of Dunlin, and 23.6% when stealth attacks were made on the flocks. These higher rates are more consistent with the study of Buchanan (1996) in coastal Washington State, USA, where success rates were 12.5% for Peregrines and 7.8% for Merlins. Arising from these studies were conclusions that were not unexpected: adult Peregrines are better hunters than juveniles, and the juveniles of prey species are more vulnerable than the adults. In further work (Dekker and Drever, 2016) success rates for Peregrines hunting Dunlin were reassessed for the period 1994-2015. 1369 attacks were observed at Boundary Bay, south-west British Columbia, Canada during October-February. The attack success rate rose from 10.8% in October to 15.0% in the winter months.

The data of Cresswell and Whitfield (1994) also noted the extreme mortality of certain prey species in studies in the UK: Redshank mortality in three consecutive winters (1989/90, 1990/91 and 1991/92) was 31.1%, 48.5% and 57.3%, that of Snipe was 28.6%, 33.3% and 25.0%, with the attrition rate of juveniles considerably higher than that of adults (Fig. 4.28). Capture by the three raptor species hunting in the area (Eurasian Sparrowhawk: attack success rate of 11.6%), Merlin (8.8%) and Peregrine (6.8%)) was not the only problem for the wader prey species as all three raptors had prey pirated by Carrion Crows (*Corvus corone*). Wader mortality rates much higher than those observed would have implications for the long-term survival of the species (at a local level), so the success rates of the three predators involved could not rise significantly before the wader population crashed.

It is also not only direct predation which affects the prey species population. In a study in Ireland's Dublin Bay, Quinn (1997) noted that although Oystercatchers were less prone to predation than the smaller waders that shared the same habitat, the time they spent foraging (as a result of taking flight when a predator was spotted, then taking time to resettle) and the efficiency of that foraging (more time spent watching the sky and a denser aggregation of birds

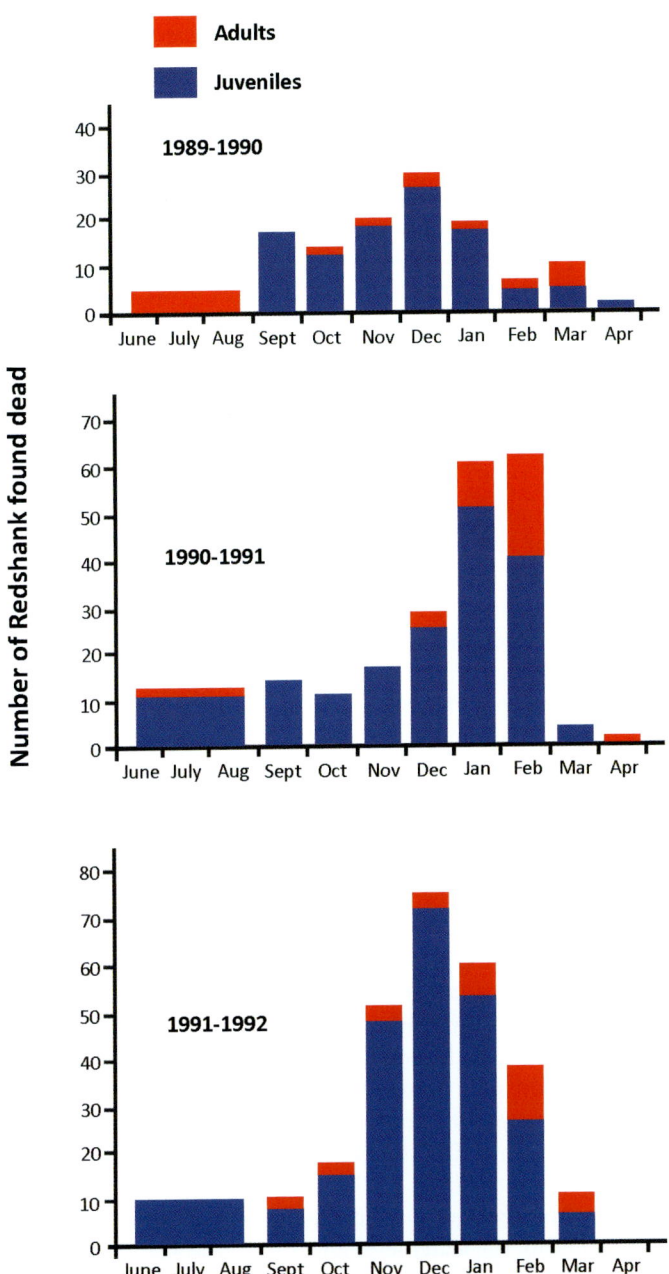

Figure 4.28 Number of adult and juvenile Redshanks found dead during successive winters at the Tyninghame estuary, south-east Scotland. Redrawn from Cresswell and Whitfield (1994).

reducing forage area) were adversely affected. This is likely to be the case for the more vulnerable species as well, and for all prey species could result in a reduction of body condition.

However, there is evidence that Peregrines can have positive, as well as negative effects, on prey species numbers. In a study on an island off USA's Washington State, Paine *et al.* (1990) found that the arrival of Peregrines on the island had resulted in a reduction in numbers of the primary prey species (Cassin's Auklet and Rhinoceros Auklet (*Cerorhinca monocerata*)), but an increase in population of other species. The increase was a result of the Peregrines also feeding on Northwestern Crows (*Corvus brachyrhynchos*) which were a significant predator of the auks, and other seabird, nests. However, there are significant differences between the two situations: in the US the Peregrines were present during the breeding season and so their attacks on crows, either killing them or curtailing their predatory behaviour, were indirectly benefiting other species, whereas the winter predation observed by Cresswell and Whitfield (1994) in the UK was a direct threat to population numbers. That said, the study of Paine *et al.* (1990) does imply, yet again, that the interaction of predator and prey is rarely as straightforward as it seems.

In his study of the hunting success of migrating Peregrines in central Alberta, Canada, Dekker (1980), mentioned above (Table 4.2) noted that the autumnal success rate for juvenile Peregrines was only 2.4% (in 42 observed hunts), consistent with the known high winter mortality of young raptors, inexperience and consequent poor hunting performance leading to starvation. In the spring migration the success rate of adult falcons was 9.8% and that of first-year falcons 7.1%, not as good, but comparable and adequate for survival. In a later study in winter in British Columbia, Canada, Dekker (1998) noted a success rate of Peregrines hunting Dunlins of 9.3% (as noted above, the overall success rate for waders was 8.8%, the falcons having better success with Dunlin).

A recent study of wintering Peregrines (*F.p. anatum* and juvenile *F.p. tundrius*) hunting shorebirds in western Mexico (Basso *et al.*, 2021) has produced interesting data, both on the success rate of such attacks with respect to flock size, and also on the number of attacks relative to the height of the tide. The study was at a semi-intensive shrimp farm in a wetland area close to a long, sinuous Pacific coastal lagoon. Shorebirds (several American wader species and terns, the main prey being Western Sandpiper (*Calidris mauri*), Black-necked Stilt (*Himantopus mexicanus*) and Willet) fed on tidal mudflats, the researchers studying the hunting close to the shrimp farm, which was well-removed from the lagoon outlet (Fig. 4.29a).

Fig. 4.29a indicates that Peregrine attacks were strongly influenced by the tides, both spring and neap, the water level affecting the availability of feeding areas for the prey and, consequently, the availability of targets for the falcons. For an individual bird, what is important is its chances of being attacked and killed by a predator. Basso *et al.* also noted the likelihood of capture by a Peregrine as a function of both tide height and flock size (Fig. 4.29b).

Diet

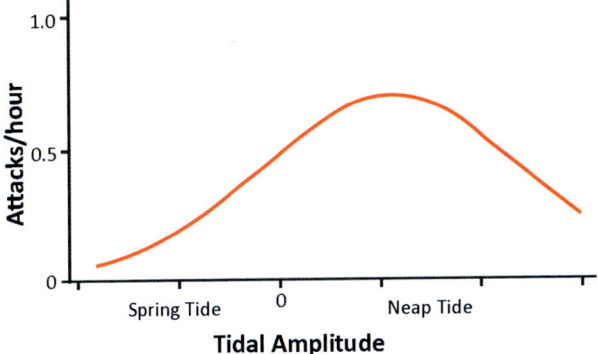

Figure 4.29a. Frequency of Peregrine attacks on shorebirds at a shrimp farm in western Mexico as a function of tide height at the nearby estuary. '0' represents mean low water at the estuary. With high spring tides there was less available foraging area for the shorebirds at the farm, so there were fewer birds and attacks decreased. With neap tides there was a much larger foraging area locally for the birds so there were fewer at the farm and attacks again decreased. The optimum time, for the Peregrines, occurred when mudflat exposure at the farm was at a maximum and the shorebird population was correspondingly high. Redrawn from Basso *et al.* (2021).

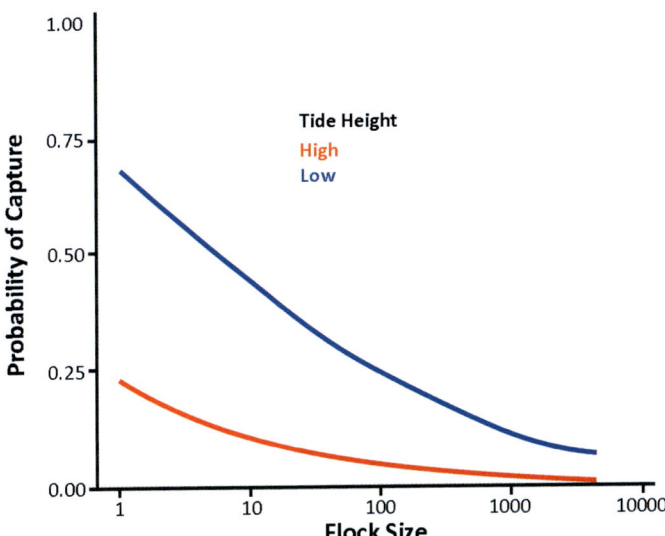

Figure 4.29b. Probability of capture of a Peregrine attacking prey flocks at low and high tide. Redrawn from Basso *et al.* (2021).

Cresswell (1994) also considered the effectiveness of flocking as an anti-predator strategy. Studying the predation of Redshanks overwintering at an estuary in Scotland by Peregrines and Eurasian Sparrowhawks noted that flocking significantly reduced the risk to an individual bird (Fig. 4.30).

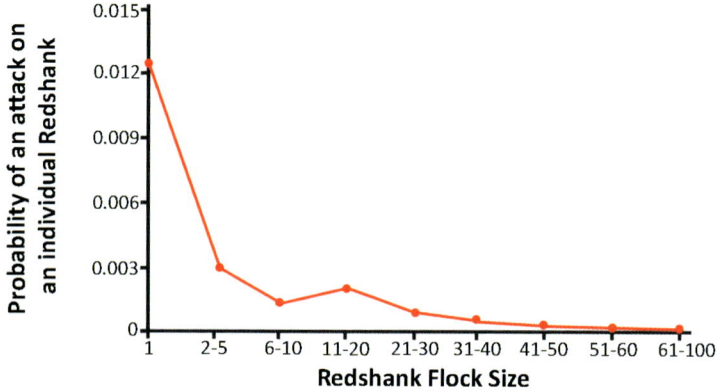

Figure 4.30. The probability that an individual Redshank in a flock of particular size will be attacked by a Peregrine, relative to the occurrence of that flock size. Each red dot represents the probability of attack x the proportion of total flocks x (1/the mid-point of each flock size). Redrawn from Cresswell (1994).

At first glance there is a significant difference between the two findings, but it is worth noting that the work of Basso *et al.* (2021) involved a limited number of attacks and captures, so that the only conclusion to be drawn from Fig. 4.29b is that a Peregrine has a much better chance of a kill if it attacks individual birds which are away from the flock, which is consistent between the two studies. Overall, Basso *et al.* found that the mean success of a Peregrine attack was 14%, which is consistent with other studies.

To augment this consideration of whether success rises or falls if larger flocks are attacked, it is worth noting that in a study of Swainson's Hawks (*Buteo swainsoni*) hunting Brazilian Free-tailed Bats in New Mexico, USA (Brighton *et al.*, 2021), the probability of success for the hawk was the same if it hunted a lone bat, away from the swarm, or attacked into the swarm, but that lone bats represented 5% of the catch, but only 0.2% of the population. The bats in the swarm therefore had a risk 25x lower than the lone bats. This again implies that flocking represents a significantly reduced risk for the individual within the flock, though the success rate for the predator is not necessarily reduced by anything approaching the same fraction. In later work, Brighton *et al.* (2022)found that when the hawks attacked the bat flock they were not fixing on an individual, but moving relative to a fixed position in space so that any bat which is on a collision course will be on a constatnt bearing and capture will occur naturally. This technique, reducing the demands on the bird's sensory system, is also likely to be used by Peregrines attacking flocks.

This *F.p. anatum*, in California has just caught a pigeon which appears to be wearing an executioner's black hood. *Gabriel Gruia*.

While it is not easy to unpick the various quoted success rates with any confidence, particularly in view of the potential for misreading some Peregrine attacks, it seems likely that Peregrines are more efficient when making stooping attacks on their primary (breeding-season) prey, and rather less successful when forced to select non-optimal prey, with probable success rates varying between *c*.10% for the latter to, perhaps, c.20% for the former, consistent with the observations of Parker (1979) in Wales, who noted success rates of 15% (female) and 17% (male) when the Peregrines were hunting pigeons.

Whichever hunting technique is employed, as noted in Chapter 1 Peregrines dispatch prey with a bite at the base of neck to break the spinal cord. If the prey is captured in the stoop the bite may be administered in the air. The Peregrine will usually bite the neck base of prey that has been killed by a stoop and knocked to the ground as well, suggesting that the bite is a reflex action when prey is clutched. Following the fatal bite, the prey is then taken to a perch for plucking and eating, the latter usually starting with the head, which is often consumed whole (sometimes the brain is picked out), then the breast muscles. The heart and liver are usually eaten, but the digestive tract is normally discarded. Though most food is eaten at perches there are records, few in number, of Peregrines eating on the wing. Cade (reported by Ratcliffe, 1993) watched a male Peregrine kill and consume a total of seven bats on the wing during a continuous 20-minute hunt over the USA's Grand Canyon. Sprunt (1951) also records a Peregrine consuming bats on the wing, reaching forward with its feet, which clutched the bat, and reached down with its head, to quickly take bites while soaring. However, Stager (1941), who also watched Peregrines feeding on bats, noted that the falcons flew to nearby trees to feed, then returned to catch more. The difference could be the sheer number of bats involved in the observation of Stager: he watched six Peregrines feed on bats for an hour, suggesting that the bats represented an almost limitless supply of prey so that time was available to land, feed unhurriedly and return to the hunt.

BOX 4.1 High intensity hunting (SW, after R.B. (Dick) Treleaven)

It's 08.45 at Symonds Yat viewpoint on June 4 2011, sunny and warm to 15°C with a southerly light wind, 10% cloud and superb visibility. The four eyasses are around 25 days old so can thermoregulate, enabling the female to leave the eyrie unattended and join the male in hunting.

9.15 Attack 1 The tiercel has been 'waiting on' for a full half hour, soaring in wide circles about 1500m away towards Coppett Hill to the north. But now his demeanour changes and a casual drift morphs into determined purposeful flight, flapping hard now, and on a straight line – '*he's onto something!*' – I hear myself shouting. Accelerating hard, wings like pistons hammering away with determined intent, he's angling down, now wings fold and he morphs into a guided missile - '*Teardrop stoop – he's stooping hard, boy look at him go*' - I can't help shouting, oblivious to all around me.

Gaining speed faster than one would believe possible he's flattening out from 45° and it's hard to keep him in focus, a long slightly downward horizontal stoop ensues with occasional lightning wing flicking flight correction, he closes in – '*Single pigeon – right in front of him!*' - a reflex shout from me - he's on his prey and then bang - two birds merge into one and a puff of white feathers drift on the wind. The guided missile morphs back into a struggling, deep flapping laden tiercel, turns into the wind and heads back to the eyrie and his four young. This kill was taken to the rock face and cached. Normally it would be plumed immediately and fed to the fast-growing eyasses, but today is different. Today hunting conditions are perfect with unusually high visibility.

9.35 Attack 2 The pair are again waiting on, high up. Now the falcon uncoils from her circular flight and heads north on a line for a full 60 seconds. A long very shallow stoop about 10° below horizontal from $c.$450m and she's immediately behind a flock of $c.$20 pigeons flying northwards. But one of them sees her coming, and they immediately scatter in all directions jinking away, the main flock disintegrating into three or four small groups. No attempt is made at tail chasing, which is rarely successful once the element of surprise is gone, and our falcon 'throws up' to regain her high station.

10.05 Attack 3 '*The tiercel's off again*' I shout to no-one in particular. '*Look at that anchor shape scything through the air!*' Flapping hard to gain attack position, pumping wings take him higher – his urgent, rangy, elastic, flapping sharp pointed wings shout 'Peregrine on a mission'.

He's climbing hard and heading due south out of sight of the viewpoint. About a minute later a large flock of pigeons flash through and above the flagpole on the 'Rock' viewpoint at a height of $c.$10-15m directly overhead. The sound is like a passing train thundering through a station, flying hard and fast but not dispersed, they maintain formation. Now the tiercel shoots through horizontally at incredible speed - little more than a blur - and roars out over our heads, wind rushing through his wings, he's twice as fast as his quarry and to see this at such close quarters is unimaginably evocative. '*Oh **** he's got it!*' I hear myself shouting. He's hit a pigeon on the outside left of the flock some 50 m from us on the 'Rock',' binding to' it, he carries the hapless prey about 10 m then unaccountably drops it. The lucky pigeon flies directly into the tree canopy, as if its life depends on it. Which of course it does, but there are Goshawks in those deep dark woods.......

No attempt is made to retrieve the pigeon and the tiercel immediately mounts into the blue sky to join the falcon slowly describing wide unpowered lazy circles in the still blue sky.

Observation: The tiercel was aware of this flock when they were probably a couple of miles away, given that pigeons can fly at around a mile per minute (60mph). He spotted them and started positioning himself for an attack around two minutes before the strike. He also flew high, directly towards, but much higher than, the incoming flock then turned 180° to attack them from behind. This might seem surprising, but it is not at all unusual (see Chapter 3 Vision*).*

10.10 Attack 4 The tiercel stoops hard on a flock, misses, and mounts up to wait on yet again. Both Peregrines have now been hunting continuously for 90mins. As if in irritation or redirected aggression the tiercel stoops hard and near vertical on a Buzzard, which flips upside down showing talons, then rights itself again only to find the tiercel powering down, wings spinning to renew the attack. Over the course of the next 20mins, another long stoop ends in failure (**Attack 5**). But now something remarkable happens.

10.30 Attack 6 A flock of *c.*15 pigeons flies hard and fast immediately above the viewpoint at an elevation of around 10m above the 'Rock'. '*What's happening here? It's almost as if they've been scattered by a...*'. I never did get to finish that sentence. Having missed his target, the tiercel 'throws up' vertically 10m above where I am standing and I'm dumbfounded. Hagar's 'ripping canvas' quote of old is made flesh and blood in this present moment and the sound of forcibly displaced air searing through the feathers of a fast-turning Peregrine coming out of a power-dive is quite extraordinary. Feathers flare and roar under extreme G forces as momentum launches him several hundred metres vertically at immense speed. I feel every bit as though I'm stood on a launchpad looking at a rocket taking off.

The tiercel renews his attack from his regained height and puts in a short sharp stoop on the scattered flock but to no avail and having lost the initiative he gives it up.

10.35 Attack 7, **10.40 Attack 8**, **10.50 Attack 9**. All three attacks have been unsuccessful long stoops from high positions, making nine observed hunts in two and a quarter hours, only one of which was successful. *'When will they next be successful, and can they keep up this remarkable high level of energy expenditure?'* I muse to myself.

11.00 Attack 10 *'She's off!'* The falcon has seen something we haven't and gains height impossibly quickly, ringing up in a tight, energy sapping, spiral flight under continuous lung bursting power, wings scything away with their characteristic deep whippy wingbeat. She's at 450m or so now, flattening out and heading due north. Still pumping her wings hard, she means business. Now a shallow stoop, wings flicking. Now a teardrop 45° stoop and she's accelerating hard...she's reached Coppett Hill and is about to disappear... *'Oh, pigeons ahead of her'...* *'Lost her'*. She's stooped the other side of Coppett, but was she successful this time?

The question is answered conclusively several minutes later as we pick her up again struggling, laden with prey in her talons, feathers cascading down the valley in a suddenly high wind. Drifting high over the viewpoint at maybe 200 m with quarry in her talons tucked up beneath her tail she finally coasts in with nary a wingbeat and alights on the rock face. I fully expect her to plume it and feed her young, but not a bit of it. She has other things on her mind.

11.25 Attack 11 The eyasses have not been fed since we arrived even though two successful captures have been made. Several minutes pass by and then *'What's this? – I think I've got a Peregrine with prey some 500 m up over Coppett Hill'* I exclaim. He flies in, slow and labouring, the hunt unseen, returning with another pigeon trussed up beneath his tail, caches it and then flies up once more to wait on with the falcon.

Observation: when waiting on, Peregrines will describe wide circles at a height, and a pair's orbit will often be different in location and elevation. Occasionally they will come together and often one bird will join the other in the lining up attack phase once prey has been spotted.

12.00 Attack 12 The tiercel flies in from the north with yet another kill. He's panting hard and stays put for five minutes or so. The eyasses are fed and both adults now 'switch off'.

15.50 Attack 13 They are 'waiting on' again at 15.00 and at 15.50 *'Big Stoop!'* He's stooping fast, steep and still quite high, around 400m I would say. Now he's flattening out, but still at electric pace he shoots forward and.... *'Single pigeon directly in front of him!'* ....*'He's on it'*, the tiercel is immediately behind his target, now he brakes sharp and hard, tail fans wide and talons shoot forward from below. In a blur the oblivious pigeon dangles, struggling, from the feet of its nemesis and feathers fly again over Coppett Hill.

Today I am left wondering how Peregrines are able to do what I have just witnessed. They combine an unparalleled adjustability of flight surfaces with an athleticism and stamina that is remarkable - the spell cast by Peregrines on the hunt always has, and always will, hold me in its thrall.

Notes on terminology

'Throw up' originates from falconry and describes an upwards vertical or near vertical flight with closed wings re-directing failed stoop momentum. Its purpose is to regain height as quickly as possible.

'Waiting on' also originates from falconry and describes a high soaring, often circular, flight with wings and tail spread, holding station without flapping. Its purpose is to enable continuous scanning of territory for prey using the high visual acuity provided by bifovial eyesight.

'Switched off' means that the Peregrine has no interest in hunting and often faces into the rock face with one foot tucked up into its feathers, apparently oblivious to everything.

The day's attack data are tabulated below.

Symonds Yat, England
Peregrine Falcon high intensity attack analysis
June 4 2011
Successful attacks in red *(minimum estimated times in italics)*

Hunt no.	Flight times			Flight analysis (seconds)						Red=capture	
	Start	Finish	Elapsed seconds	Waiting On	Approach	Stoop	Secondary Attack	Out of sight	Laden w/ prey	Attacker Male	Female
1	08.45	09.20	2100	1965	15	5			115	M	
2	09.20	09.35	900	830	65	5					F
3	09.35	10.08	1980	1882	90	8				M	
4	10.08	10.10	120	85	*30*	5				M	F
5	10.10	10.15	300	260	5	5	30			M	
6	10.15	10.32	1020	640	*60*	5	15	300		M	F
7	10.32	10.37	300	193	10	7	30	60		M	F
8	10.37	10.40	180	143	*30*	7				M	
9	10.40	10.55	900	845	10	5	40			M	F
10	11.00	11.07	420	120	*100*			200			F
11	11.15	11.29	840	540				120	180	M	F
12	11.32	12.05	1980	1560				300	120	M	F
13	15.00	15.55	3300	3000	8	5		180	107	M	
Total seconds			14340	12063	423	57	115	960	722	11	8
Total minutes			239	201	7	1	2	16	12		

Kleptoparasitism
Peregrines may also be the attacker and the victim in kleptoparasitic attacks. In his study at a coastal site in British Columbia, Canada, Dekker (2003) noted that Peregrines which were suffering such attacks from Bald Eagles during the day often flew inland at dusk, following ducks to their roosting sites. Although he did not witness hunting directly, the inference was that the Peregrines were hunting ducks in twilight to avoid piracy. In a later study at the same site, Dekker *et al.* (2012) noted that kleptoparasitism by both Bald Eagles and Gyrfalcons accounted for 36% of Peregrine Dunlin kills, and that the falcons had to increase their kill rate of the waders from 0.05/hour, when the pirating birds were largely absent, to 0.30/hour when they were present. Kleptoparasitism was therefore having a significant effect on the Dunlin population: as noted above, piracy by Carrion Crows has had a similar effect on Redshanks (Cresswell and Whitfield, 1994).

Dekker (1980) observed Peregrines being both the aggressor and victim of piracy in his migration-period studies in central Alberta, Canada. He saw adult Peregrines taking prey from juveniles, females pirating males, and Peregrines taking prey from Merlins, Northern Harriers (*Circus hudsonius*) and Sharp-shinned Hawks (*Accipiter striatus*): but he also saw Peregrines falling victim to attacks by larger raptors. In the UK, King (2009) saw an adult nominate Peregrine pirating prey from a Eurasian Hobby in Hampshire, while both Collar (2002) in Cambridgeshire, and Rees (2009) in Pembrokeshire, both in autumn, noted piracy of Merlins. In south-eastern England, Robinson (2018) saw a Peregrine dive between a pair of Marsh Harriers (*Circus aeruginosus*) to pirate prey the male harrier was dropping for his mate.

Zuberogoitia *et al.* (2002) also witnessed a Peregrine (*F.p. brookei*) apparently successfully pirating prey from a Carrion Crow in northern Spain, but also saw 'Mediterranean gulls' (presumably Yellow-legged Gulls) pirating pigeons from Peregrines. Piracy by Eurasian Buzzards of Barbary Falcons on Tenerife has also been recorded (Siverio *et al.*, 2007). One of us (SW) has observed a successful Peregrine piracy attack on a Red Kite, the kite being robbed of its mammalian prey.

Food Caching
As with the smaller falcons, Peregrines cache food, presumably for similar reasons, *e.g.* as a hedge against bad weather preventing hunting tomorrow, a strategy particularly relevant in winter. Ratcliffe (1993) notes that one UK (Cornish) nominate female Peregrine killed six pigeons during a six-hour period: one was lost, one was fed to the falcon's brood and four were cached. Male Peregrines will, of course, also cache food, particularly if the day's hunting has provided a surplus. Males in breeding pairs will also cache food for themselves: female Peregrines can aggressively take any food the male has, either for herself (during courtship) or for her brood, so the male may cache in the hope of being able to enjoy a meal later. We have seen a nominate urban

Female nominate Peregrine taking the remains of a prey away (a male Blackbird (*Turdus merula*)) for cache. The brood, in south-west England, was two young chicks. Having fed them, the female ate some herself, then took the remainder to a cleft at the far end of the nesting quarry: she was back with her brood within a few minutes.

male cache food on a building a short distance away from that on which the nesting box was situated: in this case the cache was not far enough away, as the female either saw, or was aware of, the site from courtship, and robbed the male of his meal. In their study of *F.p. macropus* in New South Wales, Australia Cameron and Olsen (1993) considered that there were several reasons for food caching during the breeding season. Firstly, it minimises wastage and so reduces the time spent hunting; it acts as a buffer against temporary prey shortages due to bad weather or, perhaps, the need for the male to rest; it may allow the male to limit his hunting if, for instance, he finds a nest with vulnerable young and so can take all of them over a short time; and finally, it separates hunting from chick feeding. Although in the case of most males this latter reason may be minimal as females tend to do the majority of chick feeding, it allows food deliveries to be spread evenly through the day, and by caching food the male can hunt at his most efficient weight, and then use cached food to eat later in the day. Figs. 4.31 and 4.32 illustrate the use of cached food in the feeding of nestlings.

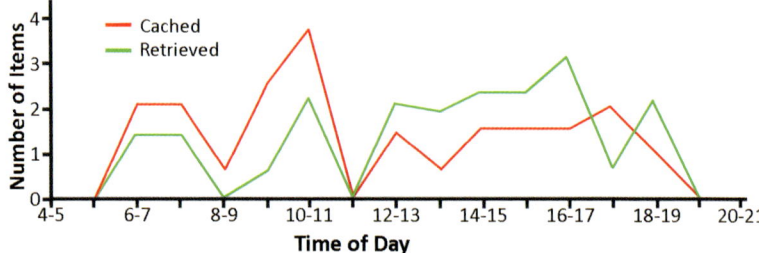

Figure 4.31. Number of items retrieved and cached by a pair of breeding Australian Peregrines at times during the day, averaged over 200 hours of observation, between 1 October and 5 November. Observations covered 13 hours on study days, with study days being every 2-3 days. Redrawn from Cameron and Olsen (1993).

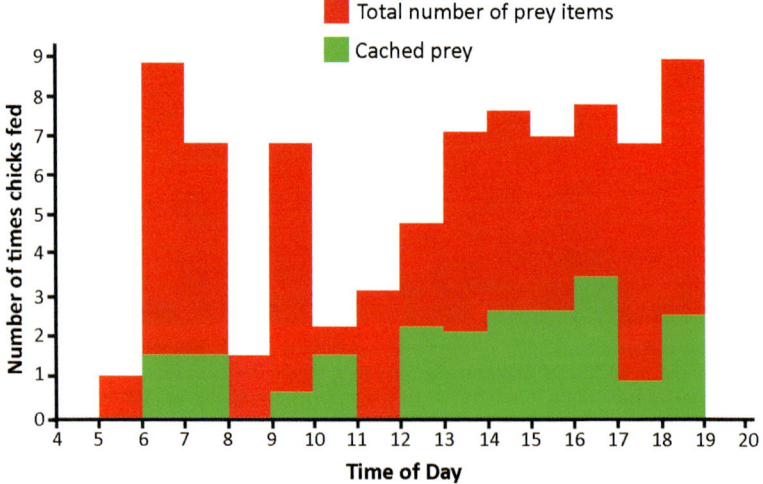

Figure 4.32. Number of times nestlings were fed per 10 hours by a pair of breeding Australian Peregrines. Redrawn from Cameron and Olsen (1993).

One additional reason for caching might be that as some prey species are too large to consume in one feeding session, they removed it to a suitable cache for further consumption at a later time. Cache sites are usually in fissures in cliffs. In his study of Peregrines on Canada's Langara Island, Nelson (1970) noted that the falcons frequently cached food. However, in this case, the Peregrines were feeding exclusively on Ancient Murrelets, which leave their nesting burrows before dawn and return after sunset. The Peregrines therefore had limited opportunities to hunt, these accentuated by the local weather, which was frequently bad, and so, not unreasonably, took every bird they could catch, caching to continue to hunt while prey remained available.

Diet

Above *F.p. peregrinator* with prey, Kannur Kattampally, Kerula, south-west India. *Afsar Nayakkan*.

Below Two *F.p. anatum* Peregrines bemused by a live Starling, San Pedro, Los Angeles, California, USA. *Gabriel Gruia*.

5   Food consumption and energy balance

For a species which is both so widespread and so iconic, surprisingly little research has been carried out on the energetics of Peregrines, and even estimates of the daily energy input are based to a large extent on data from falconers on captive birds. While the latter are obviously useful, they can only act as a pointer to the energy requirements of wild birds which do not have the benefit of being fed rather than having to hunt, and of being housed in aviaries that may also be warm and wind-free. Ratcliffe (1993) sets down data gathered in the 1950s and 1960s, noting that three captive male Peregrines in the USA (mean body weight 683g) averaged a food intake of 104g/day in autumn and winter (maximum intake 147g/day) and a single male weighing 721g having a mean intake of 83g/day (maximum 120g/day). The average intake was, therefore, 11.5% of body weight in summer and 15% in autumn and winter. Ratcliffe also quotes food intake data for nominate captive falcons of 141g (female) and 113g (male)[1].

---

[1] The data in this paragraph are set down in Ratcliffe (1993) p136, but are taken verbatim from the first edition of Ratcliffe's book (p145), published in 1980.

Food Consumption and Energy Balance

Much more comprehensive data became available in the 1980s. Lindberg (1983) recorded the Gross Energy Input (GEI) of two imprinted captive Peregrines kept in outside aviaries at Kalmar, in southern Sweden (at 57°N, close to the Baltic Sea). As they were imprints it was possible to weigh the birds regularly, and very accurately, on a triple beam balance. The two falcons were weighed every day with an empty crop before being fed daily throughout the study. Moulting was closely watched, and the falcons were not flown as falconry birds at any time. Ambient temperatures were routinely taken. Lindberg calculated GEI by assuming an energy content of the food fed to the birds of 2kcal/g wet weight: this value was converted to 8.37kJ/g wet weight for this comparison with other data – see below. For the male Peregrine Lindberg's figure equates to a gross energy intake (GEI) of 678–770kJ/day. Fig. 5.1 shows the variation of the male Peregrine's weight and GEI over the two-year study period. The female Peregrine was also weighed, as was her food intake, but these data were only taken daily for the months July-December 1976 and January-April 1977 as the falcon was breeding. Lindberg's figures equate to a gross energy intake (GEI) of 911–936kJ/day.

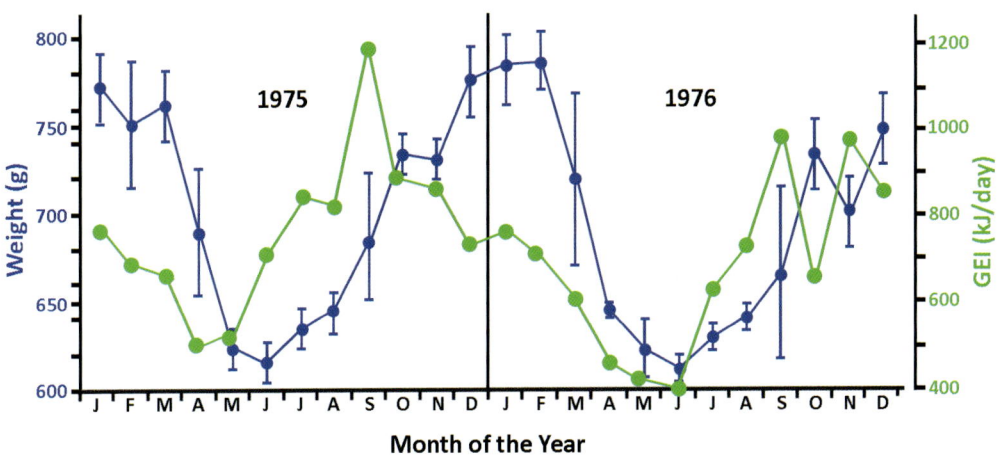

Figure 5.1 Weight and Gross Energy Intake of a male Peregrine throughout a two-year period. The bird was weighed each day before feeding, monthly datapoints being the mean weight during the month with the bars on each datapoint being 1 standard deviation (SD). Note that the SD for April 1976 was too small to be visible at the graph's scale. GEI is plotted as a point only. Each point has a large SD (comparable to, but often larger than, those associated with weight): these have been omitted for clarity of the overall figure. Drawn from data tabulated in Lindberg (1983).

Fig. 5.2 shows the variation of the male Peregrine's weight and GEI over the two-year study period.

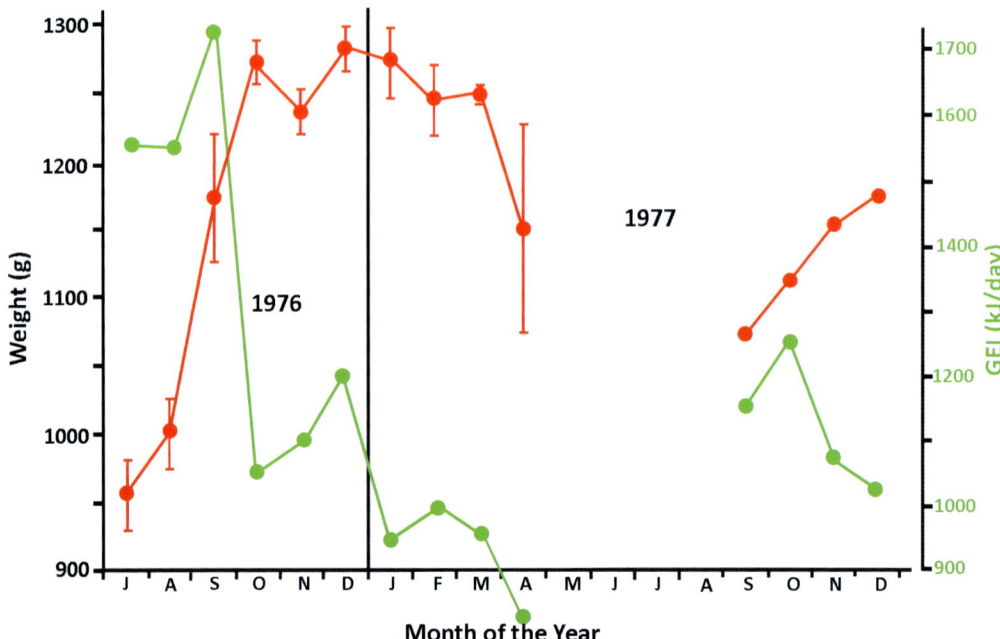

Figure 5.2 Weight and Gross Energy Intake of a female Peregrine during 1976 and 1977. In 1976 the female bred, and was laying/incubating during May-July 1977, at which times no weights were taken. During September-December weights were taken only once or twice. As in Fig. 5.1, the standard deviation on GEI points (which again were large) have been omitted for clarity. Drawn from data tabulated in Lindberg (1983).

Over the period of observation of the two falcons the mean food intake per day was 81-92g/day fresh weight for the male Peregrine, and 109-112g/day fresh weight for the female. These numbers compare reasonably well with the work of Barton and Houston (1993) who studied the dietary requirements of Peregrines (and several other captive raptor species) which were fed day-old chicks (without yolk sacs or intestines) at 20°C. When the male Peregrines were fed chicks at 0°C they required 143g/day. The difference was due, in part, to a drop in digestive efficiency (*i.e.* the effectiveness of extracting energy from consumed food), but would also be influenced by heat losses. In a separate report, Barton and Houston (1994), the researchers also found that there was a correlation between intestinal length and digestive efficiency. In general, raptors feeding on birds have short intestines and, therefore, poorer digestive efficiency. This work has already been mentioned in Chapter 1 and will be mentioned again towards the end of this Chapter.

Lindberg (1983) also calculated the Basal Metabolic Rate (BMR) for each of the male and female Peregrines from an equation of Aschoff and Pohl (1970):

$$\text{BMR (kJ/day)} = 361.5 W^{0.734} \quad (5.1)$$

where W is the weight in kg.

(Note that this equation has been converted from kcal (in the original reference and in Lindberg's 1983 work) to kJ.)

The calculated BMRs for the male and female Peregrine are given in Fig. 5.3.

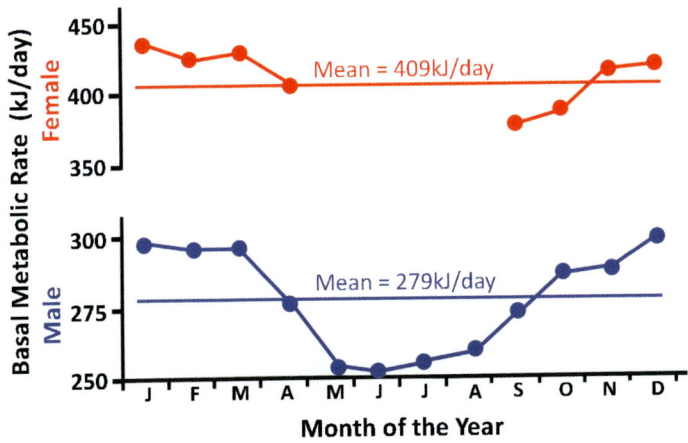

Figure 5.3 Variation of the Basal Metabolic Rate (BMR) of male and female Peregrines through one calendar year. The female bred that year, BMR data not being collected during the incubation chick/rearing period. Drawn from data tabulated in Lindberg (1983).

In a classic series of experiments on the Common Kestrel, a research team at the University of Groningen in The Netherlands investigated various aspects of the biology of the smaller falcons. One project (Masman *et al*, 1988) involved the calculation of the Daily Energy Expenditure (DEE) for the Kestrels. As the Dutch Kestrels were being flown regularly the calculated DEE included activity. The Dutch researchers estimated the DEE of their captive Kestrels using several different techniques[2]. For the purpose of a comparison with the

---
[2] The other techniques were metabolised energy intake and the use of doubly-labelled water (Masman *et al.* (1988).

work of Lindberg (1983) we have chosen a methodology based on an energy-time (Et) budget:

$$Et \text{ (kJ/day)} = B + T + A + H + S \quad (5.2)$$

where:
B is a basal component, the energy expenditure under fasting, thermoneutral conditions,
T is the energy expenditure for thermoregulation,
A is the energy expenditure for activity,
H is the energy expenditure for feeding,
S is the energy expenditure for tissue synthesis.

Lindberg's birds were kept in an outdoor aviary and so expended all these energy components apart from 'A'. Masman *et al.* (1988) calculated the values of all five components of Equation 5.2. For male Kestrels the average total energy expenditure for a year, excluding A, was 74,842kJ: for female Kestrels the value was 77,745kJ. Equation 5.2 excluding A corresponds to Lindberg's definition of BMR. For the male Peregrine Lindberg calculated the mean daily BMR in 1975 as 66.6±4.3kcal/day and a daily mean in 1976 of 65.8±4.4kcal/day. Converting to kJ/year yields 101,709±6,567 for 1975 and 100,487±6,720 for 1976. Given the size differential between the two falcon species these values are in extremely good agreement. For the Kestrel the dominant energy expenditures were the basal component and thermoregulation. These are also likely to be dominant in the Peregrine, the

For the first days and weeks the time budget of this nominate chick, hatched in Scotland, will be dominated by the conversion of prey to energy for growth.

difference in size and weather conditions in the two test centres probably accounting for the bulk of the differences. Because no values for BMR were calculated during the four months of breeding no similar comparison was possible for the female Peregrine.

Lindberg's BMR data for the male Peregrine showed a decline in summer, but the variation through the year was small, from a minimum of about 500kJ/day in June to about 600kJ/day in January, the difference being the energy requirement for thermoregulation which was obviously greater in winter. The calculation for the captive female was more difficult because of breeding which meant that for four months the falcon could not be weighed on a daily basis. However, the BMR showed a similar decline in summer (about 750kJ/day in April and September) with respect to winter (about 860kJ/day in January).

GEI is directly related to daily energy expenditure (DEE) by the falcon's metabolic efficiency, which Lindberg assumed lay between 70% and 80% (*i.e.* DEE = 0.7xGEI-0.8xGEI). The ratio of DEE to BMR varied from about 2 in winter, when heat losses were significant, to around 1.5 in summer, and then to about 3 in late summer/autumn when the falcon was moulting: the same figures applied to both male and female.

In the absence of feeding duties, the male's GEI (and, therefore, DEE) does not show the expected sharp increase during the breeding season, the peak being displaced to autumn when the captive Peregrine moulted. Instead, the variation in weight follows a relatively smooth U-shaped curve, dominated at both ends of the year by the need to increase weight to counteract the effect of dropping temperatures (Fig. 5.1). By contrast, in their study of Kestrels the Groningen team were able to assess the energy intake and expenditure of free-living falcons during the year. Fig. 5.4 shows that during the breeding period the energy expended by a male Kestrel in initially feeding his mate (in order to minimise her energy output in hunting and so prepare her for egg production) and then in provisioning the brood exceeded his energy input.

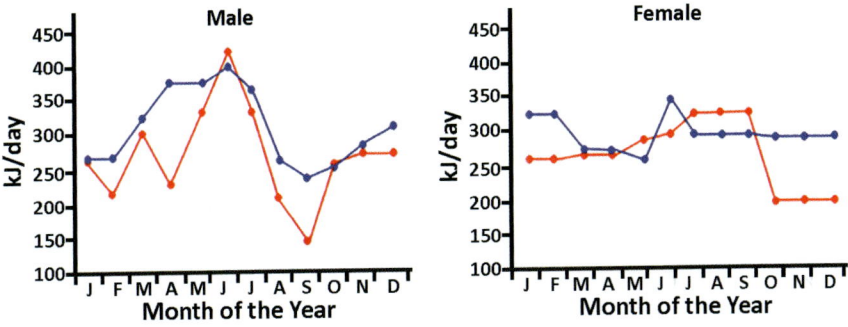

Figure 5.4 Variation of daily energy intake (in red) and daily energy expenditure (in blue) of male and female breeding Kestrels. Drawn from data in Masman *et al.* (1988).

Being unable to compensate for the energy loss because of the limited hours available for hunting, the male lost weight (Fig. 5.5). For the female Kestrel the situation was different, the bird initially gaining weight during egg production, then losing it later when she joined her mate in hunting for the growing brood, then losing weight significantly during the moult (Fig. 5.5).

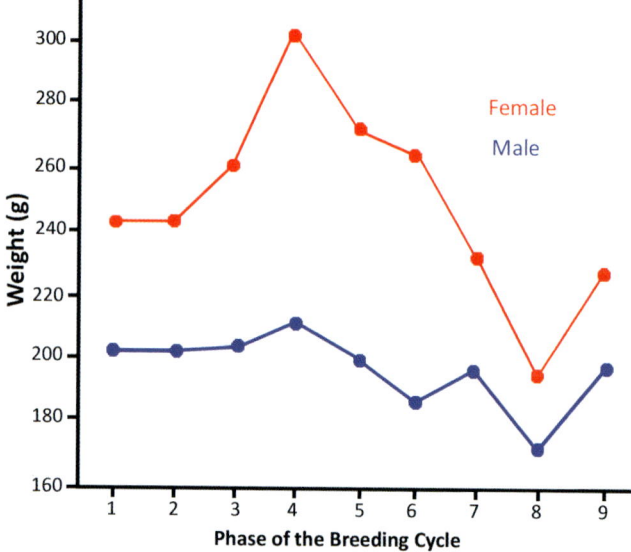

Figure 5.5 Variation in body weight of male and female breeding Kestrels by phase of the annual cycle.

The phases are:
1 Wintering unpaired	4 Egg Laying	7 Nestlings >10 days old
2 Wintering paired	5 Incubation	8 Dependent juveniles
3 Courtship	6 Nestlings <10 days old	9 Post-reproductive moult
Drawn from data in Masman *et al.* (1986).

There is no reason to believe that the situation is very different for wild, breeding Peregrines, though the percentage weight gains and losses might be different. While Lindberg (1983) was not able to assess the weight losses for breeding Peregrines, he did estimate the DEE of wild Peregrines, making the assumption that the ratio of DEE to BMR was essentially similar to that of captive birds, but allowing for the energy input of breeding (gonadal growth in males, egg production in females, *etc.*). Lindberg therefore assumed that the DEE for a wild male Peregrine was 3xBMR, that the DEE of a wild female Peregrine was 2xBMR and that the DEE for chicks was 1.7xBMR. Lindberg was then able to estimate the DEE of wild birds during the breeding season for varying brood sizes (Fig. 5.6). The DEE in Fig. 5.6 is very much higher than that observed in the captive male Peregrine (Fig. 5.1, assuming DEE ≈ 0.75xGEI), an indication of the energy demand of provisioning a growing brood. Based on Lindberg's calculations it is very likely that a male Peregrine will lose weight, just as male Kestrels do. During the later days of fledging and when the fledglings are learning to fly and hunt, a female Peregrine will also lose weight. For both sexes weight loss is probable as in addition to the rigours of feeding their brood, both also start the moult during the time when feeding demands are greatest.

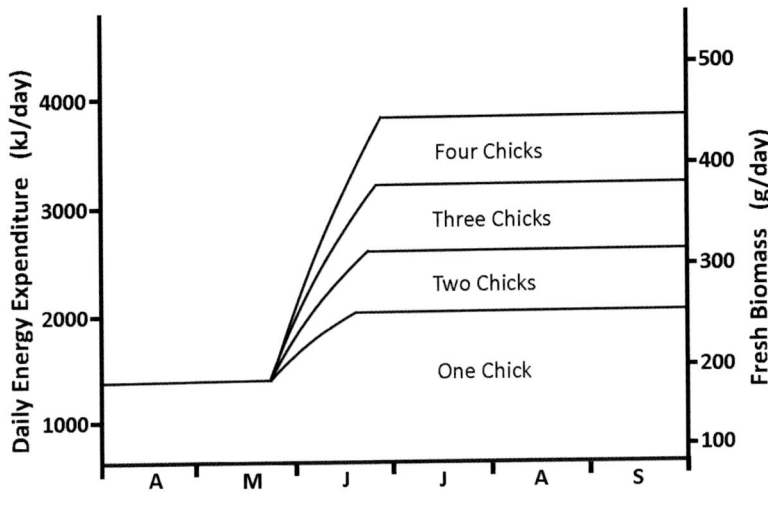

Figure 5.6 Estimated Daily Energy Expenditure (DEE) in kJ/day and in terms of food requirement in g/day (fresh biomass) of a Peregrine pair with different brood sizes. Drawn from data in Lindberg (1983).

From Fig. 5.6, the DEE of a Peregrine pair raising four chicks is 3,782kJ/day. Lindberg converted this to a food requirement of 124kg (biomass) for the six-month period April–September, which equates to the capture and consumption of 476 Black-headed Gulls (*Chroicocephalus ridibundus*).

As noted at the start of the Chapter, Barton and Houston (1994) found that in raptors there was a correlation between intestinal length and digestive efficiency: in general, raptors feeding on birds have short intestines and, consequently, poorer digestive efficiency. Working in Scotland, Barton and Houston compared the digestive efficiency of several UK raptors, simulating winter and summer conditions by feeding them with food at either 0°C or 20°C. Fig. 5.7 (overleaf) sets down the data for three of the UK's breeding falcons, together with other diurnal raptors and a number of owls: not all species were tested under both summer and winter conditions. The 'residual intestine length' was calculated as a percentage of the predicted small intestine lengths using the method of Barton and Houston (1992).

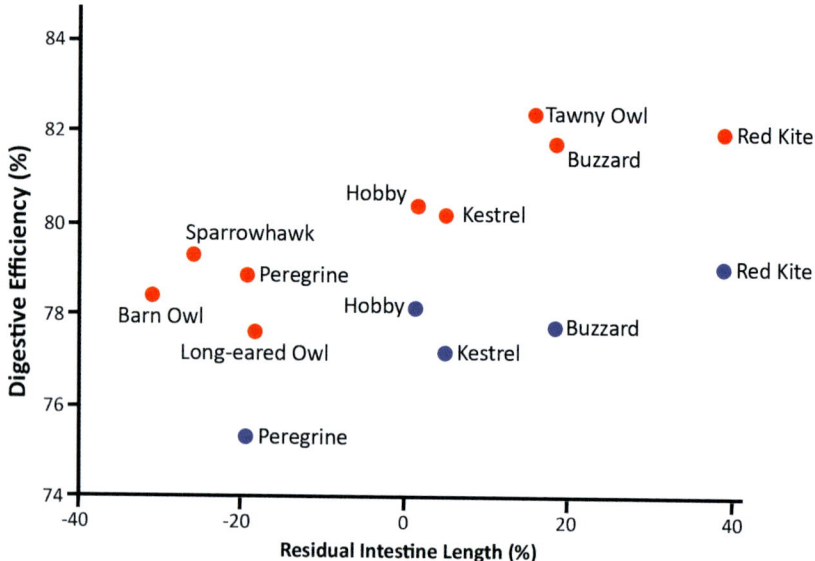

Figure 5.7 Relationship between digestive efficiency and small intestine length for various raptors in summer (red) and winter (blue). In addition to the Peregrine, the species illustrated are the Eurasian Hobby and the Common Kestrel, together with the Eurasian Sparrowhawk, the Eurasian (Common) Buzzard, the Red Kite, the Barn Owl and the Long-eared Owl. Redrawn from Barton and Houston (1993).

Barton and Houston (1993) also investigated the efficiency of the digestive tracts of the Peregrine, a specialist avian feeder, and the Common Buzzard, a generalist feeder. Four Peregrines and four Buzzards were fed on rabbit meat for eight consecutive days and then pigeon meat for eight consecutive days. The food was without fur, feathers and bones, and all fat was also removed to avoid pellet production, but roughage (day-old chicks) was fed for two days between the two test periods for the health of the birds. The mean digestive efficiency of the Peregrines on the two differing diets was 94.2% for rabbit and 94.1% for pigeon, whereas the efficiencies for the Buzzards was 96.3% (rabbit) and 95.8% (pigeon). The Buzzards gained weight (an average of 2.8% of body weight) when fed rabbit, the Peregrines lost weight (an average of 5.0% of body weight): both species gained weight when fed pigeon (Buzzards an average of 7.2%, Peregrines 1.8%). Interestingly Barton and Houston set down the water, fat, nitrogen and gross energy content of rabbit and pigeon meat (Table 5.1 opposite).

Opposite A nominate adult with a fledgling, anxious to get its talons on another meal, in tow. The shot was taken on the Isle of Man, UK. *Peter Christian.*

Food Consumption and Energy Balance

Prey	Water Content (Percentage)	Fat Content (Percentage)	Nitrogen Content (Percentage)	Gross Energy Content (kJ/day)
Rabbit	74.25 ± 1.04	5.0 ± 1.6	144	24.72 ± 0.58
Pigeon	72.21 ± 1.41	20.7 ± 2.6	130	25.67 ± 0.45

Table 5.1 Percentages of water, fat and nitrogen, and gross energy content of rabbit and pigeon meat. From Barton and Houston (1993).

The difference in fat content of the two fresh meats is considerable, and may explain, at least in part, the difference in diet between Peregrines and Buzzards. The Peregrine's attacking lifestyle is high-energy, fuelled by the (relatively) high fat content of most avian species, and particularly that of their preferred diet in most environments.

6   BREEDING PART 1

As with other raptors, territory and hunting range differ for Peregrines, the former being a relatively small area centred on the nest site and from which intruders are usually vigorously ejected, the latter a much larger area, one whose size depends on the density of prey.

Territory and Hunting Range
Competition for good nest sites and hunting areas can result in antagonism if a territory holder discovers an intruder. Generally such encounters end with one bird retreating before major conflict, but fights may occur and can end fatally. Tordoff and Redig (1999a) report two disputes, both in Minneapolis, USA which resulted in death for one combatant. In one case a female Peregrine was discovered dead after flying into a building, either having been driven into it or hitting it while fleeing, probably after a fight with her half-sister; in the second case a female died from injuries sustained in a prolonged (two hours) battle with another female. Hall (1955) also records a territorial dispute which resulted in the death of an adult male whose half-eaten remains were found on the nesting ledge of the Sun Life Building in Montreal, Canada. (Further details of the urban breeding of these Peregrines is given in Nest Sites below.) At one of our 2022 breeding study sites, Charing Cross Hospital in London, UK, a dead adult four-year-old male was recovered from the roof on 3 November 2021 following a territorial dispute. The dead bird had obvious

Breeding Part 1

Desiccated male Peregrine found at Charing Cross Hospital. It is believed this was an intruder male killed in a battle with the resident male. The red circles indicate the position of very clear puncture wounds caused by talons.

puncture wounds to the back following observed combat with the resident male. Chapter 7 has the 2022 breeding data for this site.

Mearns (1985) used telemetry to record flights of two breeding female nominate Peregrines in southern Scotland. The two breeding eyries were 4.5km apart and data were gathered over a period of about 20 days during the breeding season, ending once the fledglings had left the nests. One female had a hunting range of 22km^2 which increased by only 5% once the fledglings had left the nest ledge: this female also spent most of her time (about 85%) within 2km of the nest. The other female had a much smaller range initially (9km^2) and spent over 90% of her time within 2km of the nest, but her hunting range increased to 117km^2 once the fledglings had left the nest ledge: she then spent 40% of her time further than 2km from the nest. There was no apparent overlap of hunting range of the two females (Fig. 6.1 overleaf).

The data of Fig. 6.1 are in sharp contrast to that involving the radio-tracking of three female and two male Peregrines (*F.p. anatum*) from three pairs in Colorado, USA. There, Enderson and Craig (1997) found that the hunting ranges varied from 358km^2 (a male) to 1,508km^2 (the female of the same pair). The distances from the nest at which the adult Peregrines hunted were also much greater than had ever been seen before, 20% of female hunts being at distances greater than 23km from the nest, with one female being observed hunting 43km from her eyrie. One male was followed 14.6km from the nest site (by helicopter). There was also considerable overlap of the ranges of the

The Peregrine Falcon

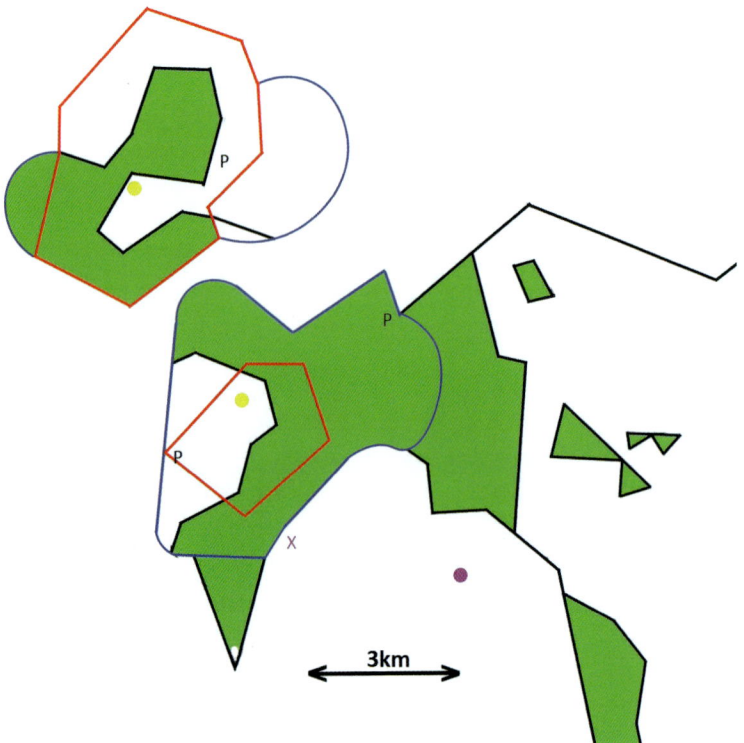

Figure 6.1 Changing hunting ranges of two female nominate Peregrines during the breeding season in Scotland. The yellow dots are the two occupied eyries. The red lines are the range limits of the adult Peregrines during chick rearing with the nestlings ≤18 days old, the blue lines are the extension of the hunting range when the chicks were more than 18 days old. 'P' represents the favoured perches of the two female Peregrines. The purple dot is a Golden Eagle with the purple X representing the position where the female Peregrine and a Merlin had an aerial dispute with a Golden Eagle from the nearby eyrie. None of the birds were injured. Redrawn from Mearns (1985).

three pairs despite the minimum inter-nest distance being 28km (Fig.6.2). Interestingly, in a study of *F.p. tundrius* on Alaska's North Slope, White and Nelson (1991), using a helicopter to follow 21 hunting forays, noted the male of a pair flying exactly the same maximum straight-line distance from the nest site (14.6km) and having a near identical hunting range (about 391km^2). Using satellite tags on four breeding *F.p. calidus* on Russia's Kola Peninsula, Ganusevich *et al.* (2004) measured the hunting ranges at 1490, 1550 and 1556km^2. The ranges overlapped, but most interestingly the fourth female had a range of only 104km^2 which was completely contained within the range of one of the other three females, and more than 50% contained within the ranges of the other two.

Breeding Part 1

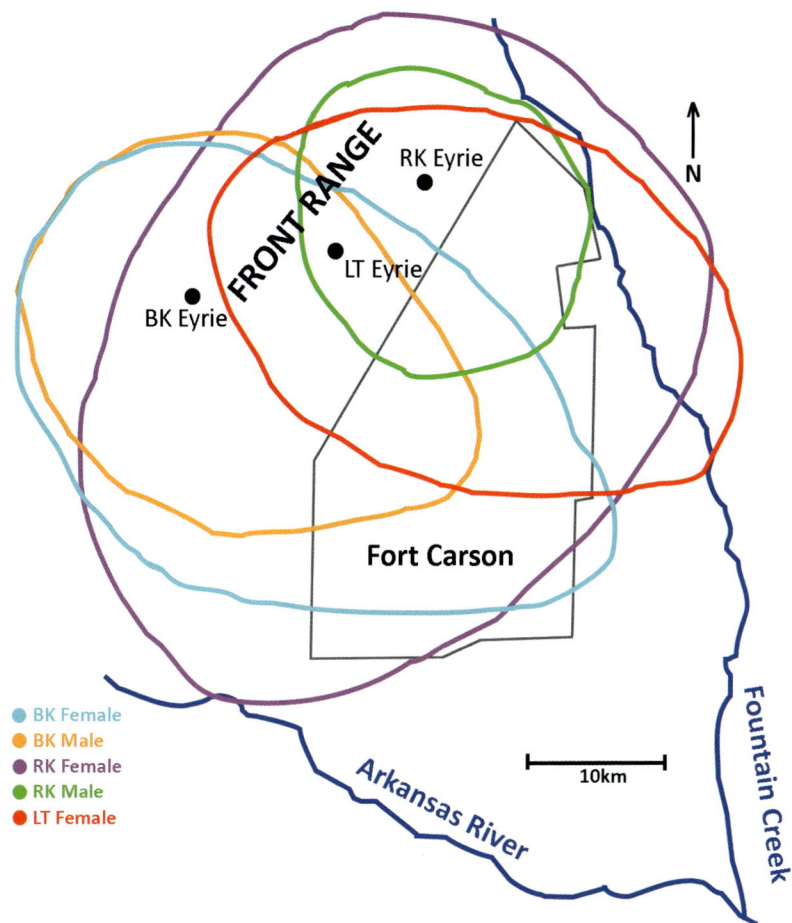

Figure 6.2 Disposition of the hunting ranges (95% contours) of adult Peregrines (*F.p. anatum*) of three pairs (BK, RK and LT) in Colorado, USA during May-August 1994. Redrawn from Enderson and Craig (1994).

As interesting as the vast difference in the size of hunting ranges, and the fact that in the Scottish study these did not overlap, was the difference in distance between eyries which implies a willingness of Peregrines to breed where there are suitable nest sites even if prey density is low and ranges must, therefore, be shared with other Peregrines.

The differences in Figs. 6.1 and 6.2 raise two important issues. Firstly, how close Peregrine adults hunt to their eyrie, and secondly, the extent to which they are willing to tolerate hunting range overlaps.

Considering the first question, some researchers have suggested that Peregrines do not hunt close to their eyries as this could identify nests to

The Peregrine Falcon

potential predators. In California, Enderson and Kirven (1983) radio-tagged a breeding pair of Peregrines (*F.p. anatum*) to measure their movements from the eyrie, collecting data from 139 flights by the female and 40 by the male. In general, the female remained within 1km of the nest (74% of flights) while for the male 65% of flights were more than 1km and most flights were along specific corridors defined by the local topography (Fig. 6.3). Prey was taken in all directions and there appears to have been no evidence for hunting at distance from the eyrie. Indeed, the return flights with prey were always directly to the nest, which would seem to negate the idea that any specific attempts at avoiding eyrie detection were being made.

In the study of Enderson and Craig (1997) at Fort Carson, Colorado (noted above and indicated in Fig. 6.2) two male and three female Peregrines (of three pairs) were tracked. Again, most flights by the five birds were within 7km of the eyrie (Fig. 6.4). However, as noted above, on some flights the

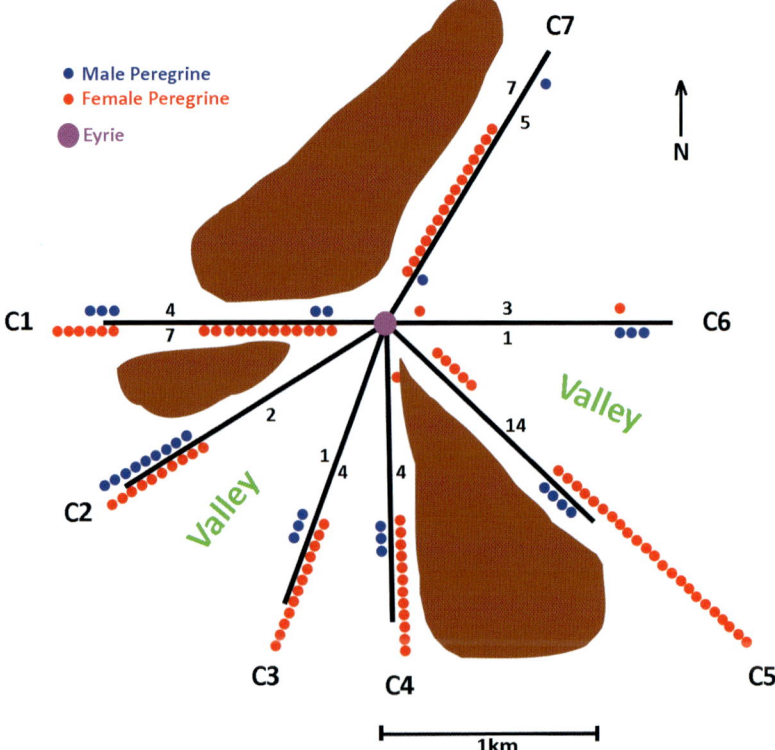

Figure 6.3 Distribution of flights around the eyrie by a breeding *F.p. anatum* pair in California. The eyrie is at the focus of the lines which represent generalised corridors used by the birds. Each coloured dot represents a flight by the male or female falcon, with dots close to the eyrie being flights of less than 1km, and dots further away being flights greater than 1km. The numbers beside each flight line show flights of unknown length along those flight corridors. Redrawn from Enderson and Kirven (1983).

Nominate Peregrine, England. The use of coloured leg rings on birds is discussed later in the Chapter. *Dan Lowth*.

falcons travelled up to 43km. On one flight a female Peregrine travelled 19km in ten minutes, an average speed of 115kph/71mph. For more than half the recorded flights the average speed of the falcons exceeded 60kph/37mph. Enderson and Craig's study was on *F. p. anatum*, but in a study of *F. p. minor* in South Africa, Jenkins and Benn (1998) found both overlapping ranges (Fig. 6.5) and a very similar spread of distances (Fig. 6.6) though with shorter distances being travelled (presumably because of higher prey densities). In each case males tended to make shorter flights. While it is clear that females can carry the same prey weight greater distances and so may range further, it is also the case that females disperse greater distances from their natal sites. In the US study it was mainly females which made much longer flights, but this was less noticeable in South Africa as few journeys of more than 10km were made. While these data deal with sub-species of the Peregrine which differ from those in other parts of the Peregrine's range, the similarity in behaviour in the USA and South Africa implies that Peregrines, particularly males, tend to forage within a short distance of the nest cliff, presumably to reduce the distance they have to carry prey, either to avoid exhaustion, or to minimise the turnaround time of prey capture and delivery in order to hunt again.

Range tolerance, if circumstances require, is also supported by evidence which suggests that the density of Peregrines has increased in areas with suitable nesting sites as the population has expanded, in addition to the falcons finding other sites – quarries, towns – in which to breed. Beebe (1960), studying

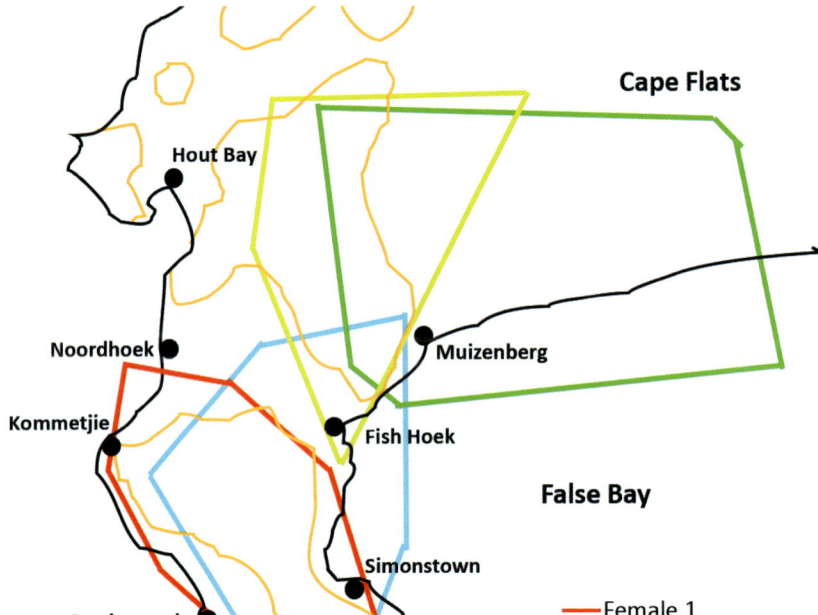

Figure 6.4 Percentage of distances flown from their eyries by hunting *F.p. anatum* in Colorado, USA. The eyrie position and hunting ranges of the falcons are shown in Fig. 6.2. Redrawn from Enderson and Craig (1994).

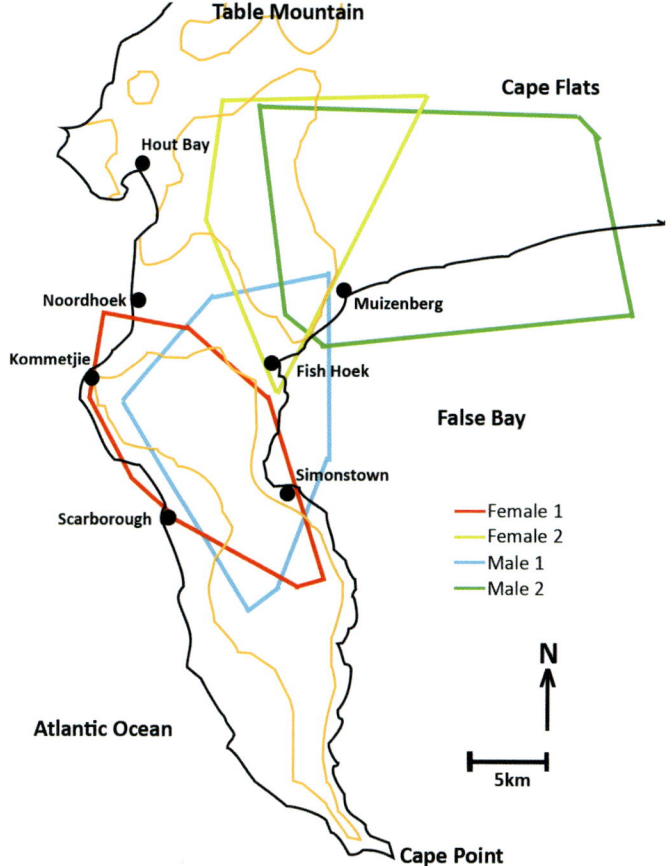

Figure 6.5 The Cape of Good Hope and Table Mountain, South Africa, showing the ranges of two pairs of breeding *F.p. minor* relative to the peninsula mountain chain, the main urban centres and the coastline. Redrawn from Jenkins and Benn (1998).

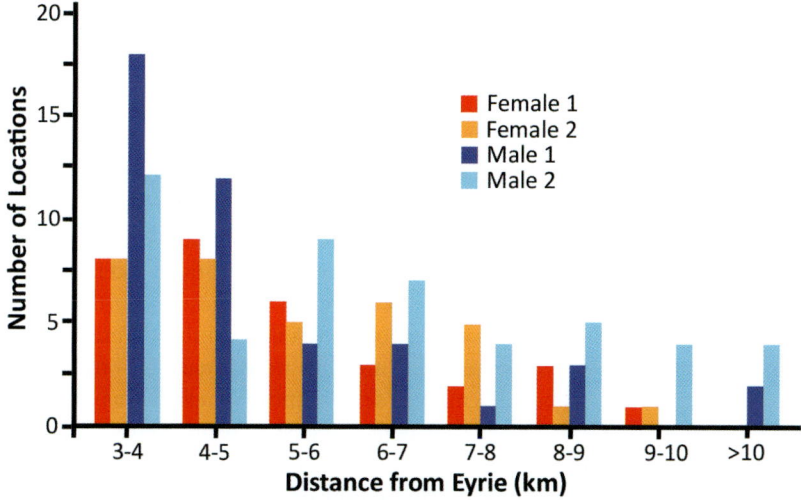

Figure 6.6 Number of locations recorded against distance from the eyrie for the two pairs of *F.p. minor* whose ranges are shown in Fig. 6.5. What is noticeable in this figure, in comparison to Fig. 6.4, is that the males of the pairs are flying further than the females, in contrast to the situation in Fig. 6.4 where the females flew the longest distances. The difference was almost certainly due to the timing of the studies. In South Africa the pairs had commenced breeding: the males were therefore courtship feeding their partners as they commenced laying. At that stage the females are more sedentary, while the study of Fig. 6.4 was much later in the breeding season when the female was less tied to the eyrie. Redrawn from Jenkins and Benn (1998).

F.p. pealei on Langara Island, British Columbia, Canada[1] found that in one particular area there were 'never less than five, usually six and sometimes eight' breeding pairs in an area (encompassing both land and sea) of 'less than two square miles', *i.e.* approximately 5km². This corresponds to a breeding density of up to 160 pairs/100km², which is extraordinary. The Peregrines were hunting Ancient Murrelet (the main prey), Cassin's Auklet, Leach's Storm-petrel and Fork-tailed Storm Petrel (*Hydrobates furcata*). According to Beebe, the availability of these prey species on Langara could only be described as 'astronomic' and the high numbers, and density, of Peregrines 'appears to be nothing more than a response' to this. Beebe notes, for instance, that although both Fox Sparrow (*Passerella iliaca*) and Hermit Thrush (*Catharus guttatus*) were present on the island and were often observed far away from cover and therefore vulnerable, the remains of neither was found among Peregrine kills. Assuming that the Peregrines had nest sites at the centre of the 5km² identified by Beebe, the observed density suggests an inter-nest distance of 2km, much less than would be expected: in a study area in south-west Scotland, Mearns and Newton (1988) noted some Peregrine pairs were within 3km of another pair, whereas other pairs were separated by up to 30km. While the shorter distance implies a breeding density of about 3 pairs/100km², the longer distance implies less than 1 pair/100km².

Sokolov *et al.* (2014) studying *F.p. calidus* on Russia's Yamal Peninsula, close to the Kara Sea, note that the hunting range of a breeding pair, hunting for growing chicks later in the breeding season altered both the size of the hunting range and the behaviour of the adult birds. In their study nine females and a single male were trapped and fitted with trackers which allowed an understanding of the size and shape of the range. The male had a small range, close to the nest site, but when the females began to hunt their range was larger, and changed annually (Fig. 6.7). A similar situation was seen in a study of *F.p. anatum* in Kentucky where, after the pesticide crisis, Peregrines are considered a Species of Greatest Conservation Need. Using satellite tags, Taylor *et al.* (2020) found that the size of the hunting range of a breeding female varied through the season, being smallest when the chicks were pre-fledge and largest after fledging. Interestingly, during the non-breeding season the female's range was smaller than it had been after the chicks had fledged but were still dependent on their parents for food.

Ratcliffe (1993) collated data from both coastal and inland Peregrine sites across Britain. For the coastal (sea cliff) areas he noted many instances where the site separation was about 2km, and several more where it was only 1km. At inland sites, Ratcliffe found that when nest sites were evenly distributed the Peregrine pairs tended to space themselves evenly, but where the sites were less uniformly distributed close nesting could be seen just as it was at coastal sites. Ratcliffe noted the mean distance to neighbours and the corresponding

---

[1] Langara is an islet off the northern coast of Graham Island. Graham and nearby Moresby Island, to the south, are separated by the Haida Gwaii channel from western British Columbia about 350km north-west of Vancouver Island.

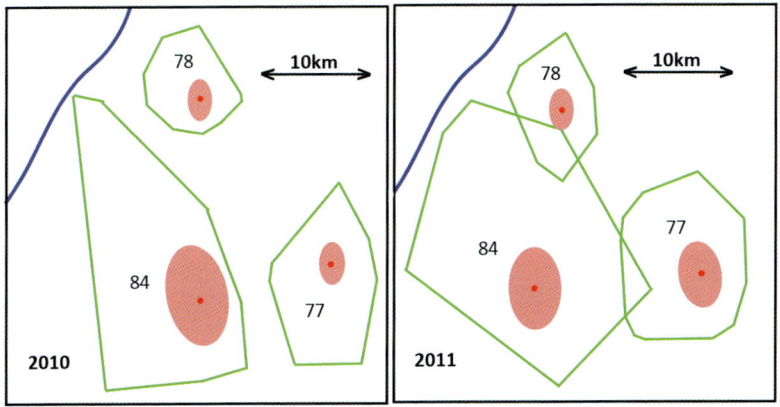

Figure 6.7 Variation of hunting ranges of three breeding female *F.p. calidus* in Yamal, Russia. The pink ellipses represent 95% of all hunting for each falcon, the green polygons encompassing 50% of the total hunting range. Redrawn from Sokolov *et al.* (2014).

particularly belligerent birds will attack and pursue intruders who come within 1.5–2km. This implies that while inter-nest separations are high if circumstances allow, they could reduce if circumstances changed.

With an increase in Peregrine population it appears that the linear separation between nest sites in Britain has decreased, but the falcons have sought to maintain an approximately equal separation to neighbours in all directions, though continuing to hunt over areas which are not defined by this inter-nest separation. The breeding density of Peregrines is, it seems, defined by food resource rather than territory size, but, adopting the human maxim that high fences make good neighbours, nest separations are kept as large as possible to minimise antagonism. This is certainly consistent with the finding of Wightman and Fuller (2006) in North America – see *Nest sites* below – that one important factor in defining 'better quality' nest sites is distance from nearest neighbour.

Mearns and Newton (1988) found no evidence to suggest that the minimal distance in their study was a factor in breeding success, but ultimately, of course, there will be a limit, one defined by a combination of prey availability and the minimum tolerable distance to the adjacent Peregrine pair. Hunt (1998) investigated the way in which populations might grow, then stabilise, in an area with a defined number of available nest sites. Though somewhat mathematical, the investigation does shed interesting light on the influence of 'floaters' (non-breeding adults) on the approach to, and achievement of, population stability. In Hunt's hypothetical population, stability was achieved in 32 years (for a nominal raptor population), which approximates the timescale actually seen in British Peregrine populations. Hunt's work notes that an equilibrium (and, therefore, limit) will be reached: that limit may (but may not) have been achieved on Langara Island, where inter-nest distances could not have been greater than about 500m, and where the combination of nest-site availability and prey resource allowed a situation approaching the maximum achievable. Hunt's work will be mentioned again in Chapter 10 *Population*.

Box 6.1 Floaters
The existence of 'floaters' (non-breeding adults) in a population derives from the idea that in any habitat the highest 'quality' pairs would occupy the prime nest sites (a situation occasionally called the 'despotic distribution model') as 'despot' occupiers would see off those of lower quality. Fretwell and Lucas (1969) laid the foundation for this idea in a paper that does not actually use the term 'despotic', later researchers building the terminology now in common usage. Despotic distribution implies that Peregrines of better 'quality' hold 'better' sites, with 'lesser-quality' pairs being forced to use poorer sites. In this context 'better-quality' implies that 'better' falcons are either more aggressive in the acquisition and defence of sites and/or more efficient hunters. The former may be true, but caution must be exercised as regards the latter. In a study of *F.p. minor* in South Africa, Jenkins (2000b) noted that falcons occupying higher cliffs (this being one definition of a 'better' site) had higher hunting success rates. However, while this implies higher efficiency, Jenkins noted that the height difference between the falcon and the prey was significantly correlated with hunting success. In other words, falcons attacking from a higher start point, and probably achieving higher attack speed as well as greater surprise, had an enhanced hunting success rate, which could imply that occupation of the 'better' site was as important, perhaps more so, than 'pure' hunting proficiency.

Assuming that first arrivals obtain and hold the better sites or, if arriving later, acquire them by evicting those in occupation, this means that a second cohort of breeders would be forced to use nests in lower quality habitat. Ultimately, a cascade of breeding pairs moving to even lower quality habitat results in some potential breeders being unable to find nest sites at all: these individuals become floaters. This idea is attractive, but there is an alternative view of floaters which suggests that some birds are of such low quality that they cannot breed in anything except the prime nest sites: such incompetence may be age-related even for non-juveniles. But either lack of available nest sites or a lack of breeding competence means that any population will have a number of floaters.

The very low level of extra-pair breeding in Peregrines suggests that competent floaters need to bide their time, waiting for suitable nest sites to become available. While juveniles may become 'third birds' at nest sites, gathering experience which will be of advantage when an opportunity for breeding arises (perhaps when one of an adult pair dies), studies in other species have suggested that there are alternatives. Third birds in species other than falcons occasionally exhibit behaviour apparently more complex than that seen in the 'helpers' of some falcon observations. In work on Purple Martins (*Progne subis*), Stutchbury (1991) noted that male floaters did not intrude on all occupied territories equally, but preferentially choose 3-5 territories. While attempts to usurp these territories usually failed in conflicts with the occupying male, the floaters fought a war of attrition which was sometimes rewarded with long term success. Another idea is that late dispersal of juveniles might explain their presence in the nest area in the following year, though this theory would only explain floaters in non-migratory species: Kimball *et al.* (2003) also note that in most cases floaters are not family members so late dispersal cannot be the sole reason for their existence. However, as Kimball and co-workers also note, since floaters are much more prevalent in raptors than in other species there must be some evolutionary process at work.

More recently, Campioni *et al.* (2010), in a study of Eagle Owls (*Bubo bubo*) in Spain, noted that the behaviour of territory holders and floaters differed, the former sitting on prominent posts to emphasise their territory ownership, the latter being more secretive. This suggested that female floaters might actually be using the ability to move through occupied territories without being subject to an aggressive response as a way of comparing them for future occupation. This is consistent with the finding of Bruinzeel and van de Pol (2004), who, in a study in The Netherlands, removed one breeding bird of a pair of Eurasian Oystercatchers (*Haematopus ostralegus*) and observed how the resulting vacancy was filled. In 80% of cases it was filled by an 'intruding floater', *i.e.* a non-paired bird which had been seen to spend time within the territory of a breeding pair. Bruinzeel and van de Pol found that intruding male floaters were more successful at filling vacancies than females. While the breeding behaviour of Oystercatchers and falcons differ, the Dutch study suggests that it is possible that both male and female floaters in falcon species are gathering not only breeding experience prior to their own first breeding, but also acquiring territorial information which might be of value later. In males this might allow a decision to be made on which territories to compete for, while for females it might indicate which males/territories would be favourable to partner.

Pair formation

Before examining pair formation in Peregrines it is worth noting here that the advent of coloured leg rings has aided a better understanding of both pairing and pair fidelity. The original use of small metal leg rings aided an understanding of falcon lifetime and, initially, migration, though the difficulty of unequivocally reading the number on these rings at a distance meant that usage was very largely confined to the finding of dead birds. Both the original metal rings and the newer coloured rings have opponents as well as adherents: as with any decision to put furniture on a bird, the potential benefits of the obtained data have to be considered against the potential harm to the bird and the possibility that the collected data will relate to abnormal behaviour. While each generation of researchers is inevitably anxious to obtain its own data, care should always be taken that data replication for its own sake is not an adequate reason for adding hardware to an individual bird. In passing, it is worth noting that Smith *et al.* (1993) suggest that individual Peregrines can be identified by visual comparison of their toe-scale patterns, though that, of course, requires very good photography which might not be available in all required cases. It is also the case that working with Eurasian Hobbies in Germany, the late K.D. Fiuczynski, an authority on the species, identified individual birds by their calls (K.D. Fiuczynski pers. comm. to RS).

Many years ago, R.B. (Dick) Treleaven noted in conversation with one of the authors (SW) that a distinguished falconer, Jack Mavrogardato, had told him that whereas falcons' plumage did not tend to vary greatly with age, the degree of wrinkling in toes and tarsus was a much surer indication of age. However, whereas this might be helpful to a falconer with an intimate knowledge of falcons at close quarters, it is unlikely to be of any use at all in the field.

Where year-round prey availability is sufficient then, as a general rule, lowland UK Peregrine pairs remain on territory, regardless of site type, throughout the year. This is also true of pairs in North America these observations implying that pairs may adopt the same lifestyle across the entire range. However, some birds definitely winter alone, particularly in non-breeding areas, and lone wintering may also be necessary for pairs if the food resource is insufficient to sustain both birds, which is often the case for upland territories. There are different views on whether the male or the female of a given Peregrine pair remains at the nesting territory in order to avoid it being usurped, both sexes having been seen to do so. Dick Treleaven in Cornwall, England was of the view that when one of a territorial pair goes missing, then the other will remain on site as territory holder, be it male or female and one of the authors (SW) has also found this to be the case based on 40 years of observations in Gloucestershire, England.

The pair bond is monogamous – bigamy has been recorded, but rarely, which is consistent with the finding of only 1.3% extra-pair copulations in a Canadian Arctic (*F.p. tundrius*) population, based on DNA analysis (Johnstone, 1998): the population in question was dense, which may account for the finding, and long-lasting, perhaps lifelong. Treleaven (1998) observed bigamy on the coast of Cornwall, England in three successive years, but apparently all breeding attempts failed. Peregrines can mate in their second calendar year, but this is uncommon in both sexes (and very uncommon in males). Wendt and Septon (1991) note that if second calendar year Peregrines pair they usually fail to produce eggs, and if they do hatchlings rarely survive to fledge. The researchers do record a pair of second calendar year *F.p. anatum* birds which bred in Milwaukee, USA, in a box mounted on top of a building and raised four nestlings to fledge. However, the male consistently attacked the fledglings when they were learning to fly, stooping at them and knocking each of the four to the ground at least once: one fledgling broke a femur on hitting the ground and died of resulting complications. Another youngster died in collision with a window, though Wendt and Septon do not say whether this

Breeding Part 1

Having paired, the two young nominate Peregrines considered further in Chapter 7 *Evesham Bell Tower*, discuss using the installed nest box for breeding purposes. *RS/SW*.

accident was due to evasive action after an attack. A third fledgling was injured twice, rescued and relocated to avoid further injury. The fourth fledgling survived, but subsequently left the city. Wendt and Septon conjecture that the breeding male's youth may have resulted in this aggressive behaviour, but note that as a breeding adult the bird continued to harass his offspring when they were learning to fly, though never to the same extent. Similar harassment by the adult male of a chick was observed during the filming of the breeding behaviour of the Peregrines at a London hospital which is discussed in detail in Chapter 7 *Charing Cross Hospital*.

In a study of nominate Peregrines in southern Scotland Mearns and Newton (1984) found 19 second calendar year females in a total of 398 territorial pairs (5%). Of these 19 only seven laid eggs, and of those seven only one raised young. The females were identified in part by trapping and in part by plumage: of six trapped males, one was a yearling male, four were two-year old birds and one was four or five years old. In a later study in the same area Mearns and Newton (1988) found only two yearling females breeding (in a sample of 62), neither of which was successful. In northern Spain Zuberogoitia *et al.* (2009), in a study (of *F.p. brookei*) over 11 years which involved ringing 426 chicks and 16 trapped adults, 3.3% of males and 22.7% of females bred in their second calendar year; the remainder bred in the third calendar year, though there a small percentage of both sexes bred later: one male did not breed until its fifth year, one female in her sixth year. The mean age of first breeding was 3.7 calendar years for males and 4 years for females.

While the evidence for assortative mating on the basis of age is therefore clear, there is some evidence that size also plays a part, Olsen *et al.* (1998), in a study in Australia (*F.p. macropus*), finding that female Peregrines with longer wing lengths tended to mate with males with similarly long wing lengths.

The situation in the London nest box (see Chapter 7 *Charing Cross Hospital* for more details) differed radically from that in Evesham (see image on previous page). In London, although the male was the significantly older of the pair, the female was a mature female who had bred successfully in the previous year. At Evesham the male took the lead in the early phases of breeding, but in London the male, in the foreground, often took up an appeasement gesture when the female arrived at the box. *NM/RS/SW*.

This is not surprising as it is known that in Common Kestrels females appear to assess plumage colour, the width of the male's sub-terminal black tail band and the length of the tarsi in mate choice (Sale (2020) and references therein). There is no reason to suppose that female Peregrines are not making similar appraisals in their choice of mates, perhaps also noting the agility of males in display flights as an indication of hunting ability, and perhaps also the male's choice of, and ability to defend, a good territory. In addition, Whittaker and Hagelin (2021) note that chemical signals from the uropygial gland may allow male birds to assess females in terms of breeding receptiveness and quality, so mate choice may be a two-way process.

If one member of a long-lasting pair dies, the widowed bird will select a new partner. In such cases the widowed male partner generally stays on territory and seeks to attract a new partner. There is one known pairing of a female Peregrine with her two-year-old son after the death of the young male's father, and other incestuous relationships are also known. In 2008 at a church in Bath, England a second calendar year male, son of the breeding female, helped raise that year's chicks when the adult male (his father) disappeared part-way through the breeding season. Even more remarkably, in subsequent years to 2013, mother and son successfully bred in an incestuous relationship producing 12 chicks before the falcon was 'replaced' in 2014. Furthermore, at the same site in 2017, one of the male offspring from 2016 remained on territory and assisted in raising that year's brood by taking on incubation duties and bringing in food for the chicks whilst at the same time begging for food from his father. This behaviour stopped in 2017 when the male offspring eventually left the area.

Pagel and Sipple (2011) report a case of sibling mating in California, while Bélisle *et al.* (2012) report a case involving male and female Peregrines that were the only hatchlings at an urban nest in Montreal, Canada (not on the Sun Life Building – see above – but on the University of Quebec Tower). The latter case was very curious. The young female had a partially disabled leg and stayed with her parents all winter. Although she could hunt, the leg disability reduced her abilities and she begged for food from her parents and stole their cached food on occasions. The following year the juvenile remained with her parents while they copulated and even displaced her father from incubating several times, though she showed little interest in incubating herself. Despite the juvenile's occasionally aggressive behaviour she was accepted by her parents. Three eggs hatched and one day the juvenile female grabbed a nestling from the unguarded nest and flew off, chased by the adult male. The juvenile returned without the nestling (which was not recovered) but was ignored by her parents and eventually left the area. The following year the juvenile female returned to Montreal and paired with her brother. The pair copulated and three eggs were laid, but these were lost when the nest they selected collapsed: on retrieval it was found that two of the three eggs were fertile. In general, as for other species, fledgling dispersal is assumed (in part at least) to be an evolutionary trait designed to prevent in-breeding. But the tendency of Peregrines to return to natal areas and the traditional use of nest sites may lead to more such relationships than might usually occur. Tordoff and Redig (1999b) report several instances from the US mid-west (*F.p. anatum*) noting sibling-sibling, half-siblings breeding (same mother, different fathers) and mother-son breeding.

While the behaviour of the juvenile female in the study of Bélisle *et al.* (2012) was clearly aberrant, there are recorded incidents of juvenile males helping at nest sites (*e.g.* Monneret, 1983, who reported incidents from the 1970s and 1980s with nominate Peregrines, and Kurosawa and Kurosawa, 2003, who watched a juvenile male *F. p. japonensis* helping to feed chicks at the nest of an unrelated pair on Hokkaido, Japan). Spofford (1947) also saw a second female at a nest in the USA. The second bird was smaller than the mated female: the male fed his mate and ignored the smaller female. The same situation arose the following year, but this time the smaller female immediately covered the eggs when the larger female left, though she stopped incubating the moment the larger bird returned.

Ratcliffe (1993) records pairings in which the death of a nominate female resulted in the male pairing with a second female who acted as foster-mother to the original brood. Ratcliffe also records one event in which an unpaired female evicted a female from her clutch, taking over not only the eggs but the male (the father of the clutch). Even more remarkably, Nelson (1970), in British Columbia, records a male (*F.p. pealei*) pairing with a foster-mother after the loss of the female, and then the male being lost, at which point the foster-mother took a new mate, the clutch being incubated by two foster-parents.

Displays
Pair formation, whether the renewal of an existing bond or the beginning of a new partnership, usually begins in January or February in the northern hemisphere[2], though earlier observations have been made. Meier *et al.* (1989) recorded courtship flight displays in autumn (fall) in two pairs (one of which comprised a juvenile male and an adult female) in Puerto Rico. The pair including the juvenile male was also observed to copulate on one occasion. Courtship behaviour, including copulation and subsequent egg laying (though the eggs were infertile) has also been seen in autumn in captive birds.

In general pair formation begins with occupation of a territory by the male, who then advertises his possession (though some observers have suggested instances where female Peregrines have arrived first). A lone male will start advertising by flying about close to the nest site while chupping, increasing both the flights and the calling if a female appears. At this stage, the occupant of a good nest site will willingly display to any passing bird of the opposite sex, being more interested in maintaining the site than in maintaining the pair bond. In cases where the female of a pair has not wintered on territory, on her return the pair will vigorously chase other falcons away. This seems to imply that pair bonds are not strong, though it may also mean that where pairs have spent the winter apart and do not arrive together, the first arrival has to cover the possibility that his/her partner may not have survived the winter. Reunited pairs will often roost at the nest site, but in the early stages of new pairings the birds tend to find separate roosts, these becoming closer together as the bond strengthens.

The pair bond is established or reinforced by courtship feeding[3]. This is often associated with cooperative hunting, the pair choosing a target prey and assisting each other in preventing it from escaping until one or other falcon makes the kill (see, for instance, Fig. 4.24). If the female does so, she will eat the prey. If the male kills he will present the prey to his mate. Partnership displays are, at first, aerial. In one display the male will fly fast and close along the nesting cliff, occasionally sweeping up to perform high, soaring circles above the site before swooping down to fly fast along the cliff again. This display seems to be the 'Z flight' of Monneret (1974), though Monneret

---

[2] Autumn is March-May for Peregrines in the southern hemisphere, winter being June-August.
[3] Nelson (1970) feels the term 'courtship feeding' is not appropriate to describe the exchange of food between male and female. Although in its earliest form it is likely that the giving and accepting of food cements the pair bond, Nelson considers the continued instances of food passing are as much related to the two birds ensuring that the female acquires the reserves necessary for egg laying and, consequently, reproductive success, and that it is therefore in the interests of both that the male continues to pass food to his mate after the initial 'courtship' phase.

Making the scrape for the eggs. In the image to the *left*, the female is scraping the substsrate to produce a cup angled to keep the eggs in the correct position for chick growth. To the *right*, the female is using her body to create the cup as she rotates in the box. *NM/RS/SW (left)* and *RS/SW (right)*.

emphasises that the male alters his body position so that the light underparts and dark upperparts are displayed, and suggests the flight can be considered a territorial display as well as advertising the nest site to the female. Monneret also describes a flight by male Peregrines in which the bird flies from the nest site and follows a horizontal figure-of-eight track. Hagar (1938) also described this flight, writing *'nosing over suddenly, he flicked his wings rapidly 15 to 20 times and fell like a thunderbolt. Wings half closed now, he shot past the north end of the cliff, described three successive loop-the-loops across its face, turning completely upside down at the top of each loop, and roared out over our heads with the wind rushing through his wings like ripping canvas.'*

Some displays are silent, but occasionally the male will chup as he flies. The two birds will also fly together, making a series of loops as they make mock stoops at each other, passing very close, then separating to loop around again. Occasionally the bird being 'attacked' will roll over and extend its talons, just as an intruder would do if attacked by a territorial male: occasionally the two birds will even make talon contact. Such displays are remarkable both for their intensity and for the sheer aerobatic ability of the birds, particularly when the pair approach each other at high speed.

Although courtship feeding appears to cement the pair bond early in the breeding cycle, it becomes more important as egg laying approaches, and it forms part of the ledge displays which precede the female making the nest scrape, and copulation. (A scrape is a cup-shaped depression in the substrate made by the female Peregrine pressing her chest into the substrate whilst slowly rotating the body.)

Wrege and Cade (1977) studied these displays in captive Peregrines, and in captive Gyrfalcons, Prairie and Lanner Falcons: the displays were essentially similar in all four species, and no differences were found across the North

The four egg clutch at London. The ridge line of the circular cup made by the rotating female can clearly be seen around the eggs. *NM/RS/SW*.

American Peregrine sub-species. Wrege and Cade define 13 behaviours, but these can be usefully compressed here. The male will bow towards the female, usually bowing low into a horizontal position, though sometimes merely from the neck, as he approaches the nest scrape, often standing up on his toes (which Wrege and Cade called the 'high-step' or 'tippy-toe' gait) and chupping. The female will also bow, but hers is often a less pronounced gesture: she will also chup (see Fig. 2.15). The two birds will also bow and chup simultaneously, and will nibble each others' bills. As well as being an indication of bonding, head bowing can also be a threat posture (usually termed agonistic behaviour, as it rarely leads to aggression). Either sex can make this gesture, which usually follows one of the pair being startled by the other: despite the pair bond the two birds are by nature aggressive and distinctly non-gregarious, and remain wary of each other. In each case the feathers are flared, though the female's bow is usually only of the head: she may also extend her wings. To counteract the female threat the male may bow extremely low, his bill almost touching the ledge, in a gesture of appeasement. Monneret (1974) in the interesting study of Peregrine displays mentioned above also suggests that some of the aggressive displays by the female to the male are used by the former to force the latter into appeasement behaviour, leading to the male hunting for food which he brings to her.

Food transfers are another ledge display, the female usually making the wailing call as she begs for food. The male transfers food from his feet to his bill, bows several times, then uses the 'tippy-toe' walk to advance on the female. She may await his approach, or walk towards him. She then snatches the food, occasionally aggressively: the male will then leave the ledge quickly.

Nest sites

What constitutes a preferred nest site if you are a pair of Peregrines seeking to set up home and raise a family? In one sense this question is being asked too late as choice of nest site will define the territory and hunting range a falcon pair will inhabit and defend. But once an area has been identified as having an adequate density of the correct prey types, the nest site is critical. Working with *F.p. tundrius* in central west Greenland, Wightman and Fuller (2006) identified what constituted a 'better' site. They defined a series of 20 variable characteristics for nest sites. The obvious characteristics were ledges that were inaccessible to predators and with minimal human disturbance. The preferred cliff was tall and set high on a hill. For the ledge itself, the Peregrines chose one which was wide, deep and relatively level. The falcons preferred a ledge which was partially protected by an overhang rather than open to the elements, though complete protection was avoided, presumably because it cut down the outlook from the eyrie, as a broad aspect was definitely preferred[4]. The ledge covering was important, of course, as it was the substrate for the eggs and chicks. Sand or dirt was preferred, vegetation being a second choice, and both proving very much better than bare rock[5]. Interestingly, while recently adopted ledges might have had stick nest remains, traditional nest sites did not, presumably because over time the Peregrines disposed of them and there was never time for other nest-building species to build new ones. The local topography was also important, if choice was possible, Peregrines preferring sites with ready access to water, though with sufficient elevation to avoid flooding. Wightman and Fuller also found that the distance to the nearest Peregrine neighbour was important, the further away the better. This finding is consistent with the data on *F.p. calidus* eyrie separations on Russia's Taymyr Peninsula where the mean separation was 23.1km, with a range of 6.3-34.5km (Quinn *et al.*, 2000).

However, nearest neighbour distance was not found to be a dominant factor in the breeding site decision of Peregrines in adjacent alpine study areas

---

[4] While aspect is important, the direction of exposure is likely to be important in some breeding areas. Studying Gyrfalcons in western Alaska, Henderson *et al.* (2021) found that exposure to the weather was negatively correlated with breeding success. It is likely that a similar effect would be seen in Peregrines exposed to bad weather in northern areas and, perhaps, for those exposed to the sun at southerly sites. In northern Spain Zuberogoitia *et al.* (2015) noted a negative correlation between high April rainfall and nest productivity.

[5] In a study of breeding *F.p. anatum* on Baja California Sur, Mexico, Ayala-Perez *et al.* (2021) found that nest sites with artificial substrates were more successful than those with natural substrates, a success rate of 83±15% with 1.91±0.12 fledglings/breeding pair against 36±21% success and 0.36±0.21 fledglings for natural substrates.

Nominate Peregrine in south-west England breeding in a vast Raven nest. *Hamish Smith.*

of Italy and Switzerland by Brambilla *et al.* (2006a). (The Peregrines in this study were breeding in a probable 'cross-over' zone from nominate falcons to *F.p. brookei*.) Brambilla and co-workers categorised nest sites by similar characteristics to Wightman and Fuller (2006) and found that in all respects, site choice was identical to those of the Greenlandic falcons. The fact that nearest neighbour distance was not deemed very relevant, even though the breeding density was low (1.43 pairs/100km^2) is interesting, Brambilla *et al.* considering that the overriding issue for the Peregrines was the distance to urban settlements with their abundance of feral pigeons. In the study area of one of the present authors (SW) in Gloucestershire, England in recent years two pairs of Peregrines have breed within 1km of each other. The local topography means that the two eyries are on each side of a triangular cliff edge around which a river makes a sharp bend. The birds are not, therefore in sight of each other, the pairs seeming to use hunting ranges which do not significantly overlap. Nevertheless, the proximity reinforces the idea that as the falcon population increases, then provided prey numbers are adequate, the mutual intolerance which appears normal is, to an extent, relaxed. One of the two sites involved in this instance of close breeding is shown on the front and back covers of this book.

In his study of Peregrines (*F.p. minor*) and Lanner Falcons in South Africa, Jenkins (2000b) noted that Peregrines used larger nest ledges on higher, more elevated cliffs than their falcon cousins and preferred an outlook over open, structurally complex terrain. Jenkins's study was across the country and so took in both temperate and tropical areas: in the former, Peregrine nests were on smaller ledges of lower cliffs overlooking vegetation that provided less refuge for prey than did the cliff sites of their tropical-breeding cousins. Temperate Peregrines also chose sites offering shelter from spring rains. Overall, Jenkins noted that cliff size and height, ledge size and the vegetation below the site accounted for 80% of the difference between Peregrines and Lanners, and between temperate and tropical Peregrines. In another study of *F.p. minor*, in this case in Kenya, Thomsett (1988) noted that Peregrines need to be careful of cliff choice when choosing nest sites as baboons often sleep on cliffs and regular destroy falcon eggs and chicks. Cliff-roaming mongooses can also be a hazard to the falcons. In Ecuador, Hilgert (1988) noted three eyries. In each case the cliffs were 100m high and at elevations of 2500m. The eyries were 5m, 12m and 35m from the cliff top and faced E, NNE and NNE respectively. In a study of Peregrine eyries in eastern France, Formon (1969) noted nine eyries facing SW, 6 S, 3 SE or E, 2 W or NW, and one facing NE. On Russia's Taymyr Peninsula, Quinn *et al.* (2000) found that most eyries were on south-facing cliffs with only 4% facing north. Given how far north the Peregrines were breeding, these data are not surprising. South of the Equator, in Zimbabwe, Hartley (2000) noted the orientation of 113 eyries as N, 7.1%; N-E (*i.e.* NNE, NE, ENE), 11.5%; E, 6.2%; E-S, 16.8%; S, 13.3%; S-W, 23.9%; W, 4.4%, W-N, 16.8%.

Bruggeman *et al.* (2018) looked at preferred territories, studying, as with Wightman and Fuller (2006) *F.p. tundrius*, but this time pairs breeding in the Colville River Special Area of north-west Alaska. Again, it was the nest site that was found to be critical. Nearest neighbour distance was not considered, but since the falcons were breeding in an area with limited competition for prey resources and minimal human disturbance, the falcons did not have to be overly concerned with nest site spacing.

In general, across their vast range Peregrines are cliff-breeders, eggs being laid on ledges large enough to accommodate four fledglings in safety. Such sites are often traditional, with usage lines going back over decades. The ledges often have sparse or short vegetation, though ledges which have formerly been used by buzzards, crows and eagles are often used. High, steep nesting cliffs are preferred, Ratcliffe (1993) noting the studies of the American Joe Hickey who pointed out that the very best sites on such cliffs would become traditional even if there was a high rate of killing of the adult Peregrines by humans. Ratcliffe also states that breeding performance is improved on higher, steeper cliffs, though if a Peregrine pair were to lose a clutch, for whatever reason, they would invariably choose a different site for a repeat clutch. However, despite the idea of nest sites being traditional, Zuberogoitia *et al.* (2015), studying *F.p. brookei* in northern Spain, suggest that contrary to what might be expected successful falcon pairs were more likely to move to a new site in the following year than to stay at the same site. In their 17-year study the researchers found that newly established females moved sites on 59% of occasions, though for older territorial females moving was reduced to 38%.

For males the switching rate was much lower. This suggests that 'win-stay, lose-switch' would seem logical for both successful pairs and for those with repeat clutches. Zuberogoitia and co-workers also note that in their study the Peregrines had an average of 3.1 eyries per territory, so it is possible that the relatively low density of Peregrines in the area meant that moving eyries in successive years was a response to other factors. Although Zuberogoitia *et al.* did not find parasite burdens at sites was a direct predictor of movements, they did find that breeding success declined with consecutive usage.

In a study of nominate Peregrines in wilder areas of the UK, Ratcliffe (1993) also noted that many sites were traditional, and that the ledges utilised often had short or sparse vegetation. In some cases there were still the remnants of stick nests, those of Ravens being the most popular for the falcons (Table 6.1). Within the nest, or in ledge soil, the female would make a scrape, usually about 20cm in diameter and 2-4cm deep (though some traditional sites have much deeper scrapes, these often combining soil, dead vegetation and the detritus of years of usage).

	Ledge (*i.e.* no nest)[1]	Raven	Common Buzzard	Golden Eagle	Crow
North Wales	28	3	0	0	0
Northern England	56	21	0	0	0
Southern Scotland	104	60	4	0	1
Scottish Highlands	58	18	1	1	0
Percentage of Total	**69.3**	**28.7**	**1.4**	**0.3**	**0.3**

Table 6.1 Nest sites used by Peregrine Falcons in Britain. From Ratcliffe (1993).

Note
1. In a study of Peregrines in south-west Scotland over a nine-year period Mearns and Newton (1988) found that stick-nest sites had a higher success rate (in terms of number of clutches which produced fledglings) than bare ledges. They considered the likely reason was that stick-nest builders were not constrained by the need to find flat ledges and so could utilise sites which offered better shelter from the weather and were less accessible by humans.

The Spanish study of Zuberogoitia *et al.* (2015) and the data of Ratcliffe (1993) suggests that the definition of 'traditional' when talking of nest sites may need to be considered flexibly: Peregrines might use the same cliff or quarry, but a different ledge, or a different section of the same ledge. It will

Nominate Peregrine heading into its inland cliff site in Germany. *Reinhold Möller*.

be interesting to see if the boxes now being extensively placed for Peregrines on urban sites (such as churches in the UK) become traditional in the true sense.

At coastal sites nest ledges are often on the upper part of tall cliffs, though, of course, the structure and stability of the face, and availability of a suitable ledge are determining factors. Most of Treleaven's studied eyries on Cornwall's north coast were situated on the lower third of the face. However, these are often high cliffs with friable, unstable upper sections (of Porthtowan Slate) where the ledges are unstable or unsuitable (P. Welsh pers. comm. to SW). It is clearly essential that any coastal eyrie site is sufficiently high to gain protection from storm-driven waves or salty spume. Ratcliffe (1993) states that 'since cliff-nesting is an adaptation against predation, the value of the nest ledge depends largely on its inaccessibility and to fulfil its function it must have a steep fall of rock above and below.' He further states that 'ledges with some degree of overhang are evidently preferred'.

At inland sites in Scotland, nominate Peregrines nest at altitudes up to about 1100m, which is about as high as cliffs can be found in Britain, though in other parts of the falcon's vast range nests are found above the 3000m contour. At sea sites the Peregrines usually have little choice in terms of the direction the chosen ledge faces, but at inland sites where greater choice is available, studies across the falcon's range suggest sites with an open view are usually favoured. That said, relatively tight gorges are also utilised, so the protection a ledge offers appears paramount in the selection process. Cliffs, being natural structures, do not have preferred orientation. There is little evidence for a preferred direction, though studies in North America suggest that if the Peregrines have a choice they prefer a position which offers shade from the sun and may choose a ledge shaded by a boulder if that is not possible. Such choices to minimise the effects of climate are likely to occur across the Peregrine's range if such effects are potentially extreme.

Peregrines in various parts of their range occasionally utilise stick nests in trees. Buchanan *et al.* (2014) has details of incidences in North America[6], but does not include the report of Whitman and Caikoski (2008) who photographed an apparently incubating Peregrine (probably *F.p. anatum*) in a stick nest built the previous year by Bald Eagles in a Quaking Aspen (*Populus tremuloides*): breeding was inferred, but not confirmed. Thomsett (1988) notes two uses of stick nests in trees by African (Kenyan) Peregrines, one being in a snake-eagle (Circaetus *spp.*) nest in a Candelabra[7], the other being in a eucalyptus (nest builder not stated). Both Pruett-Jones *et al.* (1981) and Hurley (2009), studying *F.p. macropus* in Victoria, Australia noted most falcons nesting in cliff sites, but some in tree-hollows and in stick nests in trees. Of the former almost all were in River Red Gum (*Eucalyptus camaldulensis*) trees, while in the latter the Peregrines were mostly using nests built by Wedge-tailed Eagles. That Peregrines might use a tree hollow to breed seems very surprising, but Pruett-Jones *et al.* (1981) noted that of 95 identified Peregrine nest sites in 1976-1977 62 were in cliffs or quarries, with 17 in tree hollows and 16 in stick nests.

On New Caledonia, Baudat-Franceschi *et al.* (2013) note that *F.p. nesiotes* hunting seabirds at the coast were often utilising the stick nests of Ospreys (the sub-species *Pandion haliaetus cristatus*). Tree nesting was also widespread in central-east Europe at one time but is now concentrated in Germany and Poland as a result of Peregrine reintroductions which began in 1990 and continued through the 2000s – see Langgemach *et al.* (1997) and Kleinstäuber *et al.* (2009) for details on Germany, and Mizera and Sielicki (2009) and Sielicki and Sielicki (2009) for details regarding Poland. In a study in Germany, Kirmse (2001) noted that Peregrine chicks raised in stick nests in trees may choose to use a stick nest, a cliff or a building when they begin to breed, but that chicks raised in non-tree nests had never been noted as breeding in trees. In an interesting study in Russia's southern Urals, Pazhenkov *et al.* (2018) have attempted to reintroduce tree-nesting by removing chicks from cliff sites (which were in danger of destruction by humans) and placing them on nest platforms (and artificial nests) in trees. The adult birds readily

---

[6] One of the earliest examples noted by Buchanan *et al.* (2014) was a report from 1878, although Buchanan *et al.* do not give full details. The report was by Goss (1878), who in 1875 found a pair of '*Falco communis var. anatum*' which he refers to as Duck Hawks, breeding in a cavity 50ft (15m) up in a Sycamore (*Acer pseudoplatanus*) in Kansas. Goss shot the female, collected it and sent a boy up the tree for the eggs. In 1876 he found a pair nesting in a Cottonwood (probably *Populus fremonti*), shot both but '*failed to get them*'. Then in 1877 he noticed another pair in the same tree. He shot both. Goss writes '*I now have the three birds in my collection*'. Goss claims any Peregrine seen is always shot because of '*how destructive they are to the water-fowls*'.

[7] Thomsett (1988) states the tree was a *Euphorbia candelabra*, but the species is now *Euphorbia ingens*, Candelabra being the common name.

Nominate tree-nesting Peregrines in Germany. *Torsten Pröhl*.

Single nominate Peregrine chick hatched on a peat hag in moorland country, northern England.

transferred their feeding to the new sites, though whether the experiment will be successful in the longer term is unclear. There have been a handful of observations of tree nesting in Britain. In 1999 three chicks were fledged from a clutch of four eggs in a Buzzard stick nest in Scotland (Leckie and Campbell, 2000). This seemed to be a very rare occurrence, but more recently Pictor (2013) has suggested that tree nesting might be more common than has been assumed. Pictor identified Peregrines successfully breeding (two fledglings) in a Common Buzzard nest built at a height of 15m high in a mature Larch (*Larix spp.*), and also noted instances of tree breeding in Sussex and Shropshire. It is likely there have been other instances, the rarity and unexpected nature of tree nesting, together with the difficulty of locating such nests, meaning the possibility is often overlooked.

Instances of ground nesting are also known. In the UK Ratcliffe (1993) records several, these usually being on slopes, often with broken rock outcrops and/or thick vegetation, offering a high degree of protection, they may also be on more open ground. The authors are aware of several UK ground nesting sites in 2022 where successful breeding has occurred.

Ground nesting is also seen in very remote areas where human disturbance is very unlikely, for instance in Siberia where peat mounds and tussocks are found on the tundra, and where cliffs are in short supply. Mebs (2001) also notes that it is common in Finland (90% of all nests). Much rarer are instances in the USA (Pagel *et al.* (2010) with *F.p. anatum* in California, and Boettcher and Mojica, 2016). Note that the latter was for captive-bred, mixed heritage falcons released into the eastern USA following the profound reduction in the population due to DDT.

Peregrines have also nested on electricity transmission towers (pylons) and other man-made structures such as quarries, bridges, factory chimneys, shipyard cranes and buildings of all kinds (each of which, of course, are very similar to a natural cliff). The fact that Peregrines have moved into the human world with such alacrity is an indication of their remarkable adaptability. While the falcons initially bred in old, abandoned quarry sites, there were

F.p. anatum clutch laid close to the Pacific Ocean, and across the bay from the towers of San Diego, California, USA. *Joel Paget.*

soon instances of the use of working quarries. The present authors know of many such breeding sites in operating quarries in the UK, but a more remarkable usage is noted by White *et al.* (1988b) where *F. p. macropus* pairs were breeding in seven of 11 working quarries in Victoria, Australia. In one quarry the eyrie was 50m from rock crushing equipment and other machinery. At another quarry falcons bred at the top of a gravel silo just 20m above where lorries were being loaded with gravel. A third, inactive, quarry was used by shooters: the female left her brood during target practice, returning when it stopped. White *et al.* also noted a falcon pair breeding on a hydroelectric dam site within 20m of a service road and car park for reservoir staff [8].

---

[8] In an interesting combination of natural feature and human assistance for breeding Peregrines, Pagel (1989) reports the use of explosives to enhance a natural site, enlarging an eyrie which was too small, resulting in both eggs and chicks having been lost by falls over several years. The operation was entirely successful: over five years (1984-1988) 13 chicks fledging (2.6/year). The building of artificial eyries and the improvement of natural eyries (but without the use of explosives) was also used to aid the post-pesticide recovery of nominate Peregrines in France's Jura Mountains. In 1984, 231 eyries were 'enhanced', a project which was extremely successful (Monneret *et al.* (2015).

Nominate pylon-nesting Peregrines in Germany. *Torsten Pröhl.*

Just as man-made quarries could be seen as a natural extension of the use of cliffs, the adaptability of Peregrines to humans was a distinct 'trick up the sleeve' of the falcons, allowing them to make the seemingly short step to human-constructed pseudo-cliffs – tower blocks, churches, cathedrals, hospitals, power plants *etc*. In one sense, urban breeding Peregrines is not a new phenomenon. In the UK, having been recorded on Salisbury Cathedral in the early 1860s, the Peregrine tolerating not only street noise, but the occasional bursts of bell-ringing. Culver (1919) records a pair of Peregrines in the centre of Philadelphia in January 1918 (though noting they might have been present much earlier in the winter). The pair – Culver does not state whether they were male and female, but the recorded display flights suggest pair bonding or reinforcement – probably roosted on a tall tower in the centre of the city from which they flew to capture pigeons. (Such common roosting is often seen in breeding pairs prior to egg laying.) The falcons stayed until March, then left the city. One of the more famous urban nests was that on the Sun Life Building, Montreal, Canada where Peregrines bred continuously from 1936-1947, then intermittently until 1952 (when, it was believed egg breakage due to DDT prevented yearly success) before disappearing (Hall, 1955). The first breeding attempt of the Sun Life pair was on a ledge: the eggs rolled off. A box filled with sand/gravel was then provided and used successfully. But the explosion of urban Peregrine breeding since the 1990s is astonishing, no doubt given additional momentum by the availability of webcams which have raised the interest of human populations in the falcons so that breeding boxes have been erected on an increasing number of buildings, and the adaptability of the falcons has ensured their use. An extra factor, of course, is that many

Two curious egg substrates.

Above   *F.p. cassini* chicks hatched from eggs laid in a volcanic crater in Argentina. *David Ellis*.

Below   *F.p. tundrius* egg laid on frost-shattered rock on top of a raised rock plateau, Southampton Island, Canada. The photographer apologises for the imperfect focus. In his defence it was a cold, fiercely wind-swept day, the egg was found accidentally, and the photograph was taken quickly to minimise exposure time, both for the egg and the photographer's hands. A later visit found the female incubating a clutch of three eggs.

urban sites are both inaccessible and guarded (to a greater or lesser extent) providing the best protection against unwanted human disturbance[9].

While urban breeding might be assumed to be entirely dependent on the availability of nest sites (and, perhaps, the provision of nest boxes), an interesting study in South Africa found that the most important driver was immigration (Altwegg *et al.*, 2014) – Fig. 6.8. Peregrines, it seems, have grown to love living in the city[10]. The change appears to be a worldwide phenomenon, as Faccio *et al.* (2013) noted that after Peregrines had become extinct in New England in the 1960s a reintroduction programme had caused

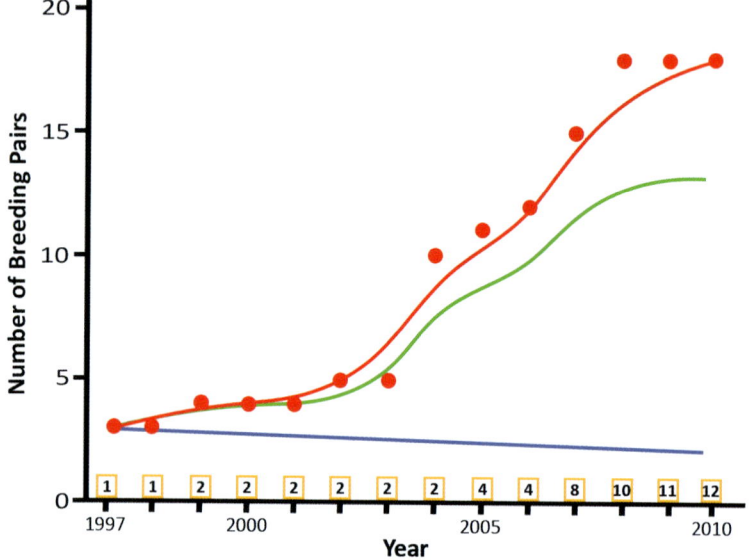

Figure 6.8 Variation of the number of breeding Peregrine pairs in urban Cape Town, South Africa, 1997-2010. The red dots are the actual observed number of pairs. The red line is a fitted curve to the observed data. The green line assumes that those Peregrines nesting in nestboxes would not have bred had the boxes not been erected. The blue line is the number of falcon pairs that would have bred had there been no immigration into the city. The numbered boxes at the base of the figure show the number of nestboxes available for the Peregrines in each year of the study. Redrawn from Altwegg *et al.* (2014).

a rapid population expansion which soon filled the original range, with the falcons utilising many urban structures – buildings, bridges *etc.* – for breeding. While most Peregrines on first breeding tended to favour nest sites similar to their own natal habitat (*i.e.* cliff or building) there was significant cross-immigration of rural and urban populations.

---

[9] While lack of disturbance is obviously important, a study in London (Mak *et al.*, 2021) notes that site selection was also influenced by other factors, Peregrines preferring proximity to water bodies, built-up areas, parks and gardens, but avoiding, where possible, agricultural land and woodland. Not surprisingly, these environment choices are clearly influenced by prey availability.

[10] For a review of the early development of urban nesting in the USA see Cade and Bird (1990). For a review of urban nesting by Peregrines (and other raptors) across South Africa see McPherson *et al.* (2021).

Above Baskets placed in trees offer an alternative to urban boxes for nominate Peregrines in Germany. *Torsten Pröhl.*

Below True urban living. Nominate female at the nest box placed on Tewkesbury Abbey, England. Behind the adult is the camera system which allows the public to watch the comings and goings at the box. One chick can just about peer over the side of the box. *Peter Rowland.*

The rise of urban breeding across the Peregrine's range suggests it is beneficial for the species, an idea which was highlighted in a study by Kettel *et al.* (2018) who collected and compared data on urban breeding raptors (falcons, hawks and owls) from across the globe and found that Peregrines were the only species which showed a statistically significant increase in clutch size (and hence fledging success) compared to their traditionally breeding cousins. In a second paper, dealing only with Peregrines breeding in the UK, Kettel *et al.* (2019) noted that in a comparison of 22 urban and 27 rural sites across the country, urban falcons raised one more chick to fledge than rural pairs, whether the comparison was between nesting attempts or successful nests (see, also, Footnote 5). Peregrines clearly benefit from urban living, though the urban environment is not without its hazards, fledglings in particular being at a higher risk of collisions than their rural cousins and being occasionally poisoned by pest control programmes. For an interesting recent review of urban raptors see Boal and Dykstra (2018).

In general, nest selection, whichever site is chosen, seems to involve both birds, with males often attempting to persuade the female to accept a site by scraping one or more potential sites. As Nethersole-Thompson and Nethersole-Thompson (1943) note, it is the female who makes the final decision, improving the scrape, and sometimes spending several days 'brooding' it before egg laying begins. While this activity would appear to be a final decision, occasionally the female will even then change her mind and choose a different site.

Before leaving nest sites, an in an excellent study on nominate Peregrines in Brittany, France (Cozic, 2019) noted an interesting change as the population

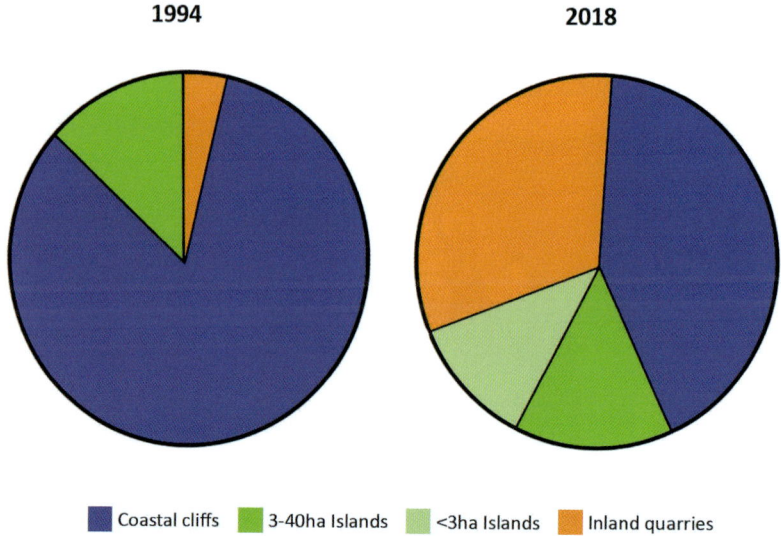

Figure 6.9 The evolution of breeding sites in Brittany as the Peregrine population increased from 1994 to 2018. Redrawn from Cozic (2019).

increased. Prior to 2012 the falcons mostly bred on coastal cliffs, with a small number on offshore islands and a single pair at an inland quarry. As the population increased, the number of suitable coastal sites ran out, the large offshore islands that had been used also became filled with falcons instead using much smaller islands. The number of pairs in quarry sites also increased dramatically, but interestingly, there were many fewer pairs on buildings than would have been expected (Fig. 6.9).

Copulation
A continuation of the 13 display behaviours noted by Wrege and Cade (1977) mentioned earlier in the chapter, relates to copulation. Male Peregrines will raise their wings whilst flying, the wings being held high above the plane of the body with short, clipped wing beats maintaining a slow-motion display flight with the tail often depressed (the oft-called 'hitched wing display'). After landing close to a potential nesting site the hitched wing is maintained momentarily adding to the ritualised nature of the display. The male will then bow from the neck while chupping quickly to solicit copulation. The female will also make the wailing call while bowing extravagantly. Copulation follows, with, as usual, the male mounting the female and flapping his wings vigorously. Throughout copulation he maintains his toes and talons in a clenched, balled position so as not to inadvertently inflict injury on his mate. After copulation the male then leaves the ledge, while the female remains, fluffing her feathers, this often accompanied by a low, soft, chupping sound. Copulation takes only a few seconds at first, but may take up to ten seconds as egg laying approaches. The rate of copulation also increases, Ratcliffe (1993) suggesting that frequencies of 3–4 per hour, or even more, have been seen. In

Headless Eurasian Teal found at the base of the Evesham Bell Tower in the weeks prior to the start of breeding. The lipids in the brain are valuable to a male Peregrine intending to breed, counting for why the head has been removed.

Breeding Part 1

general copulations cease after the laying of the third egg. In a study of *F.p. tundrius* in Arctic Canada noted the rate of copulation as 7.1±6.9/day based on the observation of 15 Peregrine pairs.

Copulation of nominate Peregrines on the Isle of Man, UK.

In the first image, *top left*, the male approaches the female. After landing on her back, *bottom right*, the male quickly folds his toes inward, *above*, to avoid damaging the female with his lethally sharp talons. In some cases the male will actually ball his talons to avoid all contact between them and the female. *Peter Christian.*

7   Breeding Part 2

A female Peregrine intending to breed and having already decided upon a mate and nest site, is faced with two critical decisions – when to start laying and how many eggs to lay. While there appears to have been little work on this issue specific to Peregrines, research on other species has shown that each of these decisions has a significant impact on the likelihood that a female's eggs will result in surviving offspring. Significant research has been carried out on these aspects of the breeding behaviour of the Common Kestrel at the University of Groningen in The Netherlands – see Sale (2020 and the large numbers of references therein). The Dutch work showed that without the constraints imposed by the physiology of both male and female Kestrels, breeding could take place at any time of year. But for the male the need to hunt in order to feed the incubating female and, after hatching, the growing brood, meant day length constrained the breeding period. For the female, the need to accumulate the bodily reserves required for breeding was also a constraint if she overwintered near the breeding grounds, a period when, again, daylight reduced hunting time. For females of migratory sub-species, accumulating the necessary reserves is potentially an even more important issue as on arriving at the breeding grounds the female has first to replenish the reserves used during the migration flight.

Another aspect which research has noted as important in raptor breeding is Lifetime Reproductive Success (LRS) which defines the number of fledglings an adult bird produces over its lifetime. LRS has significant inputs: does a juvenile raptor attempt to breed as early as possible, or does early breeding shorten reproductive life as the demands of breeding (the energy requirements of the male's hunting and of the female's egg-laying), shorten their life by imposing huge burdens on still-developing bodies? Is it best to lay more eggs and so produce more fledglings, or does the workload of raising an increased number of chicks mean that, again, breeding lifetime is reduced? All other things being equal (probable lifetime, probable prey density, probable weather conditions *etc.*), to maximise LRS, the parent birds need to maximise their reproductive effort annually. To investigate LRS the Groningen University team reduced or enlarged clutch sizes by taking chick(s) from one nest and placing them in another (Daan *et al.*, 1990). Brood size was altered 10 days after hatching. The results of these alterations are shown in Figs. 7.1-7.3.

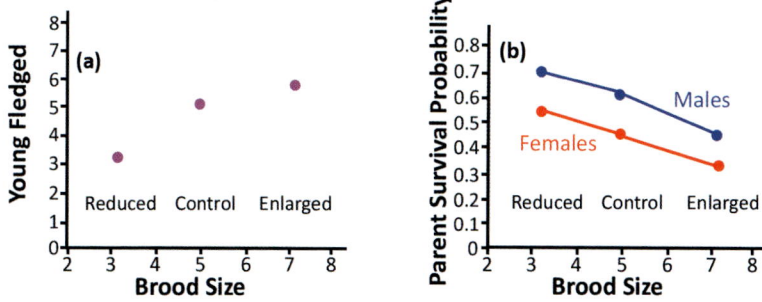

Figure 7.1 Effect of brood reduction and enlargement on the number of chicks fledged (a) and survival of the parent birds to the following year (b). Redrawn from Daan *et al.* (1990).

Although Fig. 7.1a notes that the number of chicks fledged in enlarged broods increased, the survival of an individual chick declined in those nests, *i.e.* the number of chick deaths in enlarged nests was higher than average. Fig. 7.1b notes that the effect on parent birds was equally pronounced. Parents with reduced broods were more likely to survive to breed again, their workload having declined, but the survival probability of parents (both males and females) with enlarged broods was reduced, *i.e.* parental effort raised more chicks, but at the expense of a shortened life. The Dutch team then defined 'reproductive value' for both parents (V_P) and chicks (V_C) and showed that for the chicks 'value' declined with laying date – as a chick you are more likely to survive to breed if your egg is laid early – but rose steadily with parental effort at first until the increased risk of high clutch sizes caused it to decline again

(Fig. 7.2). For parents the situation is different as their likelihood to breed again declines with the effort of chick raising.

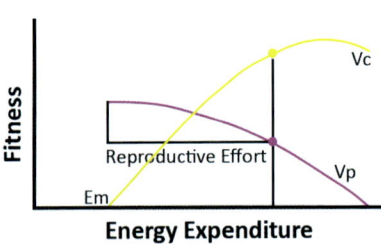

Figure 7.2 General scheme of clutch size optimisation theories. With increasing energy investment in the current season's reproduction, the value of the clutch (Vc) increases. But the rate of increase declines due to competition for food and increased risk at higher clutch sizes. The residual reproductive value of the parent birds (Vp) declines with increasing energy expenditure. The dots indicate the optimal position. The 'Reproductive Effort' is the energy requirement beyond Em, the energy requirement of non-reproduction (*i.e.* the energy requirement of self-maintenance). Redrawn from Daan *et al.* (1990).

While the solution for parents seems obvious – do not breed and live longer – that is hopeless in terms of passing on their genes. The Dutch therefore added Vp and Vc to give an overall reproductive value (Fig. 7.3) and found that evolution had already organised Kestrels to provide the optimal solution. Doubtless evolution has done much the same job with the equally successful (in terms of geographical spread and total population) Peregrine.

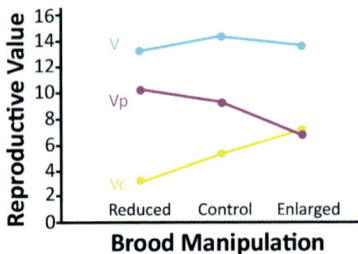

Figure 7.3 The reproductive value of the brood (Vc) and parent birds (Vp) and their sum (V). The scale of reproductive value is defined by the mathematical calculations of Daan *et al.* (1990), but is essentially arbitrary in terms of the numeric value. Redrawn from Daan *et al.* (1990).

Studies of breeding at sites in the UK

Over recent years, the choice of urban nest sites by Peregrines has allowed the use of installed camera systems to investigate the breeding habits of the falcons at multiple sites (see Dixon (2021) for a list of such sites in the UK). However, the majority of these sites are in urban settings, and not all of them allow an intimate study of breeding behaviour. In order to provide comparative data three sites were identified for comprehensive studies. In southern Scotland a trailcam was installed at a remote site (too remote to allow the use of high-quality cameras which require mains electricity). This site was chosen to provide data for a 'wild country' falcon pair. At Evesham, in England's south-Midlands, a camera system was installed on a medieval bell tower. A purpose-constructed nestbox was installed. External to it, a zoom-capable,

4k camera was set up. Within the box two cameras were installed, one to monitor adult arrivals as the constricted nature of the site meant the external camera might only monitor the back of arriving adults (and, of course, chick behaviour), so the internal camera could identify prey as it was delivered. A second internal camera monitored chick behaviour. At night infra-red lights on all three cameras allowed night-time behaviour to be recorded. This site was semi-rural, and previous experience had shown that the Peregrines hunted in the adjacent riverine country as well as taking the occasional urban pigeon. In London data from a previously-installed, infra-red enabled camera was available to study an urban breeding Peregrine pair, a replacement infra-red enabled camera being installed by the present authors as a further aid. At all three sites activity was recorded throughout the egg-laying to fledging period. The authors wish to thank Nathalie Mahieu of *FaBPeregrines*[1] (who has been monitoring the Peregrines here since 2007 and has kept meticulous breeding records for this site as well as acting as its guardian and ambassador) for her generosity in providing access to her video clips, photographs and copious detailed breeding notes and incubation charts.

Weather data was acquired from the UK MetOffice in order to investigate if rain, wind (both mean speed and gust speed) and temperature had any influence on the hunting of the breeding Peregrines. While such an influence, should it occur, would have the most significant affect during chick rearing, data were collected from first egg through to the time when either the chick(s) had fledged – at the London site – or had disappeared from continuous view – at the other two sites. Data was collected from two MetOffice sites close to (and upwind of) each of the three breeding sites. Data at these six MetOffice sites was logged hourly during the entire filmed breeding season (midnight 4/5 March-midnight 24/25 May at the London site; midnight 5/6 April-midnight 13/14 June at Evesham; and midnight 14/15 April-16/17 June in Scotland. The 2022 breeding season coincided with one of the most benign spring/early summer periods in recent UK history, with minimal instances of rain, and minimal rate of rainfall when it did rain, and minimal instances of high winds. Temperatures were also benign, although sub-zero night-time temperatures were registered during incubation at both the Scottish (to -4°C) and Evesham (to -3°C) sites. The night temperature also fell close to 0°C at the Scottish site for a short period during chick rearing. The highest wind speeds (both mean and gust) and rainfall were seen at the Scottish site. These data are shown in Figs 7.14 and 7.15 later in the Chapter. In highlighting these data it has not been our intention to suggest that Scottish weather is worse than the weather in other parts of the UK. The reason for choosing these examples is solely because the Scottish prey delivery data showed only a single prey delivery during a 65-hour period – see *Scottish Borders* below and related Figs. 7.14 and 7.15.

---

[1] The location explains Nathalie Mahieu's social media site name *FaBPeregrines* from **F**ulham **a**nd **B**arnes, and her love of the (**fab**ulous) Peregrines.

The data from these three sites in differing habitats are set down below. Later in the Chapter these UK data are compared with other studies from across the Peregrine's vast range.

Charing Cross Hospital, London

The name Charing, for this area of London, derives from the Old English word 'cierring', referring to a bend in the River Thames. The addition of 'Cross' to the hamlet's name originates from the cross erected in 1291–94 by King Edward I as a memorial to his wife, Eleanor of Castile. The cross was placed between the former hamlet of Charing and the entrance to the Royal Mews of the Palace of Whitehall (today the top of Whitehall on the south side of Trafalgar Square). Folk etymology suggests the name derives from chère reine ('dear queen') but the original name pre-dates Eleanor's death by at least a hundred years. The wooden, sculpted cross was the work of the medieval sculptor, Alexander of Abingdon, but was destroyed in 1647 on the orders of Parliament during the Civil War. A 21m stone sculpture in front of Charing Cross railway station is a copy of the original cross. Erected in 1865, it is situated a short distance east of the original cross, on the Strand. It was designed by the architect E.M. Barry and carved from Portland stone, Mansfield stone (a fine sandstone) and Aberdeen granite by Thomas Earp of Lambeth. It is not a faithful replica, being more ornate than the original.

Charing Cross Hospital is an acute general teaching hospital located in Hammersmith, west London. The present hospital was opened in 1973: the original had been established in 1818 in central London. Although rather curiously (and confusingly) still named 'Charing Cross Hospital', the building is now located in a suburban residential area (Fulham and Barnes) some 8km west of central London, and is situated 500m from the River Thames, 800m east of the London Wetland Centre. The Peregrine nestbox is set on a small balcony on the 15th floor, the very top of the building, and is therefore not overlooked.

The female Peregrine of the breeding pair is a four-year-old falcon fledged from a nestbox on the Heineken Brewery at s-Hertogenbosch in The Netherlands, ringed on 18 May 2018. Named *Azina* by Nathalie Mahieu (NM) from the darvic code AZN on her right tarsus, the female dispersed 385km from natal to breeding site. *Azina* first arrived on site on 19 December 2020 and laid three eggs in 2021, producing one chick to fledge.

The unringed male Peregrine (named *Tom*) arrived at this site in October 2011 in adult plumage and so is at least 12 years old. He has bred successfully here with 3 females and had produced 30 eggs and 14 fledged young prior to the 2022 breeding season.

In 2022 the first egg was laid at 18.48 on 7 March, the earliest ever recorded at this site. A second egg was laid at 07.51 on 10 March, 61 hours after the first, a third egg being laid at 19.48 on 12 March, 60 hours after the second. A fourth egg was laid on 15 March at 07.33, almost 60 hours after the third.

Azina with a captured (Common) Moorhen (*Gallinula chloropus*). It is notable that the London Wetland Centre, maintained to allow city dwellers to enjoy the delights of the country was also used by the urban Peregrines for the same reason, but with an entirely different outcome. *Nathalie Mahieu*.

Fig. 7.4 shows the egg laying times and incubation share of the male and female from the day of laying the first egg to hatch on 16 April. As was the case for the Evesham pair (see below), hard incubation only started once the penultimate egg had been laid. Only one egg hatched: Nathalie believes it was either egg 2 or egg 3. If it was egg 2 then period from laying to hatch was 36.9 days, if egg 3 then 34.4 days.

Figure 7.4 Incubation times of male and female Peregrines at Charing Cross Hospital, London, 2022. The times, and times of day, that each adult incubated are set down from the time of laying of the first egg.

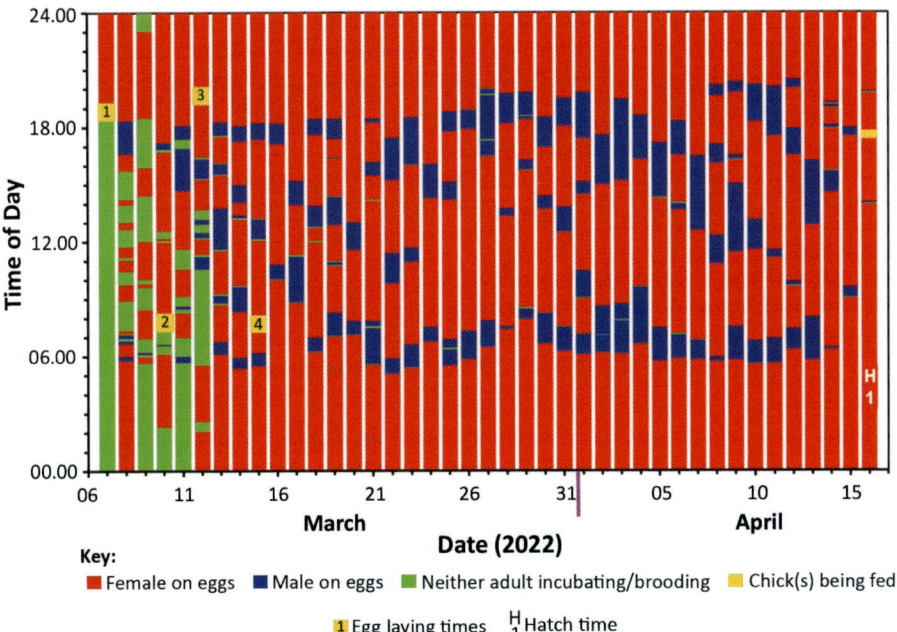

Azina laid a four-egg clutch (*left above*), but only one egg hatched. *Above right* shows that egg pipping. NM/RS/SW.

Incubation followed a regular daily pattern whereby the female always incubated overnight, averaging 11 hours 47 minutes with a minimum of 9 hours 12 minutes and a maximum of 15 hours 47 minutes. The female incubated for 84% of the total time and the eggs were only left uncovered for an average of four minutes per day.

Typically the female's overnight incubation was curtailed when the male arrived with prey, this occurring regularly between 05.00 and 07.00. The male would then take over incubation for typically 1-2hrs while the female was away feeding and preening. The male would regularly incubate for two to three daily sessions, averaging a daily total of 3 hours 48 minutes with a maximum of 6 hours 50 minutes and a minimum of 1 hour.

As is normally the case, the female was dominant at this site and determined the incubation timetable, even though the male was a much more experienced breeder. On occasions *Azina* would get off the eggs suddenly, perhaps distracted by an intruding conspecific or human activity, and in such cases the male would invariably fly in and cover the eggs almost immediately. Clearly vigilant, Tom's breeding experience showed in his behaviour.

Only one egg hatched, on 16 April at 04.45. It might be conjectured that this could be associated with the male's age. However, since 2014, *Tom* has sired no young on three occasions, one on four occasions and four in 2019, so age-related male fertility does not necessarily appear to be the problem. It is certainly the case that there is evidence that female fertility peaks at around eight years of age and then declines (see Fig. 7.13). As far as the authors are aware, no study has been done to investigate age related male fertility and therefore this can only be deemed to be speculation on our part. Chick brooding started to drop off sharply after day 14 post-hatch which is of course a function of the chick's ability to thermoregulate at that point. At 02.23,

Saturday 7 May (Day 22) the chick positively rebuffed the female's attempt to brood.

Fig. 7.5 shows the number of prey items delivered per day at this site. It is notable that this remained fairly constant throughout the hatch to fledge phase, averaging 4.6 prey items with a maximum of 8 and a minimum of 2. Note that the number of prey items delivered is distinct from the number of prey captures made in that multiple feeds may be taken from a single prey item. Typically, the male would bring the prey item to the nest ledge and the female would then immediately grab it unceremoniously, as if to say, 'what kept you?'. The male would then smartly depart and *Azina* would feed the chick immediately or fly off with the prey for a few minutes before returning to feed the chick.

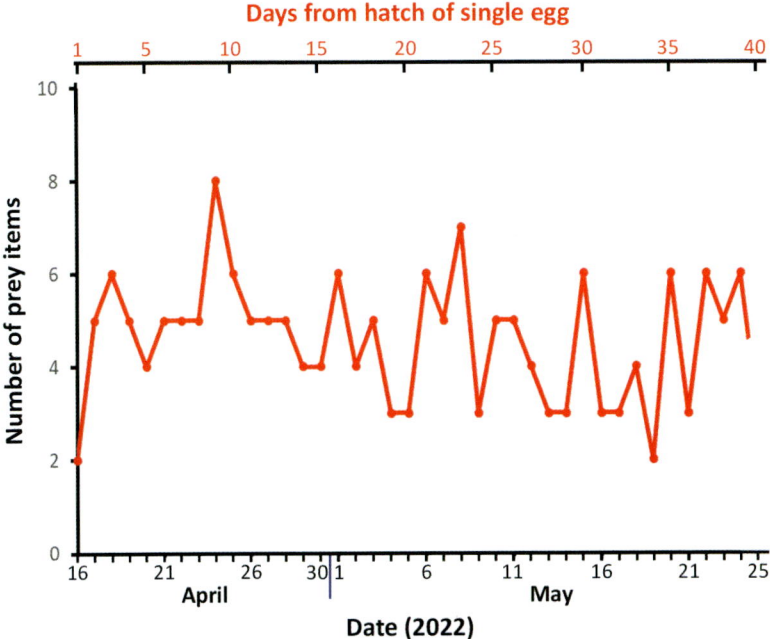

Figure 7.5 Number of prey items delivered to the single chick each day at Charing Cross Hospital, London, 2022.

Fig.7.6 (overleaf) shows the number of prey items delivered by time of day. There is a clear peak of deliveries first thing in the morning, between 05.00 and 06.00, these items often being cached from prey captures made the previous day. There is also a second peak in the evenings between 17.00 and 21.00. See also the similar figs at the other two studies sites below, and Fig. 7.17 later in the Chapter.

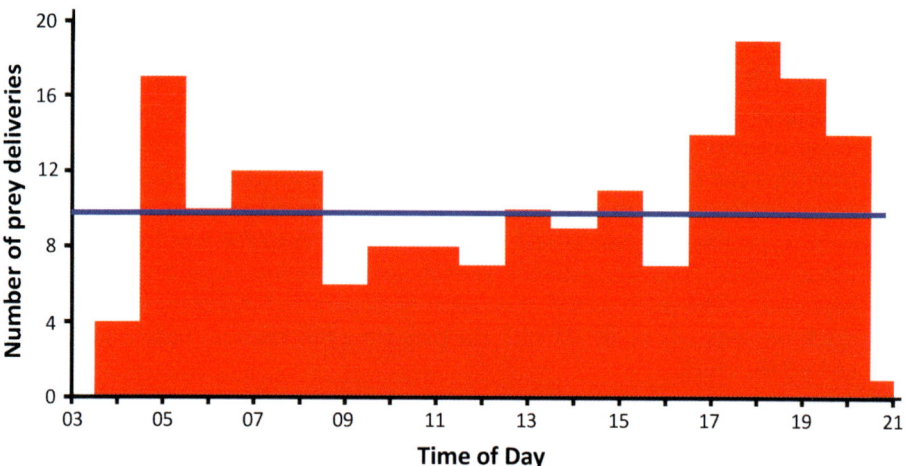

Figure 7.6 Time of day of prey deliveries to the single chick at Charing Cross Hospital, London, 2022, listed as deliveries in one hour time slots between 03.00 and 21.00. The blue line represents the mean delivery rate of 9.8 items/day between 03.00 and 21.00.

Fig.7.7 shows the prey spectrum of captures made over the hatch to fledge period of 40 days between 16 April and 25 May. 93 prey captures were made in total over this period, averaging 2.3 per day, unsurprisingly, the vast majority being feral pigeons, these representing 63.5% by number and 86.5% by biomass. Starlings were the next most favoured prey (20.4% by number, 6.4% by biomass), followed by Ring-necked Parakeets (8.6% by number, 5.0% by biomass). Other prey items included a single Swift and several small

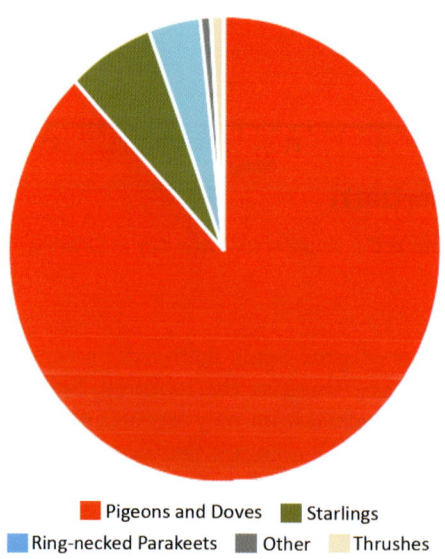

Figure 7.7 Prey spectrum by biomass percentage of nominate Peregrines breeding at Charing Cross Hospital, London in 2022. Some prey captures by both adults will not have been brought to the nestbox but will have been consumed off-site. So the spectrum will not necessarily be a complete record of prey captures during the breeding period.

Breeding Part 2

Above   The single chick, named *Indy* but with a darvic coded PDT, being fed from a captured feral pigeon. *NM/RS/SW.*

Below   *Indy* being fed a captured Ring-necked Parakeet. *NM/RS/SW.*

passerines. It is interesting to note that two of the last three items brought to the chick prior to fledge were delivered alive. As this did not happen during the previous 38-day period it does not appear to have been by chance. Both authors have observed live prey being dropped for strong flying juveniles to chase and hone their hunting skills, so this is likely to have been similar 'training' behaviour, the prey delivered alive for the chick to kill. In both cases, however, the prey was actually despatched by the adults on the nest ledge, although this does not preclude the possibility that the behaviour constituted 'training'. Fig. 7.8 shows feeding data per day split by percentage delivery by male and female, showing the age at which the chick began feeding itself, and the extent of that feeding thereafter.

SW was surprised to observe that the male was allowed to feed the chick commencing on Day 2 as previous experience suggests that this is unusual given the dominance that most females exercise at this early stage of feeding. One might conjecture that the combination of a highly experienced male and a considerably less experienced female may be part of the inter-pair dynamic in this instance, allowing the male to feed the chick very early on.

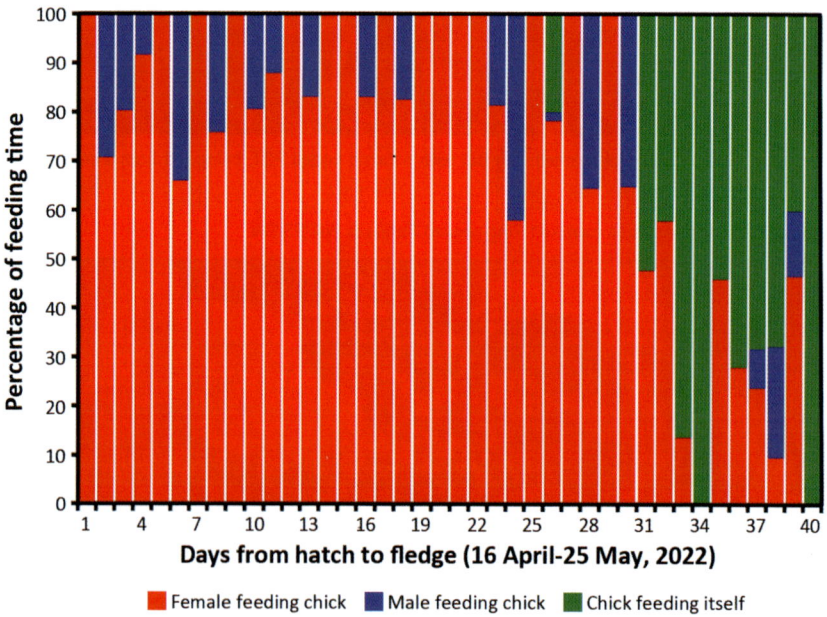

Figure 7.8 Variation with chick age of the percentage of chick feeding by male, female and self-feeding by the chick over the 40-day pre-fledging period at Charing Cross Hospital, London in 2022. The data shows that the female fed for 74.7% of the time, the male for 9.5% and the chick fed itself for the remaining 15.8%. The chick started feeding itself at Day 26 and this became the norm from Day 31. In the last 10 days before fledge the chick fed itself 66% of the time, the female 29%, and the male 5%. The average daily frequency of prey deliveries was almost identical at the London and Evesham sites. Clearly the length of time taken for each feed was greater at Evesham as this site raised two chicks whereas Charing Cross had only one.

After fledging, the London chick *Indy* was photographed in London's Richmond Park having captured a European Green Woodpecker (*Picus viridis*) fledgling. *Warren Richardson.*

Evesham Bell Tower, Worcestershire, England

Evesham is a market town, with a population of some 20,000, in England's south-Midlands. The town grew up around an abbey built in the early years of the 8th century after a local swineherd or shepherd reported seeing a vision of the Virgin Mary. The local was said to have been called *Eof*, the town that grew around the abbey being named *Eof's ham* (Eof's place) which became, over time, Evesham. After Henry VIII's Dissolution of the Monasteries in the 16th century, the abbey was largely dismantled and sold as building stone. Little now remains – some wall remnants, together with an outline of the main buildings) – apart from the square Bell Tower. Built by Abbot Clement Lichfield in the early 16th century (Perpendicular Gothic style), the tower stands alone, remote from the later town church and other buildings. The tower's 14 bells are considered one of the finest sets of change-ringing bells in the country.

Peregrines first bred behind a modern clock face on one of the tower's faces. Then, in 2022 with the permission of the incumbent, the authors installed a purpose-built nestbox and a camera system in the hope the falcons would use it. The system included a hi-res external camera and two smaller cameras mounted within the nextbox. One of these monitored activity within the box, the other monitored movements external to the box. The need for two cameras watching external movements was required because of the nature of the tower's roof structure which was a diamond of lead surrounded by narrow passageways which precluded a camera being mounted at right angles to the box entry. The cameras were switched on prior to the hoped-for arrival of the Peregrines. The authors realised the likelihood of the box being taken in the first year was limited, but on the morning of 6 April 2022 the female was recorded making a scrape in the fine gravel substrate added to the nestbox. Later that morning both birds were recorded in the box, the male also making a scrape.

The female falcon did not have a ring: the male bird did, but the full number was never adequately photographed by the on-site cameras. However, the male was photographed external to the nestbox. The full ring number was not identified, but the early section of the number suggested it had been hatched three years previously at Tewkesbury Abbey, some 20km away.

The Evesham Bell Tower male. *Peter Grimmitt.*

The first egg was laid in the afternoon of 9 April. A second egg followed during the early morning of 12 April, the clutch being completed by a third egg laid during the afternoon of 14 April. The actual time of laying could only be established when the female arrived at, and left the box revealing the absence, then presence of the egg. On that basis the first egg was laid between 16.47 and 17.25, on 9 April, the second 05.00-06.56, 12 April, the third 17.58-18.02, 14 April. From these data the times between laying were 61.7±1.3hrs for the second egg and 59.0±1.5hrs for the third. Fig. 7.9 shows

Breeding Part 2

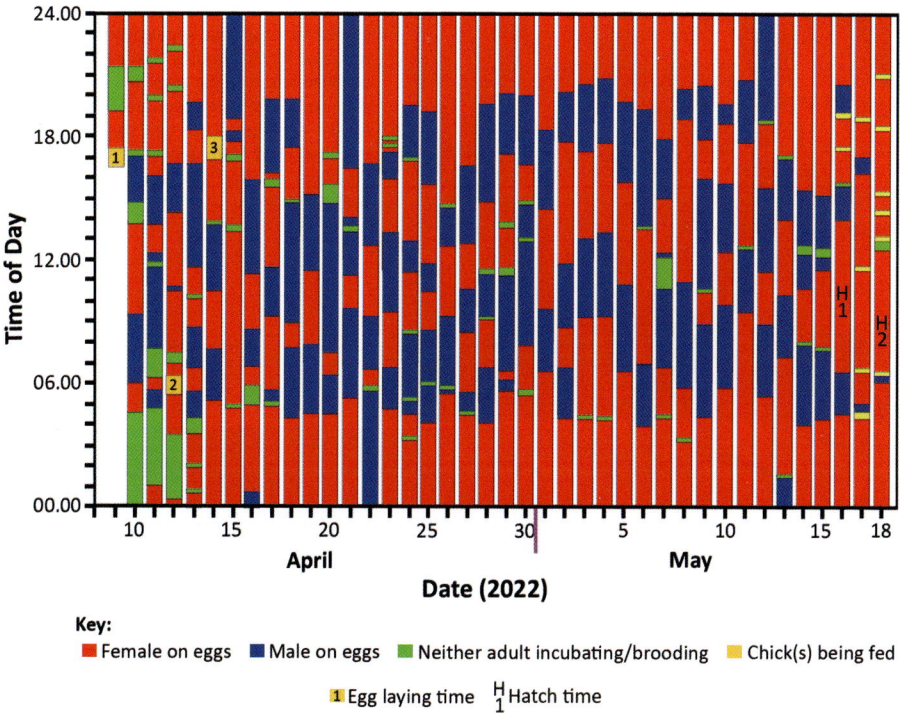

Figure 7.9 Incubation times of male and female Peregrines at Evesham Bell Tower, 2022. The times, and times of day, that each adult incubated are set down from the time of laying of the first egg. Laying and hatch times are approximate as eggs and chicks can only be seen when the female moves to expose them.

the incubation/chick brooding by both adults from the laying of the first egg to the hatching of the second egg. The third egg did not hatch. From Fig. 7.9 incubation began soon after the second egg was laid. There was a 48 hour interval between the two egg hatchings implying that either the first egg or, more likely, the third egg, was addled. The time interval between the laying of the first egg and the first hatch was 37 days. Between the laying of the second egg (and likely start of incubation) and the first hatch was 34 days. The time interval between the laying of the third egg and the second hatch was 33.5 days. It is therefore considered the most likely scenario was that the third egg was addled and the incubation time for each hatched egg was 33.5-34.0 days.

The three-egg clutch, with one addled egg suggests the female Peregrine was breeding for the first time. The way in which incubation developed certainly implies a lack of experience of the female, together with a frustration on the part of the male, despite his own young age. In total, as can be seen from Fig. 7.9, the male incubated for 35% of the time. The male's incubation included one complete night-time session (of 13.5 hours) and two shorter night-time sessions (of six hours and seven hours). Each of the latter were

Above The female, well-fed from the size of her crop, was kept waiting on this occasion for 50 minutes by the male who declined to get off the eggs. His concerns over her behaviour seem justified in an earlier photograph of her stepping on the eggs, and later when she was seen stepping on the chicks (*photographs below*). RS/SW.

The male's own behaviour was not without problems. His first attempt to feed the chicks ended when, apparently baffled by what was needed, he dropped the prey onto the youngsters and left. *RS/SW*.

terminated when the female arrived at the nestbox in the early hours of the morning (at 00.41 on the first occasion, at 01.16 on the second): the Bell Tower is illuminated at night allowing the female to arrive safely. The curious incubation behaviour seems to reflect reluctance on the part of the female rather than over-enthusiasm from the male. To support this contention, on 24 April when the female arrived to relieve the male, the latter declined to move off the eggs for 50mins during which time the female made no attempt to bully him off the clutch, as has frequently been seen in other Peregrine pairs. Similar behaviour, though for much shorter times, was seen on other days: on 7 May the male declined to move off the eggs for 6mins on one occasion and 4mins on another, and after finally leaving on this second occasion the male perched on a tower battlement for 1.5 hours staring at his partner (see photograph opposite *above*). To the human observer this looked as though the male did not trust his partner to incubate properly, and while this is

The Evesham female has taken a feral pigeon from the male and is taking it away to feed on it herself. *RS/SW*.

Incubating the eggs, the male found that the female had moved them close to the back of the box. Trying to cover them by pushing his tail up the box the male somersaulted. Quickly looking around, almost as though he wanted to make sure no one had seen what had happened, he hauled himself back over the eggs, turned over and resumed incubation. *RS/SW.*

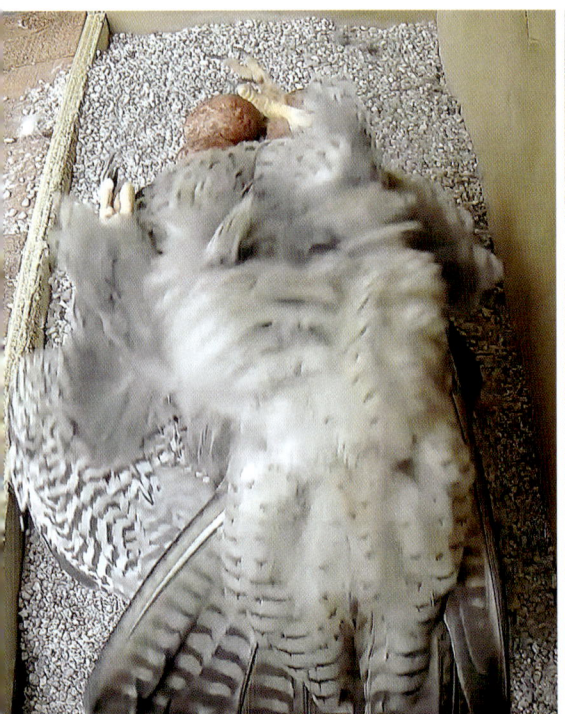

Breeding Part 2

Left The male invariably arrived at the nestbox by slaloming down the steep lead roof. But on one occasion he landed on top of the box, with a thump terrifying the female (*below*). *RS/SW.*

clearly an anthropomorphic view, it was the case that the female behaved as though agitated for much of the time. The incubation behaviour of the male is unusual, but not extremely rare as we shall see when considering incubation across the Peregrine's range below.

The first hatch occurred between 06.40 and 14.01 on 16 May, the second between 06.18 and 10.50 on 18 May. Fig. 7.10 (overleaf) shows the prey delivery/chick feeding times during the period 16 May-13 June. Although chicks were present at the nestbox after that time – the last shot of a chick was taken on 24 June – the nature of the breeding area, with four narrow gullies between battlements and roofing, separated from each other by right-angled corners, meant a chick could wander (and be fed) and be off-camera. No reliable feeding data could, therefore, be taken after 13 June. What is interesting about the feeding rate of the two chicks is that the rate did not alter significantly as the chicks grew. This is in sharp contrast with, for instance, the feeding rate of Kestrel chicks (*e.g.* Sale, 2020). The difference is the weight of prey being delivered (and, to a lesser extent, brood size). Kestrels feed their chicks on rodents which are of 'standard' size, so as the food demand rises the adult falcons must deliver more prey to compensate for growing appetites. By contrast, a Peregrine's avian prey spectrum is extremely

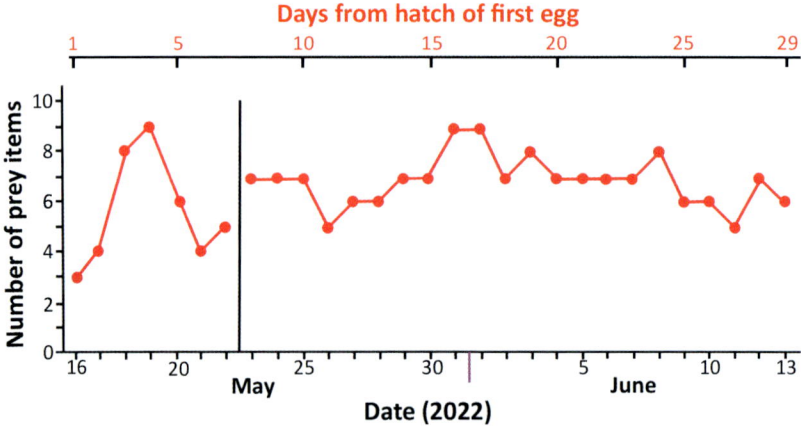

Figure 7.10 Number of prey items delivered to the two chicks each day at Evesham Bell Tower, 2022. To the left of the vertical bar dividing 22 and 23 May the chicks were being fed morsels of food from prey items. To the right of the vertical bar substantial amounts of whole prey were being fed to the chicks, though the feeding rate to 25 May did not necessarily conform to the number of prey items delivered. After 26 May whole prey items were being delivered, dismantled and fed to the two chicks.

wide (see Chapter 4) and the adult birds can catch larger prey to compensate for the chicks' growing appetite. This was very clearly seen at this nest site as the chicks aged. Initially morsels of food were fed from carcasses which had clearly been partially consumed by the adults. Later, larger prey portions were delivered, then (for instance) Starling juveniles were brought in whole and plucked before being dismantled and fed to the chicks. Next feral pigeon squabs were delivered, followed by whole pigeons.

Fig. 7.11 shows the timing of prey deliveries to the growing chicks. Note that it has not been possible to differentiate between prey caught by the two adults. It is likely that prey delivered by the male was caught by the male, but some

Figure 7.11 Time of day of prey deliveries to the two chicks at Evesham Bell Tower, 2022, listed as deliveries in one hour time slots between 03.00 and 23.00. The blue line represents the mean delivery rate of 9.6 items/day between 03.00 and 21.00. The single delivery at 22.35 on 2 June has been excluded from this calculation.

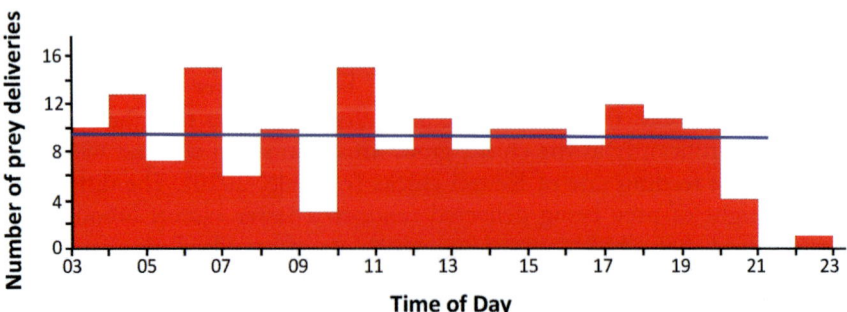

Left The male brings in a Starling. *RS/SW*.   **Right** The incubating female preens her tail feathers. *RS/SW*.

prey caught by the male may have been intercepted and then delivered by the female, particularly during the period before the chicks could thermoregulate as the female would need to be more or less continuously present. From the timing of prey deliveries, it is clear that once the chicks could be left alone for extended periods the female was hunting independently: as an example, on 6 June the male delivered prey to the nestbox at 15.50, and the female delivered prey three minutes later. While Fig. 7.11 suggests a reasonably regular supply of prey during the day, it is worth noting that during the chick feeding period from mid-May to mid-June sunrise advanced so that prey captures at both

In 2022 the breeding season was mostly good weather. The incubating male is highlighted by the evening sun (*left*). But it did occasionally rain, the now exploring chicks becoming bedraggled (*right*). *RS/SW*.

Above The male brings in prey. *RS/SW.*

Below The female poses for the camera. *RS/SW.*

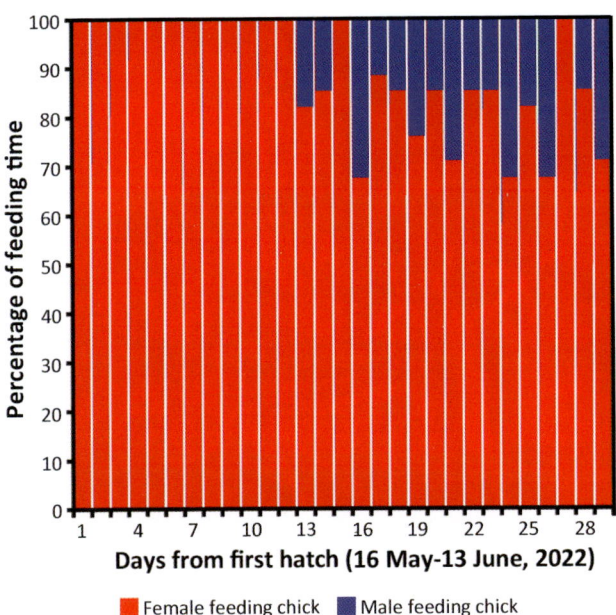

Figure 7.12 Variation of the percentage feeding by the male and female with chick age.

03.00-04.00 and 04.00-05.00 represent early morning hunting. The data suggests that feeding early in the morning was from prey cached the evening before. However, it may also suggest that the adults hunted early each day, then rested during the late morning before hunting again in the afternoon/early evening.

Fig. 7.12 shows the percentage of chick feeding by the adult Peregrine pair. The behaviour of the male once the eggs had hatched reinforces the authors' view that the male was inexperienced and, probably, breeding for the first time. The male's first successful attempt to feed the chicks was when they were 13 days old, though he had tried, and failed several times before. Even after working out how to feed the chicks, the male did not exceed 30% of a daily feed at any stage, and only managed 30% on two days.

After hatching, the female brooded the chicks until 31 May. From 1 June onwards the female did not spend the night in the box. By 31 May the older chick was then 15 days old, the younger one 13 days old, though the size difference between them was by then negligible. However, during the last three days prior to 31 May the female was not brooding in the accepted sense, the chicks being rather too big for her to cover. The female was, rather, standing close to the chicks who pressed themselves against her.

Although data collection ceased once one or both chicks were off-camera for long periods, the authors visited the site often. Both chicks fledged successfully.

The female feeding the chicks at the Evesham Bell Tower. *RS/SW*.

Scottish Borders

This site was in 'wild' country, about 1km from the nearest habitation and 15km from the nearest town. The site was an old and overgrown quarry which was also home to a large number of Jackdaws. Adjacent areas included open moorland and a small river: Snipe were relatively common as were other moorland and riverine species.

In the absence of an available power supply a modified trailcam system was used, set up to take images when triggered and powered such that it could remain in situ from installation in mid-April through to the point at which, if breeding was successful, chick(s) had fledged. The camera system was removed on 6 July, having taken many thousands of images. Data collected during incubation suggested the adult Peregrine pair behaved much as the falcons in London, with the female doing the bulk of the incubating, the male covering the eggs only when the female needed to feed, preen *etc*. Three eggs were laid, but only two hatched. The first hatch was during the afternoon of 13th May, the second hatch was during the following morning. Prey deliveries to the two chicks are shown in Fig. 7.13. The data shows a curious blank period between during 30-31 May when the chicks were 17 days old. On 30th May no prey was delivered during daylight hours, but a single item was brought to the chicks by the female during twilight at 20.57, 30 May. No prey was delivered during 31st May. Prey deliveries began again at 13.14 on 1 June. There was, therfore, a period of 65 hours during which only a single prey delivery was made to the two chicks On the afternoon of 1 June there were four prey deliveries in a five-hour period. Figs. 7.14 and 7.15 show the weather (wind gust speed and rainfall) for the filmed breeding period at the site. The absence

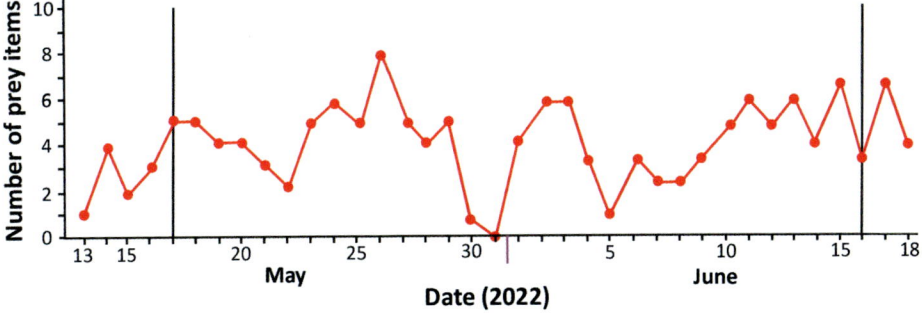

Figure 7.13 Number of prey items delivered to the two chicks each day at the Scottish Borders site, 2022. Prior to the vertical line on 17 May morsels of food were delivered to the two chicks. The second vertical line on 16 June notes that last time when both chicks were consistently visible on all images. Therefore, from the 17 June the number of prey deliveries photographed does not necessarily represent the number of prey items delivered.

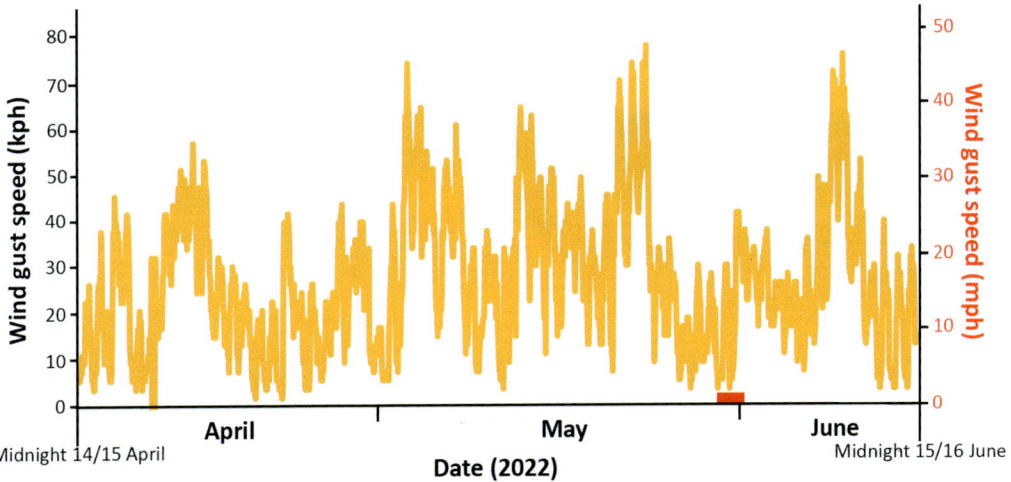

Figure 7.14 (*above*) Maximum wind gust speed at the Scottish site in kph and mph.

Figure 7.15 (*below*) Daily rainfall at the Scottish site.

In each figure the red box marks the days when the Peregrines delivered no food to their brood.

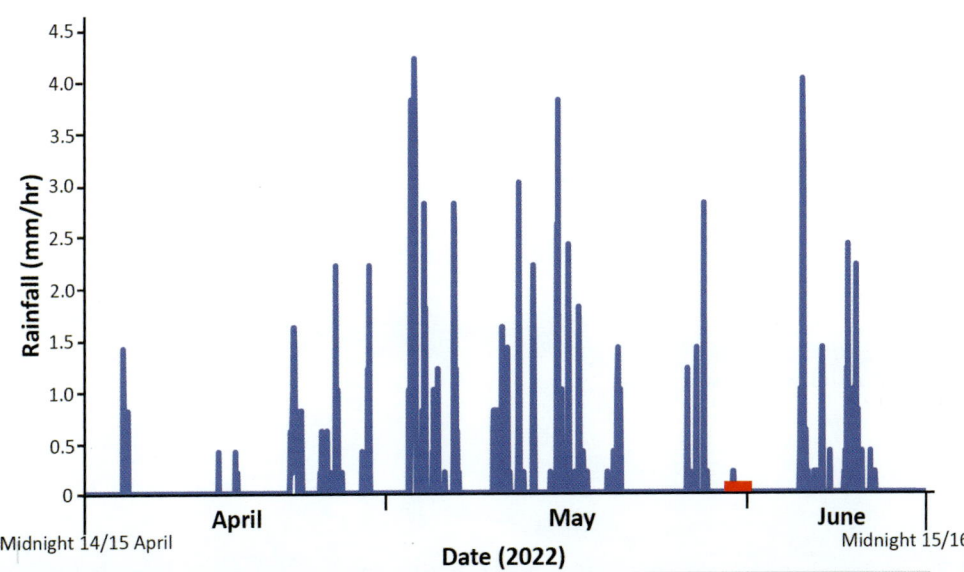

of prey deliveries is marked with a red box. Clearly the absence coincides with a period of benign weather. It is therefore conjectured that the male Peregrine was unable to hunt during this period due to a short-term illness, perhaps caused by eating cached food which had become rancid, or injury. It is further conjectured that the single prey delivery during the period (twilight (20.57) on 30 May) was by the female, either cached prey or captured by the female herself. Whatever the cause, prey deliveries were resumed on 2nd June, with the male behaving much as he had during the days prior to 30th May.

During the chick raising phase the bulk of the delivered prey which could be distinguished (much consisted of a blooded mass) was Jackdaw. One pigeon (a Stock Dove or a somewhat lost feral bird) was seen. While it is not possible to be definitive as the camera filmed only nest site activity, it seems probable that many of the hunted Jackdaws were adults breeding in the quarry and, later, fledglings from quarry nests.

From the feather mass in the photograph *above* it is clear that the Peregrines were regularly taking Jackdaws breeding in the old quarry. The photograph *below* shows a Jackdaw flying into its nest which was situated immediately below the Peregrine eyrie.

Fig. 7.16 shows the time of prey deliveries. The earliest times are significantly less than those for English Peregrine pairs, with only two deliveries prior to 04.00 and only three before 05.00. However, it must be recalled that the Scottish pair were breeding much further north and while Scotland enjoys more hours of daylight during the summer, winter daylight hours are much shorter and the sun rises later in the early months of the year. As noted above, it is possible that early morning deliveries were of prey cached the night before.

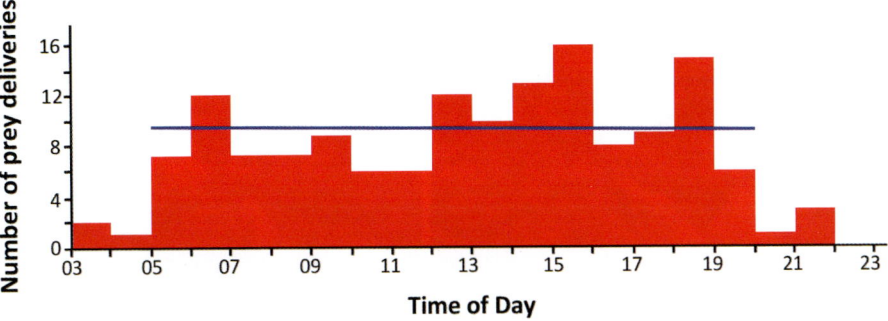

Figure 7.16 Time of day of prey deliveries to the two chicks at Scottish Borders site, 2022, listed as deliveries in one hour time slots between 03.00 and 22.00. The blue line represents the mean delivery rate of 9.5 items/day between 05.00 and 20.00.

The female stopped brooding the chicks overnight on 27-28 May despite the temperature falling to 2.2°C (36°F). The lowest recorded night-time temperature 1.1°C (34°F) was recorded early on 5th June: during the last week of May/first week of June the night-time temperature was rarely above 3°C (*c.*37°F). However, on 10th June the female brought prey to the chicks in twilight at 21.26. She fed the chicks until 21.54, then stayed with them until 03.23 when she left. Two days later the female arrived with the chicks at 21.55, this time without a prey item. She stayed with them until 03.13 the following day, then left. This curious behaviour was repeated the following evening, the female arriving at 21.08 on 13th June, without a prey item, and leaving at 00.34. The female did not visit the chicks the following night, but she returned at 21.43 on 15th June, without a prey item, and left at 02.38. She then repeated the same behaviour on four consecutive evenings arriving regularly at around 21.30 without prey and staying until about 02.30 the following morning. The female then skipped two nights, then repeated the behaviour again on the nights of 20-21, 21-22 and 22-23. After 23rd June the behaviour was not repeated. As noted in Fig. 7.13, photographs with both chicks in shot stopped on 16 June. On 21st June the male Peregrine arrived with prey (another Jackdaw) at 06.21, but there were no chicks at the eyrie and neither arrived back before he left some minutes later. Shots with both chicks in frame then became increasingly scarce: At 12.44 on 26 June the

Three images from the trailcam stream at the Scottish eyrie.

camera took the last photograph of a (single) chick. The camera equipment was removed on the morning of 6 July.

A comparison of the breeding behaviour between the two main sites covered (London and Evesham) reveals interesting differences which probably reflect the experience of the two Peregrine pairs. At the Charing Cross site we have an experienced male of at least 12 years of age partnered by a 4-year old female, while at Evesham we have reason to believe we have a three-year old male and (by observed behaviour) an inexperienced female. The younger female at Charing Cross hospital was clearly highly dominant over the older, more experienced male throughout the breeding period, the pair showing a clear understanding and mutual acceptance of their respective roles. There was very rarely a stand-off between them, and the male often adopted a subsidiary appeasement type of behaviour (see photo Chapter 6 p.270).

The Charing Cross female, *Azina*, clearly controlled incubation timing and incubated for 84% of the time. The dominance relationship in the Evesham pair was very different as the young male frequently declined to move off the eggs during incubation, the female failing to 'bully' him off. As a result, the female only incubated for 65% of the time, the 35% male incubation time being more than double that of the more experienced Charing Cross male. The time lapses between egg layings at Charing Cross were 61h 3m, 59h 57m and 59h 45m, while at Evesham were 61h 40m ±1h 20m and 59h ±1h 30m. The Charing Cross female had the endearing habit of rising after laying to admire her efforts. In addition, after observing the breeding for many years Nathalie Mahieu could pinpoint egg laying by watching the female's posture. Having worked with Common Kestrels for several years RS could also accurately assess laying, but being less familiar with Peregrines waited for more definite evidence.

The time intervals of the two females clearly indicates that egg production is instinctive with female Peregrines. However, the very different incubation behaviour at the two sites indicates that while there is an instinctive quality, experience also plays a significant role. Once hatching had occurred, instinctive female nurturing was immediately apparent in the Evesham female, while the inexperienced male seemed bemused by the chicks. His early attempts to feed were poor and failed each time, and it took him many days before he managed to work out how to transfer prey to the chicks.

At each of the three monitored sites all hatched chicks were raised to fledge.

Feeding time, as well as feeding rate, is also interesting. Figs 7.6, 7.11 and 7.16 all suggest that feeding early in the morning, perhaps with prey cached the evening before, and later in the day are the general rule. A similar feeding rate was seen by Rejt (2001) who also saw a peak delivery in the early morning (from cached prey?), then a lull in hunting prior to another peak in the evening in a study of nominate Peregrines in urban Warsaw, Poland (Fig. 7.17).

Figure 7.17 Daily pattern of chick feeding in two successive years at an eyrie of nominate Peregrines in Warsaw, Poland. Redrawn from Rejt (2001).

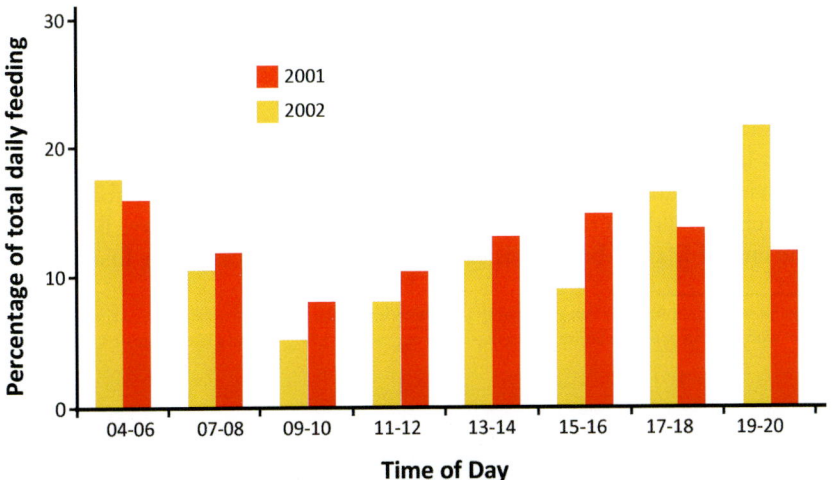

Egg laying

Data on the laying of the first egg by Peregrines are sparse, but the record earliest for the northern hemisphere seems to be that of Wimsatt (1940a) who notes that in 1939 three, four-week old, *F.p. anatum* fledglings were removed from a nest close to the Potomac River in Maryland, USA by falconers on 10 April. Allowing for incubation, Wimsatt considers the first egg must have been laid on, or about, 10 February. He also states that this was two weeks earlier than the previous year, so that the 1938 first egg was also laid in February. Wimsatt notes that other nests in the area did not have first eggs until late March/early April, implying that the Maryland female or the pair were remarkable in their ability to breed early. Wimsatt calculates his first egg date on the assumption of a 28-day incubation and four days to complete the clutch. Since both these timings are less than the generally accepted data (see below), Wimsatt's 10 February date could well have been 4-6 days earlier[2].

Collecting data for nominate Peregrines in the UK, Ratcliffe (1993) found that there was consistency in the date both by area and also by female. Ratcliffe found that the first-egg date was dependent on the temperature in February (probably because cold Februaries delay the onset of mating behaviour), latitude (though the spread of mean dates from southern England to central Scotland was only about seven days) and altitude (though again the change was small, only about three days in mean date between nests at about 200m to those above 500m). There was also an apparent change in first-egg date from data collected in the period 1905–1924 and that for 1925–1940 of about three days which coincided with an increase in mean annual temperatures in the UK during the 1930s. That the Peregrines were responding to mean daily temperature is in accord with data from the British Trust for Ornithology (BTO) which suggests that since 1970 the laying of the first Peregrine egg has advanced by about 14 days as atmospheric Carbon Dioxide levels and, therefore, global warming have increased.

Overall, there seems to be a remarkable consistency in the first-egg date of individual female Peregrines, with instances of the same bird laying her first egg on the same date in successive years: Ratcliffe (1993) quotes his own experience of a female in Galloway, Scotland who laid her first egg between 3 April and 6 April in seven successive years. Captive Peregrines are also very consistent in laying dates, though in this case, of course, the vagaries of climate and food supply are largely eliminated. Mearns and Newton (1988) in their study of Peregrines in southern Scotland over a nine-year period noted that females tended to lay their first egg earlier as they aged, birds five years of age or older laying an average of eight days earlier than two-year old birds, laying date having advanced 2-3 days each year. Mearns and Newton also found that the clutches of older birds were larger. In the UK, full clutches have been seen by the end of March, but this is exceptional. More normal is for first eggs to

---

[2] It must be borne in mind that Peregrines in the southern hemisphere breed at very different times, Jenkins (2000c) studying *F.p. minor* in South Africa lists a first egg date of 5 August and a latest date of 4 November.

Breeding Part 2

appear in the last week of March or the first week of April. However, as noted earlier in the Chapter, data collected from the London breeding site implies that warmer ambient temperatures in cities is reflected in earlier first egg dates.

The data above is for a resident population. For migratory Peregrines, in addition to acquiring the reserves necessary for laying, the female falcon has to recover reserves lost during the migration itself. This was investigated by Lamarre *et al.* (2017) for *F.p. tundrius* which may have been involved in some of the longest migratory flights of any sub-species. In addition, *F.p. tundrius* are constrained to a short breeding window by the limited Arctic summer: recovery from migration must, therefore, be swift or breeding may need to be abandoned. Lamarre *et al.* captured breeding females at Rankin Inlet and Igloolik in Arctic Canada and took blood samples which were analysed for two metabolites (β-hydroxybutyric acid and triglyceride) known to reflect short-term changes in fasting and fattening, and the hormone corticosterone which is associated with energy allocation. The results of the study confirmed that lower rates of pre-laying fattening delayed laying, with corticosterone levels rising to compensate for the energy cost of egg production. Fig. 7.18 neatly illustrates the effect that migration recovery and the energy requirements of egg production have on clutch size.

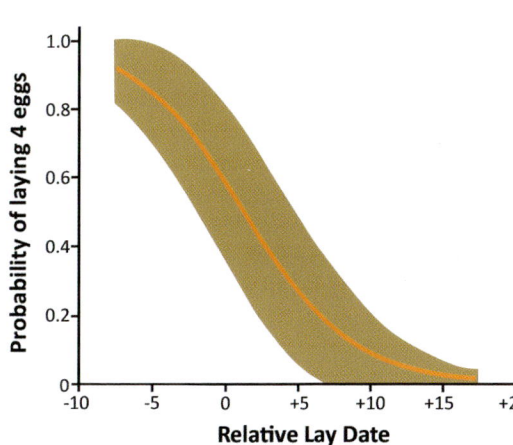

Figure 7.18 Probability of a female *F.p. tundrius* laying a clutch of four eggs in relation to the relative lay date with values standardised relative to the yearly mean first lay date at Rankin Inlet and Igloolik. The standardising was required because on average the falcons at Rankin Inlet laid six days earlier than those at Igloolik. The solid line is a regression fitted to the data from the two sites. The shaded area is the 95% confidence interval. Redrawn from Lamarre *et al.* (2017).

Egg-laying Intervals

It is generally assumed that female Peregrines lay eggs at 2-2½ day intervals and at any time of day, though usually in the early morning. Ratcliffe (1993) quotes Cade's view, based on North American captive Peregrines, that the laying interval was longer, 52-62 hours. As noted above the time lapses between egg layings at Charing Cross were 61h 3m, 59h 57m and 59h 45m, while at Evesham were 61h 40m ± 1h 20m and 59h ± 1h 30m.

In general the period between eggs increases after the third egg, but this was not seen in the interval to the fourth egg in our London study. Repeat clutches are laid if the first clutch is lost, but only if incubation is not significantly under way: if incubation has already reached 7-10 days repeat clutches will not be laid. Repeat clutches require the breeding cycle to start again, with the full spectrum of displays, though these are significantly shortened as the repeat clutch usually starts within 14 days of the loss of the first. Instances of third clutches being laid after the loss of the first and repeat are known, but are very rare. There are no recorded instances of second clutches being laid if the first clutch is successful.

Eggs
The avian egg is a remarkable object, one whose intimate details, in terms of construction and ability to aid and protect a developing embryo, could fill a book of their own. Since the thinning of the shell of the Peregrine egg during the DDT crisis has been of such importance in the population dynamics of the species (and will be considered again in Chapter 10 *Population*) it is worth noting here that shell thinning is a natural process as the developing embryo removes calcium, continued protection being afforded by an internal membrane whose importance is often overlooked (see Castilla *et al.*, 2010, for further details with specific reference to Peregrines).

Peregrine eggs are elliptical, measuring 47-59mm along the major axis (mean 52mm) and 36-45mm along the minor axis (mean 41mm) – data from a sample of 300 nominate eggs (Cramp and Simmons, 1980). The mean values quoted by Cramp and Simmons accord well with data in Ratcliffe (1993), who gives values of 51.5×40.8mm based on measurements of 2,253 British (*i.e.* nominate) eggs. Ratcliffe notes that while there is no significant variation of egg size within a clutch, there is a tendency for females to lay smaller eggs as they age. Though this trend is slight, Ratcliffe does quote data for one Welsh female whose eggs 'shrank' from 50.0×39.5mm to 46.5×32.5mm over a seven-year period. Ratcliffe also assembled egg size data by area and found no significant differences across the British Isles. In their study of egg sizes in North American Peregrines (both wild and captive, unspecified sub-species) Burnham *et al.* (1984) found significant variation in size, quoting differences of 37% in length among eggs from various sites, and 17% within clutches; and 31% in breadth (but only 0.3% in clutches). There was also a considerable variation across years – 14% in length and 48% in breadth. In Australia, Olsen (1982) found that the size of the eggs of *F.p. macropus* increased with the latitude of the breeding pair, and with the mean temperature at the time of laying (in September): the differences were small, but significantly different. Olsen also found that eggs laid in tree hollows were larger than those from other nest types.

Fresh eggs weigh 39–46g (Cramp and Simmons, 1980); 38.5–52.6g, mean 45.5g (Ratcliffe 1993), of which the shell contributes about 3.8g[3]. A clutch of

---
[3] Hoyt (1979) notes that the volume of an avian egg can be estimated to within 2% from the equation $V = 0.51LB^2$ where L is egg length and B the maximum diameter. Fresh weight can also be estimated within 2% from $W = KxLB^2$ grams. K varies with species. Hoyt does not quote a value of K for Peregrines, but the variation across a large number of species is 0.527-0.567 (excluding the Ostrich (*Struthio camelus*) for which it is 0.597).

four eggs therefore weighs a mean of 182g, or about 17% of the weight of an average nominate female. These data are consistent with those of Burnham *et al.* (1984) for North American Peregrines of unspecified sub-species.

The ground colour of the eggs is variable, from buff through shades of red to crimson, with blotches or speckles of red-brown to dark brown, and even with greys and shades of purple. As with other falcons, the variation in speckling is considerable both between birds and within clutches, and in general the later eggs of a clutch are the least coloured. Village (1990) suggested for Common Kestrels that this resulted from depletion in protoporphyrin, the pigment which creates the speckling, and that is also likely to be the case for Peregrines.

Clutch size

Hickey and Anderson (1969) recorded a North American Peregrine laying a clutch of seven eggs and laying a repeat clutch of six when the first clutch was lost: this seems remarkable, but it is known that if eggs are repeatedly removed from a laying female she may lay up to 16 eggs. However, 'true' (as opposed to manipulated) clutches, as seen by Hickey and Anderson, are rare. Ratcliffe (1993) presents data for a total of 1,920 clutches from across Britain (*i.e.* nominate Peregrines) which suggest no significant difference between regions, but do show a decline in clutch size from the period prior to 1980 to the following decade (Table 7.1). There is no obvious reason for this decrease, which was reflected in the data from all British regions, particularly as the health of the Peregrine population would have been expected to improve following the organochlorine contamination problems of the 1950s/1960s. Indeed, the mean clutch size during the time when shell thinning was causing egg breakage and severe depletion of the Peregrine population remained remarkably consistent with pre-contamination figures, which makes the later decline even more perplexing.

Period	Number of Clutches	Clutch size[1] (as a percentage of total number of clutches)				Mean
		2	3	4	5	
Pre-1980	622	19 (3.1)	205 (33.0)	386 (62.0)	12 (1.9)	3.63
1980-1991	1298	81 (6.2)	600 (46.25)	600 (46.25)	17 (1.3)	3.43

Table 7.1 Clutch sizes for clutches observed across Britain. From Ratcliffe (1993).

Note:

1. These data are consistent with those of Mearns and Newton (1988) who studied Peregrines in south-west Scotland over a nine-year period and found single eggs on 2.0% of occasions, with 2 eggs (4.5% of occasions), 3 eggs (34.8%), 4 eggs (56.3%) and 5 eggs (2.4%).

Single egg clutches are also seen – Mearns and Newton (1988) found five such single eggs in 247 observed clutches during their nine-year study period – as are clutches of six. Data collated from captive falcons suggest that females lay clutches of consistent sizes throughout their breeding years, with no decline in clutch size until they cease to breed at an age of about 20 years. However, this suggestion is not supported by more recent data as we shall see in Breeding Success below.

If repeat clutches are laid, these are significantly smaller than the first clutch. Ratcliffe (1993) notes that in the pre-1980 period the mean size of repeat clutches was 3.21 (*cf.* 3.63 for first clutches: Table 7.1), while in the following decade the mean declined even further, from 3.43 to 2.91.

One interesting aspect of the increased (and increasing) urban population of Peregrines is the increase in reported clutch sizes. In the USA, (Caballero *et al.* (2016) and reference therein) clutches of 4 and 5 eggs are regularly seen in urban nests. Caballero *et al.* also note that the provision of nest boxes for urban falcons aids chick survival, while the occasional rescue of a fledgling from the ground by humans also aids the number of successful fledglings.

Incubation
Incubation normally begins with the third egg, though occasionally with the fourth, so that hatching is almost synchronous. Wiebe *et al.* (1998) studied the incubation behaviour of 17 wild Kestrel pairs to assess the impact on hatching times. The researchers found that in general hatching occurred in the order eggs were laid (*i.e.* asynchronous hatching) and that both this, and the total time span of hatching, corresponded well with incubation behaviour. The Finnish team identified three incubation modes (Fig. 7.19). In 'rising' incubation the female steadily increased the time spent incubating as the clutch grew, though the starting time for commencement, and the intensity of, incubation could vary with egg number. The effect of these variations in incubation rate was to increase the time span of asynchronous hatching (Fig. 7.20) with the total incubation time before the last egg was laid being linearly related to the hatch time span. The combination allows the female to reduce or increase the asynchronous level of hatching.

In Peregrines breeding at more northerly latitudes (*e.g. F. p. tundrius* in northern Canada) incubation may start earlier, resulting in asynchronous hatching (Court, 1986, Court *et al.,* 1988). The reasons for asynchronous hatching are not clear, and Court (1986) suggests that the studies of *F.p. tundrius* do not support the brood-reduction hypothesis (*i.e.* that broods of mixed ages offer insurance against poor food availability, older birds outcompeting younger siblings for scarce resources, so ensuring the survival of some, as synchronous hatching might result in all siblings receiving less, with consequent endangering of all). Court (1986) found that 7% of all chicks died in asynchronous broods, with 50% of last-hatched chicks dying within five days. Deaths were not due to direct fratricide, though dead chicks may be fed to

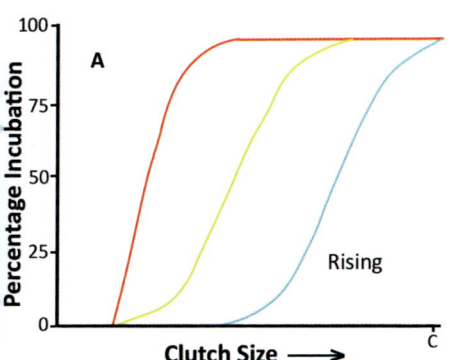

Figure 7.19 Possible incubation schedules. The curves show the percentage of the day spent incubating as the clutch size increases to its final size (C). 'Rising' incubation with differing start times and rates of rise, allows the female some control over the synchronicity of hatching. Redrawn from Wiebe *et al.* (1998)

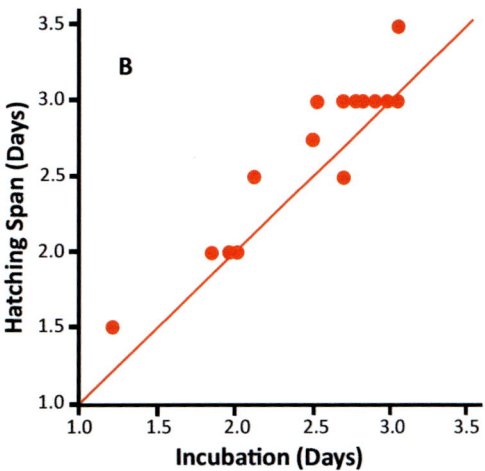

Figure 7.20 Relationship between total incubation time prior to the laying of the last egg and the hatching span. Data based on 15 clutches in Finland. The line is perfect correspondence between incubation and hatching times. Redrawn from Wiebe *et al.* (1998).

siblings by the female. Rather, deaths were due to the inability of youngest chicks to compete with their siblings – a newly hatched chick weighing 36g may find itself in a nest with siblings weighing up to 140g if they are up to five days older, and may be trampled to death as a consequence – but parental effort was also involved, as all chicks in some asynchronous broods survived (see photo overleaf). Last-chick deaths were also influenced by female care in brooding and feeding. In those nests where the last-hatched chicks survived they received enough food and grew at the same rate (in terms of weight gain per day) as their older siblings[4].

The incubation period varies from 28 to 33 days according to Ratcliffe (1993), though there is disagreement over this figure from North American researchers, who claim that 33–35 days is more usual, based on data from Hagar (1938) – see, also, Nelson (1972) who quotes 32-34 days, exceptionally 35 days for *F.p. pealei* – and from captive falcons. In a study on nominate Peregrines in southern Sweden, Lindberg (1983) found the period varying from 28 to 33 days depending on ambient temperature and the efficiency of incubation. Razafimanjato *et al.* (2007), noted 33 and 35 days for two breeding pairs of *F.p. radama* on Madagascar.

---

[4] Studying Barn Owls, Dreiss *et al.* (2015) suggested that the chicks used vocal signals to compete for the next food item which would be delivered by a parent, and that these signals were not related to either age hierarchy or hunger level, but rather defined a turn-by-turn feeding regime. While there is evidence in some avian species for direct competition for food, the findings of Dreiss *et al.* imply that if resources are not overly scarce, siblings may behave non-competitively. It would be interesting if future work were to discover that the lack of overt competitiveness in the species is also related to vocal signalling.

Above   A female nominate Peregrine feeding the larger two chicks of her three-chick brood. The third, smaller chick tried very hard to reach the female, but failed to do so. It is not known whether the smaller chick hatched later or was sick in some way.

Below   A few days later the two larger chicks looked well-fed, but there was no sign of the third chick. It had probably died of starvation and may have been fed to its larger siblings. In his study of Common Kestrels RS was watching a brood of four chicks. The male Kestrel disappeared and over time the female fed the smallest, weakest chick to the remaining chicks, on one occasion doing so while the starving chick was still alive (though, obviously, only briefly). Ultimately only one chick was left, at which point RS added food to the nest box to ensure its survival.

As noted above, in our studies at London and Evesham defining the time between laying and hatching was complicated by the fact that eggs at both sites were addled (three of four at London, one of three at Evesham). At the London site, if egg 2 was the 'live' egg, then period from laying to hatch was 36.9 days, if egg 3 then 34.4 days. At Evesham the variation in time interval was 33.5-37 days depending on which egg was addled. It is our view that the most likely scenario was that the third egg was addled and the incubation time for each hatched egg was 33.5-34.0 days, consistent with the best estimate time at London.

In his study of *F.p. anatum* in Ontario, Beaupre (1922) noted, charmingly, that '*the female attends to all the domestic duties of the falcon home, but is spared the task of seeking for food. This is the duty of the male.*' That division of labour seemed to fit with the lifestyle of other falcons. But for the Peregrine it is not entirely true. Both sexes have twin lateral brood patches, though these are less well developed in males. In general, females are responsible for the majority of incubation, and for incubation through the night. However, Johnson (2011) observed a male Peregrine incubating at night at an urban (London) site: our data also shows a male incubating through the night, for one complete night only, but for periods through to the early hours of the following day on other occasions, and taking a greater share of incubation than would be expected. Ratcliffe (1993) says that in his observations (based on the number of times the male or female bird was flushed off eggs on his approach) males incubate for 12% of the time. Formon (1969) studying nominate Peregrines in eastern France raised this fraction, noting that in his study males incubated for 16–25% of the time. In a study in North America, Enderson *et al.* (1973), using time-lapse photography of Peregrines at five nests on Alaska's Yukon River (probably *F.p. anatum* given the location) and analysing the resulting 69,461 frames (which constituted 4,208 hours of filming), suggested a greater proportion of male incubation. The male incubated for periods of 2-4 hours, taking 40% of the total incubation time between 15 and 11 days from hatching, this fraction decreasing as hatching approached, though remaining at about 25% or so of the total time. The female usually spent four hours sitting. Such fractions seem consistent with the fact that the male, being smaller, will have more problems covering the clutch, particularly if it is large, than the female, and also has less efficient brood patches. The total time when neither bird incubated was minimal (<1% of total time and never for more than three minutes at a time). After hatching, male brooding was minimal (<2% of total time).

Other North American studies suggest even longer fractions of incubation by males than recorded by Enderson *et al.* (1973), with Nelson (1970) suggesting 30-50% of the total time (though a decreasing fraction as the time to hatching approached) and an extreme example of a pair of Peregrines (*F.p. anatum*) in northern New Mexico where Clevinger (1987), observing for 202 hours over 18 days of an incubation period which lasted 33 days, noted that the male incubated for 63% of daylight hours, with a range of 27–87% (Fig. 7.21). The male incubated on average for 2 hours 34 minutes, the female for 1 hour 10 minutes. A check on local sunrise and sunset times during incubation by the present authors notes that at the mid-point of incubation sunrise was 06.00 and sunset was 18.00, meaning that daylight lasted 12 hours. Clevinger suggested that the female might have been a first-time breeder, her inexperience accounting for her reduced fraction, or that she might have had an abnormal lack of inclination to sit on her eggs. The latter suggestion and the data from all studies suggest that male incubation is variable, and dependent on decisions made within the Peregrine pair, but that males may spend significant time incubating.

During these periods, of course, the female is free to maintain her plumage, but also to hunt, reducing her dependency on the male and so maximising his energy resources for the critical time of chick feeding immediately post-hatching. However, most studies indicate that the male still does a significant amount of the hunting for the female during incubation, his times for incubating usually following his arrival with food. The male will call to the female from a usual perch and transfer food when she arrives, leaving him to fly to the nest while she eats. Males will also chup, seemingly as a question to the female, asking if he may incubate: the female may or may not respond. However, if the male is incubating and the female returns and wishes to take over, the male is typically not asked if he wishes to depart, but ordered to do so. Nelson (1970) did observe an occasion when the female returned and the male declined to leave the eggs, the female ultimately departing: however, this seems to have been very unusual behaviour.

Similar long incubation periods were also seen in a pair of *F.p. minor* in Zimbabwe, Hustler (1983) noting that the eggs were covered 98.8% of the time, with the male sitting for an average of 72mins and the female for an average of 101mins. Also studying *F.p. minor*, though in South Africa's Transvaal, Tarboton (1984) noted a mean covering by the male of 90mins and by the female of 145mins, for daylight hours during five consecutive days. In a study of *F.p. brookei* in northern Spain, Zuberogoitia *et al.* (2018)

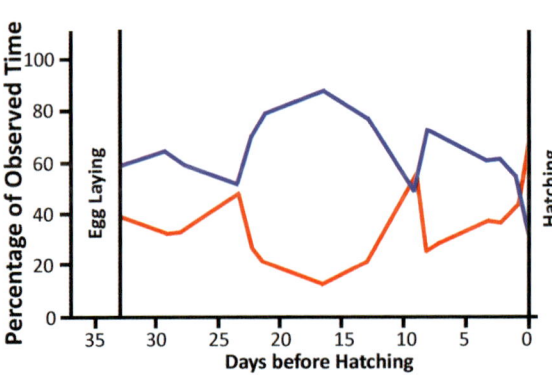

Figure 7.21 Percentage of observed time that the male (blue line) and female (red line) Peregrines of a breeding pair were engaged in incubation. Redrawn from Clevinger (1987).

found that male incubation time was significantly related to their breeding at a territory, but not related to prior breeding success. Males also tended to incubate for less time during the first and final weeks of incubation and spent longer periods on the eggs in warmer weather, presumably because the lack of efficient brood patches makes less difference on warmer days: the female takes advantage of this to devote slightly more time to herself.

The figure of 40% for male incubation quoted by Enderson *et al.* (1973) is higher than, but of the same order, as the value of 35% for male incubation in our study at Evesham. The behaviour of the breeding pair at Evesham is also consistent with the view of Clevinger (1987) that inexperienced females spend a reduced time incubating. By contrast the London male incubated for only 15% of the time.

The incubating bird will turn the eggs at intervals using its bill, and is usually careful on leaving the nest, rising gently and stepping clear of the clutch to avoid moving an egg out of the scrape. Equal care is taken when the bird moves onto the eggs. However, King (2008) noted one nesting female Peregrine fly away from the nest with an egg attached to her breast feathers. When she perched locally she pecked at the egg and detached it: it fell to the ground where it smashed. It must be assumed that the female damaged the egg and it became attached when she sat over it. It seems it was the first egg to be laid and no more were laid at the nest. Nelson (1970) includes a comprehensive description of the incubating bird, covering arrival at the eggs, settling on them, the incubating bird preening, stretching, fidgeting with nest site debris, and finally dozing and sleeping; and covering brooding as well as incubating. Both activities, which were regularly seen at London and Evesham, would, it seems, try the patience of a saintly falcon.

Chick growth

As with other species, Peregrine chicks start to cheep before hatching, usually from about 2-3 days before. Hatching takes some time – Nelson (1970) records an average of 72 hours between first starring of the eggshell and full hatching – though much of this time seems to be spent with the chick recuperating from early attempts to begin the shell-breaking process. Nelson noted the death of one chick when the shell membrane dried and the chick was unable to force a way through it, suggesting that atmospheric humidity was critical to hatching, something which has been proved in artificial incubators. Ratcliffe (1993) notes that some researchers consider that the parent birds leaving their eggs early on the morning of hatching is intentional, the eggs being exposed to the atmosphere from which they would otherwise be shielded. Nelson (1970) notes that females become more aggressive as hatching approaches, conjecturing that hearing the chicks cheeping within the shells triggers this, as the change in female behaviour does not occur if the clutch is addled. As for other avian species, the chick uses the egg tooth to break open the shell. Shell remains may be eaten by the female, to replace calcium loss. Peregrines often

Development of the chicks at Evesham Bell Tower. *RS/SW*.

neglect to remove spent eggshells from the nest area, these being trampled to join the other detritus which accumulates during the nestling stage: some parent birds will remove prey remains from the nest area when the chicks are very young, but in general once the nestlings are mobile there is little housekeeping.

In various studies, the sex ratio of hatched chicks is 0.52:0.48 in favour of females, a bias which is often seen in the Falconiformes (see Olsen and Cockburn (1991) studying *F.p. macropus* in south-east Australia and references therein). Olsen and Cockburn noted three conclusions from their study. Firstly, Peregrine pairs that breed early not only had a higher female bias but were more successful; secondly, that the early-laid eggs in clutches were more likely to be females, later eggs being more likely to be male; and thirdly that heavier female chicks than expected (≥10%) were only found in female-biased broods.

The newly hatched chick weighs 35-40g, the egg having lost about 17% of its weight during incubation. The chick is semi-altricial and nidicolous, with a thin covering of white down. The eyes are closed, but open within a few days, occasionally as soon as the day of hatching, though in that case only as slits initially. Brooding is mostly by the female, as the male has even more problems covering chicks than he did the eggs, particularly once the youngsters are a few days old. Brooding is more or less constant (about 90% of the time) during the early days of chick life when they are unable to thermoregulate, and is certainly most intense if there is bad weather (rain or cold), but also if there is very bright sunshine, as the lack of complete down covering means the chicks are at risk of sunstroke. Brooding reduces after about ten days, falling more or less linearly with age to reach about 15% of total time at age 20 days, and declining to zero time at age 25 days, though there are still exceptions at times of rain or low temperatures. At that age the nestlings are also able to recognise the adults when they are still a considerable distance away. Once the chicks can be left for extended periods the female may start hunting, though the extent of her hunting varies with individual pairs: in some the male continues to do the majority of the hunting, the female merely spending time away from the nest, but close to it, guarding the chicks. In other cases, particularly with large broods and as the nestlings approach fledge, the female will take a larger share in provisioning the young.

The chicks can feed on their first day, gaping instinctively even though they may not be able to see their mother. The female feeds by taking small morsels of food from prey delivered by the male and dropping it into the open gapes. By the second day the chicks can make the chup call, though it is more of a begging whine for food, and as soon as they can see they will often nibble the female's bill to stimulate her to feed. Both male and female feed the chicks, but the female is the main provider and may even prevent the male from doing so on occasions: both Nelson (1970) and Treleaven (1977) suggest that female Peregrines are much more likely to prevent the male feeding when the chicks are young (less than two weeks old), male feeding being more often accepted with older nestlings.

However, instances of males both feeding and brooding young chicks are known (*e.g.* Treleaven, 1977) and were certainly seen at each of the authors' study sites.

In a study throughout the breeding cycle in south-west France, Carlier (1993) noted the times five pairs of Peregrines[5] were at the nest (Fig. 7.22). The percentage of total time was highest during courtship, remained high during incubation (unsurprisingly), but then declined as the nestlings grew.

The chicks beg for morsels, then consume what they are given before begging again. Nelson (1970) noted the chicks forming a neat semi-circle in front of the feeding parent, while Treleaven (1977) stated that feeding involves

---

[5] Carlier (1993) states that the Peregrines which he studied in the Massif Central were *F.p. brookei*. Current opinion believes that southern France is a cross-over area between that sub-species and nominate Peregrines.

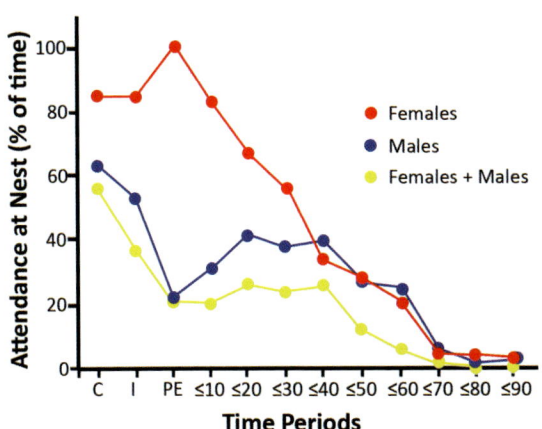

Figure 7.22 Proportion of observation time during which the male and female Peregrines, and both male and female at the same time, were close to the nest site. C is Courtship, I is Incubation, PE is Pipping Egg, ≤ is age of the chicks in days. Redrawn from Carlier (1993).

each chick being fed in turn until its lack of begging calls suggests satiation and the parent moves on to the next. Our observations suggest that a more haphazard regime is as likely as these controlled events. When the chicks are very young, say up to 14 days old, they cannot mantle prey as they have no wings to speak of and require the adults to feed them. But as they age, they can move more freely around the eyrie, and can therefore grab food and mantle it before siblings can take it. However, while mantling occurs, Peregrine chicks are sociable and often huddle together for warmth and mutual comfort. That said, Tordoff and Redig (1998) report an apparent case of siblicide, with three chicks seemingly pecking at a weaker one and causing its death: this behaviour might have been precipitated by hunger. In a rather more horrifying case, Franke *et al.* (2013) report the first-ever record of infanticide in Peregrines. Studying *F.p. tundrius* near Rankin Inlet, Arctic Canada, they were using a motion-sensitive camera to watch the breeding of a Peregrine pair with two chicks. During a period of intense rainfall which clearly prevented hunting, the two chicks became increasingly distressed. When the rain-sodden female returned without prey, she killed and partially consumed the weaker chick, but did not feed the remaining chick. Shortly after, the female returned and killed and partially consumed the second chick. This is the only case of infanticide we have been able to find in the literature. SW has observed similar behaviour, a female eating her chick after three days of continuous rain – but in that case the chick was lifeless. RS, studying Common Kestrels, also observed that after the adult male had disappeared, the female fed dead (starved) chicks to her remaining brood: one chick appeared to be still (but barely) alive when the female began to dismember it.

This behaviour contrasts starkly with the maternal care shown by female Peregrines on other occasions. At a coastal site in central Italy, Casini and Morelli (2008) observed the smallest chick of a brood of four fall from an eyrie 115m high on a 135m sea cliff. An adult (probably the female of the

Adult nominate Peregrine brings prey to her brood in Scotland.

Adult nominate Peregrine feeding her brood in Scotland. The disappearance of the smaller chick is commented on in the photograph at the top of p326.

pair) swooped down and seized the chick in mid-air and took it back to the eyrie. But almost immediately the chick fell again: the adult swooped once more and gathered up the chick, this time installing it safely in the eyrie. Soon after the adult returned to the eyrie and fed the chick. Similar behaviour has been seen in Cornwall (P. Welsh pers. comm. to SW). There, the smallest of three chicks fell from a sea cliff eyrie and was retrieved by the diving female. But the chick struggled and fell again. Again the female caught it, but while flying out over the sea prior to returning to the eyrie the struggling chick fell from her grasp. The chick landed in the sea, but was again retrieved by the female. It was taken back to the eyrie where the female placed the, now wet, chick between its larger siblings, then brooded the three: the chick survived the ordeal.

The nestlings will attempt to grab food from an early age, but are unable to feed themselves, as opposed to swallowing what they get hold of, until they are about 14 days old, or a little older, and are not fully capable of dismantling prey until they are about 25-30 days old. Once the parent birds realise the nestlings are capable of this, prey is just left for them rather than being broken up completely. Feeding frequency and the times of food delivery seem variable, but in general feeding starts very early in the day, is less frequent around midday, then increases again prior to sunset. Parker (1979), observing an eyrie on the Welsh coast, noted 4-8 feeding visits daily on average throughout the period of chick growth.

Olsen *et al.* (1998b) studying *F.P. macropus* in south-east Australia noted that during the nestling period male Peregrines provided 85% of prey deliveries (71% of biomass) and that during the main period of nestling growth males provided 93% of prey (86% of biomass). In the first 29 days post-hatch, males with a brood of three or four provided up to three times the prey deliveries and twice the biomass of males with a brood of one or two. Larger broods were fed more regularly than were smaller broods, and were fed at peak rate longer into the nestling period. By contrast, Olsen *et al.* found that female prey deliveries were not related to brood size, fledgling success or the male's delivery rate. The females paired with the better male providers did not benefit by having their own hunting efforts reduced, but did benefit from more successful breeding. In a related observation, Zuberogoitia *et al.* (2013), studying *F.p. brookei* in northern Spain, compared the average body mass of prey delivered to broods with different numbers of chicks. The study included 15 nests with single chicks, 48 nests of 2 chicks, 72 of 3 chicks and 29 of 4 chicks. The respective mean prey masses were 114.7g, 107.7g, 134.9g and 129.2g. Given the large standard deviation of each mean, the data suggested that there was no significant difference in prey mass, *i.e.* the adult falcons were not adjusting their hunting, in terms of prey spectrum, to compensate for differing numbers of bills to feed. However, studying prey deliveries by *F.p. tundrius* to a brood of four chicks at an eyrie in west Greenland Hovis *et al.* (1985) noted that the number of prey deliveries varied with chick age (Table 7.2).

Chick age (days)	3-5	6-10	11-15	16-20	21-24
Prey deliveries/hr	0.8 ± 0.1	1.3 ± 0.2	1.2 ± 0.2	1.4 ± 0.3	0.7 ±0 .6

Table 7.2 Feeding rates for a brood of four *F.p. tundrius* in west Greenland. Data from Hovis *et al.* (1985).

Studying breeding *F.p. macropus* in south-east Australia, Cameron and Olsen (1993) noted that the size of prey delivered to a brood varied in terms of biomass, both during the day and as the chicks aged (Figs. 7.23 and 7.24). In a related study, Olsen *et al.* (1993c), noted that the diet of the chicks was related to breeding success. During the breeding season the adult falcons were preferentially hunting Galahs, this increasing the condition of their chicks.

Similar data was recorded at all three sites in the authors' studies which are detailed earlier in this Chapter. There, the rate of prey deliveries stayed remarkably static, but the biomass of delivered prey increased. This was most apparent at London where identification of delivered prey over the entire period, hatching to fledging was possible. Prey identification at Evesham was not always possible as usually the prey had been partially consumed by the adults, particularly while the chicks were young. However, it was possible to note that when whole prey was delivered this was, initially, juvenile Starlings then, later, feral pigeons. At the Scottish Borders site the nature of the filming did not always allow prey identification, but again it was possible to

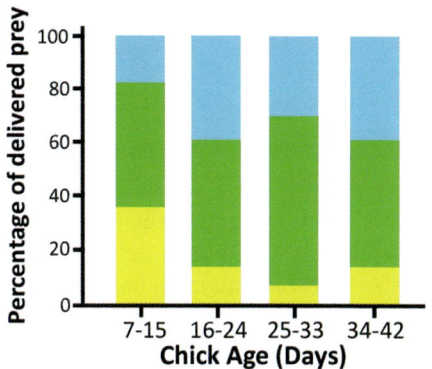
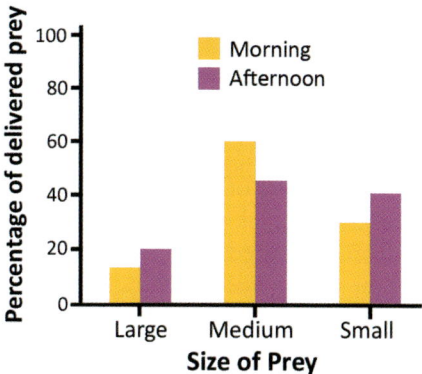

Figure 7.23 Percentage of small (turquoise), medium (green) and large (yellow) prey delivered to a brood of Peregrine chicks as a function of chick age. Redrawn from Cameron and Olsen (1993).

Figure 7.24 Percentage of the size of prey delivered to a brood of Peregrine chicks as a function of time of day. Redrawn from Cameron and Olsen (1993).

discern that as the chicks aged, the number of Jackdaws delivered to the eyrie increased. One interesting question when viewing an eyrie with several young Peregrines is can we tell how the brood is composed – how many males and females? And what is their age? In a very interesting report on the development of Peregrine (*F.p. macropus*) chicks in nests near Melbourne, Australia, Boulet *et al.* (2001) found that the food intake of male and female chicks and their growth rates were more or less equal until they were 20 days old, at which point male growth slowed, while that of females did not. By the time the chicks fledged, female weight and food intake had exceeded that of males by 45% and 25% respectively. Boulet *et al.* found that the growth efficiency of females (*i.e.* the efficiency of converting food intake to growth) was greater than that of males, explaining the apparent discrepancy between food intake and final body weight. This differential weight gain was also seen in another study of *F.p. macropus* in the same area of Australia where Olsen and Tucker (2003) manipulated Peregrine broods. Fifteen days after hatch the chicks of five broods were manipulated – Table 7.3.

Nest	Pre-experiment Days 1-15	Experimental Period Days 16-30	Post-experimental Day 31 onwards
A	1M, 2F	1M	1M, 2F
B	3F	4F	3F
C	1M, 2F	1M, 3F	1M, 2F
D (Control)	1M, 2F	No change	No change
E (Control)	2F	No change	No change

Table 7.3 Brood manipulation experiment on *F.p. macropus* near Canberra, Australia. Two female chicks were removed from nest A (leaving a single male chick), one each being added to two nests which already had three chicks, to produce two broods of four. There were also two control nests of two and three chicks. After 31 days, the added chicks were returned to the original nests.

In each case of brood manipulation there was minimal effect on the growth rate of the chicks (males or females), but when a chick was added the parent falcons increased the size of the prey (and, therefore, the biomass) delivered to the nest, while if a chick was removed, the size (and biomass) was reduced. The parent Peregrines were obviously aware of food needs of the chicks, though whether they were counting the chicks or were being stimulated by chick hunger is unclear. Fig. 7.25 and Fig. 7.26 show the change in delivered biomass across the nests and the growth rate of the chicks.

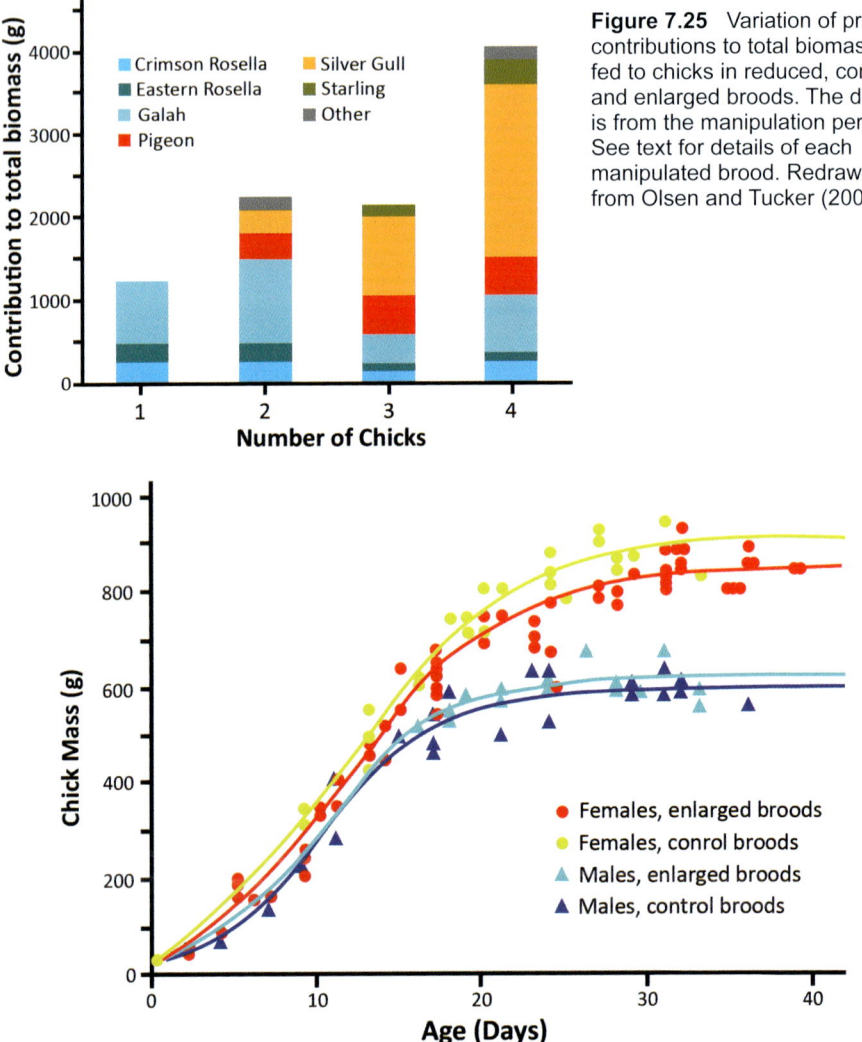

Figure 7.25 Variation of prey contributions to total biomass fed to chicks in reduced, control and enlarged broods. The data is from the manipulation period. See text for details of each manipulated brood. Redrawn from Olsen and Tucker (2003).

Figure 7.26 Variation in weight of 22 nestling Peregrines with age. Overlapping standard errors for the mean growth curves indicate that there is no significant difference between the control and enlarged groups for either females or males. Redrawn from Olsen and Tucker (2003).

Chick growth is rapid, both in terms of feather growth and mobility. At ten days the sheaths of remiges and rectrices can be seen. At 14 days the second down is very dense, and the first primary (P9) emerges from its sheath. Contour feathers begin to emerge at about 17 days and are visible beneath the down a couple of days later. At 21 days all primary and tail feathers have emerged, and wing flapping begins in earnest. At 30 days the chick is half-down, half-feather: about five days later the chick is mostly feathered, but patches of down are still visible.

At 6-8 days the chicks can preen and stretch their infant wings and legs. The second down erupts at this time and is complete by about ten days, at which time the primary sheaths break the skin, the sheaths of the outer rectrices following a day or so later. At 12 days the chicks are able to move to the nest edge and to turn around to excrete.

Although the chicks wander about the nest ledge within a short distance of the nest scrape, they spend most of their time within it, often huddled together, but after about Day 24 they spend much more time away from the scrape and alone, frequent wing flapping removing the final down feathers. For the observer the nestlings are a fine sight at about Day 30, with a delightful 50:50 mixture of down and juvenile feathers. At about 40 days the last traces of down are still visible, but juvenile plumage is essentially complete, though the rectrices are not fully formed. The fledgling can now fly, but not well, as some primaries and rectrices are not fully grown: Weir (1978) suggests that males fledge 1 or 2 days earlier than females. Lindberg (1983), studying nominate Peregrines in southern Sweden, found that the age of first flight was 41±3 days; in his study of *F. p. pealei*, Nelson (1970) quotes 41-44 days.

Even when they have begun to fly the young Peregrines stay close to the nest site at first, often roosting with siblings (and occasionally with adults if roosting spots are few), but eventually choosing their own roosts. The fledglings are also fed close to the nest site by the adults, though the latter begin to teach the rudiments of prey capture by making food transfers in mid-air, the youngsters catching dropped prey or taking it from the adult's talons. Prey dropping seems to occur too frequently to be a chance event rather than a deliberate strategy – see, for example, Hewitt (2013) – and can be highly entertaining for the observer.

Observers have described seeing adult Peregrines catching, then releasing, prey, the prey by now injured, for fledglings to pursue. During his 40-year study at a Gloucestershire, England eyrie SW has frequently observed similar behaviour. This occurs once the fledglings have attained strong flight capability and are attempting to capture prey themselves. At this time the juveniles will increasingly join the adults in hunting forays, though they still lack the flight capability and experience to make a capture. Adults will sometimes capture prey, but deliberately not despatch it as they normally would. Instead, the prey, still alive, will be dropped into the path of inexperienced juveniles to allow them to hone their skills. SW has observed a dropped capture outfly chasing juveniles only to be swept up again by the adult Peregrine, and then re-released: one of the juveniles then managed to make the capture. RS witnessed a further example of the adult birds teaching their brood. The nest site was a quarry in southern England into which the adults drove a pigeon. The two adults then circled above the quarry, preventing all attempts by the frantic, and increasingly exhausted, pigeon to escape, while the young Peregrines attacked it.

Sherrod (1983) in his highly illuminating book on the behaviour of fledgling Peregrines, notes that most leave the nest in the early morning. The reason for this timing is usually stated to be that it enhances survival, but there is scant evidence in the literature to support this hypothesis. Recently, Santema *et al.* (2021a) collected data for over 1500 Blue Tits (*Cyanistes caeruleus*) leaving 230 nests to test this and other hypotheses. The conclusion was that early departure did not enhance long-term survival, and was more likely to be physiological – that reaching the developmental threshold for departure was more likely to occur overnight for near-fledging birds. A second possible hypothesis, that the timing was related to parental provisioning, was not supported by the evidence of prey deliveries. In a second study (Santema *et al.*, 2021b) the researchers noted that fledging was also influenced by siblings. By transferring Blue Tit chicks into nests of chicks that were two days older and near-fledging, the fledging of the younger chicks was advanced. Sibling pressure was modulating individual behaviour.

Early flights by the juvenile Peregrines also help hone their hunting skills as they chase each other and make mock attacks on any large bird which happens into the area. Most young Peregrines hunt insects at this stage, but true attacks soon start, though the majority of these are unsuccessful at first. By early September in the northern hemisphere fledglings have to have mastered hunting as adult feeding ceases (the period of adult feeding after their offspring had left the eyrie thus lasting up to about nine weeks in total): in migratory populations this period is much less, usually 5-6 weeks). Whether or not the fledglings are then driven away from the area by their parents is still debated, with evidence both for and against being available. But whether by choice, through adult intimidation or simply because the adults cease to provide food, the young Peregrines are now ready to disperse. A further interesting note on the timing of fledging was raised by Olsen and Georges (1993) in their study of *F.p. macropus* in south-east Australia who note that contrary to received wisdom, Peregrines time the hatching of clutches when prey numbers, particularly juveniles, are abundant to aid chick rearing, the timing of Peregrine young fledging is synchronised with prey juveniles to give the best opportunity for young falcons to hone their hunting skills.

On 23 August 2011 SW observed some interesting post-dispersal behaviour by the adult female Peregrine at his Symonds Yat, Gloucestershire, England study site. The falcon pair bred successfully, producing four juveniles which fledged between 23-25 June. The fledglings remained on site until around

Adult nominate Peregrine passing food to a fledgling, England. *Jon Watson*.

15 August, then dispersed. Eight days later, the adult female came in very high (around 300m) holding a white dove. Instead of gliding down into a pitch close to the eyrie (as expected) the female held station, circling at the same height. The female then plumed the prey, white feathers scattering in the wind. Unprecedentedly, in the author's experience, the female maintained height for 15 minutes, continuing to scatter feathers into the wind, then glided slowly down, landed and sat with the prey for 20mins before starting to feed. Was this a form of prey display, suggesting that she is back in control of her territory following the recent dispersal of her juveniles? But, intriguingly, was the female making a prey availability display in case her hungry juveniles were still in the area? Certainly, white feathers in the wind at height on a bright day would be a highly visible sign of prey availability for any juveniles within several miles of their natal territory.

Breeding success
Defining the breeding success of any avian species is not trivial. Breeding pairs will hold a territory and may or may not breed. If breeding is attempted, it may or may not be successful (as egg clutches may be lost to predators or due to bad weather). Breeding may be successful. Distinguishing between these definitions of success in the literature is not always straightforward. Jenkins (2000c), studying *F.p. minor* in three areas of South Africa, noted that the overall success rate was 1.31 fledglings/territorial pair, 1.97 fledglings/breeding pair and 2.37 fledglings/successful pair. For comparison in the northern hemisphere, Cozic (2019) found success rates of 1.0-1.6 fledglings/territorial pair and 2.35-2.71 fledglings/successful pair for the period 1995-2018 for an increasing population of nominate Peregrines in Brittany, France.

In the data given below we have attempted to compare like-with-like in all cases.

Mearns and Newton (1988) studied the breeding ecology of nominate Peregrines in south-west Scotland and noted both the productivity of the falcons (*i.e.* the number of young produced per territorial pair), a measure of the overall breeding success of the species, and the mean number of young produced per brood, a measure of the success of breeding pairs. The study area included both coastal and inland sites, which allowed a degree of comparison between them, though as the number of territories was not equal in each case direct comparison must be treated cautiously. The overall performance of the population is set down in Table 7.4.

From Table 7.4 it is clear that coastal sites were significantly more successful than inland sites across the years of study. Mearns and Newton considered the likely cause was the accessibility of inland sites by humans. In general, sites which were occupied earlier (both coastal and inland) were more successful than those occupied later. Mearns and Newton considered that this was probably because less vulnerable sites were occupied first and were more likely to be occupied by experienced Peregrine pairs (who might also be better

	Number of territorial pairs	Number (%) of territories at which eggs were laid	Number (%) of territories at which young were fledged	Mean clutch size	Mean number of fledglings/ brood	Mean number of fledglings/ territorial pair
Inland sites	297	246 (83)	131 (44)	3.53	2.40	1.06
Coastal sites	100	83 (83)	58 (58)	3.53	2.17	1.26
Total	397	329 (83)	189 (48)	3.53	2.33	1.10

Table 7.4 Mean clutch size, fledglings per brood and productivity of Peregrines in south-west Scotland, 1974-1982. Data from Mearns and Newton (1988).

hunters), the combination of these factors leading to improved success. The use of more vulnerable sites by Peregrines due, in part, to the fact that the Peregrine population increased during the years of the study, resulted in some pairs having to occupy sites on lower, and therefore more accessible, cliffs: the success rate of nest sites, measured as the percentage of nests producing young, was 42% for cliffs <10m high, rising to 56% for cliffs 11–20m, 58% for cliffs 21–40m, and 71% for cliff >41m. Mearns and Newton also defined three classes of cliff – those accessible without ropes, those accessible with ropes, and those impossible to access. While these are subjective definitions (and rock climbers would possibly disagree), the resulting data gave success percentages of 51%, 60% and 73% in the three classes. Interestingly, although sites below overhangs were more successful, as might be expected, those in recesses or caves were the most successful, suggesting that shelter from poor weather as well as accessibility (by humans and predators) played a part in the final success. However, Mearns and Newton were not able to disassociate the influence of 'better' sites and 'better' falcons (the latter in terms of hunting skills) and so could not judge which was the more important.

In a study covering south-central Canada and the north-eastern USA, where Peregrines of mixed heritage have been reintroduced, Gahbauer *et al.* (2015) noted that overall there was no significant difference between the breeding success of rural and urban Peregrines, but that nest sites with overhead cover, including human-erected boxes, had higher productivity (in terms of number of nestlings which fledged). This result is consistent with a study in the Jura Mountains of France (Monneret *et al.*, 2015) where the breeding success in artificial nests was 2.2 fledglings per breeding pair, compared to 1.85 on natural sites. However, the success rates quoted by Ayala-Perez *et al.* (2021) for *F.p. anatum* in Mexico's Baja California were much lower. There the number of egg-laying pairs each year (2012-2016) was 6.8±1.6 of which 4.4±1.8 were successful (≈65%) with a productivity of only 1.23±0.06 fledglings/egg-laying pair. The researchers also found that the substrate of nests was an important contributor to breeding success (see Footnote 5 of Chapter 6). It is not clear why this particular breeding population has such a low success rate, though the number of breeding pairs has doubled in the area in recent years, this perhaps increasing the pressure on hunting males to provide adequate food resources to nestlings. In an eight-year study of Black Shaheen (*F.p. peregrinator*) in west Maharashtra, India (south of Mumbai) the breeding season covers both winter (November-February) and spring (March-April) and so encompasses a temperature range of 5-40°C, together with high humidity (45-50%). In this environment Pande *et al.* (2017) noted large variations of breeding success, the mean success being 1.27±0.81 fledglings per breeding attempt (range 0-3, N=122) and 1.56±0.59 fledglings/successful pair (range 1-3, N=99).

While the position of the Barbary Falcon is, as noted in Chapter 1, contentious, Shafaeipour *et al.* (2016), who studied the falcons in south-west Iran (and consider them a Peregrine sub-species) found that seven clutches (all in cliff cavities at an average height of 80m) had a mean clutch size of four and a mean breeding success of 3.43 fledglings/eyrie, the highest productivity we have found in the literature.

Mearns and Newton (1988) also investigated the causes of egg and chick loss (Table 7.5) and found that most chick mortality involved death soon after hatching, though some deaths did occur later, usually during spells of rain or mist when the adult birds were unable to hunt (Fig. 7.27).

Ratcliffe (1993) gives several other examples of adverse weather affecting nominate Peregrine breeding success, from Scotland, the Lake District and Wales, noting in particular 1981 when snow and frost in April, followed by a cold, wet May led to at least 55 pairs losing eggs or young. In a study in the Republic of Ireland, Norriss (1995) noted that the breeding range of Peregrines was limited by weather, this being seen in the interaction of cliff height and orientation, and spring rainfall. The highest (and therefore most attractive) cliffs were not occupied in high-rainfall areas or if they were

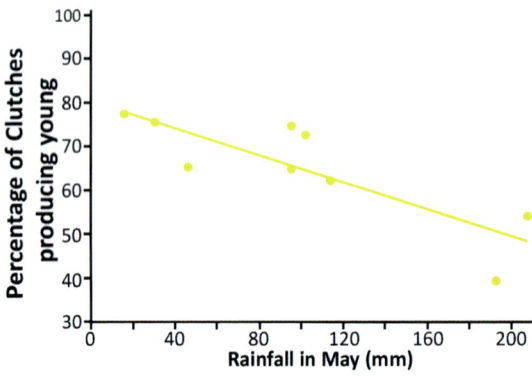

Figure 7.27 Variation of percentage success of Peregrine clutches with rainfall in May in nine successive years (1974-1982) in southern Scotland. The correlation between clutch success and May rainfall is significant at the 99% level. Redrawn from Mearns and Newton (1988).

Breeding Part 2

Cause of failure	Number	Percentage of all failures (n = 208)	Percentage of all breeding opportunities (n = 397)
Eggs not laid	68	33	17
Eggs broken/addled	30	14	8
Eggs deserted	3	1	1
Eggs disappeared	8	4	2
Eggs robbed[1]	23	11	6
Chicks disappeared at hatch	13	6	3
Chicks died	8	4	2
Chicks disappeared	20	10	5
Chicks robbed	13	6	3
Unknown	22	11	6

Table 7.5 Causes of failure of eggs and chicks of Peregrine nests in south-west Scotland 1974–1982. From Mearns and Newton (1988).

Note:
1. Known to have been robbed by humans. Humans may also have been responsible for some eggs which disappeared.

orientated north-west to north-east, while lower cliffs were unoccupied in a greater range of orientations and at lower rainfalls. No significant correlation was found between clutch success and rainfall over a longer period, either April–July (*i.e.* the entire breeding period) or May–July (*i.e.* the post-hatch period), indicating that rainfall in May was a critical factor in hatchling survival.

Rain was also an issue for Peregrines (*F.p. minor*) in South Africa, Jenkins (2000c) noting that spring weather had a very significant effect on the breeding success of Peregrines on the Cape Peninsula with egg and nestling survival being lower in wet years, and fledgling survival being higher in warm years (Fig. 7.28 overleaf). In the Orange River area, Peregrine productivity was also positively correlated with the height of the river when breeding began, while at Soutpansberg Peregrine productivity was higher in seasons following years of high rainfall.

Emison *et al.* (1993) noted very similar weather-related issues when studying *F.p. macropus* near Melbourne, Australia, where both rainfall and

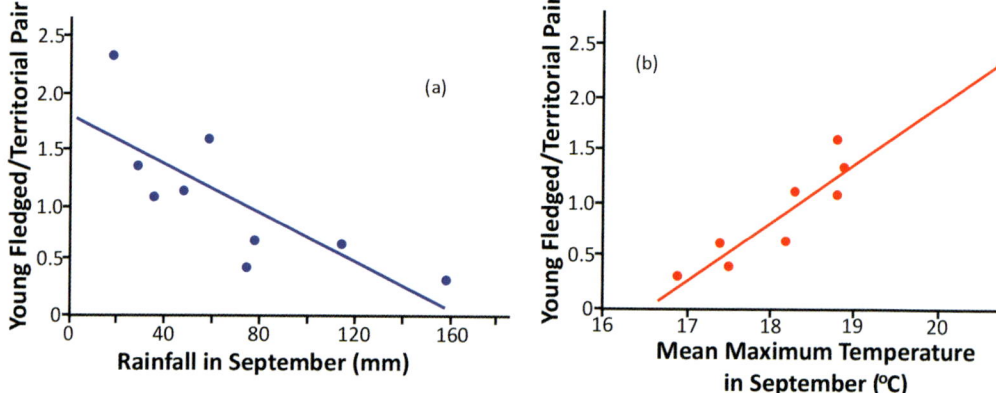

Figure 7.28 Mean annual productivity of Peregrines on the Cape Peninsula, South Africa, in relation to spring weather in nine successive years (1989-1997). In each case the lines are linear regressions. The correlation of breeding success and rainfall (Fig. 7.28a) is significant at the 98% level. The correlation of breeding success and mean maximum temperature (Fig. 7.28b) is significant at >99.9% level. Redrawn from Jenkins (2000c).

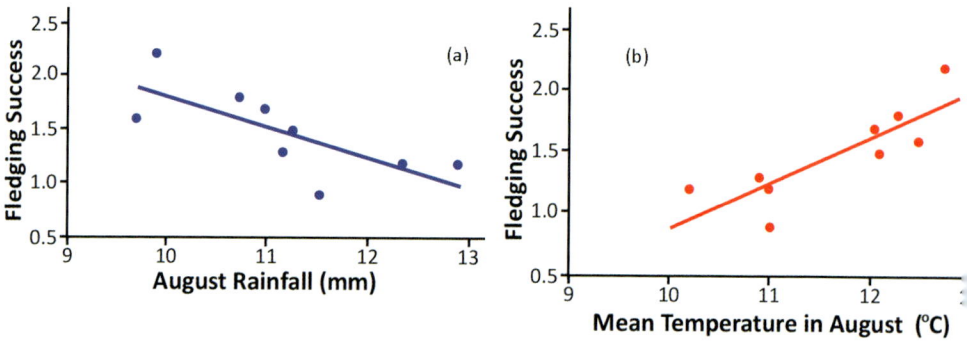

Figure 7.29 Fledgling success of Peregrines near Melbourne, Australia in relation to August rainfall (a) and temperature (b). Redrawn from Emison et al. (1993).

the cold of the austral winter reduced the breeding success of Peregrines (Fig. 7.29.

In the climatically much harsher Arctic environment of Rankin Island, Canada, a cross-over area between *F.p. tundrius* and *F.p. anatum*, Anctil *et al.* (2014) found a very similar correlation between rain and Peregrine breeding success, noting that success was negatively dependent on both the amount of rainfall and the hours of rain in July and August. An earlier study in the Rankin Inlet area, Bradley *et al.* (1997), had also found heavy rainfall reducing breeding success. In 1987 three broods totalling seven chicks died when 50mm of rain fell during a seven-day period. Then, in 1990, 17 chicks in five broods died during a four-day storm which combined 69mm of rain and 75kph winds, while in 1992 30mm of rain fell in three days and killed

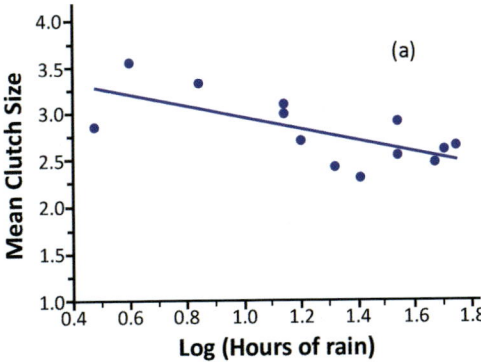

Figure 7.30 Variation of Peregrine clutch size in Arctic Canada with (a) total hours of rainfall and (b) total snowfall during the pre-egg laying period across the 13-year study 1982-1994. In each case the lines are linear regressions. The correlation of mean clutch size and rainfall is significant at the 99% level. The correlation of mean clutch size and snowfall is significant at the 97% level. Redrawn from Bradley *et al.* (1997).

14 chicks in four broods. Bradley also found that rain (and snow) affected breeding success much earlier in the breeding season, each reducing the clutch size laid by breeding females, though there was no correlation between weather and egg mortality. It seems the females were dutifully incubating their eggs, but had laid small clutches because the weather had reduced their breeding condition (Fig. 7.30).

In another study near Rankin Inlet, Robinson *et al.* (2017) noted another effect of the uncompromising Arctic weather. Those chicks which did not succumb to high rainfall or snow, did not gain weight if the weather was cool and wet in comparison to those raised when the weather was warm and dry (Fig. 7.31). As Fig. 7.31 indicates, the normal growth rate of nestlings

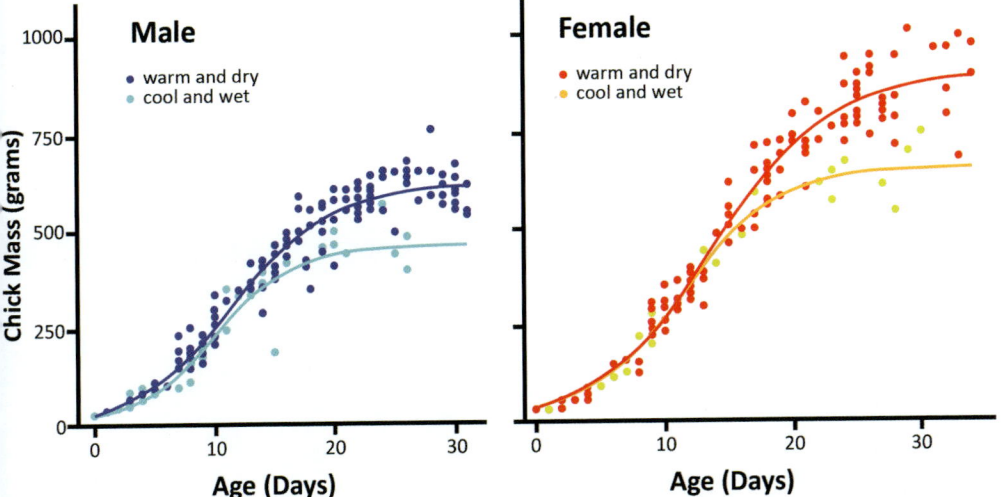

Figure 7.31 Growth rates of male and female Peregrine nestlings in Arctic Canada in relation to breeding period weather. The lines are best fit curves to the data of the raw dots. Redrawn from Robinson *et al.* (2017).

Nominate Peregrine food pass, but in this case with an aerial adult passing prey to a land-bound recently fledged youngster. In its haste to get to the adult quickly, the second, recently fledged youngster on the right attempted a hasty flight and went crashing into the rock ledge. Fortunately it was unhurt.

was curtailed after about 14 days of age if the weather was cool and wet, presumably because the adult birds were unable to hunt as successfully in those conditions. Robinson *et al.* surmise, but were not able to prove, that underweight fledged falcons were likely to have a higher mortality than those of 'correct' weight. Reduced weight could also, of course, prejudice breeding attempts, particularly for female Peregrines, for those fledglings which achieve adulthood.

At a coastal site in north-east Washington State, USA Wilson *et al.* (2000) noted a further weather-related influence on Peregrine breeding, in this case *F.p. pealei*. In years when the El Niño-Southern Oscillation produced warm surface waters in the Pacific Ocean the falcon's brood size fell by a third (Fig. 7.32). The exact reason for the decline was not established, but it was almost certainly the lack of prey. During warm water periods it is known that the numbers of breeding Brandt's Cormorants (*Phalacrocorax penicillatus*), Double-crested Cormorants (*P. auritus*) and Common Murres (or Common Guillemot) fell sharply, the inference being that other local auk and seabird species, that formed the bulk of the Peregrines' prey, would see the same decline.

The lifetime breeding success of wild Peregrines is not well documented. Hall (1955), studying the urban falcons which bred on Montreal's Sun Life building, states that one female raised 22 young over a 17-year period with three male partners. Since the female lost clutches in three of those years and did not breed in each of her first two years, her breeding period was 12 years. However, in a study of captive Peregrines Clum (1995) investigated female

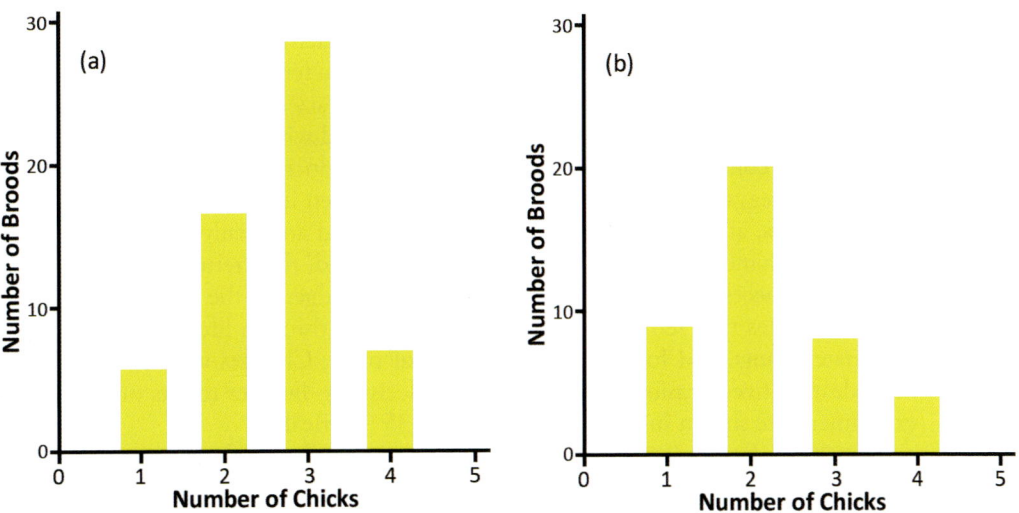

Figure 7.32 Variation of the number of Peregrine broods on the outer coast of the Olympic Peninsula, Washington State, USA during years without (a) and with (b) the El Niño Southern Oscillation. Redrawn from Wilson *et al.* (2000).

F.p. anatum food pass, adult bill to juvenile talons - a sequence which seems very risky. Shot taken at San Pedro, California, USA. *Gabriel Gruia*.

reproduction in two ways. Firstly, by removing eggs that female Peregrines had laid, and so forcing the falcons to lay again, Clum measured the maximum egg production attainable over the breeding life of the birds (Fig. 7.33a). Clum also measured six factors indicative of reproductive success – clutch size, egg fertility, 'hatchability' (*i.e.* the likelihood that a fertile egg will hatch), brood size at hatching, survival of nestlings and brood size at fledging. For the experiment 21 female Peregrines were kept year-round with males chosen on management considerations. As noted above, egg production was maximised by removing eggs to induce re-laying. The females then incubated their eggs for seven days, after which the eggs were removed and artificially incubated. Chicks were hand-raised until they were 7–14 days old, then returned to the falcons (not necessarily their parents) for raising to fledge. Of the 21 females, nine had mates that remained with them constantly through life, eight had one mate change and four had two changes of mate. Changes were due to male deaths (five occasions) or management decisions. Further results of the experiment are shown in Fig. 7.34, and Fig. 7.35 (overleaf).

From these figures it can be seen that all measures of reproductive success except nestling survival peaked at a female age of seven years, and then declined. Nestling survival is curiously anomalous: in the wild it is usually assumed that experienced parents are more likely to have hatchlings which survive, and this factor would not, therefore, decrease with female age. However, in this

Breeding Part 2

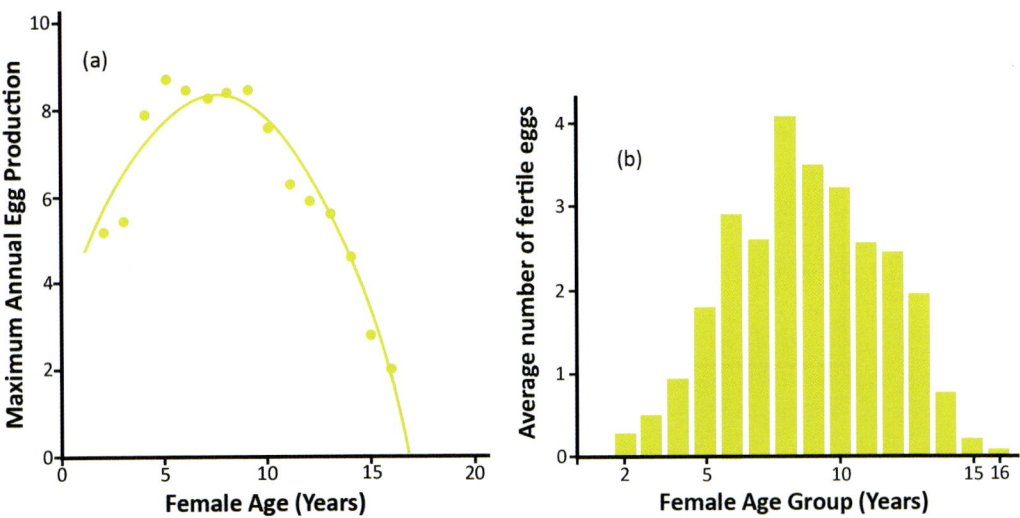

Figure 7.33
(a) Variation of maximum egg production with age for female Peregrines. Redrawn from Clum (1995).
(b) For comparison, the data here shows the variation of the mean egg production with age for 20 captive female Peregrines in Sweden. Redrawn from Lindberg and Sjoberg (2009).

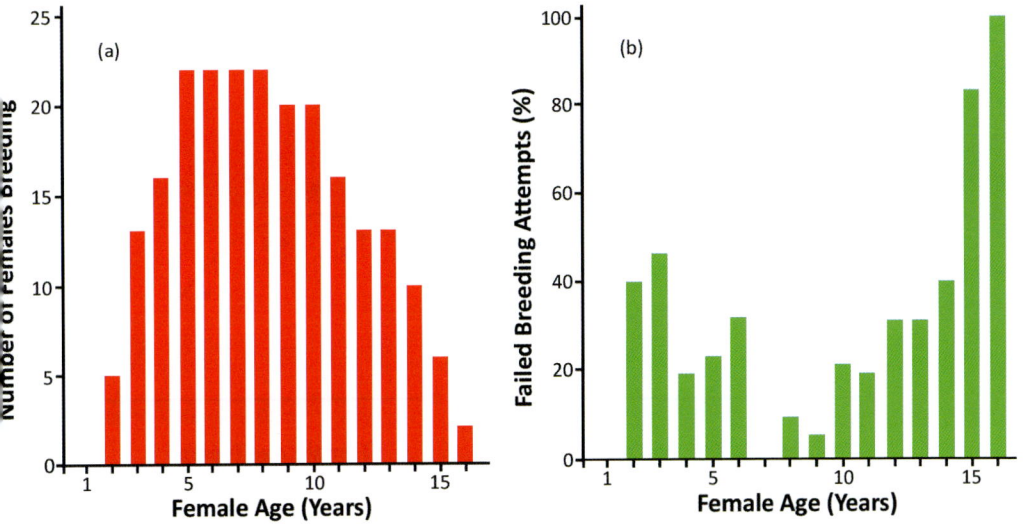

Figure 7.34 (a) Number of captive female Peregrines breeding as a function of age, and (b) number of failed breeding attempts, also as a function of age. Failures tend to be due to infertility in the female's first two years and from age 10 years onward, and from embryo or nestling deaths in other years. Redrawn from Clum (1995).

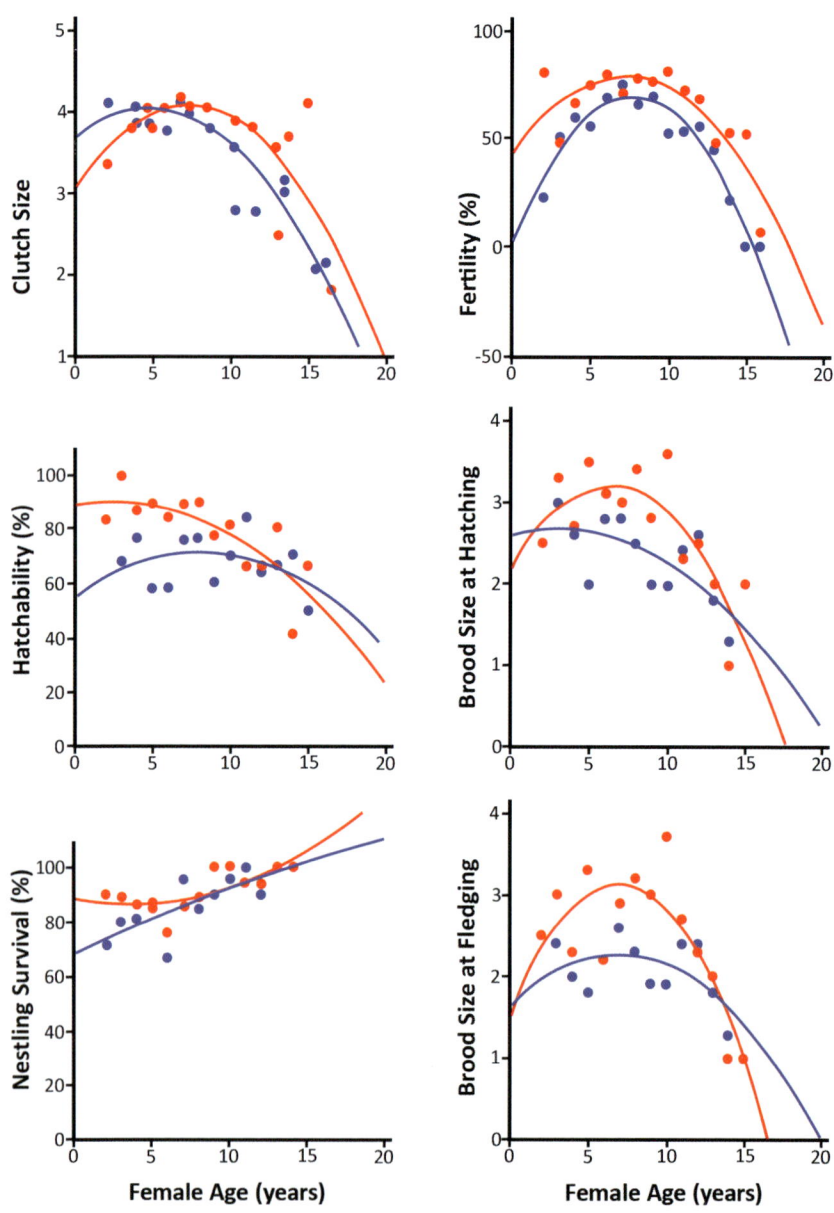

Figure 7.35 Variation of six measures of annual reproductive performance with the age of the adult falcons. The red circles and lines represent females retaining their mates throughout breeding life. The blue circles and lines represent females who experienced at least one change of mate during their breeding lifetime. Redrawn from Clum (1995).

experiment hatchlings were not necessarily raised by their parents, suggesting the experimental result is spurious.

Clum was able to conclude that birds with prior breeding experience had higher productivity than those of the same age which had not previously bred, and that increased experience resulted in increased productivity. By moving the pairs, Clum was able to show that productivity was not affected by such movements. However, changing mates did affect productivity, causing a 53% fall in the change year, though there was an increase in subsequent years. The results of Clum were not entirely borne out in a study of wild Peregrines in Spain's Basque Country, by Zabala and Zuberogoita (2015), who studied 37 breeding territories over a 16-year period. They found that breeding performance stabilised once the falcons had acquired adult plumage (usually by the end of their second calendar year), though age did eventually mean a fall-off in performance. Zabala and Zuberogoita noted that the performance of yearling (*i.e.* second calendar year) breeders was poorer than that of adults, but considered this was due to a lack of maturity, as the performance of inexperienced breeders (*i.e.* third calendar year) did not differ from that of experienced adults. Zabala and Zuberogoita consider that the differences between their study and that of Clum might be associated with wild birds being able to choose partners rather than having them assigned. This suggestion has implications for breeding fidelity which are considered in the following section.

Nest-site, territorial residency and mate fidelity
Site Fidelity
In a study in southern Scotland, Mearns and Newton (1984) found that during a five-year period (1977-1981) of 66 retrapped females, 60 (91%) were on the same territory as the previous year, while six (9%) had moved territory. Of eight retrapped males, six (75%) were on the same territory. In a later study covering northern England and southern Scotland, Smith *et al.* (2015), looked at two distinct periods, 1974-1982 when the population was still in recovery from the pesticide-induced decline (a low population density phase) and 2002-2011 when the density of falcons was relatively high. Observing site fidelity for adults (birds >2 years of age), sub-adults (birds 1-2 years of age) and juveniles (birds <1 year of age), the researchers found that site fidelity was very significantly higher for sub-adults and adults than for juveniles, and was higher for females than for males. Interestingly, when the population was in the low density, pesticide recovery, phase, the fidelity of juveniles was much higher than during the later, high density, phase. That was also true for adults (particularly males) though to a much lesser degree. The fidelity of sub-adults was the same for both the low and high density phases.

The critical nature of nest site fidelity in the breeding success of Peregrines can be gauged by the fact that the highly migratory *F.p. tundrius* falcons of North America show a similar degree of nest site fidelity to the sedentary

British population. In a five-year study in Canada's Northwest Territories, Court (1986) noted that four females were in the same territory in four successive years, four others were in the same territory for three years, and seven females were in the same territory for two years. For males three were seen for four years, three more for three years and one for two years. In terms of partnership, one pair stayed together for four years. In a longer study (seven years) in the same area Court *et al.* (1989) noted that mortality appeared to account for virtually all the observed changes in male territory holdings. There were changes by females, but these were few, and chiefly associated with nest failures. Overall, territories were held by individual birds of each sex for at least 6 years, a time comparable to the actual breeding lifetime of the birds. One falcon pair remained together for four consecutive years.

One interesting example of nest-area fidelity which did not involve nest-site fidelity was reported by Ponton (1983) studying *F.p. anatum* in northern New Mexico, where the falcons were nesting in eroded potholes in volcanic tuff (ash) along a 1km cliff. The female of one pair at the cliff used ten different potholes in ten consecutive years. After the female failed to lay during the next two years a new female continued the process by changing potholes in consecutive years. Interestingly, the latter female laid two repeat clutches and in each case laid the second clutch in the pothole used (successfully) in the previous year, a process which Ponton referred to as 'fall-back-one' behaviour.

Also interesting was the experience of Wimsatt (1940b) in Pennsylvania. Wimsatt removed an incubating *F.p. anatum* female from a clutch of four eggs, hooded the bird and took it 100km to Cornell University where it was kept in darkness. Wimsatt returned to see if the male would continue to incubate. He did, but only for a few days before deserting. Wimsatt therefore removed the clutch, and then released the female. The bird headed off in the direction of the nest site. Four days later Wimsatt found the female at a new nest site 3m from the old one: she had already laid an egg and would go on to complete a clutch of four, three of which hatched. Wimsatt has no information on whether the male responsible for the second clutch was the original partner.

The authors have explored site-type fidelity of nominate Peregrines by examining population data since 1930. Table 7.6 sets down the Peregrine territorial pair population for the period 1939-2014 split by country into four broad, site type categories: traditional natural inland and coastal sites, together with recently occupied quarries and man-made structures. Table 7.6, and Fig. 7.36 overleaf, reflect the temporal and spatial changes occuring since the pre-war baseline data of Ratcliffe (1993). Note that the site type categories in the Table show no quarry breeding prior to 1981, and no human stucture breeding prior to 1991. This is simply because such breeding sites were not mentioned in the BTO Bird Study papers. We do know that buildings have been used sporadically: Ratcliffe (1993) notes specific historical records including Corton church tower, Norfolk (in ruins from *c.*1840) and Salisbury Cathedral used occasionally according to records going back to 1864.

UK	1930/39 Baseline	1962	1971	1981	1991	2002	2014	Sparkline Inland Coastal
Coastal								
England (Incl. Isle of Man)	128	16	16	60	123	184	207	
Scotland	233	150	107	175	189	200	201	
Wales	62	8	11	43	72	86	106	
Northern Ireland	22	6	10	14	26	23	20	
Coastal totals	445	180	144	292	410	493	534	
Inland - Natural								
England (Incl. Isle of Man)	50	20	29	64	131	121	150	
Scotland	282	197	223	297	410	316	211	
Wales	65	19	16	69	132	111	82	
Northern Ireland	32	10	25	34	32	9	26	
Inland Natural totals	429	246	293	464	705	557	469	
Inland - Quarry								
England (Incl. Isle of Man)	0	0	0	6	44	152	228	
Scotland	0	0	0	7	26	49	89	
Wales	0	0	0	9	53	81	80	
Northern Ireland	0	0	0	5	38	50	56	
Inland Quarry totals	0	0	0	27	161	332	453	
Inland - Human structure								
England (Incl. Isle of Man)	0	0	0	0	5	42	258	
Scotland	0	0	0	0	1	8	26	
Wales	0	0	0	0	1	5	12	
Northern Ireland	0	0	0	0	0	0	1	
Inland Human structure totals	0	0	0	0	7	55	297	
UK and Isle of Man totals	874	426	437	783	1283	1437	1753	

Table 7.6 The number of territorial pairs of nominate Peregrines by site type in the UK and Isle of Man (excluding the Channel Islands) 1939-2014. The 'heat map' gradient varies from low (white) to high (red). The sparkline also indicates the change in population with time.

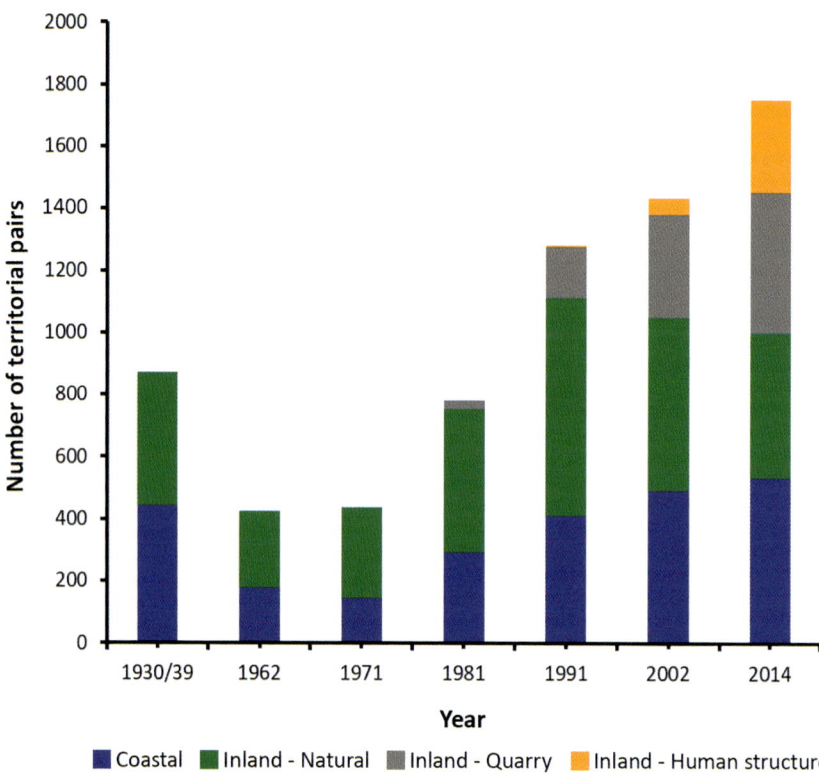

Figure 7.36 Estimated number of territorial nominate Peregrine pairs in the UK 1930-2014 by site type.

Note that these data are caveated on the basis that the authors have been required to make some broad assumptions in respect of the site type allocations of unchecked estimates and, for 2014, of random square estimates. The unchecked and random square estimates have again been allocated proportional to actual counts. It is likely that this methodology will understate the values for human structures and quarries (M. Wilson pers. comm. SW). It is also difficult to differentiate, with confidence, new population site types as between quarry and human structure within the 2014 data because for the first time random square territorial pair numbers were included in the estimate based on sample data. Random square estimates in 2014 accounted for 520 pairs in a total of 1753 *i.e.* 30% The rationale for this change in estimation basis was that the known large increases in the numbers of new quarry and human structure breeding Peregrines since the 2002 survey made it difficult to assess where to monitor and also how to allocate the required manpower. Nevertheless, the authors feel that it is worth showing the historical data in this format to identify broad regional site type trends over the whole survey periods 1939-2014.

As the UK Peregrine population expanded, the falcons initially took to quarries (N. Dixon pers. comm. to SW) both disused and working, and began to breed successfully while at the same time becoming habituated to breeding close to both the presence of humans and human activities. It has been suggested, reasonably, that quarry nesting could thus be a precursor for an increased level of breeding on man-made structures because of the habituation of some pairs to humans. However, an analysis of the available natal site type fidelity data, obtained from BTO and other resightings on breeding territories, has, surprisingly, found no evidence for this. Once Peregrines are breeding successfully at a particular site type we contend that the young produced are highly likely to perpetuate and extend breeding at that site type. In our experience, the interaction with humans at quarries is generally very positive, quarry workers often referring to 'their' Peregrines and taking active measures to protect and safeguard the birds, often denying access to 'unknown' visitors to prevent disturbance[6]. At these sites the Peregrines become inured to adjacent heavy machinery workings, the dynamiting of rock faces and general human comings and goings: it is instructive to note that the breeding success in working quarries (as defined by either occupation levels or fledglings produced/breeding pair) is greater than it is for disused quarries, presumably because of the reduction in malign disturbance.

Once the trend for Peregrines to breed on man-made structures became established, the general public took a keen interest, sometimes casual, sometimes more formal, to the point where if Peregrines failed to breed through lack of appropriate location and/or substrate, concerted efforts would often be made to provide suitable, sometimes tailor-made, nest boxes for subsequent seasons. Such boxes often become the basis for a passionate and enthusiastic social media following, particularly if webcams were installed relaying activity in the box. As such cameras occasionally pick up the rings on breeding adults, further dispersal information is made readily available to Peregrine researchers and armchair enthusiasts alike.

Urban environments are also welcoming in other ways. Typically, installed nestboxes offer protection against sun, wind and rain, while most urban boxes inherently deter human persecution. In addition, cities and towns are 2-3^{0}C warmer than 'out of town' sites and invariably offer a dense supply of readily available prey, *e.g.* feral pigeons, Starlings (and, increasingly, Ring-necked Parakeets whose population is increasing in, and which are spreading out from, London). Urban prey is also unlikely to be heavily contaminated with industrial or agricultural chemicals. In their study of the breeding performance of Peregrines in urban and rural nest sites between 2006 and 2016 (22 urban sites and 58 rural sites) Kettel *et al.* (2019) found that breeding success was higher in urban sites (94% against 78% in rural sites) and that the number of fledglings produced per site was also higher, urban pairs on average producing approximately one more fledgling than their rural cousins. Kettel *et al.* also compared available prey biomass within a 2km radius of each nest site (an area of 12.57km^2) - Figs. 7.37 and 7.38 overleaf.

---

[6] Disturbance has to be seen in context. Peregrines very quickly become inured to local human activity, ignoring dynamite in quarries and bells on church towers, but they dislike being observed, particularly from above, in the early breeding season. This would seem to be an increasing problem at coastal sites where paths adjacent to breeding cliffs are increasingly used by humans.

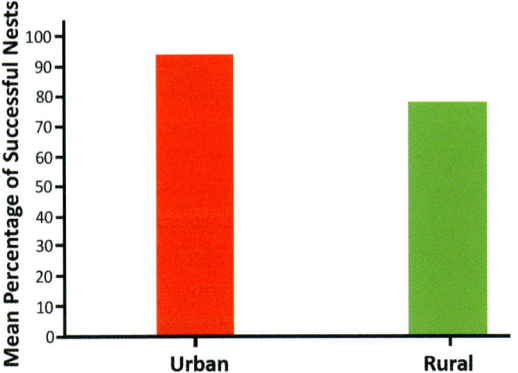

Figure 7.37 Mean percentage of successful nesting attempts (i.e. fledging ≥1 chick) at urban and rural sites in the period 2006-2016. Urban nests had a mean success of 94%, while the mean success for rural nests was 78%. Redrawn from Kettel *et al.* (2019).

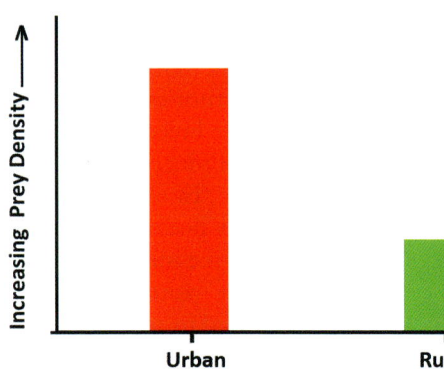

Figure 7.38 Relative prey density for urban and rural nest sites. The density was measured as mean/km² within a 2km radius of the nest site by the researchers but the actual values calculated have not been transcribed here, the arrow noting the direction of increasing prey density. Redrawn from Kettel *et al.* (2019).

When considering future potential urban range expansion it is interesting to note the current maximum breeding pair densities in London and to compare that to the total UK urban landscape (see Chapter 10 Fig 10.15). One of the authors (SW) has monitored breeding habitat changes in Gloucestershire, south-west England, since 1984 and observed rapidly increasing breeding in towns and villages on relatively low, but still prominent buildings, within the local (essentially rural) landscape. If a building is relatively high and holds breeding potential then it is increasingly likely to be tenanted, particularly if prey availability is good.

It is probably the case that Peregrines have bred in small numbers in inland quarries (>1 km from coast) for many decades, but this site type population first appeared in BTO survey data in 1981 when 27 pairs were counted and were represented in all UK countries: by 1991 the territorial pair population at this site type had increased six fold to 161. By 2002 this figure had doubled, and by 2014 had increased by a further 36%. Over the 33 years to 2014 UK territorial pair numbers in inland quarries had therefore increased from 27 to 453 pairs, a 17-fold increase (or 9%/year). Over the same period the number of UK human structure breeding site types has increased from 7 in 1991 to c.300 in 2014, a 43-fold increase (17%/year). If we combine these two manmade site types (so as to eliminate any misallocation between them) then for the 33 year period 1981-2014 the territorial Peregrine pair population has increased from 27 to 750 pairs, *i.e.* 28-fold (10.6%/year). Over the course of these surveys a handful of sporadic and occasional tree and ground nesting breeding attempts have been monitored in the UK and are not specified separately in this analysis – they are included in 'Natural Inland' data.

Breeding Part 2

Site type analysis was assessed in greater depth than before in 2014 (Wilson *et al.*, 2018). Table 7.7 is derived and reformatted from that study and correlates the number of detailed site types **monitored** with percentage occupancy for each in 2014 (providing an occupation indication as this metric will change year on year).

Country split	England (incl. Isle of Man)		Scotland		Wales		Northern Ireland		UK	
	No. monitored	Occupancy	No. monitored	Occupancy	No. monitored	Occupancy	No. monitored	Occupancy	No. monitored	Occupancy
Site types										
Human structure - pylons	62	90%	4	25%					66	86%
Human structure - buildings	74	96%	27	63%	9	78%	2	25%	112	85%
Human structure - other	18	94%	11	64%	3	50%			32	80%
Quarry - working	46	86%	51	81%	13	88%	60	58%	170	75%
Quarry - disused	88	79%	68	54%	50	61%	53	36%	259	60%
Coastal	207	86%	363	42%	119	61%	29	66%	718	59%
Natural Inland - large cliff	62	66%	181	44%	86	49%	25	60%	354	50%
Natural Inland - small crag	100	42%	280	29%	49	31%	26	31%	455	32%
Natural Inland - other	22	43%	117	13%	4	38%	14	18%	157	18%
Totals - all site types	**679**	**77%**	**1102**	**39%**	**333**	**55%**	**209**	**47%**	**2323**	**53%**

Table 7.7 UK Peregrine occupancy by site type based on sites monitored in 2014. The data have been sorted by percentage occupancy. Data reformatted from Wilson *et al.* (2018).

It is interesting to note that occupancy levels at human structures and quarries are significantly higher than at the equivalent natural and coastal sites. For instance, the occupancy rate at UK buildings is 70% higher than that at natural inland large cliffs, and 44% higher than at coastal sites. Extracting 2014 data for monitored sites with known breeding outcome from Wilson *et al.* (2018) has allowed us to construct Table 7.8 overleaf.

Combining the site type average for percentage fledged young with average breeding success rates shows an interesting comparison between site types split by country. Human structures, on average, produced more fledged young per successful nest in 2014 than all other site types, being 36% higher than traditional coastal sites and 24% higher than natural inland large cliffs. It is also of note that there is a consistent proportion of successful breeders across all site types with the notable exception of those at large natural cliffs in inland England. Traditional type sites are just as successful (in terms of percentage of breeding pairs that fledge one or more juvenile) as those on human structures and quarries.

Country split	England (incl. Isle of Man)		Scotland		Wales		Northern Ireland		UK	
	Average number of fledged young	Successful	Average number of fledged young	Successful	Average number of fledged young	Successful	Average number of fledged young	Successful	Average number of fledged young	Successful
Site types										
Human structure - buildings	2.6	68%	2.7	60%	2.0	75%			2.6	67%
Quarry - working	2.2	61%	2.4	74%	1.5	44%	2.8	66%	2.3	65%
Natural Inland - small crag	2.3	52%	2.4	69%	2.0	63%	2.3	60%	2.3	63%
Natural Inland - other	2.3	43%	2.2	69%			3.0	50%	2.3	60%
Human structure - other	1.9	63%	2.6	100%					2.2	71%
Quarry - disused	2.3	56%	2.2	64%	1.7	68%	2.6	47%	2.1	61%
Natural Inland - large cliff	1.9	35%	2.3	58%	2.0	63%	2.5	79%	2.1	52%
Coastal	2.1	59%	1.7	73%	1.7	60%	2.9	84%	1.9	67%
Human structure - pylons	1.8	66%							1.8	66%

Table 7.8 UK successful breeding attempts and average fledged broods from those successful attempts, sorted by UK average number of fledged young.

Note that this dataset should be viewed cautiously as site types with low pair numbers can give distorted fledged young averages and success rates – this applies to human structure data for Scotland, Wales and Northern Ireland and also Natural inland – and 'other' for Wales).

The increase of the urban Peregrine population in the UK is considered further in Chapter 10.

Territorial residency

Residency, as with site fidelity, is in the interests of a Peregrine, as secure nest sites are essential for breeding success, and moving away means the possibility that on return in the spring the territory will have been claimed by another pair. As a consequence, if pairs are able to stay at a site, sufficient food resources being available, then they will. Juveniles also tend to spend the winter within a short distance of their natal site if territories are available, often moving to coastal sites where they can feed on waders if they are not, or if prey resources are inadequate. These birds, assuming they survive the winter, will then move back towards their natal sites unless they have moved to areas where territories and nest sites prove equally, or more, attractive. Enderson and Craig (1988), studying *F.p. anatum* in Colorado, USA, noted 83% of males and 77% of females on the same territory in a subsequent year. In their study of *F.p. brookei* in northern Spain, Zuberogoitia *et al.* (2009) found that territorial fidelity for males was at least 3.4 years, while for females it was at least 3.7 years.

Mate fidelity
There is also evidence that female birds prefer philopatric males because occupation of a territory suggests a willingness to defend it (and, therefore, an ability to withstand competition) and also suggests that local resources are good. Philopatric males are also likely to have some knowledge of local predators which is also advantageous for successful breeding. (As an aside, Greenwood (1983) noted that in birds the tendency is for males to be more philopatric than females, but the reverse is the case for many mammals.)

In most cases falcon mortality was the primary reason for birds not to return to a territory, though females are more likely to switch territories than males. Breeding success underpins site fidelity, and mate fidelity is presumably similarly predicated, given that both sexes would be inclined to stay on a good territory, and, as noted above, the work of Clum (1995) indicates that pairs with prior breeding experience together have a higher productivity (in terms of fledglings raised per breeding cycle) than newly-paired birds. While evidence from captive birds cannot be considered conclusive, this strongly suggests that mate fidelity is an important criterion.

Dispersal
Mearns and Newton also looked at the dispersal of ringed nestlings, finding that females had moved further (an average of 68km, maximum 185km) than males (an average of 20km, maximum 75km). Perhaps dispersal within the UK is defined by the distance, east or west, to the sea, as greater distances, to about 500km, are seen in continental Europe, though the ratio of dispersal distances of the two sexes remain the same. Interestingly, in Australia Mooney and Brothers (1993) noted a mean male dispersal distance of 24km, with females travelling a mean of 120km. As Australia is a vast country (though with a desert heartland that would not suit Peregrines) it might be expected that distances would be longer, but this study was in Tasmania. In the USA, comparable distances to those seen in Europe have been observed. In the Midwest Tordoff and Redig (1997) found mean distances for males of 176km and for (*F.p. anatum*) females of 320km. However, far greater distances have been seen. Telford (1996) notes a female Peregrine fledged in Virginia in 1989 which bred in Vermont, 781km away, in 1991. Dennhardt and Wakamiya (2013), in Midwest USA noted one female which moved 876.08km, but also noted a male which moved just 190m. Faccio *et al.* (2013) record a male Peregrine which in 2004 moved 1009.72km from Maine to Virginia. In a study of urban breeding Peregrines in Midwest towns (involving reintroduced Peregrines of mixed heritage) Caballero *et al.* (2016) found mean dispersal distances of 124km for males and 226km for females which seems to be more the usual pattern.

Other very long dispersal distances are often quoted, but in some cases the timings of these and the subsequent history of the bird, make it hard to differentiate dispersal from migration. A good example of this is a Finnish male Peregrine ringed close to Oulu, central Finland, which was photographed eight months later in Devon, England, 2300km away. In the absence of further information on the later whereabouts of the bird it is difficult to know if the male had dispersed permanently to the UK or had migrated there and subsequently flown back to Scandinavia.

Robinson *et al.* (2019) record a nine-month old female Peregrine ringed in May 1997 in Northumberland, north-east England which was found dead in March 1998 in Lanzarote (in the Canary Islands), a straight-line distance of 3065km, while Drewitt and Sutton (2021) record a male Peregrine ringed in Taunton, south-west England in late May 2019 being found dead in Tiznit, Morocco 155 days later having travelled a straight-line distance of 2435km. Are these birds dispersing or migrating? Certainly the Finnish bird might have been avoiding the definitely harsh north-Scandinavian winter, but while the winter in northern England can also be harsh, most UK Peregrines are resident. Doolittle *et al.* (2013) record the tarsus of a female Peregrine being found in Ticino, Switzerland in 2008. The tarsus had a ring which had been placed on a female juvenile in Wisconsin, USA, near Lake Superior in 1993: it was not clear when the falcon had died. This was the third American Peregrine known to have made it across the Atlantic, but the find was the furthest east of the three (the two others made it to Lisbon, Portugal and Brighton, England).

Dispersal is beneficial to the species as it reduces the possibility of incest (see, for instance, Ponnikas *et al.* (2017) on the maintenance of genetic variability of a specific population of falcons due to dispersal). However, the recruitment of juveniles into the local breeding stock does occur. Court *et al.* (1989), studying *F.p. tundrius* in Canada's Northwest Territories noted that a small fraction (<4%) of local juveniles did join the local breeding stock. As might be expected, given the mean dispersal distances being significantly greater for females than males, males were much more likely to be locally recruited than females.

An irate nominate female finds her mate feeding the chicks and demands to take over. *Torsten Pröhl.*

An adult *F.p. anatum* chases a young American White Pelican (*Pelecanus erythrorhynchos*) off its territory near San Diego. *Will Sooter.*

F.p. pealei with prey. *Nick Dunlop.*

8   Movements and Wintering Grounds

Peregrines have both residential and migratory populations, the latter being the northern breeding species who travel south to avoid the cold, the lack of daylight for hunting and the sparse/non-existent prey at their breeding grounds, most of the prey species utilised during chick rearing also being migratory. True migration is, therefore, concentrated in the Arctic breeders: *F.p. tundrius* and *F.p. calidus*; the northern breeders of *F.p. anatum* and, perhaps, *F.p. pealei* in North America; nominate Peregrines in Scandinavia and Russia; and *F.p. japonensis* in north-east Siberia, although several other sub-species also move from breeding areas, occasionally to lower elevations for the winter (*e.g. F.p cassini* in South America and *F.p. peregrinator* in central Asia).

The evolution of migration in birds has long been studied, with the idea that some northern breeding species hibernated during winter (as Aristotle conjectured) not being finally dispelled until the mid-19th century. Although the reasons for migration seem straightforward, the minutiae of observed differences between species are still debated, particularly the existence in some species of both resident and migratory individuals. Any such debate must also take into consideration whether local conditions offer a breeding success that outweighs the stresses long migratory flights impose on the individual bird[1]: it is known, for instance, that some individuals fed by humans during

---

[1] Klaassen *et al.* (2014) used satellite tagging to study the danger to life of migration in raptors. Although Peregrines were not one of the studied species, the findings of the study probably apply to all migratory species. Klaassen *et al.* found that the mortality rate was six times higher during migrations, with more than half of all mortality occurring during that time. Total mortality across the whole year was however comparable: it is the migration flights which are dangerous for an individual bird, the 'resident' periods at each end being relatively safe. Spring mortality was higher than that in autumn – migration is hard on young birds.

the winter will choose not to migrate as others of their kind do. Nevertheless, it is now generally thought that migratory behaviour is inherent in many, perhaps most, bird species, being repeatedly evolved and lost over time, an idea supported by the speed with which migration to northern areas was re-established following the last Ice Age. To test this idea, Winger *et al.* (2012) looked at 108 of the 110 species (including recently extinct members) of the Parulidae (New World warblers (or wood-warblers)). The collected data are illustrated in Fig. 8.1. Winger *et al.* conclude that the ancestral wood-warbler was migratory and that throughout history its descendants have both lost and gained migration behaviour. Sedentary wood-warbler populations (some 58% of the total) have clearly lost migratory behaviour while the behaviour in others has been extremely variable, from 1° to 60° in terms of the latitudinal difference in migratory flights.

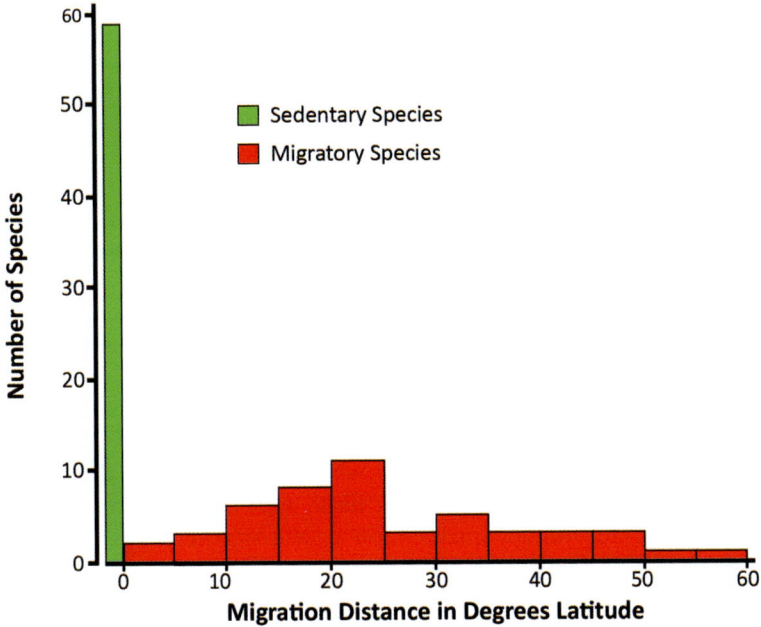

Figure 8.1 Analysis of the 108 Parulidae species considered in terms of distance travelled, in degrees of latitude, by migratory species (in red) as opposed to sedentary species (in green). Redrawn from Winger *et al.* (2012).

The ability of birds to fly is clearly an advantage when a species is seeking better feeding areas in breeding and non-breeding periods, *i.e.* is choosing to migrate. Flying consistently at, say, 80kph allows an individual to travel almost 2000km in a day, a distance way beyond the abilities of any terrestrial mammal. But while flight is critical, there are other factors which are equally important. To find either its breeding or wintering grounds a bird must carry the equivalent of a map, a compass and a clock, just as a walker out for a day's birdwatching must (though the clock is optional). In addition, the bird may need to carry something our day-walking birdwatcher can leave at home – a calendar.

Since it was known that juveniles sometimes migrated after adults (as they in general do in Peregrines – see Fig. 8.6) it was thought that the direction of, and time taken for, migratory flights must be inherent. Early experiments in which large numbers of juveniles of several species were caught and moved to a release area remote from their natal site suggested that flight direction was indeed inherent, the birds setting off in the correct direction and landing in an area remote from normal wintering grounds by a distance equal to the initial displacement. Most striking was the migration of the European (Common) Cuckoo (*Cuculus canorus*) which finds its way to the wintering grounds of its parents without ever having met them even when the wintering grounds of its 'foster' parents lie in an entirely different direction/area. In one experiment (Rabøl, 1993) suggested that flight time was also inherent: Pied Flycatchers (*Ficedula hypoleuca*) and Garden Warblers (*Sylvia borin*) shipped from Denmark to Kenya at what would have been the time to set out on migration, and then kept caged, continued to show a restlessness in their migratory flight direction for the duration of what would have been their transit time. However, to confuse the issue, Lesser Whitethroats (*Curruca curruca*) did not. There were also other anomalies: in some experiments, birds finding themselves at displaced wintering grounds readily made it back to their breeding areas the following spring, while some juveniles did not follow an inherent directional vector, adjusting their direction to compensate for displacement. The exact nature of the inherent directional vector was, therefore, still debated.

Nevertheless, it appeared that in general both the direction and time of travel were inherent, raising the obvious questions of how a juvenile bird acquires the information it needs, and what information an adult bird uses to reassess direction and time when it finds itself off its normal route, due either to deliberate human intervention or the weather. The first question seemed initially to involve mysticism, so early efforts were concentrated on the second.

Obvious navigational clues are celestial – the position of the sun during the day and the stars at night, as each follows a defined path which varies with latitude. For some Peregrines, the flight paths are more complicated than for migrants whose travels are contained within either the northern or southern hemispheres. For Peregrines north of the Equator the sun lies to the south, while south of the Equator it lies to the north: in each case the sun moves east-west. At night, north of the Equator the stars rotate anti-clockwise around the north celestial pole marked by Polaris (the North or Pole Star): south of the Equator the stars rotate clockwise around the south celestial pole, a point not marked by a single prominent star (although Polaris Australis (occasionally called Sigma Octantis) is very close to the pole it is a very dim star) human

Migrated *F.p. calidus*, photographed against the setting sun at the Little Rann of Kutch, Gujarat, western India. *Dhairya Dixit*.

stellar navigation using a line drawn through stars of Crux, the Southern Cross. A Peregrine using celestial clues therefore needs an understanding of its position relative to the Equator and an understanding of the changing star map if it crosses that line (which some, though few in number, Peregrines do, both in the eastern and western hemispheres). The changing length of day and night with time of year is also an issue, though the bird's circadian rhythm would still define a 24-hour day, as it does for humans, even those that venture to the poles and so experience days without nights (or vice versa)[2]. That the sun acted as a compass for migratory birds has been known since the 1950s when German ornithologist Gustav Kramer[3] carried out experiments on caged Starlings. Using the sun has the obvious disadvantage of days when it is not visible, though some birds can read the polarisation pattern of the sky which would allow information to be gathered even if the sun itself is hidden (see Box 8.1). Experiments with birds placed in planetaria where the stellar pattern could be readily manipulated, have also shown that birds do indeed read the night sky for orientation: the prominence of the polarisation pattern at sunset and the fact that both the setting sun and the emerging night sky may also be visible probably explains why nocturnal migrants usually begin flying at that time.

---

[2] The 2017 Nobel Prize for Physiology or Medicine was awarded to three American scientists for the discovery of the genetic basis of the circadian rhythm in *Drosophila* fruit flies. There is no reason to suppose the rhythm is not similarly controlled in all animals.
[3] Gustav Kramer (1910-1959) was trying to take Rock Dove chicks from a nest in southern Italy when he slipped and fell. The fall was fatal, a tragic lesson to all would-be field workers.

Box 8.1 The polarisation of light
A book on Peregrines is not the forum for a discussion on quantum physics and whether light is wave-like or comprises a stream of particles (photons). Suffice it to say that experiments have shown that in specific circumstances each has its merits. Assuming light is a wave, it is a transverse wave, *i.e.* the oscillations are perpendicular to the direction of travel – as in the wave that can be produced by flicking a length of string at one end when the other end is tightly fixed. Sunlight is unpolarised as it comprises a whole series of random individual light waves. But the scattering of light by particles or water droplets in the air can create polarisation, in the same way as polarising lens in sunglasses can, reducing glare and, for photographers, enhancing the blue of the sky. Since scattering enhances polarisation it is prominent on cloudy days and towards twilight. Many people can see the polarisation of light by viewing a blue sky with their backs to the sun. The faint, visible image is known as Haidinger's Brushes after Wilhelm Haidinger (1795-1871), an Austrian mineralogist who first described the phenomenon. The brushes (Fig. 8.1.1) are formed in the macula of the eye and can aid navigation as the yellow 'bow-tie' points towards the sun.

Figure 8.1.1   Haidinger's Brushes. The 'brushes' are so-called because their fuzziness are said to resemble brushes. To the present authors the brushes better resemble crossed bow-ties. The yellow axis points towards the sun.

For a better understanding of polarisation patterns under different sky conditions and possible navigational methods they create see Cronin and Marshall (2011) and Zhao *et al.* (2018). In view of the use of polarisation to aid navigation in avian migration, it is interesting to note that Haidinger's Brushes has been postulated as a navigation tool for those long-distance mammalian migrants, the Vikings, whose sea voyages reached Iceland, Greenland and North America, though it could be argued their journeys were dispersal rather than migration. Viking usage of polarisation patterns is controversial: for a discussion on this see Horváth *et al.* (2017).

Opposite   Migrating North American juvenile Peregrines resting on a ship moored 40-45km off the Mexican coast in the western Gulf of Mexico. *Amanda Stafford.*

The Earth's magnetic field was thought to offer another useful navigational cue. The longitudinal force lines of the Earth's field are similar to those produced by a standard dipole, the lines dipping vertically at the North and South Magnetic Poles (which are close to, but not coincident with, the geographical poles), and lie parallel to the Earth's surface at the Equator. The angle of the force line (known as inclination) thus varies from +90° at the North Magnetic Pole, through 0° at the Equator, to -90° at the South Magnetic Pole (Fig. 8.2a). The intensity of the magnetic field also varies from pole to pole though in a much less controlled way as the Earth has several magnetic anomalies, in particular a minimum intensity area over central South America and the adjacent South Atlantic, and higher intensity areas in north-central Siberia, north-central Canada and coastal Antarctic near Australia. (Fig. 8.2b). While these anomalies preclude a general use of geomagnetism to provide a two co-ordinate mapping system for a bird, the longest migration paths of Peregrines are across areas of minimal deviation from a uniform change and so offer a possible world map – if the falcons can read it.

In a seminal experiment Wiltschko (1968) showed that caged European Robins (*Erithacus rubecula*) exposed to a magnetic field more powerful than that of the Earth changed their direction of migratory interest, *i.e.* they were reading what they assumed was the Earth's field: when the imposed field was removed, the birds re-aligned to the Earth's field. Later experiments showed that other species were also reading the field, implying that all migratory birds could read both celestial information and the Earth's magnetic field (Fig. 8.3 overleaf). In later work, Wiltschko and Wiltschko (1999) noted that the Robin's detection of a magnetic field was dependent on the wavelength of the light to which the birds were subjected. In green light (wavelength 565nm)[4] the Robins aligned with the imposed magnetic field, but with yellow light (wavelength 590nm) they did not. In later work, Wiltschko *et al.* (2002), covered each eye of a Robin separately and found that the magnetic compass was associated only with the bird's right eye (Fig. 8.4 overleaf). This was initially baffling, but later it was discovered that birds had not a single, but two magnetic detection mechanisms (Wiltschko and Wiltschko, 2007). The first mechanism used specific flavoproteins called cryptochromes (specifically CRY4) located only in the retina of the right eye that are sensitive to blue light (and also involved in circadian rhythm[5]). The light-sensitivity creates a magnetic compass by a process known as radical-pair formation[6]. A very

---

[4] A nanometer (nm) is a thousand millionth of a metre, *i.e.* 0.000000001m, usually written as 10^{-9}m.

[5] Specifically this is cryptochrome CRY1 – see Pinzon-Rodriguez and Muheim (2021) – or CRY2 – see Einwich *et al.* (2022).

[6] The story is told that in 1978 Klaus Schulten, a young physicist in Germany working on influence of magnetic fields on chemical processes and the production of radical-pairs (a highly complex process described by quantum physics), submitted a paper to a leading scientific journal in which he suggested this could be used as a navigational tool by migrating birds. The journal suggested he put the paper in the waste basket. Altering the paper, Schulten, now working with colleagues, submitted a revised paper to a German journal (Schulten *et al.* (1978)), again mentioning the potential use in avian migration. The paper was ignored for many years until the discoveries of recent years showed that Schulten's idea was excellent rather than dustbin material.

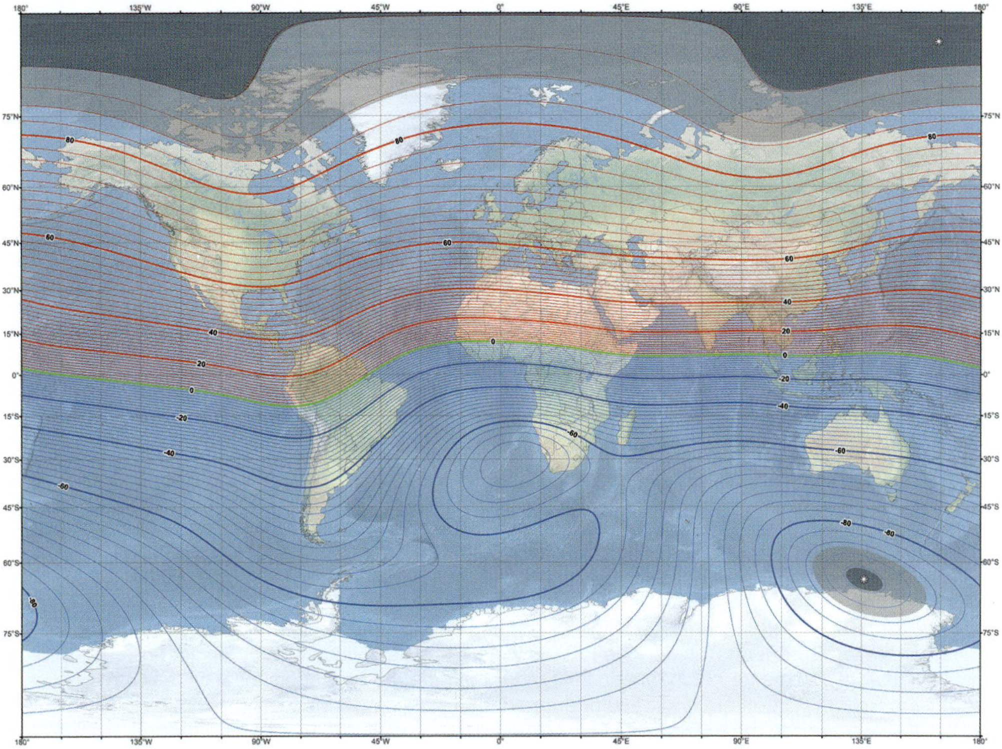

Figure 8.2a Earth's Magnetic Field: Main Field Inclination 2020. Angles are in degrees: red (positive) contours are 'down', blue (negative) contours are 'up'. Figure from NGDC.NOAA, US Department of Commerce.

Figure 8.2b Earth's Magnetic Field: Main Field Intensity 2020 (nanoTesla - nT). Contours are at 1000nT intervals. Data 0-2000nT is an 'unreliable zone'; data 2000-6000nT is a 'caution zone'. Figure from NGDC.NOAA, US Department of Commerce.

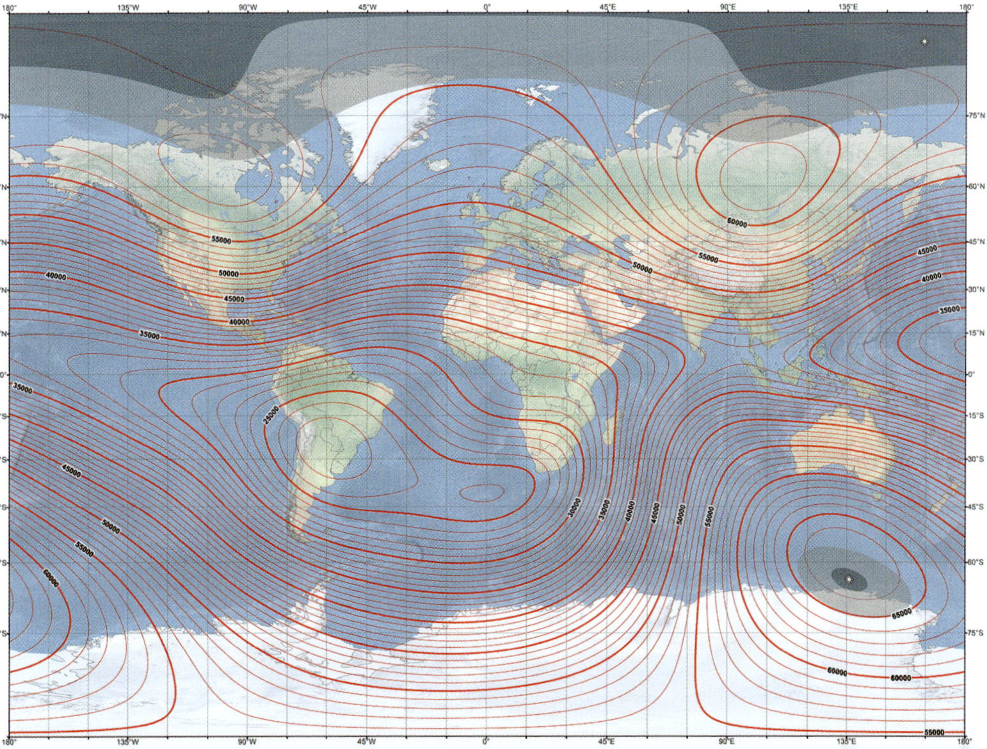

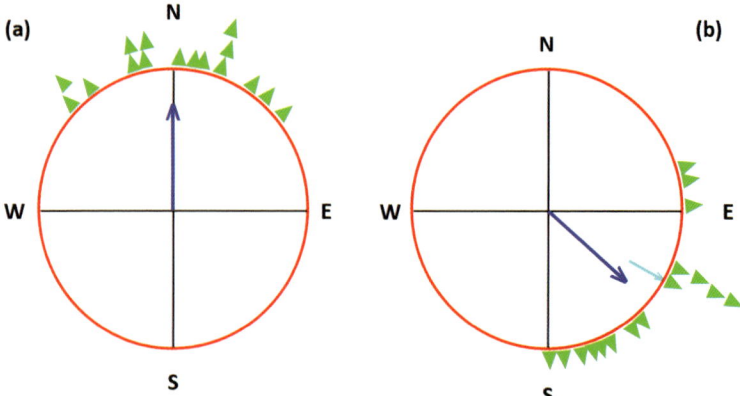

Figure 8.3 A demonstration of the use of a magnetic compass orientation in 16 migrating European Robins. The Robins were placed individually in an enclosure, initially with no external magnetic field other than that of the Earth's local field. In (a) the Robins (Green arrows) have orientated themselves with the local magnetic field (Blue arrow). In (b) a Helmholtz coil was used to produce a magnetic field around the bird's enclosure more powerful than the local Earth field. The Robins orientated themselves preferentially to this field. The small turquoise arrow is the mean direction of the Robins, the large blue arrow is the imposed magnetic field. Redrawn from Wiltschko and Wiltschko (2007).

good review article on magnetic navigation, avoiding the more complex mathematics is Wiltschko and Wiltschko (2022).

The second mechanism is the use of magnetite, an iron ore (Fe3O4) which is a permanent magnet. Magnetite is found in many animals. In birds it tends to be found close to the olfactory nerve and in bristles in the nasal cavity, *i.e.* close to the bill. As magnetite is sensitive to the Earth's magnetic field strength it can provide latitudinal information, while the radical-pair photoreceptors provide directional, *i.e.* longitudinal information[7]. Together the magnetoreceptors can therefore provide the bird with a basic world map.

But those birds which cross the Equator also have to overcome two further problems. As noted above, close to the Equator the field lines lie parallel to the Earth's surface and so no longer offer an unambiguous directional cue, while beyond the Equator the bird has to understand it must now follow an increasing, not reducing, field strength. However, as can be seen in Fig. 8.2b, this problem does not exist for Peregrines in northern Eurasia heading for northern Africa as the curious minimal intensity area in the South Atlantic means the Earth's field strength continues to reduce all the way to South Africa. For Peregrines in America heading for northern South America the

---

[7] The basis of the radical-pair model in terms of its use in bird migration is neatly set down in Ritz *et al.* (2000) and Heyers *et al.* (2007). For a later (but highly complex) assessment of the quantum physics involved in the process see Adams *et al.* (2018). The pair production involves entanglement, a process Albert Einstein dismissed as 'spooky action at a distance' as it requires two entities separated by potentially huge distances to react in unison.

Movements and Wintering Grounds

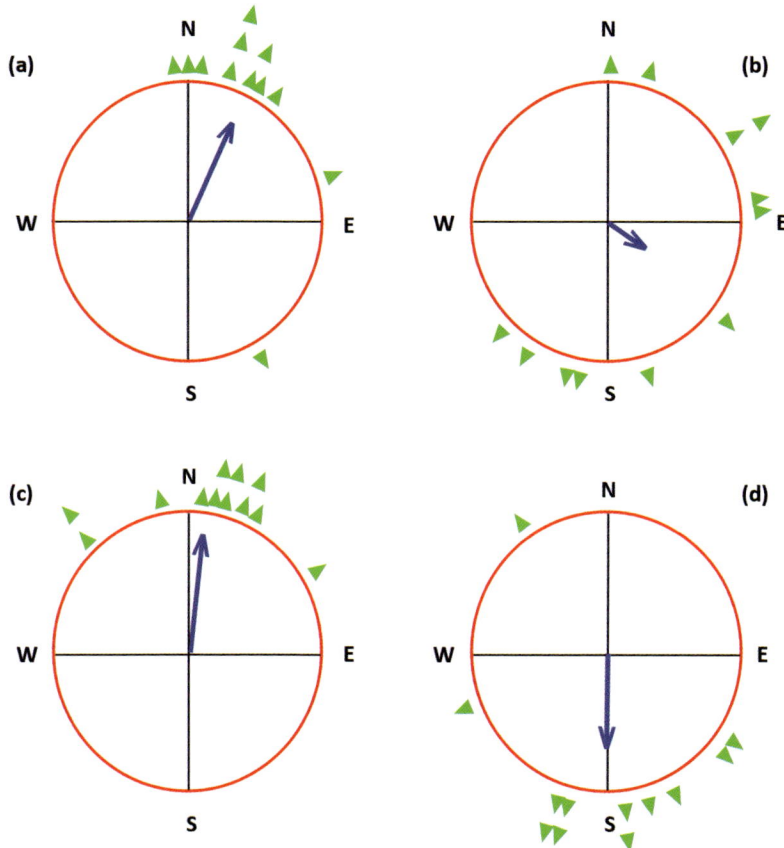

Figure 8.4 Orientation behaviour of Eurasian Robins under monochromatic green light and a magnetic field. The mean headings of the 12 birds are indicated by the green triangles in each diagram with the blue arrows being the overall mean heading in terms of direction and arrow length as a proportion of circle radius. The four diagrams are:

(a) Binocular control with the geomagnetic field.
(b) Monocular left eye with the geomagnetic field.
(c) Monocular right eye with the geomagnetic field.
(d) Monocular right eye in a magnetic field with the field's vertical component inverted.

Redrawn from Wiltschko *et al.* (2002).

same South Atlantic anomaly means that again the Earth's field strength reduces, unless the bird reaches northern Argentina or northern Chile when the field strength starts to increase all the way to Tierra del Fuego.

Given that migratory birds have an array of navigational detectors, which do they prefer if local weather conditions allow all of them to be in play?

Experiments have found that magnetic information trumps celestial (stellar and sun) except at sunset when both information from the sun's position and from polarised light may supersede magnetic signals. It is also possible that the lack of absolute values deriving from magnetic information (because of the non-linear way in which field intensity varies) means that given a plethora of celestial signals at sunset the bird uses them to recalibrate its magnetic sensors. Is it also the case that vision (seeing the sun and stars, polarisation patterns and 'seeing' the Earth's magnetic field) is the only sense used by migratory birds? In principle senses other than sight should be considered, though of these only smell and hearing would seem useful. But the olfactory system is relatively poor (though definitely not non-existent as is often suggested) in birds, and while hearing might aid a bird to detect changes in weather systems, the approach of a storm for instance, it would seem of limited value in migrations of several thousand kilometres.

Another aspect of migration which has become apparent only in recent years is the way birds prepare for long over-ocean migration flights. The advent of small satellite tracking units which can be fitted to migrants has shown, for instance, that Bar-tailed Godwits fly continuously from Alaska to Australia and New Zealand, one bird in September 2021 travelling 13,049km in 239 hours of continuous flying (averaging 54.6kph). To accomplish such feats long distance migrants put on weight (to a point where humans would be regarded as clinically obese), have a blood chemistry which would classify a human as diabetic, shrink all the now-unnecessary digestive organs, and increase the size of their hearts and pectoral muscles (see Guglielmo (2018) for further details). And they do this twice each year and for many years. Birds are also very much better at converting fat to energy than humans, who, as with other mammals, chiefly use carbohydrates and proteins for the same job the energy in them being much easier to release. But fat has little free water, and it is remarkable that long distance migrators do not die of thirst. Gerson and Guglielmo (2016) and Guglielmo (2018), working on songbirds, found that birds compensate for respiratory water loss by the catabolism of protein – the birds cannibalise their own muscles and organs to produce water.

Peregrines do not have to go through such traumatic rigours as their migrations are shorter and invariably over land. Over sea migration has been observed, but this is rarely very far from shore. Kerlinger *et al.* (1983) studied data on autumn migration of raptors observed from cruise ships in the waters off north-eastern North America, from Nova Scotia south to Cape Hatteras, North Carolina. While in general Ospreys were seen further from land than other raptors, there were sightings of Peregrine Falcons at distances which meant that the birds were not able to see either the coastline to the west or the coast in the direction in which they were heading. Kerlinger (1985), studying autumnal migrating Peregrines at Cape May Point, New Jersey, USA, noted that the falcons were using dynamic soaring – harvesting energy from air updrafts close to wave tops – in the manner of albatrosses, a flight method very different from that seen in soaring birds on land. On occasions the falcons were within 2m of the waves, and rarely higher than 10m above them. Nourani *et al.* (2021) confirmed the use of ocean uplift by using the temperature difference between the sea surface and the air as a proxy for uplift and analysed the data for sea crossings of five raptors including the Peregrine in both America and Eurasia-Africa. All five raptors maximised wind support and uplift when choosing to cross seas.

As noted in Chapter 2, the name Peregrine derives from the Latin for wanderer, and the falcons have often turned up in the most unlikely places. Chapter 4 *Hunting Strategies* notes a Peregrine making a temporary home of two ships in the Pacific Ocean over 2000km from the coast of Costa Rica, and of another finding a temporary home on a ship in mid-Atlantic. In the latter case (Voous, 1961) the bird arrived when the ship was 1,300km west of Africa and left when it was still 1,100km from South America. As mentioned in Chapter 7 *Dispersal*, a North American Peregrine (probably *F. p. anatum*) ringed in northern Wisconsin in 1993, was recovered in Switzerland in 2008, a straight-line distance of 7,030km from its natal site (Doolittle *et al.*, 2013). Though the Atlantic crossing may have been weather-assisted and the falcon may have rested on ships, this still represents an extraordinary journey, particularly as the bird was 15 years old. Such journeys indicate a willingness to move if circumstances require, but probably relate to birds blown off-course by storms or high winds, proving that Peregrines are adept at making the best of circumstances which might challenge other species. While this is undoubtedly true, Santos *et al.* (2020), studying Black Kites (*Milvus migrans*) migrating across the Gibraltar strait noted that adult birds used less energy in the sea crossing than juveniles by choosing weaker crosswinds to minimise height loss and drift. By analysing the likelihood that a bird would abort a sea crossing, Santos *et al.* noted that the height of the bird when it set out was a critical factor, birds starting at an altitude of only 100m being three-times more likely to abort the crossing than those starting at 1000m. From the study it was also clear that birds can learn to minimise the risk and energy loss of sea crossings. In the study of Nourani *et al.* (2021) mentioned above, it was also clear that when making long cross-sea flights, the studied raptors were choosing routes which offered the least uncertainty in terms of the wind conditions they would face.

As noted in Chapter 4, Peregrines do occasionally reach remote islands, but this invariably seems to have been storm-driven rather than by design. For the migrating Peregrine a more important question seems to be when to stop and refuel prior to setting out on short sea crossings, or over other food-poor areas. In particular, how does a juvenile bird know where to stop since, as we shall see below, adult Peregrines migrate before their offspring? In experiments in Sweden (see, for example, Fransson *et al.*, 2001) it was shown that manipulating the magnetic field around caged juvenile Thrush Nightingales (*Luscinia luscinia*) so the birds believed they were in northern

Egypt, caused them to increase their fat deposits just as they would have needed to do before crossing a desert to reach wintering grounds in southeast Africa. Similar studies have shown that manipulating the magnetic field around caged birds so that they believe they have reached their wintering grounds causes them to cease migratory behaviour. A related topic is how birds decide to return from wintering grounds to breeding areas – what stimulates the decision to begin the spring migration. This was studied in the case of Asian Houbaras (*Chlamydotis macqueenii*) by Burnside *et al.* (2021). The Asian Houbara is a bird of the Bustard family, breeding eastwards from Kazakhstan to Mongolia and making relatively short migratory flights (up to 1200km) to Pakistan and countries around Saudi Arabia. While both the family and migration differ markedly from Peregrines, it is likely that similar spring migratory signals apply to each species as the imperative in each case is to reach breeding grounds in time to mate, breed and raise nestlings. In their study, Burnside *et al.* attached satellites tags to 48 birds and tracked them across multiple years. The study suggested that while day length and wind speed/direction influenced departure date, the significant effect was temperature, with those individuals which overwintered furthest south choosing to start the migration at higher temperatures than those wintering to the north (Fig. 8.5a). Temperature also influenced the autumnal departure date (Fig. 8.5b), though in this case day length and wind speed/direction were equally important. This is unsurprising as the imperative to breed early and so maximise the survival of the year's brood is far more important than arrival time at wintering grounds. As interesting as the departure date and local conditions was, even more remarkable was the repeatability of the departure cues for individual birds. Once learned, a departure date appeared fixed for

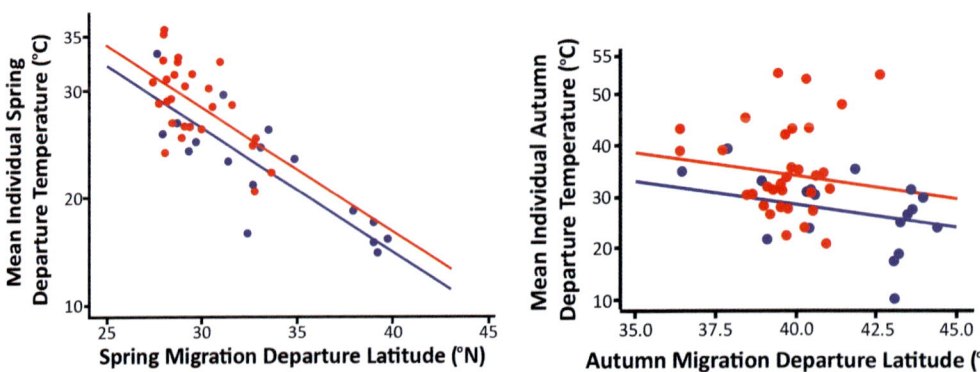

Figure 8.5a (*left*) Mean individual temperatures at spring migration departure sites related to the mean latitude from which an individual bird departed. Redrawn from Burnside *et al.* (2021).

Figure 8.5b (*right*) Mean individual temperatures at autumn migration departure sites related to the mean latitude from which an individual bird departed. Redrawn from Burnside *et al.* (2021).

Two of the migrating North American juvenile Peregrines (probably *F.p. tundrius*) resting on the ship in the western Gulf of Mexico. From the head coloration these two hatched in Alaska rather than Canada. *Amanda Stafford.*

an individual (though the arrival time at the breeding grounds is dependent upon the weather *en route*). For those Peregrines wintering north of the Equator, and for the few wintering to the south, similar cues for migration may apply. But for those birds wintering close to the Equator (for instance US Peregrines which overwinter in Peru – see below) the differences in daylight and temperature will be very much less pronounced. For those falcons, wind direction or annual circadian rhythm may be the cue.

One final aspect of migration is assessing the route which birds take. For a human walker crossing a flat landscape, route choice usually involves picking a straight line as this is the shortest distance between two points. But that is true only if the surface being traversed is a 2D-plane (which is the case for our short-distance walker). For a long-distance migrant such as the Peregrine the fact that the Earth is not flat becomes important. On a sphere, the shortest distance between two points is defined by a Great Circle (or orthodrome), a circle whose diameter equals that of the sphere. Any two points on a sphere are uniquely linked by a single Great Circle (with the exception of antipodal points, *i.e.* those diametrically opposite each other, which lie on an infinite number of Great Circles). Given that the Great Circle represents the shortest distance a bird must travel it would be anticipated migrants would choose to follow one. But experiments show that in general they do not, presumably because the computation involved, in terms of heading changes, are too complex for the bird's navigational system as it involves compensating for time shifts on east-west (or vice versa) transits: such changes require an onboard computer or full-time navigator on an aircraft.

Having established how Peregrines gather the information for successful migration, the first question raised – how do juveniles know how to migrate without being taught by their parents? – still needed to be addressed. Until just a few years ago when satellite tags for attachment to birds could be made small and lightweight enough for use, such information as existed on bird migratory routes was obtained from the recovery of rings from dead birds attached to them as nestlings, or by watching the masses of migrating species at known bottleneck sites. Over the years many such sites have been identified, for instance the Straits of Messina between the toe of Italy and Sicily for birds *en route* to north Africa; the Défilé de l'Écluse in the French Haute-Savoie; Batumi, on the eastern shore of the Black Sea in Georgia; and numerous places in the USA such as Hawk Mountain, Pennsylvania and South Padre Island, Texas. Mueller *et al.* (2000) trapped 23,000 raptors of ten species at Cedar Grove, Wisconsin, between 1953 and 1996. Most of the trapped Peregrines were *F.p tundrius*, though Mueller *et al.* do not specify a percentage split between the Arctic breeders and *F.p. anatum*. Adult Peregrines migrated earlier than juveniles. No difference was seen in the timing of adult male and adult female migration, though the sample sizes for both were too small for this to be definitive. A similar result regarding adult migration prior to that of juveniles was also found for Peregrines observed migrating at Falsterbo, Sweden, by Kjellén (1992) – Fig. 8.6.

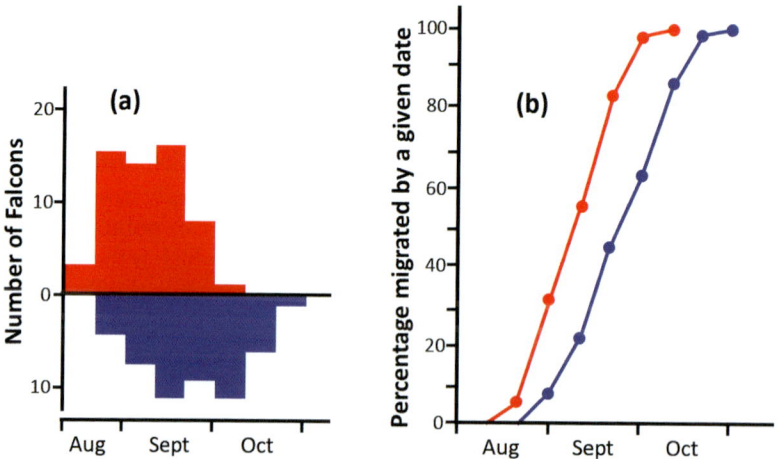

Figure 8.6 Migration counts of Peregrines heading south at Falsterbo, Sweden 1986-1990 – (a) is by number, (b) is by percentage of the total. Red rectangles/dots/lines are adult falcons, blue rectangles/dots/lines are juveniles. Redrawn from Kjellén (1992).

In a classic paper on avian migration Pulido (2007) pointed to a genetic basis for differences in the migratory behaviour within species, with Delmore *et al.* (2016) later identifying a specific chromosome cluster critical to migratory behaviour (and plumage colour) of populations of Swainson's Thrush (*Catharus ustulatus*). More recently, there have been further studies which indicate a genetic basis for both the timing of migration and the choice of route. Bossu *et al.* (2021), studying American Kestrels, noted three genes related to the timing of migration, as well as confirming the migratory input of the specific gene involved in the migration of Swainson's Thrush. In separate work, Gu *et al.* (2021) fitted satellite trackers to 56 Peregrines of six different populations breeding in northern Russia (from the Kola Peninsula to the Kolyma River) which migrated in different directions to wintering areas varying from Spain, to the west, and Indonesia to the east. The team found that a specific gene (ADCY8) was associated with migratory choice. The gene is also associated with long-term memory. Parent birds are passing their genes to their offspring and, in so doing, pass information on the direction of migration and time of travel to/position of wintering grounds, data which can be used by the juvenile to offset the effects of weather conditions or the need to modify the journey because of food scarcity. In some of the displacement experiments noted above, juveniles who found acceptable wintering quarters returned to those the following year – they had modified their genetic 'map' and were able to pass that information to their own offspring. Migratory behaviour, it seems, is capable of rapid change within populations. Gu *et al.* consider it likely that northern Peregrine populations will respond quickly to global warming as, in effect, it is just another spatiotemporal change, perhaps moving their breeding grounds further north and choosing new routes to different wintering areas. Such changes alter the genome of individual birds and hence those of their offspring, aiding an explanation of the mystery of migration. The findings of Gu *et al.* will be considered again below.

The Peregrines of the Nearctic tundra (*F.p. tundrius*) probably make the longest migratory flights travelling from breeding grounds as far north as the southern islands of Canada's Arctic archipelago to Mexico and the Florida Keys, and into South America, reaching at least as far south as central Chile. While the assumption might be that all these long-distance migrants are indeed *F.p. tundrius*, White *et al.* (2013a) noted that it is possible that some birds, probably those arriving in the north of South America, might be *F.p. anatum*. Later studies have shown that this is the case, both North American Peregrines being long-distance migrators. Beingolea *et al.* (2020) collected data from eight Peregrines which had been banded (ringed) in North America and retrapped/recovered in Peru (Fig. 8.7a overleaf). Three of these birds were juveniles on their first migration; four birds were aged 2-6 years with one being 9 years old. Interestingly of the six birds whose age could be positively identified five were male. Beingolea *et al.* also trapped a further 213 Peregrines

The Peregrine Falcon

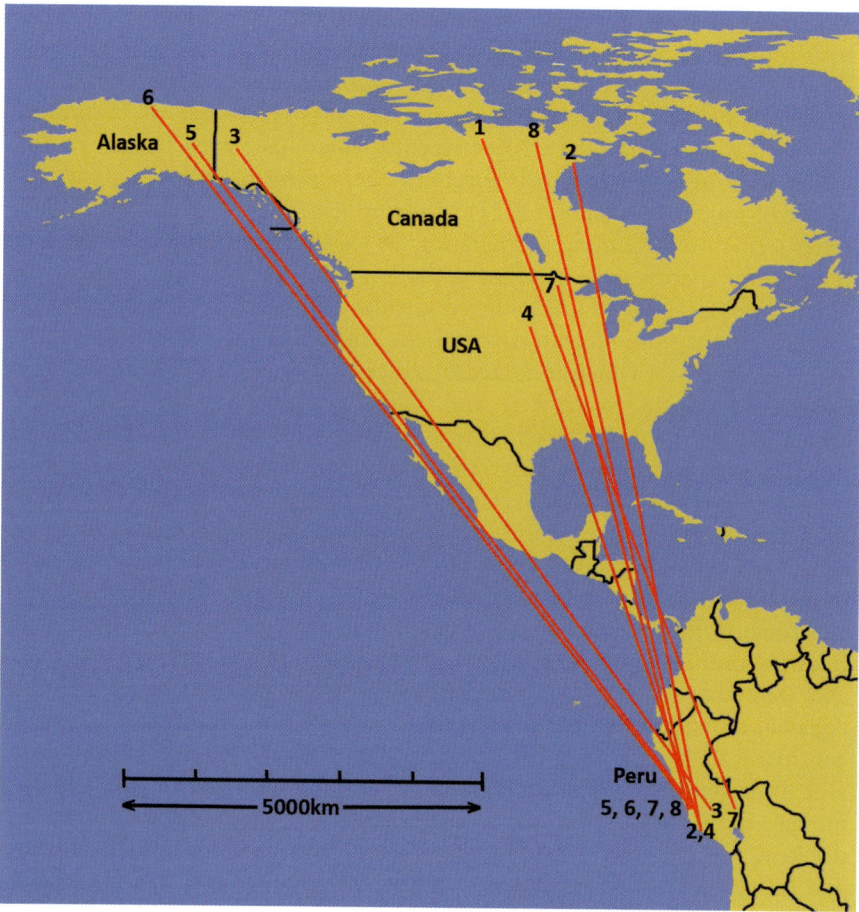

Figure 8.7a Peregrine Falcons recovered in Peru with hypothetical flight lines linking them to their natal sites in North America. Redrawn from Beingolea and Arcilla (2020).

in Peru over the period 1988-1995, of which 97.7% (208 birds) were from North America. Of that fraction, 72.1% (150 birds) were male. It seems that in general male Peregrines migrate further than females (a curiosity given that, as we have noted in Chapter 7 *Dispersal*, female juveniles, on average, disperse further than males). What is also clear from Fig. 8.7a is that both far north (probably *F.p. tundrius*) and Lower 48 (*i.e.* contiguous) States of the USA (*F. p. anatum*) falcons are migrating to wintering grounds south of the Equator in South America. In the absence of any migration route information Fig. 8.7a assumes straight-line flights, the distances travelled by the birds between banding and recovery varying from 6430km to 10,671km, the furthest being a juvenile less than one year old. But Beingolea *et al.* also collated data on birds which had been recovered in Peru after being banded at migratory stop-overs at coastal sites/off-shore islands in Texas (7 birds), New Jersey (3), Virginia (2)

Figure 8.7b Peregrine Falcons recovered in Peru with hypothetical flight lines linking them to known stopover sites in southern USA. As noted in the text, it is possible that those Peregrines migrating along the east coast of North America may island-hop to reach South America. Redrawn from Beingolea and Arcilla (2020).

and North Carolina (1) – Fig. 8.7b. There is, of course, no natal information for any of these birds and the ages of many could not be determined. Five of the birds were juveniles, others being estimated at ages of up to six years. Of the 13 birds, only four were male.

Paganini *et al.* (2018) noted a mass gathering of between 5 and 50 Peregrines (probably *F.p. anatum*) on a ship 40-45km off the Mexican shore in the Gulf of Mexico in October 2017. Most of the birds were juveniles, and they were hunting from the ship, probably on their way to wintering grounds further south. The falcons were with the ship for 13 days, the largest number being seen on 8 October.

Records exist of Peregrines reaching as far south as Argentina and Chile, with other falcons reaching Colombia, Venezuela, Ecuador, Peru, Brazil and Uruguay. White *et al.* (2013a) note one adult female reaching 41°77'S in

The Brahmaputra River in south-east Tibet is followed by some *F.p. calidus* migrating south from Siberia.

Chile. Such journeys involve huge distances, Kuyt (1967) noting one falcon ringed on 29 July 1965 close to the Thelon River in Canada's Northwest Territories having been killed in January 1966 at India Muerta, Chaco Province, Argentina, having travelled around 14,000km in about 5 months. Other journeys in excess of 12,000km have also been recorded. All the *bona fide* long journeys appear to have been of *F.p. tundrius*.

Better migration data has been collected since satellite tagging of Peregrines has been possible. Fuller *et al.* (1998) tagged 61 *F.p. tundrius* trapped in the northern boreal forest/southern Arctic and followed them as far south as central Argentina, though few of the falcons were travelling south of the Equator in comparison to those wintering on Caribbean islands, coastal Mexico, the countries of central America and northern South America. The Peregrines logged a mean travel of 8624km southward, and 8247km on the return journeys, with mean daily travelling of 172km southward and 198km northward (Fig. 8.8).

Movements and Wintering Grounds

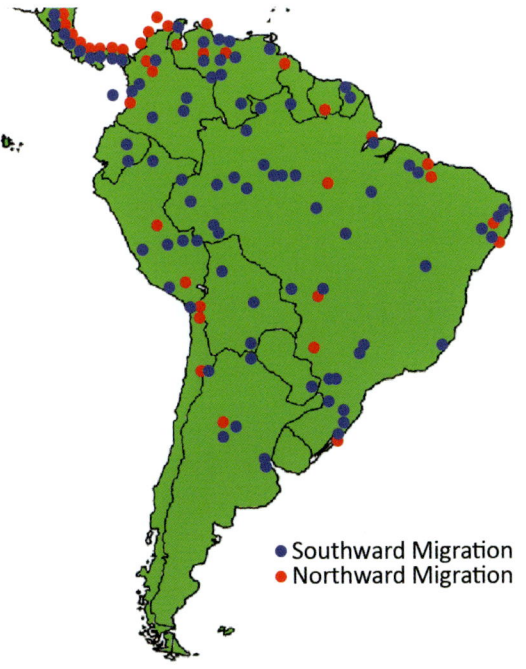

Figure 8.8 Migration of Peregrines in the Americas. Some of the Peregrines were tagged in west Greenland, in Canada's Northwest Territories and Quebec and so would likely have been *F.p. tundrius*. Others were tagged at Padre Island, Texas and so could have been either *F.p tundrius* or *F.p. anatum*. The plotted data is from individual birds, with snapshots of positions. The disparity in southward and northward data is due to the data being from 22 complete and 12 incomplete southward migrations, but only 7 complete and 11 incomplete northward flights.

The data in Central America is notional: there were many positional data points, the coloured circles indicating only that both falcons heading south and north used the land corridor to reach South America.

The study also included tagged Swainson's Hawks (*Buteo swainsoni*) whose migration was significantly different. The hawks travelled close to the east side of the Andes and converged in an area of east central Argentina. The northward route reversed the southerly tracks. Redrawn from Fuller *et al.* (1998).

McGrady *et al.* (2002) tagged 11 female and a single male in northern Canada and followed them to wintering sites in Mexico, on the coast of the Gulf of Mexico with travel distances of 4580 to 5844km[8]. The Siberian tundra Peregrine (*F. p. calidus*) has similarly long migration flights to its Nearctic cousins. In early satellite-tagged flights from Russia's Taimyr Peninsula, Eastham *et al.* (2000), tags were attached to a number of falcons in 1996-1998. The failure rate of the tags was high, but data from two falcons were of interest. One female flew almost due south, reaching northern Pakistan before the tag failed. The bird had travelled 5500km by then (during 16 September-12 November), travelling 250km on one day, but then back-tracking to almost the same point before turning south again. A second female flew 4815km (3 August-15 April) to central Uzbekistan: the tag on the falcon showed no more movement before it ceased transmitting in April of the following year.

In 2011 10 *F.p. calidus* were trapped by the Popigai River at the south-eastern edge of Russia's Taimyr Peninsula Dixon *et al.* (2017). As with the earlier tagging, the Peregrines migrating south along the Central Asian Flyway from this area face formidable obstacles. Initially the falcons must cross the

---

[8] Further migration data were found by Sivakumar *et al.* (2021) who used radar data collected from seven stations in Alaska over the period 1995-2018 to identify timings and direction of birds breeding in the area. While no specific Peregrine information could be obtained from the radar scans, the data showed that peak spring bird migration ranged from 3-30 May across the 24-year period, with autumn migration peaking between 18 August and 12 September.

Central Siberia Plateau to reach the Sayan and Altai mountains. Beyond is the Mongolian steppe/desert, the Tien Shan Mountains and the high Tibetan Plateau. Peregrines attempting to avoid these obstacles by migrating to the west meet the Hindu Kush and the high peaks of the Karakoram. Heading east the Tibetan Plateau is still an obstacle, while at its southern edge are the Himalayas which continue east through Bhutan and northern India to meet the mountains of western China. Figs. 8.9a-d illustrate the routes taken by the falcons to reach their wintering grounds. As can be seen from Figs. 8.9, the Peregrines avoided central Nepal and the high Himalayas. To the west the falcons cross the barrier via passes in Kashmir to reach northern India or followed the route of the Karakoram Highway between the Hindu Kush and the Karakoram to reach northern Pakistan. To the east the falcons rounded Bhutan to reach northern India. Some Peregrines do traverse Nepal, being seen at Khare (to the west of Everest) and Thula Kharka (to the east of Everest) in central Nepal, but are rarely seen taking the route of the Kali Gandaki between Dhaulagiri and Annapurna (the world's 7th and 10th highest peaks). That valley, one of the deepest in the world, is renowned for its katabatic (downhill) winds after sunset, and anabatic (updraft winds and thermals) after sunrise. The latter are readily utilised by soaring birds, but offer little advantage to the Peregrine which is, in general, a low-level migrant, usually seen using flapping flight at heights of around 100m. Peregrines have also been seen at Lawala Pass in west Bhutan.

Dixon *et al.* (2017) note the difficulties and dangers of the routes south, noting particularly the journeys of a male Peregrine which, having crossed the Tien Shan Mountains was now faced with the huge Taklamakan Desert, 337,000km^2 in area (about 1000km west-east, and 300km north-south). In October 2011 the male falcon made two abortive attempts to cross the desert, retreating each time to farmland at its northern edge to recuperate before succeeding on its third attempt. In 2011 the male crossed the Tibetan Plateau, flying 1100km in two days. It wintered in Sunderbans National Park of West Bengal, India. The following year the male was attempting a route further to the west. It flew about 430km in two days, then struggled to complete 300km in 7 days before stopping for the last time and, presumably dying.

From 2009 to 2015 attempts were made to follow breeding Peregrines (*F.p. calidus*) from northern Russia as they migrated southwards prior to winter and then northwards the following spring. A total of 56 birds were fitted with satellite tags within six breeding areas: the Kola Peninsula to the west (7 birds); Kolguev Island (in the Barents Sea, off the Russian mainland north of Cheshskaya Guba – (10 birds); the Yamal Peninsula – (12 birds); eastern Taimyr (Popigai) – (10 birds); the Lena Delta (8 birds); and the Kolyma Delta (9 birds). The falcons from the Kola Peninsula and Kolguev Island both migrated south-west towards Europe, reaching France and Spain (Gu *et al.*, 2021)[9].

---

[9] In an earlier study in 1994, Henny *et al.* (2000) attached satellite tags to four adult *F.p. calidus* trapped near the Ponoy River in the central Kola Peninsula. One bird lost its tag (or died) soon after. The other three wintered in The Netherlands, northern France and southern Spain.

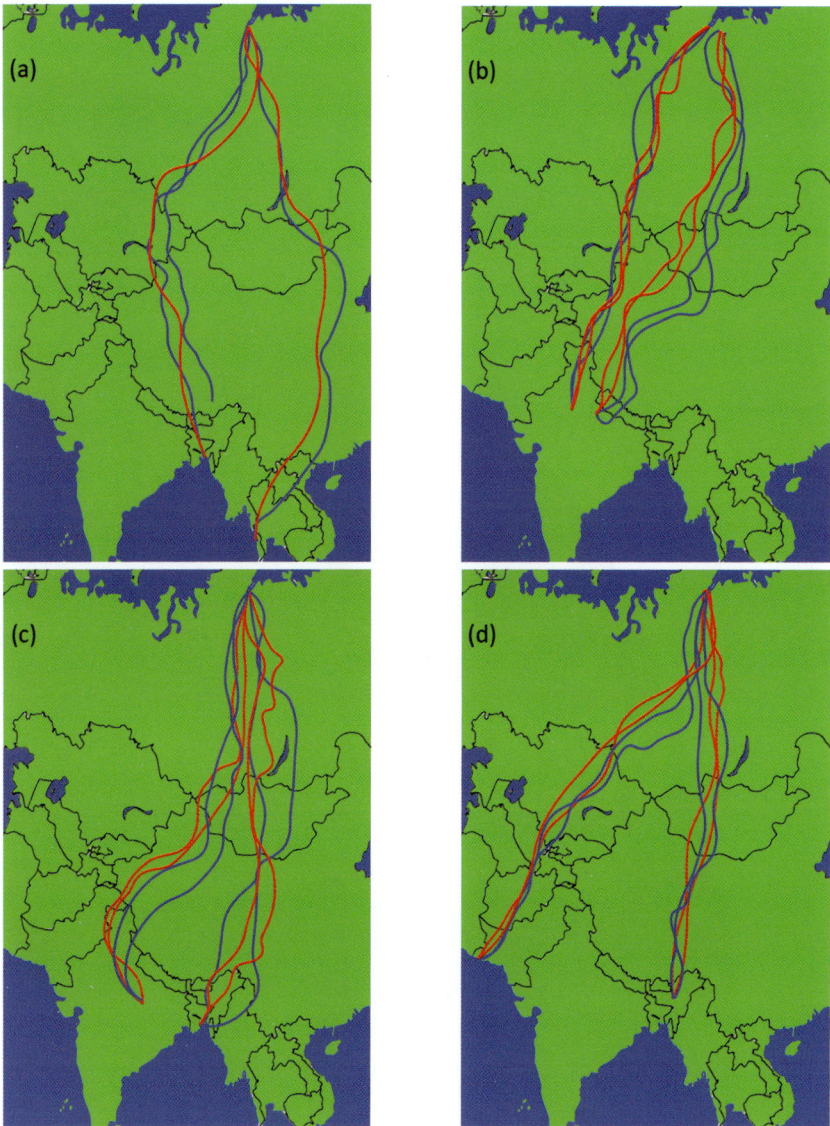

Figure 8.9a-d Migration tracks of *F.p. calidus* breeding in south-eastern Taimyr. As in Fig. 8.8, southward flights are in blue, northward flights are in red.

Fig. 8.9a shows, to the left, one complete flight and one incomplete southward flight of a male Peregrine which died in Tibet. To the right the single complete flight represents the outer flight limits (*i.e.* all three flights are encompassed by the single flights shown) of three complete flights by a female (the mate of the male in two years).

Fig. 8.9b, shows the tracks of two flights each of two female Peregrines. In each case the flights shown encompass three flights by each of the birds.

Fig. 8.9c shows two 'round-trip' flights by two female Peregrines.

Fig. 8.9d shows, to the left, two complete flights by a female (left), and, to the right, two flights which encompass four complete flights by a different female.

Redrawn from Dixon *et al.* (2017).

Adult   *F.p. calidus*, Little Rann of Kutch, India. *Annet Patel.*

Opposite   Juvenile Alaskan *F.p. tundrius*, Gulf of Mexico. *Amanda Stafford.*

The 56 birds included both adults and juveniles and of the total, 41 birds completed 151 migration paths. Autumnal migration paths varied from 2280-11,000km. On average the journeys took 27 days (the 95% confidence interval on this mean was 14-46 days) with the birds covering a mean distance of 213km/day (the 95% confidence interval on this mean was 49-420km/day). The migration paths are shown in Fig. 8.10. There is a clear difference in the paths from those observed in North America where the longitudinal spread in destinations is much less pronounced. There are obvious reasons for this, of course: Eurasia and northern Africa represent a rectangular block of land with a vast baseline on both the northern and southern edges. In North America, Mexico and central America represent a migratory bottleneck: North America is less a rectangle and more a triangle – its top chopped off, the truncated triangle then inverted. This basic difference in the area available to migrating Peregrines is readily shown in Fig. 8.11. As noted earlier in this Chapter, Gu *et al.* found that a specific gene (ADCY8) was probably responsible for the development of the five distinctly different migration directions after the re-establishment of breeding pairs and migratory behaviour (both in latitude and

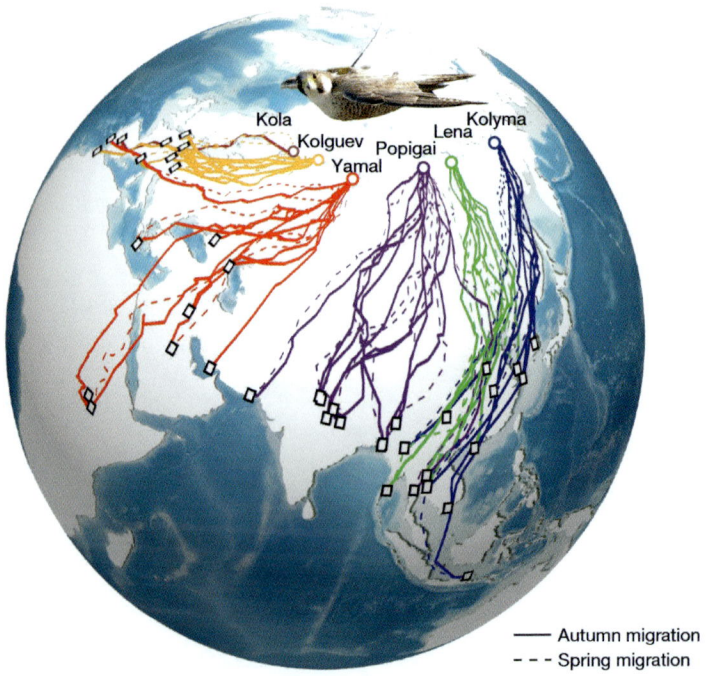

Figure 8.10 Migration routes of the Nearctic breeding *F.p. calidus*. Only 41 complete migration paths are shown. The colour code is as Fig. 8.11. opposite.

Reproduced from Gu *et al.* (2021) with kind permission of the journal *Nature*.

longitude terms) following the retreat of the ice at the end of the last Ice Age. *F.p. calidus* now migrate to wintering grounds as far west as Portugal and north-west Africa and as far east as the islands of the Java Sea. Interestingly, Gu *et al.* looked at the potential change in both breeding and wintering areas as a consequence of global warming, using model simulations to predict the position in 2070. The researchers predict that there will be no breeding population on Kolguev Island at that time, and that all the other five areas will see reductions in breeding populations, varying from 37.2% on the Yamal Peninsula to 86.2% on the Kolyma Delta and 92.7% on the Kola Peninsula.

While, as noted above, *F.p.tundrius* probably, in general, make the longest migration flights, it is possible that some *F.p. calidus* may cover comparable distances. In October 1995 a female *F.p. calidus* (weighing 1115g) was purchased from a local trapper close to Jedah, on the Red Sea coast of Saudi Arabia, and fitted with a satellite tag (Meyburg *et al.*, 2018). The bird was tracked to Cape Town, South Africa. The tag signal was then lost: the bird had travelled 6346km in 22 days. Meyburg *et al.* note that the appearance of *F.p. calidus* birds in southern Africa is rare, but has been seen several times. Given the work of Gu *et al.* (2021) mentioned above, it is probable that this bird (and perhaps other *F.p. calidus* seen in southern Africa) came from the Yamal Peninsula. To the east, White *et al.* (2013a) note a falcon photographed on the Ashmore Reef off Australia's north-western coast (about 150km) south of Timor). Of course, some birds do not make it as far.

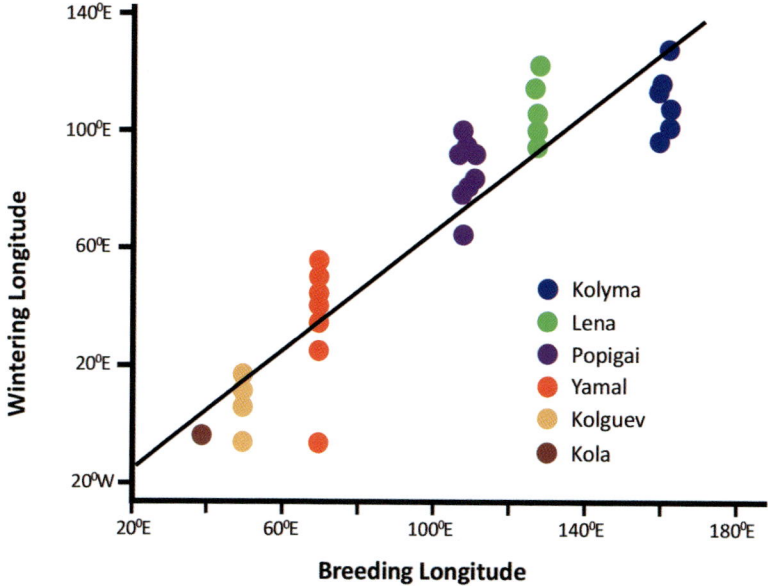

Figure 8.11 Connection between breeding grounds and wintering grounds of *F.p. calidus* for the six investigated areas of Arctic Russia. The regression line is significant at >99.9%. Redrawn from Gu *et al.* (2021).

The migration of Peregrines from north-east Siberia (see Chapter 2 for brief notes on whether these birds are *F.p. japonensis*) is poorly understood but they are believed to go as far south as Borneo and the Philippines

These shots were taken in Jurong, Singapore in mid-November. The photographer was 80 years old and his camera was not top quality. But he managed two shots of a Peregrine consuming a Javan Myna (*Acridotheres javanicus*) in Jurong's Japanese Garden. The Peregrine is widely believed by local experts to have been a *F.p. japonensis/harterti* migrated from Chukota or northern Kamchatka. *Johnny Wee.*

Nominate birds are known to migrate from Fennoscandia to avoid the harsh, dark northern winter, arriving in the UK and various countries of southern Europe, and occasionally travelling all the way to north Africa and the northern Arabian Peninsula (White *et al.*, 2013a). White *et al.* also claim that the data of Ganusevich *et al.* (2004) refers to nominate falcons. Ganusevich *et al.* (2004) recorded the data from satellite tags fitted to four female Peregrines breeding on the eastern Kola Peninsula, Arctic Russia, in 1994. One tag stopped transmitting soon after the carrying female started her migration. The other three falcons headed south-west, taking almost straight-line paths to their wintering grounds. One bird wintered on the northern coast of France after a flight of 2909km, while the second flew 4262km to winter in southern Spain, spending most of its time close to Gibraltar. The third bird reached northern Netherlands, but the signal from it then failed. While the data does indeed suggest these were nominate falcons, the suggestion of eastern Kola breeding offers the possibility of the birds being *F.p. calidus* which are known to overwinter in France, Denmark, The Netherlands and Germany. Crombach (2020) notes the retrieval of a dead adult female *F.p. calidus* ringed on Kolguev Island in June 2018 from a bridge in Nijmegen, The Netherlands on 21 July 2019. The falcon also carried a transmitter, the last signal from which was from a position about 100km away from the bridge on 18 October 2018. Crombach speculates that the bird had died in a territorial dispute with resident Peregrines.

In addition to 'standard' (latitude/longitude) migration, the two Asian shaheen sub-species (*F.p. peregrinator* and *F.p. babylonicus*) are known to practice altitudinal migration, leaving their breeding grounds high in the mountains of central Asia and the Himalayas to winter at lower altitudes where their prey species also move to escape the cold and heavy snowfalls of the high lands. Williamson and Witt (2021) studied altitudinal migration, though the 91 species they studied included only a single falcon, the Eurasian Hobby, which shows changes of altitude between breeding and wintering grounds in some areas of its range which are comparable to the two shaheen Peregrines. Willamson and Witt note that in such cases changes in the altitude of hunting areas of ≥2000m involve a ≥20% change in air density and oxygen partial pressure[10] that result in a decline in arterial blood oxygen which is accentuated in flapping flight. That these species make the change twice annually suggests inherent genetic adaptations and a metabolic flexibility human mountaineers would envy.

The computed travel speeds from the data set down above are consistent with the theoretical speed calculated in Chapter 3 *Migration Flights* (about 12m/s = 43kph) and also consistent with the tabulated data for migration speeds seen in the wild (Tables 3.1 and 3.2). These data suggest that on average a migrating bird will fly for about 6 hours each day and migrate on most days unless the weather is poor, and will always take advantage of thermals to gain height so as to reach favourable winds. White *et al.* (2013a) note that one female Peregrine migrating north to the Arctic flew 708km in a day, while another bird heading south, assisted by a hurricane, reached Florida from New Jersey in a day, a distance of 1526km. Equally interesting is the data from Dixon *et al.* (2017) who, as noted above, tagged *F.p. calidus* near Russia's Taimyr Peninsula. The researchers recorded a mean speed of 9.4kph (range 8.4-10.4kph) for falcons crossing the Central Siberian Plateau, a mean of 11.5kph (range 9.9-13.1kph) for the Tien Shan/South Siberian Mountains; 9.8kph (7.6-12.0kph) for Mongolian steppe/desert; and 15.5kph (13.5-17.5kph) for the Tibetan Plateau/Himalayas. Clearly the falcons speed up when crossing the latter, hugely significant, barrier.

Although they prefer level flight during migration, Peregrines will soar and glide if wind conditions are favourable, using wind energy to avoid flapping flight while still gaining distance. Flapping flight is energy intensive, and it is known that some migratory species which largely depend on flapping fly much higher during migration using the cold air of altitude to prevent

---

[10] In a gaseous mix the partial pressure of each individual gas element is the pressure it would actually be if the same amount filled the entire volume on its own.

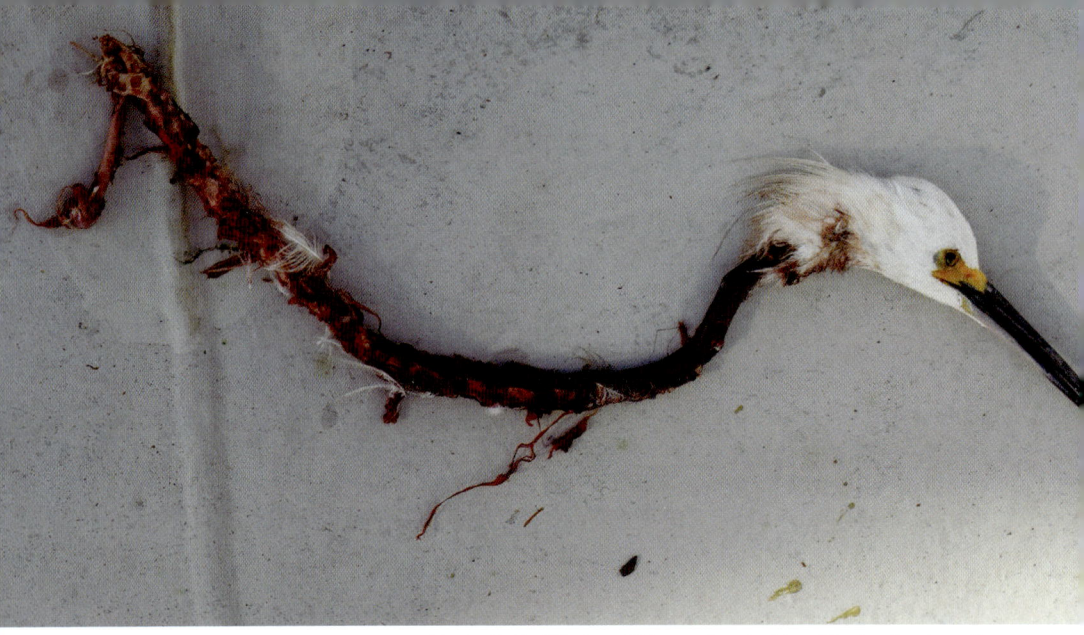

Another shot from the ship moored in the Gulf of Mexico which was 'invaded' by *c.*50 migrating juvenile Peregrines. The falcons had caught this Snowy Egret (*Egretta thula*) and stripped it clean of flesh, an indication of how hungry the birds were. *Amanda Stafford*.

overheating (see, for instance, Lindström *et al.* (2021) who tagged Great Snipe (*Gallinago media*) migrating from Sweden to East Africa: the pressure sensors on the birds noted that they flew at about 5000m during the day, but at 2000-2500m at night). It is possible that Peregrines forced to wing flap for long periods may do the same.

British Peregrines are much more sedentary than their northern cousins. As already noted, Mearns (1982) found over 80% of both coastal and inland breeding territories were occupied in winter in southern Scotland. He also noted that the coastal Peregrines had a smaller hunting range than inland falcons, presumably because of the availability of seabirds and waders. Data on the recovery of ringed British Peregrines indicate that most were found close to where they had been ringed: of 554 recovered birds the median distance travelled was only 45km, with 55% travelling less than 50km, and 78% less than 100km (Ratcliffe in Wernham *et al.,* 2002). Only 6.7% of the recovered birds had travelled more than 200km. Of the long-distance travellers, *i.e.* those travelling more than 200km, most were falcons moving from Scotland to England or Wales, or northern England to southern England or Ireland. The Peregrines in northern Scotland make the longest journeys and are the most migratory of the British population, those at coastal sites lacking prey once the seabird colonies that are their main resource disperse, as local estuaries hold smaller populations of waders and the inland moorlands are equally devoid of potential prey.

UK data shows that females travel further than males on average, but that the wintering grounds of both sexes are similar. The data also indicates that in general first-year falcons travel greater distances than older birds, leading Mearns and Newton (1984) to conclude that Peregrines make their longest movements during their first year, remaining faithful to territories thereafter. As a consequence of the minimal movements of adult Peregrines, the winter density of the birds is comparable with the breeding density and is just as lacking in uniformity. Juvenile Peregrines move out of their natal sites to secure suitable (*i.e.* vacant) territories or to establish themselves between existing territories and are chiefly to be found in lowland coastal sites where prey availability in the form of wintering waders, particularly at estuaries, is high. They will also, of course, be joined by older falcons whose breeding grounds do not hold a sufficiently high prey density to allow overwintering. Kelly and Thorpe (1993) note a communal roost of Peregrines (regularly five falcons, occasionally nine) together with five Common Kestrels and four-six Eurasian Sparrowhawks on the Isle of Man, UK.

However, it is worth noting again that prey density is critical in defining the wintering ranges of all Peregrines (just as it is during the breeding season). In Mexico, at coastal sites on the Gulf of Mexico McGrady *et al.* (2002) measured the hunting ranges of 12 wintering adult Peregrines fitted with PTT satellite trackers. The ranges varied from $17km^2$ to $700km^2$ with a mean of $83km^2$, giving an indication of the effect of prey density as the primary prey, shorebirds, were dependent on feeding grounds (mudflats) which were distributed and varied in size, and were subject to tidal inundation.

One interesting aspect of migration which may become more important in future years if the current global climate crisis does not receive the attention it deserves, is the possible effect on the distances birds choose to travel when heading to their wintering grounds, and the extent to which that distance is affected if either (or both) wintering and breeding areas shift as a consequence of warming. In a fascinating study of 29 species of Finnish birds using data collected over 55 years, Potvin *et al.* (2016) note that in general both breeding and wintering ranges were shifting northward, and that migration distances were, again in general, decreasing. For Peregrines the wintering ranges were moving northward at a rate of 7.5km/year, but the northward movement of the breeding range was less pronounced, being about 2km/year. The effect of these changes was, of course, a decrease of migration distance of about 5.5km/year.

Finally, it is worth noting one aspect of urban living which has altered some Peregrine pairs' wintering behaviour. The authors have observed a local Gloucestershire, UK female regularly moving to a local quarry to breed, but habitually then returning to the nearest town, some 4.7km away during the winter to take advantage of the higher temperatures and the abundance of feral pigeons. In another case, a coastal pair near Brixham, Devon, UK, breed and roost on their 'home sea cliff' in summer, but hunt regularly close to, and roost on, a church in town over the winter. There is a box on the church tower, but the Peregrines have shown no interest in it, presumably because the site is sub-optimal in comparison to their 'home cliff' (less protection, less visibility of prey and of other Peregrines pairs, and increased disturbance at this holiday-resort town in summer?) (Seb Loram pers. comm. SW). The authors know of other sites with similar dynamics.

9   FRIENDS AND FOES

Friends

There are many instances of small birds nesting close to breeding Peregrines, too many it would seem, for all to be chance, the inference being that the small birds gain a degree of protection from potential egg and chick predators which the Peregrine will not tolerate. In exchange it is possible that the Peregrines benefit from the alarm calls of their neighbours acting as an early warning of approaching danger. One suggestion was that the smaller birds gained protection by the reluctance of the Peregrine to hunt close to its nest for fear of giving away its position, but the many observations of the falcons hunting in proximity to their sites means that this theory is now largely discounted. However, in their study of *F.p. tundrius* nesting near Sondre Stromfjord in west Greenland, Meese and Fuller (1989) found no evidence that the falcons were preferentially hunting local prey. However, Meese and Fuller did find that, contrary to the general view regarding such nesting, most prey species had lower nest densities near the Peregrine nests: while Snow Buntings were found in relative abundance, the nesting populations of the Greenlandic sub-species of the Common Redpoll (*Carduelis flammea*), Lapland Bunting and Wheatear had lower than average abundance. Because Meese and Fuller did not see preferential falcon hunting close to their eyries, they assumed that such hunting, prior to their study, was not the cause of the lower prey densities. As well as seeing lower prey densities, Meese and Fuller questioned the hypothesis that Peregrines offer enhanced protection of local passerines by attacking Arctic Foxes searching for eggs and nestlings, as they could find no evidence for this happening (though Quinn *et al.*, 2003 do suggest that both Snowy Owls and Peregrines offered enhanced protection against the same predator in their study in Siberia – see below for further details. As mentioned in Chapter 4, Meese and Fuller (1989) noted that Snow Buntings nested in crevices in the same cliffs on which Peregrines bred, but used crevices which were too small for an Arctic Fox to enter. The buntings were therefore gaining protection from fox predation irrespective of any attacks by the Peregrines. As Meese and Fuller note, the possible reasons why the behaviour seen in the Arctic is not seen with Peregrines elsewhere is both interesting and worthy of further study.

In Britain instances of Wrens (*Troglodytes troglodytes*), Twites (*Carduelis flavirostris*) and even species as large as Ring Ouzels (*Turdus torquatus*) (much more likely to be potential prey to the falcon than the smaller birds) are known to nest close to Peregrines. At urban sites Pied Wagtails (*Motacilla alba*) are often seen breeding close to the falcons, though this is more likely because of the chances of collecting insects attracted by rotting prey remains than true commensal breeding. Such commensal nesting seems more common in the Arctic, perhaps because both Peregrines and other species are more likely to be forced into using nest sites vulnerable to terrestrial predators, much of the Arctic being short of cliffs or large rock outcrops. In the Nearctic Canada Geese (*Branta canadensis*) have been seen to nest close to Peregrines, and other raptors, while the association of Red-breasted Geese with not only Peregrines, but also Snowy Owls and Rough-legged Buzzards in Siberia, is one of the best-known examples of such behaviour (*e.g.* Quinn *et al.*, 2003 with *F.p. calidus*). One interesting aspect of these associations is that geese tend to land away from, and walk to, their nest sites rather than flying in, probably a useful precaution against attack, though in general the Peregrines seem to ignore these neighbours, even when the chicks are hatched. However, as we shall see below, Canada Geese can also be a significant foe to Peregrine pairs attempting to breed.

Ratcliffe (1962) noted a female nominate Peregrine incubating the eggs of Common Kestrels at a site in Scotland and assumed the larger falcon had ousted the smaller pair having lost their own clutch. The following year, again in Scotland, Ratcliffe (1963) saw Peregrines which had raised Common Kestrel chicks to fledge and had continued to feed them after they had left the nest. In each case it is likely the Peregrines had lost their clutches (due to DDT-induced egg thinning?) and supplanted the smaller falcon by force. Whether such commandeering of chicks can be classified as friendship, rather than kidnapping is a matter of debate. Smith (1992) saw nominate Peregrines and Common Kestrels nesting only 3.5m apart on a 93m chimney at a mill site in northern England, which suggests tolerance at least, and Johnson (2008) noted Common Kestrels and Peregrines nesting within 6m on a Devon, England sea cliff. The chicks of each falcon pair hatched at very similar times and the Peregrine adults were seen to be feeding Kestrels chicks when the latter begged them for food. This behaviour certainly cannot be explained by the Peregrines having lost their chicks. However, not all large falcon/small falcon interactions are so amicable, Wallen (1992) saw Peregrines and Merlins locking talons in the air (in England), only coming apart when they were 8m from the ground.

Above   The Peregrines breeding in the Avon gorge, Bristol, England have become a local attraction. In recent years, juveniles on their frst flights have landed on seats erected for the purpose of viewing the Gorge, to the delight of visitors. The huntin and prey passing flights of both adults and juveniles also keep the usual array of visitors with cameras enthralled. the gorge is also a well-known rock climbing centre with hard and coveted routes on the sheer limestone walls. To ensure the safety of the breeding falcons the British Mountaineering Council erect signs annually to keep climbers off the section of the Gorge the Peregrines have chosen to raise their family, throughout the breeding season.

Opposite   Red-breasted Goose family (*top*, Taimyr Peninsula, Russia), male Snow Bunting (*centre*, Bylot Island, Canada) and female Snow Bunting (*bottom*). The irony of the female Snow Bunting shot is that it was taken in Iceland, arguably the most baffling island on which Peregrines do not breed.

In south-east Arizona, USA, Ellis and Groat (1982) observed a fledgling Prairie Falcon intruding into an *F.p. anatum* eyrie and being fed by the adult Peregrines. It seemed that the adult Peregrines were aware of the intruder and occasionally stooped at the eyrie entrance, causing it to scuttle away from the edge. However, no concerted effort to remove the intruder was made by either Peregrine adult and the Prairie Falcon fed on three delivered prey items: the Peregrine nestlings seemed unperturbed. Given the fact that Peregrines have predated Prairie Falcons (see below), it is surprising that this very bold (or very foolish) behaviour was tolerated. Peregrines will also both tolerate and attack their own kind depending on circumstances. As already noted in Chapter 8 Kelly and Thorpe (1993) noted a communal roost of Peregrines (on the Isle of Man, UK) and other raptors during the winter, but the situation is very different if potentially rival birds trespass close to a nest site during the breeding season.

In a study of *F.p. calidus* breeding close to the Artic Ocean in Russia, in the Nenetsky Nature Reserve and, further east, at the base of the Yamal Peninsula, Pokrovsky *et al.* (2020) found a remarkable association of breeding falcons with breeding Rough-legged Buzzards (Rough-legged Hawks). The Peregrines aggressively defended their territories against incoming mammalian predators (foxes and mustelids, the latter including Wolverine (*Gulo gulo*). As a consequence, rodent numbers increased close to Peregrine nests. In Yamal as rodent numbers declined, the buzzards nested closer to Peregrine eyries to take advantage of the increased rodent density in those areas (Fig. 9.1). By contrast, in the Nenetsky Reserve, as rodent numbers declined, the buzzards moved further from Peregrine nests as they were now in competition with the falcons for avian resources (Fig. 9.1). However, those buzzards which nested in association with Peregrines had a higher breeding success (2-4 fledglings, with a mean of about 2.8/successful breeding pair) than those nesting 'out of' association with the falcons (0.5-1.3 fledglings, mean about 0.9): for the buzzards, choosing to alter their prey spectrum comes with consequences. Of course, the aggressive nature of Peregrines means the buzzards may also suffer conflicts when nesting close to them, so, as Fig. 9.1 shows, they may choose to reduce their breeding success when rodent numbers are high to avoid aggressive neighbours.

In later work in the same area, but also including Kolguev and Vaigach islands, Curk *et al.* (2022) investigated how Rough-legged Buzzards identified good breeding areas in which to settle. Given that the population of lemmings on which the Buzzards chiefly fed was not always homogeneous and that the number of rodents was cyclical, hoping to strike lucky on arrival was risky. What Curk *et al.* found was that the buzzards were staying on at the breeding grounds longer than expected, and were prospecting areas to determine where might be the best place to settle the following year. Pairs whose breeding was unsuccessful prospected, and therefore stayed, longer in the autumn. Those buzzards which preyed mainly on avian prey were, not surprisingly, much

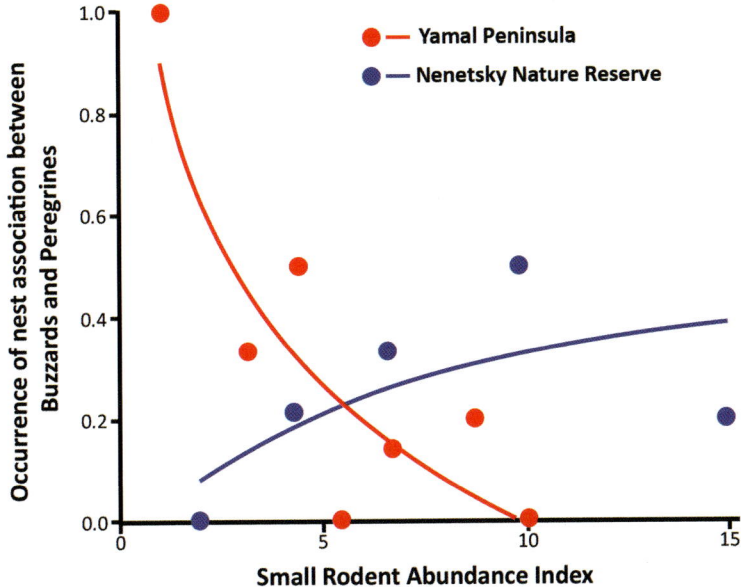

Figure 9.1 Relationship between Rough-legged Buzzard and Peregrine nest association, and rodent abundance in two areas of Arctic Russia. The lower the rodent abundance index, the lower the rodent density. Redrawn from Pokrovsky *et al.* (2020).

more philopatric as avian densities were much more regular on an annual basis. Curk *et al.* note that global warming was disrupting the lemming cycle, which might influence the buzzards to widen their prey spectrum. This raises the thought, one Curk *et al.* do not suggest, that some buzzards are prospecting the breeding sites of Peregrines. As the falcons tend to be highly philopatric, knowing where their breeding sites are could be useful to the buzzards in a poor lemming year.

Foes

Peregrines at coastal sites sometimes nest within colonies of their prey species, but this close proximity seems to arise rather less from symbiosis than from necessity, as the seabird colonies are densely packed, occupying almost every available ledge space. There is also little to be gained by either side, as the likelihood of predation by the falcons at such sites is minimal (for an individual bird) and the protection offered by the falcons in seeing off potential nest thieves seems equally small. At some sites the Peregrines may also be at risk from their neighbours. If Fulmars (*Fulmarus glacialis*) are disturbed at their nest sites they defend themselves by vomiting the contents of their stomachs at whoever approaches. Fulmars' aim is poor, the vomiting being initiated by disturbance and going in the direction the bird happens to be facing at the time rather than being directed at the intruder. But some vomits hit, the sticky, fishy mess of stomach oil being both foul and difficult to remove. Clarke (1993) notes that an examination of four dead nominate Peregrines (collected 1971-1975 in northern Scotland, Orkney and Shetland) showed conclusively that they had died as a consequence of Fulmar oiling[1]. The oil clogs the feathers, reducing insulating properties, affecting the ability to fly and reducing waterproofing, all of which will, in time (and the oil is both persistent and difficult to remove) cause the affected bird's health to decline. In addition to the four deaths noted by Clarke, there are other observations of Peregrines being fouled by Fulmar stomach oil (*e.g.* Mearns, 1983; Dennis, 1970 also notes the death of a White-tailed Eagle (*Haliaeetus albicilla*) on Fair Isle and a European Honey-buzzard (*Pernis apivorus*) on Orkney due to Fulmar oiling).

Not surprisingly Peregrines and Ravens share nesting areas: Nethersole-Thompson (1931) records an instance where the two had nests only 5m apart in the UK, while Ratcliffe (1993) notes a separation of 10m at one site, a buttress of rock shielding direct viewing between the nests and so allowing the two pairs to maintain a (presumably uneasy) truce. Relationships between the two species are poor, the falcons usually being the aggressor, and there are instances in which Ravens have been not only 'buzzed' but struck, and occasionally killed by Peregrines. Ravens have also been found in Peregrine prey remains, though this is more likely to have occurred when the corvids were killed in territorial disputes than actually hunted. However, battles between the two species do not always go the falcon's way, and there have been instances of a Raven turning onto its back in mid-air as an attack is made and using its bill and/or talons to inflict damage on the attacker, with records existing of Peregrine deaths as a consequence.

The likely reason for the antagonism between Peregrines and Ravens is that the latter are potential predators of the falcon's eggs and young. Peregrines will also attack other corvids, probably for similar reasons. Corvids will also mob Peregrines, and Cocker (2007) records an interesting observation in the UK during which a juvenile female Peregrine on prey was surrounded by 18 Carrion Crows which seemed intent on pirating her meal, to the extent of pulling at the tail and wings as the falcon mantled her catch. Eventually the Peregrine took off, and many of the crows pursued her. However, far from fleeing the scene (and the prey) the Peregrine stooped on the remaining crows, who all took off in alarm. The crows then dispersed, and the Peregrine returned to the prey. Interestingly, after Cocker's note had been published, Combridge (2008) reported a similar incident in which a Peregrine (an adult?) took no notice of 12 encircling Carrion Crows and finished its meal before flying off with the remaining carcass.

Unlike Ravens, Buzzards may have a similar diet to Peregrines, which may be a factor in the antagonism between them. There are known instances of nominate Peregrines killing Common Buzzards, which have then formed

---

[1] Stomach oil production is common in some members of the Procellariiformes and is believed to be produced as additional feeding for chicks. In Fulmars, vomiting of the oil appears to be an automatic reaction to certain external stimuli and has even been seen in unhatched chicks and in chicks when approached by parent birds.

Fulmar.

part of the falcon's diet. An astounding, and lethal, relationship between a nominate Peregrine pair and Common Buzzards was seen at Exeter, UK where the falcons were breeding on a local church tower. Once the chicks of the Peregrine pair could thermoregulate, the adult female joined her mate in the air in behaviour which suggested they were preoccupied with searching the sky for Buzzards which often crossed the city while dispersing or possibly foraging. In 2009 a new female arrived on site joining the male who had been resident since 2005. She was seen to attack Buzzards solo that year, but co-operative joint attacks on Buzzards were first observed in 2010 and continued until 2019. Why the female Peregrine attacked Buzzards with such single-mindedness and ferocity is unclear, but it is conjectured that prior to arriving in Exeter an altercation with a Buzzard, perhaps while the falcon was juvenile[2]. Whatever the cause the falcon had a remarkable antipathy toward the larger raptor, sufficient not only to underline her own behaviour, but for her to bully the male into cooperation, this occasionally resulting in behaviour which appeared detrimental to their brood. By 2012 the number of reported attacks had reached 10 with 4 Buzzards being seen to be knocked to the ground. In 2013 there were 45 reported attacks with 10 Buzzards knocked out of the sky. The behaviour is reported in several papers (Dixon, 2013, Dixon and Gibbs, 2015a, 2015b and 2016). What is reported below and in the accompanying photographs was witnessed by the current authors, by Nick Dixon who has been studying the Exeter Peregrines for many years, and several others in June 2018. Over a six-hour period ten attacks were made on single Buzzards, *i.e.* an attack every 36 mins. In each case the two Peregrines attacked cyclically, stooping at a Buzzard until it was knocked to the ground (two occasions) or driven past an invisible boundary marking the outer border of the falcons' territory. The number of stoops in each attack varied, with a maximum of 15, but in all cases the ferocity and relentlessness of the attacks seemed wholly out of proportion with what was necessary to clear the Buzzard out of the falcons' airspace and seemed to defy the logic of energy conservation. These Peregrines had, by now, perfected their attack strategy which entailed a repeated series

---

[2] Peter Welsh, who has studied raptors in Cornwall, England for many years, notes that Buzzards often attempt to pirate prey from Peregrines and that such attacks may leave the falcon injured: occasionally juvenile falcons may be killed. Such attacks on a juvenile bird may be the basis for lifelong antipathy (P. Welsh pers. comm. SW).

of 'cycle stoops' where the first stoop would invoke normal Buzzard defence tactics (*i.e.* present talons to an overhead attacker whilst inverted), then once the Buzzard had righted itself the second falcon would stoop immediately, clearly with the intention of striking the Buzzard's back or head. Peregrines will often attack Buzzards in the breeding season if they fly too close to the eyrie but this is normally to harry them out of the territory. The Exeter Peregrines, by contrast, were clearly determined to injure or kill any Buzzard that approached their eyrie, at whatever height, behaviour out of all proportion to the threat they posed. Three Buzzards were knocked to the ground on the day the authors attended, two having been killed in the air. The third grounded Buzzard was taken into care and later released. From the image sequence of the second mid-air kill, the female Peregrine travelled 10 times her length in the time it took to take three images (0.25s at the shooting rate of 12 shots/s). Assuming the female is a 'standard' size (0.45-0.5m in length), her speed was 18-20m/s, 54-60mph. Her speed did not vary significantly across the sequence. One of the dead Buzzards fell onto a, thankfully empty, table on the pavement outside an Exeter High Street coffee shop, to the consternation of the guests enjoying their drinks at the adjacent table. The other dead Buzzard was not retrieved. Interestingly, whereas all Buzzards passing close to the eyrie were attacked, *de riguer* it would seem, Red Kites however did not elicit the same behaviour.

First Exeter kill.

Second Exeter kill.

After the second kill, the plunging Buzzard was observed closely by Herring Gulls. After one gull flew upside down to make sure it was dead, a small flock of gulls followed it down as it crashed into a street in Exeter.

In Europe, nominate Peregrines are also known to have killed (and consumed) Eurasian Sparrowhawks, Merlins, Common Kestrels and Eurasian Hobbies. Rudebeck (1951), studying the diet of Swedish nominate Peregrines observed one falcon that specialised in hunting Eurasian Sparrowhawks, noting '*the day before when there had been a dense fog, I observed a male Peregrine flying by with a screaming Sparrowhawk in its claws*'.

Peregrines will also attack other Peregrines intruding on their nest area, or even during territorial disputes prior to breeding. As noted in Chapter 6 *Territory and Hunting Range*, Tordoff and Redig (1999a) report two disputes which resulted in death for one combatant. Hall (1955) also records a territorial dispute which led to death in the saga of the Sun Life building Peregrines in Montreal. The increase in urban breeding in the UK has meant that such encounters have increased in recent years and conflicts have also been known to result in deaths (see photo p.257). Older Peregrines will often come under attack by younger, fitter birds early on in the breeding season when hormone levels are high and the availability of nesting sites is limited. Such incidents may lead to death or serious injury, and may last for days or even weeks, to the detriment of foraging and breeding activities, sometimes resulting in failure of the breeding site.

Peregrines will also harass Goshawks if they come into contact. That is also the case with Golden Eagles if the two species interact. Such interactions are rare, as their ranges do not have a significant overlap, but where they do it is clear that the eagles are dominant, particularly when it comes to nest-site selection. MacNally (1979) found the plucked, partially eaten carcass of a nominate Peregrine at a Golden Eagle feeding perch in Scotland, the talon marks to the falcon suggesting capture rather than carrion feeding. In a study which included both Golden Eagles and Eagles Owls as well as Peregrines (and also included Bonelli's Eagles) Martinez *et al.* (2008) looked at the nearest neighbour distances between breeding raptor pairs to investigate the relationships between them. In general, body mass defined dominance, with Golden Eagles at the top of the raptor hierarchy. Both Bonelli's Eagles and Eagle Owls dominated Peregrines. This is not a surprising result, but the biggest surprise was that Eagle Owls were very much less dominated by the larger eagles than would have been expected – the owls were preying not only on the Peregrines (see below), but the two eagle species as well.

At Gibraltar Peregrines (*F.p. brookei*) regularly attack raptors on their spring migration from Africa, these attacks being especially aggressive when the raptors are close to the falcons' Rock eyrie. Short-toed Eagles (*Circaetus gallicus*) seem to be a particular target for these attacks, several eagles having been seen to be knocked into the sea and drowned (Garcia, 1978). Peregrines were also seen to attack Booted Eagles, Black Kites and even Griffon Vultures (*Gyps fulvus*).

Hunt *et al.* (1992) note many instances of Peregrines (*F.p. anatum*) harassing Bald Eagles (weighing about 4.7kg) in Arizona, USA: in one case a falcon

F.p. minor and a Verreaux's Eagle (*Aquila verreauxii*). The photograph was taken in the Augrabies National Park, South Africa in October so it was probably a territorial dispute as the area is well-known for the high concentration of Rock Hyrax (*Procavia capensis*), a favourite prey of the eagles. There is no direct evidence that the eagles kill Peregrines, but as in all cases, the falcon is taking a chance when arguing with a formidably-armed opponent. *Peter Warne.*

was believed to have chased an eagle which flew into power lines and was killed, and in another a falcon hit an eagle in the head. The eagle subsequently behaved erratically, indicating a brain injury, and was found dead a month later apparently from the injury it sustained. The falcon attack was not thought to be a hunt for prey, but part of the general antagonism between the two species during the breeding season. Hays (1987) also records an attack by an *F.p. anatum* pair on a Golden Eagle in Colorado, USA which caused the eagle to collide with a cliff. Now disorientated, the eagle was struck in the head by the stooping male Peregrine. The eagle fell, dead weight, over 180m into the forest below. Though the eagle was not recovered, survival seems very unlikely, and Hays notes it was not seen again. Beebe (1960) recorded Peregrines (*F.p. pealei*) mobbing Bald Eagles on Langara Island, off the eastern Canadian (British Columbia) Pacific shore. When attacks were over the ocean away from the nesting cliffs, the eagles ignored the falcons, unless they stooped. Then, the eagle would turn and offer its talons at the incoming Peregrine which rapidly aborted the stoop. Beebe saw hits on eagles only if two Peregrines attacked cyclically. However, there was a marked difference if the falcons approached the eagles' nest, such indiscretion bringing an immediate response. Beebe saw one male eagle attack a Peregrine which strayed too close to its nest: the falcon was lucky to escape with its life. Also on Langara Island, Nelson (1970) considered that Bald Eagle interference with nesting falcons caused a delay in the fledging of the Peregrines' young: Nelson saw four unsuccessful attempts by the eagles to pirate Peregrine prey, from which it can perhaps be concluded that further attempts might have been successful.

In Arctic Canada, pairs of *F.p. tundrius* and Gyrfalcons bred at each end of a river canyon, keeping interaction to a minumum by avoiding flying through the canyon. Occasionally adults would meet accidentally, but such meetings rarely resulted in serious conflict.

In the USA Peregrines are also known to usurp Prairie Falcon nests and even to kill the falcons, even though the two species are comparable in size. Gyrfalcons have been known to pirate Peregrine prey (Dekker, 1995), and Gyrfalcons predating Peregrines has been surmised. Booms and Fuller (2003) found traces (<1% of total remains) of Peregrine in Gyrfalcon pellets in Greenland, but this does not necessarily imply predation rather than feeding on a carcass. Lindberg (1983) found no evidence of Peregrine in the diet of Swedish Gyrfalcons. However, Pokrovsky et al. (2010) report a case in which an immature Gyrfalcon had apparently taken a fledgling Peregrine from close to the nest and then, while eating it, had been attacked and killed by an adult Peregrine. Burnham and Burnham (2011) studying Gyrfalcons and Peregrines breeding near Kangerlussuaq, west Greenland noted that once Peregrines had arrived in the area they had rapidly colonised it at the expense of the resident Gyrfalcons, the ratio of 1:1 breeding pairs in the early 1970s increasing to about 8:1 within a decade and to >10:1 by the late 1990s. The Gyrfalcons bred earlier and hunted adult Ptarmigans (as well as other birds, such as Snow Buntings, and they also hunted hares). The Peregrines bred later and heavily predated Ptarmigan chicks which reduced the adult Ptarmigan population the following spring. The Peregrines were also much more aggressive than the Gyrfalcons, and more agile. Quoting an earlier study, Burnham and Burnham note that when attacked by a stooping Peregrine the Gyrfalcons '*looked clumsy and confused*'. In the North American Arctic, RS has seen Gyrfalcons and Peregrines nesting in close proximity, and while in general the two ignored each other – the nests were at each end of a canyon, the falcons choosing to enter and exit the canyon at the end closest to their nest, rather than flying along its length so as to avoid interaction – aerial conflicts were seen, though these were perfunctory and none involved contact.

Interestingly, a study in Italy by De Rosa *et al.* (2019) has strongly suggested that Peregrines (*F.p. brookei*) are out-competing Lanner Falcons in the central area of the country. Habitat loss is restricting breeding options for

Prey-carrying nominate Peregrine being harried by a Herring Gull. *Chris Skipper*.

both falcons, but the Lanners are being pushed into sub-optimal habitats, resulting in lower breeding success, as the Peregrines have recovered following the DDT crisis. Again, it is the Peregrines aggression towards species which are in competition, in this case for prime breeding sites, which has led to its increasing dominance.

At coastal sites Peregrines will also attack, and potentially kill and consume, both Long-tailed Skuas and Arctic Skuas (*Stercorarius. parasiticus*), each of which is a predator in its own right, and, as noted in the section on diet, Great Black-backed Gulls.

All five British owls are also known to have been killed and eaten by Peregrines. However, the Peregrine does not have it all its own way. In Europe

Eagle Owls predate both Peregrine chicks and adults (Mikkola, 1983 – see also Krekels (2017) who reported the predation of four Peregrine chicks by Eagle Owls in The Netherlands), while the same is true for Great Horned Owls (*Bubo virginianus*) in North America (though in that case falcons killing owls is also known). Studying Peregrines in France's Jura Mountains, during the period 1964-2009, Monneret (2010) noted a significant reduction in breeding success from 1990 onwards. Monneret noted that the reduction could not be readily explained by site saturation, human disturbance or climate change, but appeared to coincide with the influx of Eagle Owls (Fig. 9.2 and 9.3). While definitive proof was lacking Monneret believed the owls were the principal cause of the abnormal reduction in both adult and juvenile Peregrine numbers. In a later study in the same area (Kéry et al., 2022) the researchers found that while the Peregrine population had risen significantly after the pesticide crisis it had peaked in different areas between 2002 and 2007, the Peregrine population had then declined. The chief cause appeared to be the increased number of Eagle Owls, but illegal persecution and other human disturbance were also involved. Studying Peregrines in Germany, Lindner (2018) came to a similar conclusion, noting not only that the falcons formed part of the owls' prey spectrum, but that the owls were also evicting Peregrines from cliff nesting sites and were moving into urban areas at the falcons' expense.

However, while the proximity of owls might therefore be expected to influence Peregrine breeding sites, particularly as the owls are prey competitors,

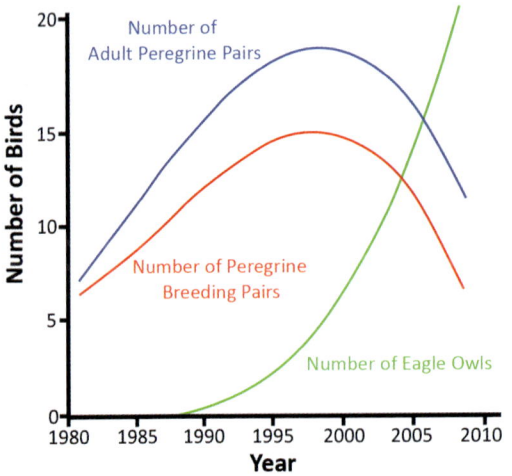

Figure 9.2 Variation of the number of adult nominate Peregrine pairs, the number of breeding Peregrine pairs and the number of Eagle Owls in the Jura Mountains of France in the period 1964-2009. Redrawn from Monneret (2010).

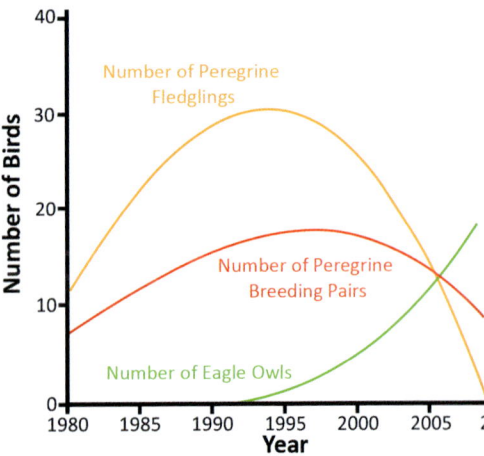

Figure 9.3 Variation of the number of breeding nominate Peregrine pairs, the number of fledglings produced and the number of Eagle Owls in the Jura Mountains of France in the period 1964-2009. Redrawn from Monneret (2010).

Eagle Owl with a Peregrine it has killed. *B. Zoller.*

Brambilla *et al.* (2006a), studying Peregrines in an alpine area on the border between Switzerland and Italy (an area which could be considered the dividing line between nominate and *F.p. brookei*) found no evidence that distance to owl nests influenced falcon nest selection. However, the same authors did find that proximity to an owl nest lowered Peregrine productivity, though it was not clear whether this was due to predation of falcon chicks, or because nest sites near owls are obviously poor quality and therefore likely to be occupied by sub-optimal falcon pairs (Brambilla *et al.*, 2006b).

Interestingly, Lindner (2018), who had noted Eagle Owls evicting Peregrines from cliff sites, also noted that Egyptian Geese (*Alopochen aegyptiaca*) had usurped Peregrine nesting sites at several locations. This news was particularly interesting to the present authors as at Symonds Yat, Herefordshire, England, close to where both authors live, Canada Geese had evicted Peregrines from nest sites they had occupied for many years. The nest sites were in small caves in vertical rock faces. When the geese first arrived they used rock ledges from which they were easily evicted by the falcons. But soon the geese realised the advantages of the caves and took them over early in the breeding season[3]. Faced with a seated goose armed with a long neck and a formidable bill, the Peregrines were unable to evict the geese and had to move to weather-prone ledges. Table 9.1 shows the detrimental effect on Peregrine breeding of the arrival of the geese.

Canada Geese at a breeding cave, Symonds Yat, England.

---

[3] Interestingly, Newton (1979) tells an identical story about Canada Geese and Prairie Falcons in western Canada, the falcons evicting the geese from ledges on earth banks, but being unable to evict them from holes.

Year	Male Adult	Female Adult	Total Juvs	1st face 'Letterbox'	1st face 'Raven's nest'	1st face 'Bees' nest'	2nd face 'Ash hole'	2nd face '7'	2nd face 'Lozenge'	Midway 'Cyclops'	3rd face Unused
1990	M1	F1	4				4				
1991	M1	F1	2	2							
1992	M2	F1	1	1							
1993	M2	F2	0				0				
1994	M2	F2	3				3				
1995	M2	F2	3				3				
1996	M2	F2	4	4							
1997	M2	F2	3	3							
1998	M2	F2	4				4				
1999	M2	F2	4	4							
2000	M2	F2	4				4				
2001	M3	F3	1				1				
2002	M3	F3	4		4						
2003	M3	F3	2				2				
2004	M3	F3	2				2				
2005	M3	F3	3		3						
2006	M3	F3	4	4							
2007	M3	F4	3				3				
2008	M3	F4	0				0				
2009	M3	F4	0		0						
2010	M4	F5	1					1			
2011	M4	F5	4						4		
2012	M4	F5	2		2						Goose
2013	M4	F5	3		3						Goose
2014	M4	F5	3			3					Goose
2015	M4	F5	3						3		Goose
2016	M4	F5	1	Goose			1				Goose
2017	M4	F5	3	Goose			3				Goose
2018	M5	F5	0	Goose	Goose	Goose	Goose		0		Goose
2019	M5	F5	0	Goose	Goose	Goose	Goose			0	Goose
2020	M6	F6	0	Goose	Goose	Goose	Goose			0	Goose
2021	M6	F6	2	Goose		Goose	Goose	2			Goose
2022	M7	F7	3	Goose		3	Goose				
Totals			76	18	12	6	30	3	7	0	
% eyrie use			100%	24%	16%	8%	39%	4%	9%	0%	
Eyrie Location Count			33	6	5	2	13	2	3	2	

Table 9.1 Eyrie location and breeding success of nominate Peregrines at Symonds Yat, Gloucestershire, England prior to, and subsequent to, the arrival of breeding Canada Geese.

Notes
1. M1/F1 refers to the sequence of observed changes in territory-holding males and females. There have been 7 males and 7 females holding territory from 1990-2022.
2. Alternating shaded and unshaded rows indicate a change in territory-holding pairs.
3. Eyries '7' 'Lozenge' and 'Cyclops' are too small to accommodate Canada Geese.
4. Breeding failures in the period 2018-2022 were at sub-optimal eyries.

Both Arctic and Red foxes have been known to take Peregrine chicks, and other terrestrial predators would almost certainly do so opportunistically, but the combination of the inaccessibility of nest sites and the fierce defence of the parent falcons is a strong deterrent. However, all records suggest that humans are the most persistent and dangerous mammalian predators of the falcons. In Britain it is likely that at most non-urban nest sites humans are the greatest potential threat to Peregrines. Since rock climbing increased in popularity as a sport the potential for human disturbance of Peregrine sites has also increased across the Peregrine's range. An evolution in climbing techniques and equipment has allowed steeper cliffs to become accessible, while competition between adherents has meant that cliffs that had long been ignored have become popular. Persistent disturbance stresses the birds. Brambilla et al. (2004), in a study in northern Italy and southern Switzerland, noted that the productivity of falcons nesting on cliffs popular with climbers had declined, some pairs raising no chicks. In Britain, sensible cooperation between the climbing fraternity and bird enthusiasts has resulted in restrictions being put in place which limit access to certain parts or entire cliffs for the duration of the breeding season – see photo p.396. In all cases, and particularly at remote cliffs, goodwill is required to ensure disturbance is eliminated.

Human disturbance in general, rather than rock climbing specifically, will always be a threat to nesting Peregrines. In a study in the Balkan Mountains of Bulgaria Djorgova et al. (2021) studied the response of three raptors, the Golden Eagle, the Long-legged Buzzard (*Buteo rufinus*) and the Peregrine (*F.p. brookei*) to increased human intrusion into their breeding areas using the ArcGIS Cost Path technique of spatial analysis. They found that the buzzards were least affected, the eagles the most affected. The effect on breeding Peregrines was intermediary, but in all cases, excessive disturbance is detrimental to breeding success and the health the of the birds. In a related study of North American raptors Morrison et al. (2006) classified Peregrines as Category 4 in terms of nest defence. This, the highest category, means very aggressive defence, and the recorded aggression was against any intruder, not only humans. Similar aggression has been seen in Britain, with records of attacks against humans, though individual falcons react in different ways, these varying from high circling of one or both parent birds, each of which will make repeated alarm calls, through dummy stoops that pass close by, to strikes, though the last have to date been reported only in the USA.

Closer to home, netting put on the top of the Parkinson Building, University of Leeds, England in order to prevent pigeons from breeding sadly trapped and killed three adult Peregrines which were either territory holders or had been seeking an urban breeding spot (Paul Wheatley pers. comm. SW). One dead adult tiercel was found on 11 June 2022 while another of the year's juveniles was being rescued, after the alarm had been raised by local bird enthusiasts, who had observed the trapped bird flapping under the netting. A second juvenile was rescued from the netting a week later. It had injured a leg in its struggles to get free: the leg was treated and the bird was released. Both recovered birds flew strongly after release. Following a social media outcry and adverse media publicity, the university took the decision to take down the netting. Two further dead adult female Peregrines were found at this stage. Both were desiccated: one was thought to have been the resident female of at least one year previously, the other was thought to have died approximately one month before. Happily, the two 2022 trapped juveniles were untangled from the netting: one was released immediately, the other had an injured leg which was treated before subsequent release. Both juveniles flew away strongly.

The effect of placing netting (to deter feral pigeons) on the roof of a building at the University of Leeds, England.

Left   One of the dessicated Peregrines trapped in the netting. *Paul Wheatley*.

Below   A luckier falcon. This one was rescued from the netting and returnd to the wild. The netting was removed to prevent a re-occurence. *Paul Wheatley*.

Overleaf are a series of images showing a juvenile *F.p. anatum* taking an Osprey chick from a nest in the presence of an adult Osprey, at Vero Beach, Blue Cypress Lake, Florida, USA. The falcon must have been extremely hungry to take such a risk. In the air Peregrines are more agile than the larger raptor but, as with the eagle of p.405, Ospreys carry significant armament in talons and bill, and an irate adult seeking to protect a chick is always a formidable opponent, particularly when both birds are grounded. *Mark Smith*.

Friends and Foes

10   POPULATION

Survival

For wild birds, an age spectrum can only be assessed from recovered ringed birds whose birth date is known with certainty. Within that cohort birds with known ages in excess of 15 years are seen. Studies of Peregrines in North America and Europe have allowed a reasonable assessment to be made of the survival of breeding adult birds – see Table 10.1 overleaf.

Opposite page *F.p. anatum* juveniles, San Pedro, Los Angeles, USA. *Gabriel Gruia.*

This page Nominate Female feeding chicks, southern Scotland.

Study Area/ Sub-species	Survival of 1st year birds (%)	Survival of Adults (%)	Source
North America F.p. anatum	30	75	Enderson (1969)
Alaska, USA[1] F.p. anatum/F.p. tundrius		77	Ambrose and Riddle (1988)
Washington State, USA F.p. pealei[3]	42.4	73.8[2]	Varland et al. (2020)
Colorado, USA[4] F.p. anatum	54.4	80.0	Craig et al. (2004)
California, USA[4] F.p. anatum	28-65[5]	85.0	Kauffman et al. (2003)
NWT, Canada F.p. tundrius		85 (male) 81 (female)	Court et al. (1989)
NWT, Canada[6] F.p. tundrius		72.5	Johnstone (1998)
NWT, Canada (1998) F.p. tundrius		73[7]	Franke et al. (2011)
West Greenland F.p. tundrius		78.8	Good and Fuller (1995)
Scotland Nominate	44	89	Newton and Mearns (1988)
Northern UK[8] Nominate	60.0	81.0[9]	Smith et al. (2015)
Sweden Nominate	41	68	Lindberg (1977)
Finland Nominate	29	81	Mebs (1971)
Australia F.p. macropus	45	95	Olsen and Olsen (1988)
South Africa F.p. minor		85.2	Altwegg et al. (2011)

Population

Table 10.1 (*opposite*) Survival estimates for Peregrine Falcons of different ages across their distribution[11].

Notes:
1. The study was in four areas, two occupied by *F.p. tundrius* and two by *F.p. anatum*. Data was combined across the four areas and, therefore, the two sub-species.
2. The study, between 1995 and 2018 at coastal sites in Washington State also measured the survival rate of falcons in their second year as 66.3%.
3. The study was carried out in winter, but identified birds ringed during the breeding season. The area is known to hold a population of breeding *F.p. pealei*.
4. As well as quoting figures for first-year and adult birds, Craig *et al.* note a survival of 67% for birds in their second year.
5. The survival of first-year birds varied from 28% in rural areas to 65% in urban areas. The mean value, corrected for numbers in each of the locations, was 38%.
6. The quoted value is a mean for males and females. Johnstone (1998) quotes survival of males as 76.5% and of females as 70.1%.
7. The study period was 1982-2008. During that period the annual survival of adult Peregrines varied from 43-100%. The weather (particularly the influence of the North Atlantic Oscillation) was a significant factor in the survival of the falcons, bad weather affecting both survival on spring migrations and the availability of prey during the breeding period.
8. The study covered northern England and southern Scotland.
9. Smith *et al.* (2015) quote 81.1% survival for sub-adults (*i.e.* second calendar year) birds and 81.0% for adults.
10. Altwegg *et al.* (2014) note a survival of birds in their second year as 71.6%.
11. For an interesting article on the methods of acquiring some of the information in this Table, and for similar information for other raptors and owls, see Newton *et al.* (2016).

The data trend of Table 10.1 is consistent with those for other raptors. The mortality of first-year birds is high, the rigours of a first winter testing their hunting experience to, and beyond, its limits. The data also suggests that those species which undertake significant migration journeys (*F.p. tundrius*, northern *F.p. anatum*, *F.p. calidus* and northern breeding *F.p. japonensis*) have lower adult survival than sedentary populations, implying the combination of migratory flights and the rigours of the Arctic/sub-Arctic breeding areas where (relatively) high prey numbers are occasionally offset by inhospitable weather, has an adverse effect on survival. One aspect of adult mortality which has been investigated, but with mixed results, is whether brood size influences adult survival, both of males (because of the stresses of hunting to provision the brood) and of females (because of the stresses of egg production, incubation, raising the brood and nest defence). Nelson (1988, 1990) found that on Langara Island, Canada, adult *F.p. pealei* survival was halved if the falcons had three chicks rather than one or two (though, curiously, was only marginally further reduced for broods of four). Further studies are clearly required before a definitive understanding can be reached.

Fig. 10.1 (overleaf) shows the variation of survival of adult and first-year birds with time based on ring recoveries from the sedentary Peregrine population of Great Britain. Over the 40-year period 1975-2015 the mean survival of adult Peregrines was essentially constant at 83.0% (95% confidence

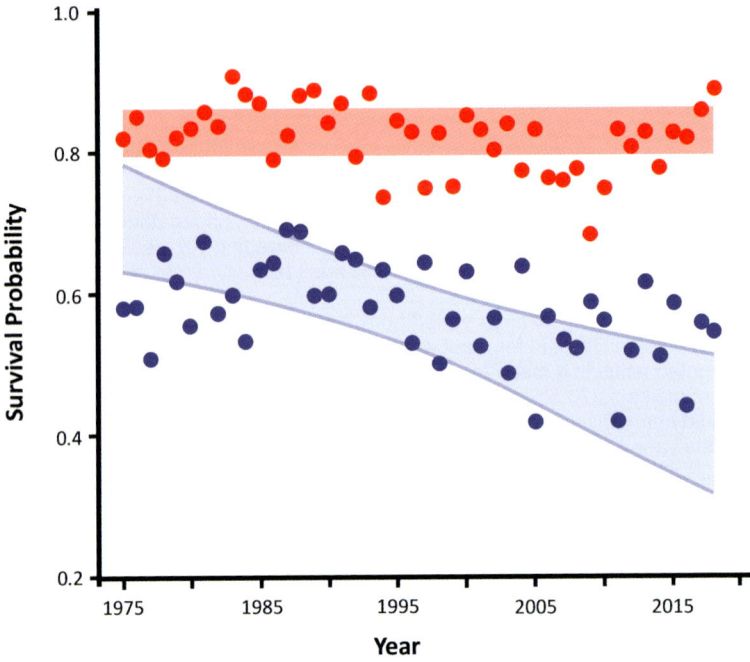

Figure 10.1 Survival of adults (red) and first-year (blue) Peregrines in the UK based on the recovery of over 700 ringed birds during the period 1975-2015. The red and blue lines represent best estimate fits to the data, the paler polygons representing 95% confidence limits. Redrawn from Robinson and Wilson (2021).

limit 79-86%, individual years varying from 68-91%). However, the mean survival of first-year falcons was not constant, varying from 42-69% across the 40 years. The reasons for the decline in mean survival are unclear. The calculated mean adult survival compares reasonably well with data from other parts of the world (Table 10.1), though as noted above, adult survival tends to be lower in migratory species. The position on first-year survival is less clear as data are sparse and inconsistent, but suggests that, on average, it is no better than 50%. Overall, it seems reasonable to suggest that for Peregrines breeding in areas with minimal human disturbance adult survival is 80-85%, with that for migratory species being lower, perhaps 75-80%: the data of Table 10.1 suggests that by 'adult' here we mean Peregrines in their second year of life and older. The survival of first calendar year birds is of order 50%.

Assuming a 50% survival for fledglings, and a mean, 82.5%, survival for adults, the probability that fledglings born in Year 1 will still be alive at 10 years of age is 8.9% (*i.e.* a 50% chance of surviving to the second year and an 18% chance of surviving a further nine years). The chances of an individual fledgling reaching 20 years of age is 1.3%. Peregrines in captivity (*i.e.* falconry or breeding birds) do survive to this age and beyond: White *et al.* (2020) quote ages up to c.25 years. In the wild, such ages are rare, though a male in the UK, ringed as a nestling in Cumbria, England in June 1994 was spotted in southern Scotland in May 2016 when it was 21 years 10 months 24 days old. White *et al.* (2020) note one female ringed as a nestling in Alaska, USA in 1981 being trapped alive in California, USA in September 2000. The bird, by then more than 19 years old, was released. Another bird mentioned by White *et al.* ringed in Minnesota, USA in June 1992 was seen alive in Virginia, USA in March, 2012 aged 19 years 9 months. The famous female breeding on Montreal's Sun Life Building was at least 18 years old when last seen. Other birds with quoted ages of up to 17 years have also been noted. However, the mean age of death of adult birds, from both recovery records and statistics is 6-7 years.

Ringing data collected in the UK allows a life table of nominate Peregrines to be drawn up (Fig. 10.2). Defining a mean life for Peregrines on the basis of Fig. 10.2 is difficult for two fundamental reasons. The first, and most obvious, is the very high death rate for first year birds (and, also, the high rate for second- and third-year birds) and how these should be treated. The second problem is the (relative) paucity of data for birds older than three years. To allow a mean age to be determined the shape of the age histogram must also be decided. At first glance the distribution of ages seems to follow a Poisson distribution with a mean age of *c.*5 years if all data is included, or a mean age of *c.*7.5 if the first three years are excluded.

Figure 10.2 Age at recovery of 428 ringed birds in the UK. The yellow bars are for birds less than three calendar years, the number recovered at that age being given at the top of the bars. The red bars are for birds older than three calendar years. The oldest recovered bird was 20+ years at the time of recovery. With thanks to the BTO for the ringing recovery data.

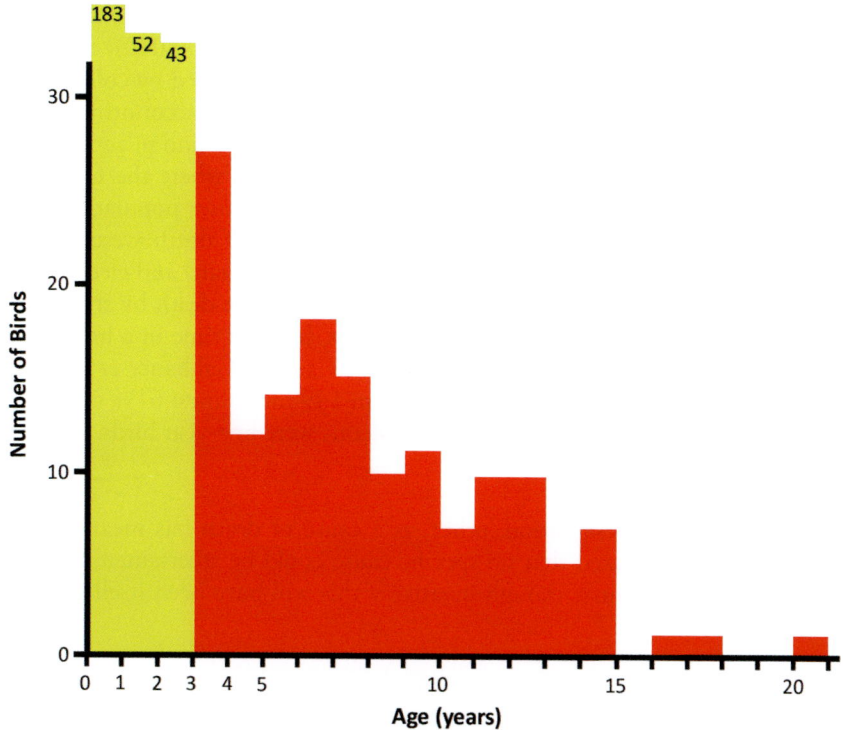

In their study of *F.p. macropus* in Tasmania, Australia, Mooney and Brothers (1993) studied the age spectrum of breeding birds. Females tended to breed in their second calendar year, males in their third. The average breeding period was 3.3 years for females and 3.2 years for males. The mean lifetime of females was, therefore, 5.3 years and 6.2 years for males. Based on the data of Fig. 10.2 these data appear to under-estimate the lifetime by 1-2 years, but the data sizes are different in the two cases, the nominate data being based on over 400 birds, while the Australian lifetimes is from a much smaller data set.

Causes of Death
Ratcliffe (in Wernham *et al.,* 2002) presents data on the causes of death of recovered ringed Peregrines in the UK, but the analysis allows only a very broad-brush assessment, with 21% taken by humans, 30% human-related (which would encompass collisions with man-made structures), 29% dying of natural causes, and 20% succumbing to 'other'. In an assessment of the deaths of 1000 wild birds (across all UK avian species) Jennings (1961) noted that members of the Falconidae (there are four UK breeding species, and which of those contributed to the eight cases noted is not detailed) three had been poisoned; one had succumbed to unspecified parasites; one to an unspecified bacterial infection; one to an unspecified fungal infection; one to an unspecified disease; and one to an unspecified trauma. In a study in the USA, Sweeney *et al.* (1997) found that 81% of 168 Peregrines brought to a veterinary care centre had sustained collision injuries: sadly 45% of the birds died or were euthanised, but 41% were successfully treated and released. In a later study, Hager (2009), looking into the causes of death of urban raptors in the USA (where the number of birds of all species which die annually as a result of anthropogenic causes is estimated at 1,000,000,000), found that falcons (all urban species) were susceptible to injury caused by collisions and electrocution, with vehicle collisions and window strikes accounting for 39% and 12% respectively of all deaths. Similar results were found in another study in the USA, this time of Peregrines in Pennsylvania, where the falcons had been re-introduced following the extirpation of the native population during the pesticide crisis (see below). Again, the causes of death were collisions (with buildings, vehicles and two falcons killed by aircraft) and electrocution, but with many deaths unexplained. There was a single death by shooting. In another study of the reintroduced Peregrines, but this time in a broader area of north-east USA and southern Ontario, Canada, Gahbauer *et al.* (2015) looked at 107 known causes of death from 149 mortalities: 61% of the birds had died from collisions. Many of the falcons were first-year birds.

Disease and Parasites
Historically, the designation 'other' as a cause of death has meant that the bird was so emaciated that no specific cause could be ascertained, starvation being the likely cause, though invariably this will have been itself caused by

an underlying illness or injury. Here we present details where specific causes of death have been identified.

One of the most important avian diseases in recent times has been avian influenza (HPAI) which has devastated chicken farms and led to restrictions on the flying of falconry birds. Bertran *et al.* (2012) have shown that it is possible for raptors to contract the disease from infected prey. Bertran *et al.* studied the disease in captive hybrid Gyr-Saker cross falcons where it was usually fatal. More recently (2022) the disease has struck with devastating results across all species in the northern hemisphere. In an article on 8 August 2022 (*www.birdlife.org/news/2022/08/08/an-unprecedented-wave-of-avian-flu-has-been-devastating-bird-populations-across-the-northern-hemisphere/*) BirdLife International note the discovery of more than 400,000 dead birds from over 2,600 known outbreaks (more than double the number of fatalities than in the 2016/2017 outbreak). Nearly all western Europe's raptor species have been impacted, with 30% of the Peregrine population of The Netherlands having been wiped out. A worrying aspect of studies of the disease is that many predatory mammal species are carriers and so may contribute to the spread in raptors.

Trainer (1969) gives a list of the bacterial, viral and parasitic infections which may be found in US raptors. For Peregrines infectious diseases include trichomoniasis, aspergillosis, coccidiosis, botulism and cestodes, with parasites including tapeworm, mallophaga, mites, fleas and ticks. In a study in Poland (Woźniakowski *et al.*, 2013) note that Peregrines hunting pigeons could be infected with Columbid Herpesvirus-1 (CoHV-1) which has been known to cause death in the falcons. Cooper (1993) gives a brief *résumé* of diseases which are known to affect captive Peregrines and may, therefore, be found in wild falcons, these including *Mycobacterium avium* (avian tuberculosis), protozoa-based diseases such as 'frounce' (see below) and avian malaria, salmonella and Escherichia coli infections, and parasites such as hippoboscidae (louse flies) and nematodes[1]. Cooper *et al.* (1980) also note that *Corynebacterium xerosis, Enterobacter cloaca, E. coli, Staphylococcus aureus, Pasteurella* (now *Riemerella*) *anatipestifer* (associated with diseases in waterfowl which the adult Peregrines were hunting), Streptococcus *spp.* and Proteus *spp.* were isolated in swabs taken from the pharynx and cloaca of wild nestling Peregrines at the time of ringing in both the USA and UK: swabs were also taken from seven captive falcons. Each of the identified organisms is associated with an avian disease. *Staphylococcus aureus*, which can cause 'bumblefoot' (ulcerative pododermatitis)[2], was only

---

[1] Ibarra *et al.* (2019) reported the nematode *Serratospiculum tendo* in the air sac of an adult male South American Peregrine Falcon (*F.p. cassini*) see photo overleaf. Air sac nematodes are known in Peregrines in North America, Eurasia and Australia, but this was only the second time one had been seen in Argentina.
[2] Bumblefoot is a well-known condition in birds of prey, being seen as inflammation and infection of the pads that grip the perch surface, and potentially leading to lesions and badly deformed feet. Those who feed garden birds may see the condition in Chaffinches who seem particularly prone to it. It affects both wild and captive (falconry) Peregrines. Müller *et al.* (2000) note that wild birds captured to enter breeding programmes are prone to the disease if they are not allowed adequate chances to fly, while Barboza *et al.* (2020) suggest certain perch surfaces and foot pads to lessen the effect in falconry birds.

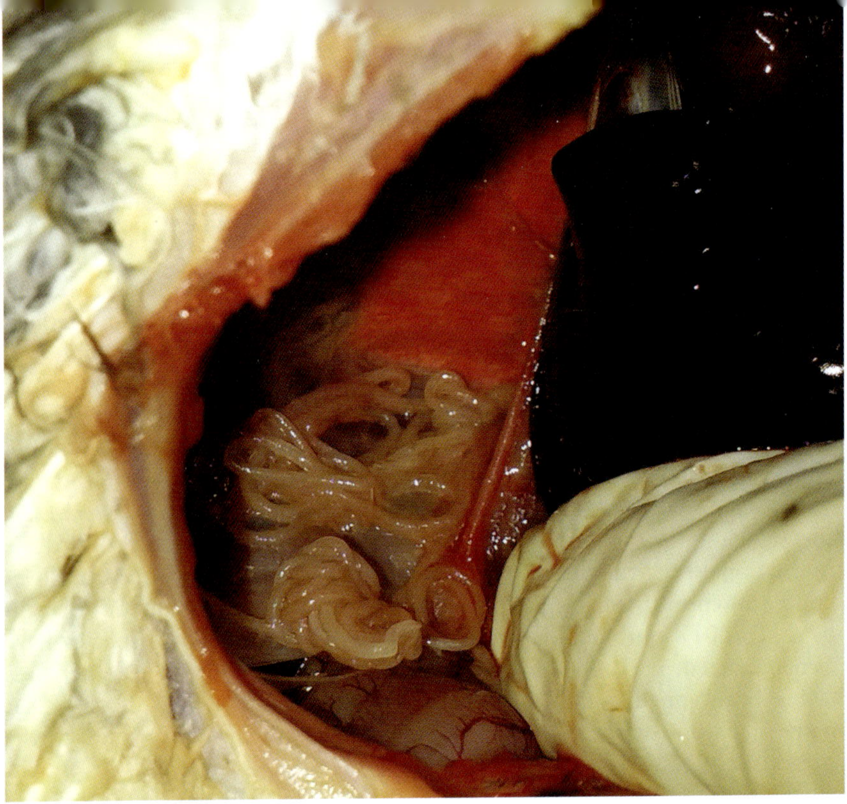

The nematode *Serratospiculum tendo* in the air sac of an adult male South American Peregrine Falcon (*F.p. cassini*). Jennifer Ibarra.

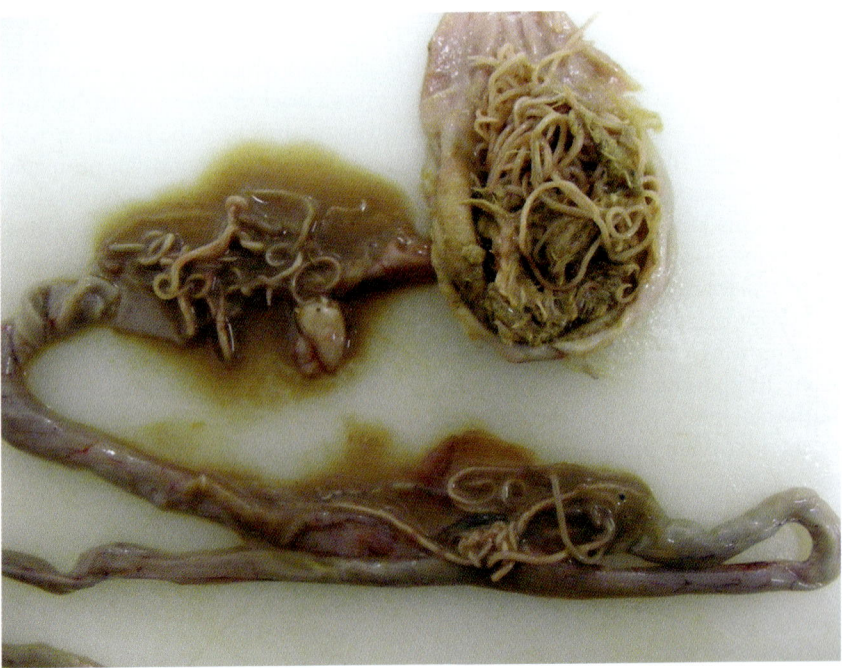

Roundworms in the gizzard and intestines of a Common Buzzard. The same species is also found in Peregrines. *Tom Pennycock.*

found in the captive falcons, suggesting that humans might have been the vector, though *S. epidermidis* was found in the wild birds. However, *S. aureus* has been positively identified in Peregrines (*F.p. brookei*) in Spain (Vidal *et al.*, 2017). Gargiulo *et al.* (2017) added Campylobacter *spp.* to the list of enteropathogenic bacteria which may affect raptors.

Just as with other animal species, birth defects can be seen in infant Peregrines. Cooper (1984) noted developmental abnormalities of a leg and wing in a two-month-old female Peregrine collected in Scotland. There was duplication of a hind toe and extra vestigial digits on a wing carpus. Such problems are rare in wild birds, perhaps because the bird does not survive long after fledging: in this case the bird was not released into the wild as the flight issues with the abnormal wing probably meant it would be unable to feed adequately. Moore *et al.* (2019) also found visual defects (photoreceptor dysplasia and secondary retinal degeneration) in juvenile Peregrine Falcons in the USA. Such visual defects would very likely limit life chances in the wild.

In captive/falconry birds, Applegate *et al.* (2017) noted four cases of gastric impaction (a Peregrine, together with American Kestrel, Golden Eagle and Barn Owl). In all cases the birds had ingested foreign material (perch substrate, newspaper and grass) which had caused obstructions. All four were treated, but only the Barn Owl survived. While it can be argued that these deaths were unlikely in wild birds, the possibility of impaction due to accidental or the unfortunate deliberate ingestion of foreign (human-created?) material cannot be entirely ruled out.

Parasite burden may also reduce a falcon's ability to hunt effectively. Body parasites in falcons include fleas, screw-worm flies, chewing bird lice and louse flies, many forms of which are associated with falcons (Maa, 1969). The louse flies are Hippoboscidae, diptera, the usual one seen in the UK being *Carnus hemapterus* which is often seen in the 'wing pits' of falcons. Other diptera have also been seen on Peregrines (see, for example, Bequaert (1954) for the USA and Levesque-Beaudin and Sinclair (2021) for Canada and Europe). See overleaf for photographs of various louse *etc.* flies.

There are also micro-organisms that degrade feathers, the annual moult offering only a partial protection against them in terms of feather efficiency. Philips (2000) lists 86 species of mites which attack not only feathers, but quills, as well as the bird's skin (both the surface and mites that burrow sub-cutaneously), and also infest the respiratory system. In Australia Raidal *et al.* (1999) found parasitic infections in the brain and eyes of a single adult male *F.p. macropus* (and in several Nankeen Kestrels) with associated parasitic cysts, possibly of Apicomplexan protozoans, which ultimately led to the death of the Peregrine.

In a study of raptor nests as a habitat for invertebrates, Philips and Dindal (1977) noted that nests are a potential home to parasites which may infect adult birds and nestlings, as well as animal saprovores (invertebrates which feed on carrion, excreta, pellets *etc.*), and humus fauna (invertebrates which

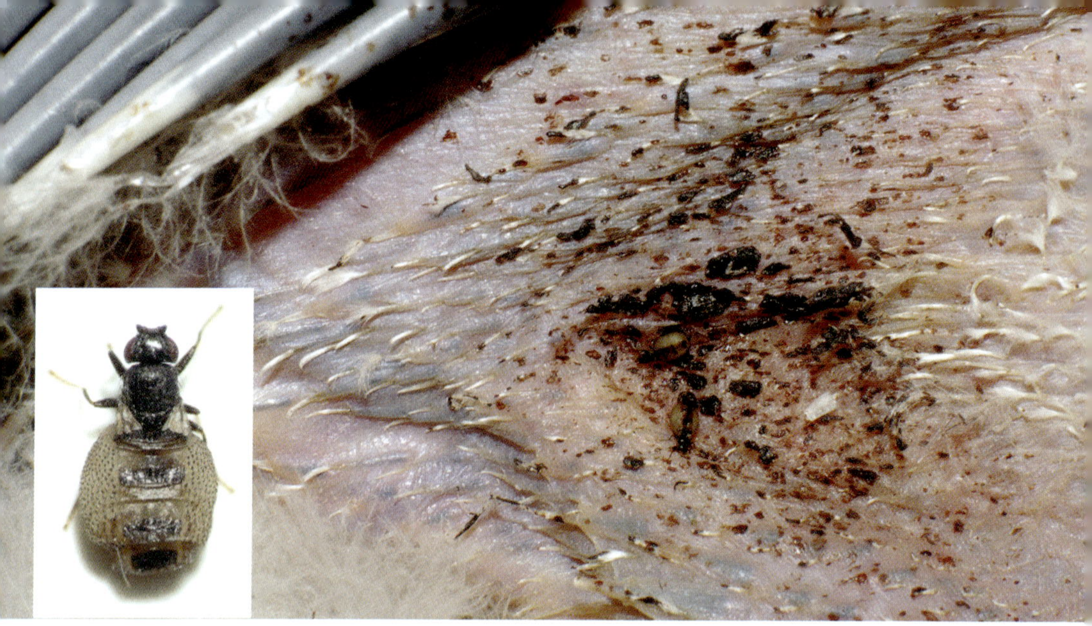

Carnus hemapterus infestation in the wing-pit of a Peregrine nestling. *Phil J. Belman.*

Inset above   *Carnus hemapterus.*

Right   Male *Ornithomya avicularia.* Found on a Eurasian Sparrowhawk, but also seen on Peregrines (see Thompson (1954) and Hill (1962)) for UK records, and O'Connor and Sleeman (1987) for Irish records. *Denise Wawman.*

Below left *Crataerina pallida,* the Swift Louse Fly. Photograph of first UK record on a Peregrine. *Denise Wawman and Nathalie Mahieu.*

Below right *Icosta americana*, which infests North American Peregrines. The out-of-focus circle on the fly's thorax is the specimen mounting pin. *Valerie Levesque-Beaudin.*

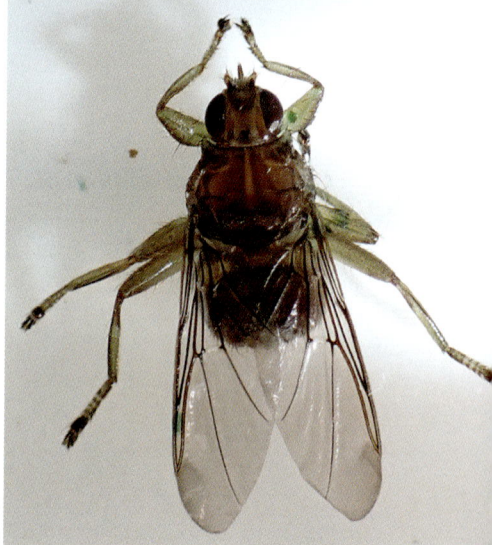

feed on nest material). Philips and Dindal noted 22 species of arthropods from 15 families which have been found in North American Peregrine nests (though they did not specifically name them), as well as listing the insects commonly found in raptor nests. It is probable that British nests will similarly carry a significant burden of parasites.

Internal parasites can also infect Peregrines, with known conditions including trichomoniasis, caused by *Trichomonas gallinae*, a motile single-celled protozoan which lives in the mouth, throat and crop and causes frounce (an upper digestive tract disease) which can be fatal; coccidiosis, (caused by caryospora (Eimeria *spp.*) protozoan parasites of the intestinal tract; the fungal infection aspergillosis; filariasis, caused by roundworms; and myiasis, caused by fly maggots growing beneath the skin. Krone (2007) provides details of the protozoan endoparasites known to infest raptors. While the potentially harmful effects of internal parasites are obviously significant, Barton and Houston (2001), in a post-mortem study of 379 birds of six raptor species, including 23 Peregrines, found that only 20% of the specimens had one or more internal parasites.

The Pesticide Crisis

While starvation will lead to a falcon being emaciated, it is possible that there are underlying problems which contribute to ill-health and the bird's inability to hunt. One possible explanation would be a contaminant burden. But the journey to the discovery that the Peregrine population across most of its range was in decline as a result of contamination began in the most unexpected way. In the UK the building of dovecotes in which pigeons could breed, and then be used as a source of fresh meat, has a long history, going back to medieval times. Over the years of semi-domestication these birds (Rock Doves which, over time, had become an essentially feral sub-species) were bred for colour and also as homing birds which could be used to race against those of fellow breeders. The sport of pigeon racing evolved in the mid-19th century and by 1900 the fact that Peregrines were taking racers had become an issue for the owners. Throughout the early 20th century pigeon breeders were administering rough justice to known Peregrine eyries. Then, during the 1939-45 War around 200,000 pigeons were employed by the British military, the birds being used for duties which ranged from bringing back information from agents in occupied Europe (in an irony doubtless lost on the birds themselves, the pigeons – more than 16,000 in total – were often carried into enemy-held lands by parachute) to being used by the crews of downed aircraft to alert the authorities of their position. Thirty-two pigeons were awarded the Dickin Medal (the animal equivalent of the Victoria Cross, the UK's highest award for military personnel for bravery in the face of enemy action) as a result of activities during the war, seven related to downed aircraft, the remainder for flights with messages from enemy-occupied lands. In order to protect the pigeons an order was made for the killing of Peregrines and the

Two important historical documents in the work to discover the reason for the decline in the UK's Peregrine population. To the *left* is a photograph taken by, and annotated by, Derek Ratcliffe of a Peregrine egg clutch. The photograph was taken in 1947, two years before Ratcliffe realised that the falcons' eggs were breaking at an alarming rate. To the right is a sketch map prepared, and annotated by, Ratcliffe when he had been asked, in 1960, to review the status of the Peregrine in Great Britain. The map defined the probable distribution of Peregrines at that time based on Ratcliffe's work and that of his co-workers and correspondents. *With sincere thanks to the Ratcliffe family and Lesley Hindley, BTO.*

destruction of their nests, eggs and young. The order covered only certain counties (a rather odd collection), but the effect was to virtually eradicate the falcons from southern England and to deplete the population in other areas. Around 600 adult and juvenile falcons were killed in total with an unknown number of eggs and chicks destroyed. Following the end of the war, the government-sanctioned persecution ceased[3]. In 1946 protection was returned to Peregrines, and in 1954 the falcon was added to a list of rarer, specially protected species.

Perhaps the Peregrine persecution during 1939-45, then the special protection afforded the falcons, or perhaps better amalgamation by pigeon

---

[3] Ratcliffe (1993) tells an insightful story of the lighthouse keepers of Ailsa Craig, an island in the outer Firth of Clyde, Scotland, who persecuted Peregrines because they took the pigeons they used to carry messages to and from the mainland despite the fact that the rock was a vast seabird colony. The story is undated, but presumably dates from a time before the advent of radio.

breeders (or a combination of these events) meant that in the late 1950s the pigeon fanciers petitioned the government to reduce the number of Peregrines. The argument was that Peregrine numbers were increasing, and that the falcons were interfering with their sport. The pigeon fanciers therefore requested removal of legal protection for Peregrines based on perceived increased predation (Box 10.1). The government set up an enquiry into Peregrine numbers as a means of deciding what action might be necessary. The investigation was led by Derek Ratcliffe (1929-2005), a leading British environmentalist and ornithologist. Ratcliffe was the ideal choice as he had already gathered copious data on Peregrine breeding (Ratcliffe, 1958) and had identified that eggs were breaking in falcon eyries, his evidence stretching back to 1949. There were also inputs from others – Treleaven (1977) had noted and reported on severe declines in both eyrie occupation and breeding success in Cornwall from 1955. Treleaven also recounted to SW that he once watched a Peregrine fall out of the sky at a Cornish coastal eyrie in 1961. He subsequently located the bird and took it home. The falcon was in tremors and died overnight. The carcass was sent away for toxicology but was sadly 'lost' (much to Treleaven's chagrin) before any tests were completed.

Box 10.1 Peregrines and Pigeons
In 1960 pigeon fanciers from south Wales petitioned the UK Home Office (an arm of the UK government) for the removal of legal protection for the Peregrine, stating that large numbers of pigeons were being predated by an increasing population of Peregrines. It is therefore ironic that their complaint led to the discovery of the decline in Peregrine numbers due to pesticides. Based on the number of rings issued by the Royal Pigeon Racing Association, the British population of racing pigeons increased by about one-third between the mid-1970s and 1991 before declining. Dixon and Richards (2003) investigated the probable kill rates of domestic pigeons in south Wales, concluding that the success rate of adult Peregrines was 0.19 kills per hour, but that adults with dependent young increase this to 0.29 kills per hour. Given the fraction of domestic pigeons in the diet of Peregrines, this represents a large number of birds, though, of course, in terms of the overall losses of pigeons from other causes, it is much less significant. For the Peregrine, pigeons are a highly significant food resource, and a decline in popularity of the sport, coupled with a change in release points of racing pigeons, has been suggested as the main factor in the decline of Peregrines in central Wales (Dixon *et al.*, 2010). Dixon and co-workers examined the diet of Peregrines during the racing pigeon season (April–September) and outside the season (October–March) and found that the percentage of domestic pigeons fell from 31.4% in season to 5.5% out of season. While the fraction of most other species remained essentially constant, the fraction of starlings and thrushes rose to compensate for the lack of pigeons.

Starting in 1960 Ratcliffe gathered data on Peregrines and Peregrine territories and found that, far from increasing, the population was actually in steep decline. The irony of the story is that without the prompting from pigeon fanciers Ratcliffe's survey simply would not have been commissioned and the true extent of the severe decline would not have become so quickly apparent. A second irony in the story is that Ratcliffe was able to use contacts among egg-collectors, an enthusiasm which by that time had been outlawed, to access data on egg-shell thinning as this seemed the mostly likely cause of the breakages he had seen: their input was to prove pivotal. In the immediate post-war period, to encourage greater crop production a new chemical, DDT, had been used as a seed dressing and Ratcliffe was able to show that its use had coincided with the shell thinning that was causing egg breakages in Peregrine eyries[4].

The third irony is that while DDT is now the name most often associated with the catastrophic decline in Peregrine numbers (and an equal decline in other species) in the late 1940s/1950s, but it was actually later pesticides (such as aldrin, dieldrin and heptachlor – cyclodienes (chlorinated methylenes) which had quickly followed DDT onto the market) which were far more dangerous to the seed-eating birds on which the Peregrine preyed that aided the sharp decline in falcons numbers (though it is worth noting that DDT was largely responsible for egg-shelling thinning: the later pesticides killed adult birds). Evidence for a potential cause was in the large number of seed-eating birds which had been found dead during 1959–1960. The new pesticides were highly effective against the insects and fungi they were supposed to target, but were far more dangerous to birds and animals than DDT had been. Peregrines were accumulating pesticide toxins in fats and brain tissues and were therefore highly susceptible to small doses ingested by prey species: the falcons were a proxy for the underlying health of the ecological systems in which they foraged. DDT and the cyclodienes are environmentally stable and persist for long periods with a half-life of 2-15 years in terrestrial environments.

Analysis of unhatched eggs in Peregrine eyries showed that there were high levels of DDE (Dichloro-Diphenyl-Dichloro-Ethylene[5]) a residue of DDT produced by chemical breakdown of the original compound, and the newer pesticides. More importantly, it was found that the number of broken eggs in Peregrine nests was increasing. Ratcliffe (1970) noted that of 100 clutches examined in the period 1905–1950 only three had contained broken egg(s), while of 163 clutches examined during 1951–1966, 51 had broken egg(s), an increase from 3% to 31%. While it was clear that pesticides were killing adult birds and fatally contaminating some eggs, it also seemed that even viable eggs might be being lost as a consequence of changes to the eggshell

---

[4] DDT (Dichloro-Diphenyl-Trichloroethane), an organochlorine insecticide, was first synthesised in the late 19th century, but whose insecticidal properties were not discovered until 1939. The discoverer of that usage, Swiss chemist Paul Müller, was awarded the 1948 Nobel Prize for Medicine/Physiology.
[5] In British Peregrines a mean level of 13.6 parts per million (wet weight) DDE in eggs corresponded with eggshell thinning of 21% on average (Ratcliffe, 1993).

itself which promoted breakage. Ratcliffe therefore needed a measure of the shell thickness, as, clearly, thinner shells were more vulnerable to breakage. He developed the equation (Ratcliffe, 1993):

$$\text{eggshell thickness index} = \frac{\text{weight of shell (mg)}}{\text{shell length (mm) x shell breadth (mm)}}$$

which is now usually known as the Ratcliffe index.

In a report on pesticide residue (and mercury) in the eggs of British Peregrines, Newton *et al.* (1989) set down the change of the Ratcliffe index for falcons breeding in inland southern Scotland during 1961 and 1986, and the consequential change in the mean number of Peregrine fledglings over the period 1961-1979 (Figs. 10. 3a and b).

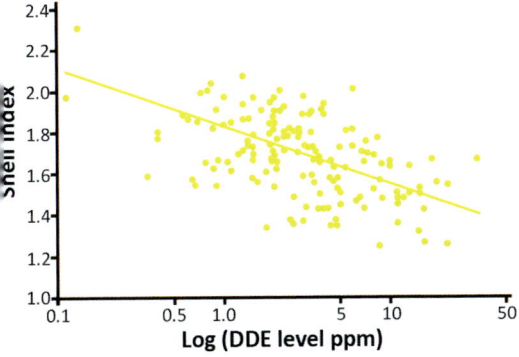
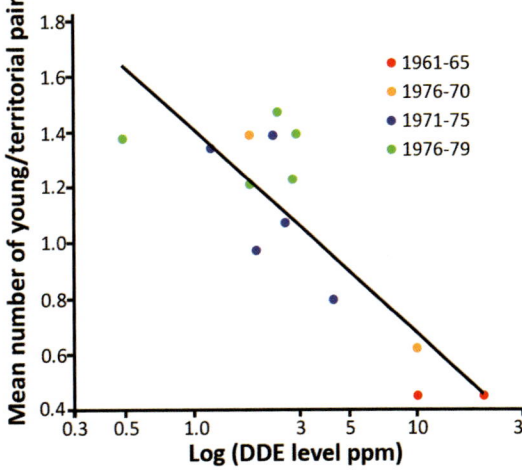

Figure 10.3a The relationship between the (Ratcliffe) shell index and the content of the DDT residue DDE in Peregrine eggs from inland sites in southern Scotland during 1961-1986. Newton *et al.* (1989) set up a correlation matrix relating the egg concentrations of DDE, PCB and the Dieldrin residue HEOD (Sterically-Hindered Epoxides Dieldrin). The researchers concluded that only DDE was significantly related to shell index. The line on the graph is the best fit regression. Redrawn from Newton *et al.* (1989).

Figure 10.3b Relationship between the geometric mean level of DDE residue in eggs across Britain and the mean number of Peregrine fledglings per territorial pair for various time periods 1961-1979. The line on the graph is the best fit regression. Redrawn from Newton *et al.* (1989).

Use of the index allowed the shell thickness of museum specimens to be examined without damaging the eggs, so that changes over time could be established and the possibility of pesticide-induced thinning identified. Use of the Ratcliffe index is now widespread and has been subject to testing by other researchers. Weighing eggs in the field, coupled with knowledge of the mean and range of embryo weights, allows the index to be calculated without the need to sample a clutch. However, there have been modifications. Nygård (1999) noted that variation in the shape of eggs of different species meant that a correction to the standard Ratcliffe index was required in some cases. Nygård noted, for instance, that cormorants had long elliptical eggs which therefore had high eccentricity (defined in terms of egg length and breadth), while owls had short eggs with low eccentricity. Nygård also noted that the size of the blow hole in museum eggs affected the Ratcliffe index, and therefore proposed a correction based on species egg eccentricity and blow-hole size. For 585 eggs of six European raptor species the correction was small (about 3%), but it could be larger for other species.

For a comprehensive review of Ratcliffe's work and the pesticide contamination story, see Ratcliffe (1970) or Ratcliffe (1993). As the authors of this book are British, and the first reports of the potential crisis came from Britain it was inevitable that the work of Derek Ratcliffe would form the basis of a discussion on it. But researchers in the northern hemisphere rapidly began confirmation studies of their own and to define the extent of the problem in their own countries. For an understanding of the position of other field workers in the northern hemisphere see Hickey (1969). Particularly salient points are also covered in more detail in the paragraphs below.

Similar egg breakages and adult deaths were being seen in the USA where Rachel Carson's book *Silent Spring*[6] brought the environmental damage being wrought by modern pesticides to the fore. Carson's book title captured the problem of the rising toll on avian populations in such a poetic way that it also captured the public imagination which aided her campaign against the use of DDT and other new generation pesticides. As already noted above, and has been often noted elsewhere, it was DDT, a simple acronym, that took hold as a rallying cry (and was the chief culprit as far as shell thinning and consequent embryo death was concerned – see Figs. 10.3a and b) despite the fact that it was later, more potent pesticides which did most of the damage (Fig. 10.4). For a comprehensive review of the relative effects (and, therefore, harms) of DDT and the cyclodiene pesticides see Nisbet (1988).

In North America there were few systematic surveys of Peregrines prior to the 1960s. Hickey and Anderson (1968) investigated eggshell thicknesses in museums (having been alerted to the problems in the UK), seeing a decline of up to 26% in shell weight in eastern USA, compared with only 1-2% in California. The two researchers also noted catastrophic declines in the populations of Golden Eagles, Red-tailed Hawks and Great Horned Owls. But only when Fyfe *et al.* (1976) made the first effectively full-scale survey (one which included Greenland) was it found that to the east of the Rocky Mountains and south of the Boreal Forest (*i.e.* eastern USA and southern Canada) the effect of post-war pesticide use to increase grain yields was an

---

[6] Rachel Carson (1907-1964) was an American marine biologist who became an environmentalist and author. She had become concerned over the effect of recently-developed pesticides on the natural world, and specifically the deaths of birds which were the basis of her book *Silent Spring* which was published in late 1962.

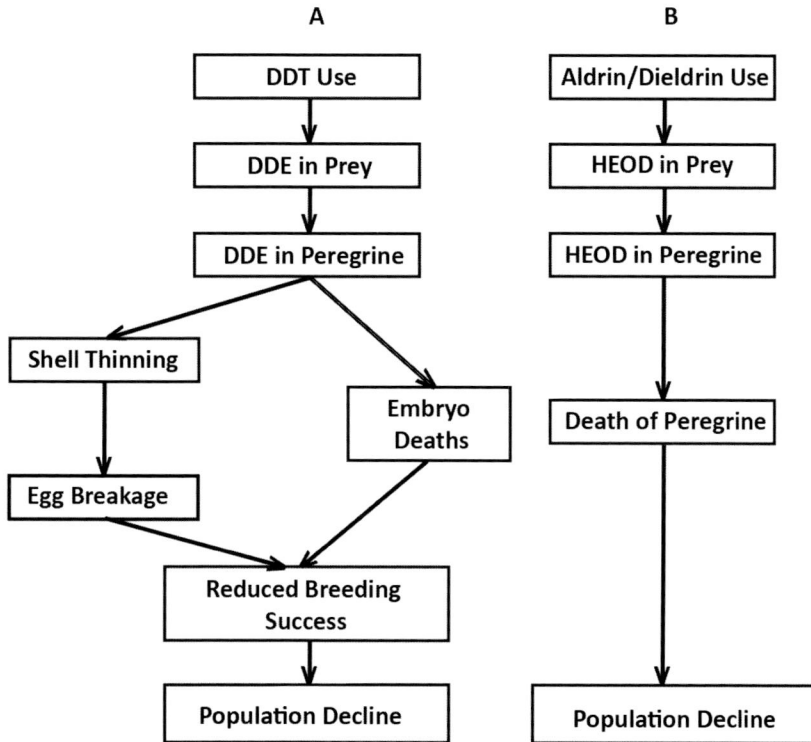

Figure 10.4 The effect of the organochlorine pesticides on bird populations. DDT was used as a pesticide from the late 1940s, the cyclodienes from about 1955. DDT was responsible for eggshell thinning in Peregrines and the name is often associated with the population crisis of the post-War years, in part because of the public success of Rachel Carson's book *Silent Spring*, but while the pesticide did contribute to population declines, it was the adult-killing cyclodienes that hastened the decline. Redrawn from Nisbet (1988) – see, also, Newton (2013).

effective extirpation of Peregrines as a breeding species: under the provisions of the US Endangered Species Act of 1973 the falcon had become an official endangered species[7]. Because of the limited understanding of the Peregrine populations of Alaska, northern Canada and Greenland, the position of *F.p. tundrius*, *F.p. pealei*, and *F.p. anatum* west of the Rockies and in northern Canada was unclear in as much as populations seemed reasonable, but there were few data to judge against. Later surveys were to identify that these populations also showed evidence of shell thinning and pesticide residues in eggs and adult birds. Studying *F.p. tundrius* at Rankin Inlet, Canada,

---

[7] It was this discovery that led to the decision to repopulate the area with a mix of non-American sub-species, which has already been touched-upon in Chapter 2. Ironically, sampling of unhatched eggs in North Carolina (Augspurger and Boynton, 1998) indicated that reintroduced Peregrines (largely from the Peregrine Fund Inc. and private breeders), showed residues of both organochlorines and Dieldrin, as well as mercury. No habitat, it seemed was safe. Similar dramatic population declines were also seen in Scandinavia and eastern Europe, but there, reintroductions were of nominate falcons which maintained the original sub-species map (see Chapter 2, specifically Footnote 2).

Court *et al.* (1990) found DDE levels marginally below the level deemed critical, but still saw egg breakages in 10% of clutches. In general, however, the residual quantities (though some were above levels deemed critical and showing a marked reluctance to reduce across the years of study) had not led to significant declines in populations – see, for instance, Johnstone *et al.* (1996) for *F.p. tundrius* in Arctic Canada; Schick *et al.* (1987) who collected data on the contamination of shorebirds in coastal Washington State and considered the uptake of *F.p. pealei* and *F.p anatum* along the US west coast from Washington to California. Equally encouraging was the finding of Falk *et al.* (2006) that the eggshell thickness of *F.p. tundrius* in both south Greenland (Narsarsuaq, Narsaq and Qaqortoq) and west Greenland (Kangerlussuaq, Manitsoq and Sisimiut) had increased over the period 1972-2003. Thinning had been measured at 13.9% in 1972, but was 7.8% in 2003, indicating an increase of 0.19%/year. In a second report (Vorkamp *et al.*, 2014) the researchers found continuing declines in pesticide residues in Peregrine eggs in the same area.

Similar population declines were also seen in Europe. Barnaby (1975) noted that in Sweden the Peregrine was in danger of becoming extinct, having declined from 350 breeding pairs in 1945 to 35 in 1965, 9 in 1974 and 6 in 1975. In Norway Nygård (1983) noted eggshell thinning of 19-23%, DDT and Dieldrin residues in eggs, and a Peregrine population that was a 'mere fraction' of its former size. In Finland, Salminen and Wikman (1977) noted a similar decline in the Finnish population. Reintroductions, chiefly from Finland, Scotland and northern Sweden have resulted in a different genetic make-up from the past stock, but in this case all the reintroced birds were *F.p. peregrinus* (see Jacobsen *et al.*, 2008). Similarly, reintroductions in Poland after the DDT crisis were from western European nominate stock (Puchała *et al.*, 2021). Wegner *et al.* (2005) studied pesticides and mercury in German Peregrines over the period 1955-2002 and noted the collapse of the population to 1975, followed by an encouraging rise. In their conference report, the European Peregrine Falcon Working Group (Mizera and Sielecki, 2009) the decline of Peregrines across Europe is enumerated.

In 1994 Henny *et al.* (2000) found low concentrations of DDE (and no traces of PCBs) in the blood of *F.p. calidus* trapped in the central Kola Peninsula. A similar result was found in *F.p. calidus* sampled on the Taimyr Peninsula (Quinn *et al.*, 2000). However, Henny *et al.* (1998) collected eggs (one per clutch) from several raptor species in areas of southern Russia. In general, the areas were heavily farmed, principally for wheat, other grains and vegetables. No Peregrine nests were found in three of the four study areas. The only egg the researchers found was an addled one in a nest near Novogorod, the furthest north site (about 200km south-east of St Petersburg): a fledgling Peregrine was seen in the area. The addled egg was heavily contaminated with DDE. The results of the study led Henny *et al.* to express deep concerns for the Peregrine population of southern Russia.

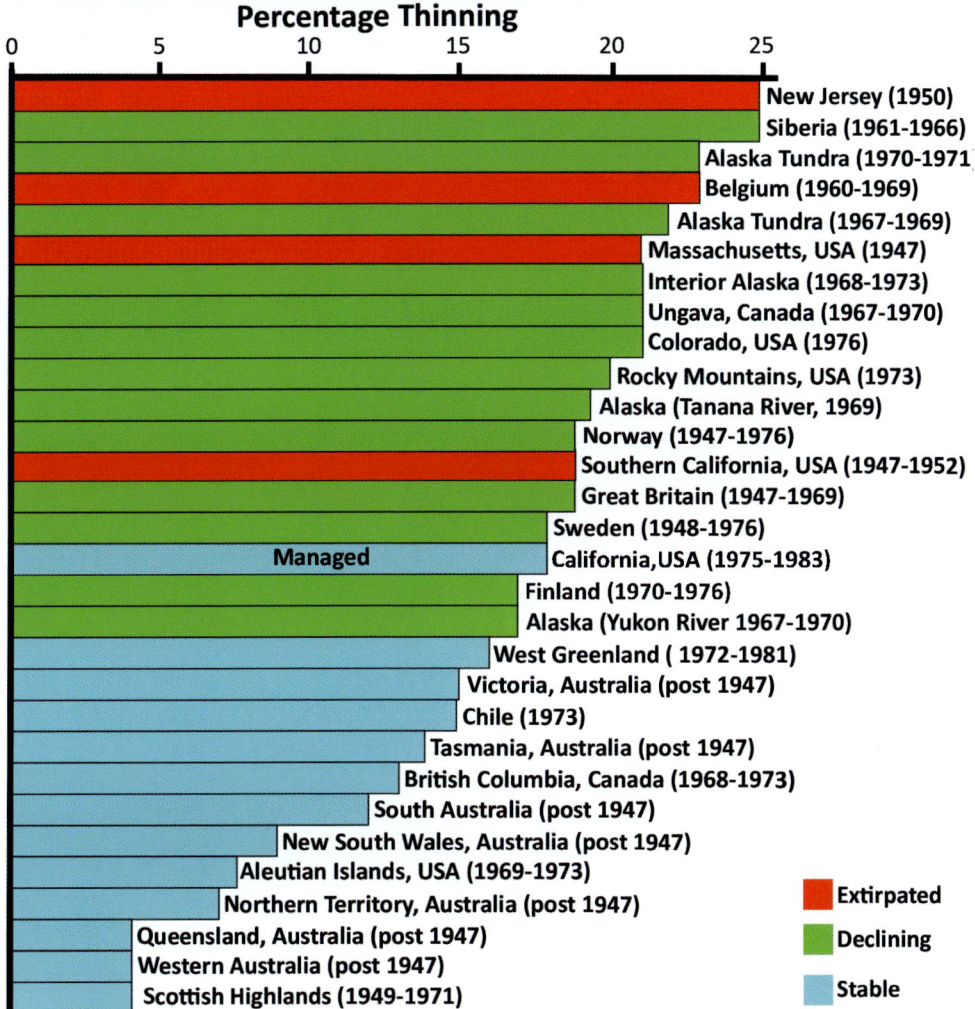

Figure 10.5 Percentage of eggshell thinning in Peregrines across the world. For references to each bar of the histogram see Peakall and Kiff (1988). The population in California, USA was 'managed' by removing contaminated eggs from nests and incubating them in the laboratory so that embryos were not lost due to within-in nest breakages. The hatched chicks were raised in the laboratory and hacked back to the wild to reinforce the population. Redrawn from Peakall and Kiff (1988).

Hartley *et al.* (1995) found eggshell thinning in *F.p. minor* eggs in Zimbabwe, as did Curtis and Jenkins (2002) in two widely separated breeding sites in South Africa. However, the use of DDT to control tsetse flies and, therefore, the prevalence of the potentially-fatal sleeping sickness (trypanosomiasis) in the human population was one of the main bones of contention in the early days of usage, *i.e.* human welfare .v. avian welfare. The fact that the value of DDT as an insecticide waned with time aided the phasing out of its usage[8].

In all countries where the early pesticides were used to improve harvests the local Peregrine populations went into serious decline (Fig. 10.5).

This figure is also reproduced in Ian Newton's excellent 'anniversary' review article (Newton, 2013) on pesticides and birds.

---

[8] The value of DDT in terms of human life is incalculable, but in a study by Hilbers *et al.* (2018) it was calculated that the cost of reintroducing the Peregrine Falcon (*F.p. anatum*) in California to a minimum viable population level after its decimation by DDT was $3,023,000.

Above   *F.p. tundrius* eggs broken as a result of shell thinning, Rankin Inlet, Canada. *Gordon Court.*

Below   *F.p. pealei*, north-west USA. Although the sub-species in remote, wild areas did not suffer from pesticides as badly as those in cultivated areas, they were not completely immune. *Nick Dunlop.*

Other Contaminants
Even though the levels of organochlorines and cyclodienes in Peregrines fell significantly after the banning of certain pesticides in the 1960s, the persistence of the contaminants means that residues continue to be found in the species. Newton *et al.* (1989) considered the levels were no longer causing shell-thickness reductions, and consequent reductions in brood sizes, but found levels of mercury (partly naturally occurring, but also derived from industrial sources) which may have been responsible for reduced brood sizes. Mercury is a persistent contaminant globally. Its organic form is highly toxic to living organisms and is known to impact humans and wildlife. In some coastal Peregrine populations, there is evidence of bioaccumulation within individuals and between age classes. Mercury methylates in water, this process increasing its toxicity at the aquatic food web entry point.

Mercury became a contaminant of note when many inhabitants of Minimata, a city of south-west Kyushu island, Japan, showed severe neurological problems, many of which were fatal, after eating fish and shellfish contaminated by methylmercury run-off from a local industrial chemical plant: the toxic effects became known as Minimata Disease. Mercury exists naturally in the environment, but can also be introduced by industrial processes (not only chemical-based industries but plants such as coal-fired power stations) and some agricultural products (*e.g.* alkyl-mercury pesticides). Once mercury has found its way into the environment it can be readily converted to methylmercury by bacteria, particularly in anaerobic conditions, *e.g.* wetlands, rivers and seas. Raptors can acquire mercury in their livers (and in the eggs females then lay) because of their position as apex predators. Newton *et al.* (1989) noted the metal in 550 addled eggs of nominate Peregrines collected from across the UK during the period 1963-1986, with some evidence that higher concentration reduced brood size in the falcons. However, Walker *et al.* (2016), studying (in the UK) mercury levels in the liver, kidneys and brain of the Eurasian Sparrowhawk (an avian predator), the liver of the Common Kestrel (primarily a rodent feeder) and the eggs of the Golden Eagle (which in the UK takes both avian and mammalian prey) showed concentrations that had remained steady over the period 1990-2014.

A study in Nevada, USA in which the mercury contamination of Peregrines (*F.p. anatum*) was assessed from feathers indicated a much wider range of mercury levels, and also some significantly higher values (Barnes and Gerstenberger, 2015). The study also showed that the higher levels were associated with the falcons taking aquatic avian prey, the mercury levels in a large lake being high, presumably as a result of agricultural organomercury run-off, causing consequent high levels in waterfowl and waders. Barnes *et al.* (2018) also found higher mercury levels in Peregrines (*F.p. pealei*) at coastal sites in Washington State where the Peregrines were likely feeding on seabirds. Barnes *et al.* noted that over their 15-year study period the level of mercury in hatch-year falcons had declined. However, rather different results came from the study of Arctic shorebirds at the Yukon Delta, Alaska where Perkins *et al.* (2016) noted that climate change was releasing mercury sequestered in thawing permafrost: the mercury was accumulating in Artic breeding shorebirds. Since these birds contribute to the dietary spectrum of Arctic Peregrines, the falcons will inevitably accumulate higher levels of mercury.

In the latest of a series of reports on studies into environmental pollutants in the eggs of birds of prey in Norway – a work programme which became part of the Program for Terrestrisk naturovervåking (TOV) (Programme for Terrestrial Monitoring) in 1992 – Nygård and Polder (2012) used data gathered over the previous 40–50 years to produce long-term trends in pollutants. The study showed that the levels of legacy pollutants continue to decline, and that the majority of eggs show concentrations below the believed critical levels, but that levels of PCBs (PolyChlorinated Biphenyls[9]) have stabilised in some species. In the eggs of Peregrines, Merlins and White-tailed Eagles the by-products of DDT, particularly DDE, and the persistent organic compounds PCB and HCB (HexaChloroBenzene), are still the dominant pollutants. Fig. 10.6 a-d shows the variation of some of these persistent contaminants, and also of mercury in the eggs of Peregrines over the period 1975–2010. In this study Peregrines, Merlins and White-tailed Eagles also have the highest levels of mercury.

Later work in Norway (Nygård *et al.*, 2019) showed that the levels of DDE and mercury had remained constant during the period 2010-2016, but that the level of PCB had risen. No HCB was found in any sampled egg. However, the number of eggs sampled was small, so the results must be treated with caution.

More worrying, as it took many years for the effect of organochlorine contamination to be understood, both brominated flame retardants (PBDE – PolyBrominated Diphenyl Ethers) and heat resistant coatings PFAS – PolyFluoroAlkyl Substances – are now being seen in raptor eggs – see, for instance, Chen *et al.* (2008) and Park *et al.* (2009) regarding the flame retardants. Chen *et al.* (2008) studied the contaminants in Peregrines in north-eastern USA. Park *et al.* (2009) studied Peregrine eggs from California and found that PBDE levels had more than tripled each decade, whereas PCB levels had seen no significant change. In general levels are low, but little is known about the biological effects of these compounds, particularly in the longer term. Nygård and Polder (2012) measured the value of these compounds in Peregrine eggs in Norway (Fig. 10.7 a and b overleaf). UK production of, and import of, PBDEs ceased in 1996.

Opposite **Figure 10.6** Variation of levels of contamination in Peregrine eggs in Norway for (a) DDE (to be absolutely accurate p', p' DDE); (b) PCB; (c) HCB; (d) Mercury. Redrawn from Nygård and Polder (2012).

---

[9] PCBs are chlorinated hydrocarbons and have a range of toxicities. Due to their non-flammability, chemical stability, high boiling point and electrical insulating properties, PCBs were used in hundreds of industrial and commercial applications. They were banned in UK in 1981.

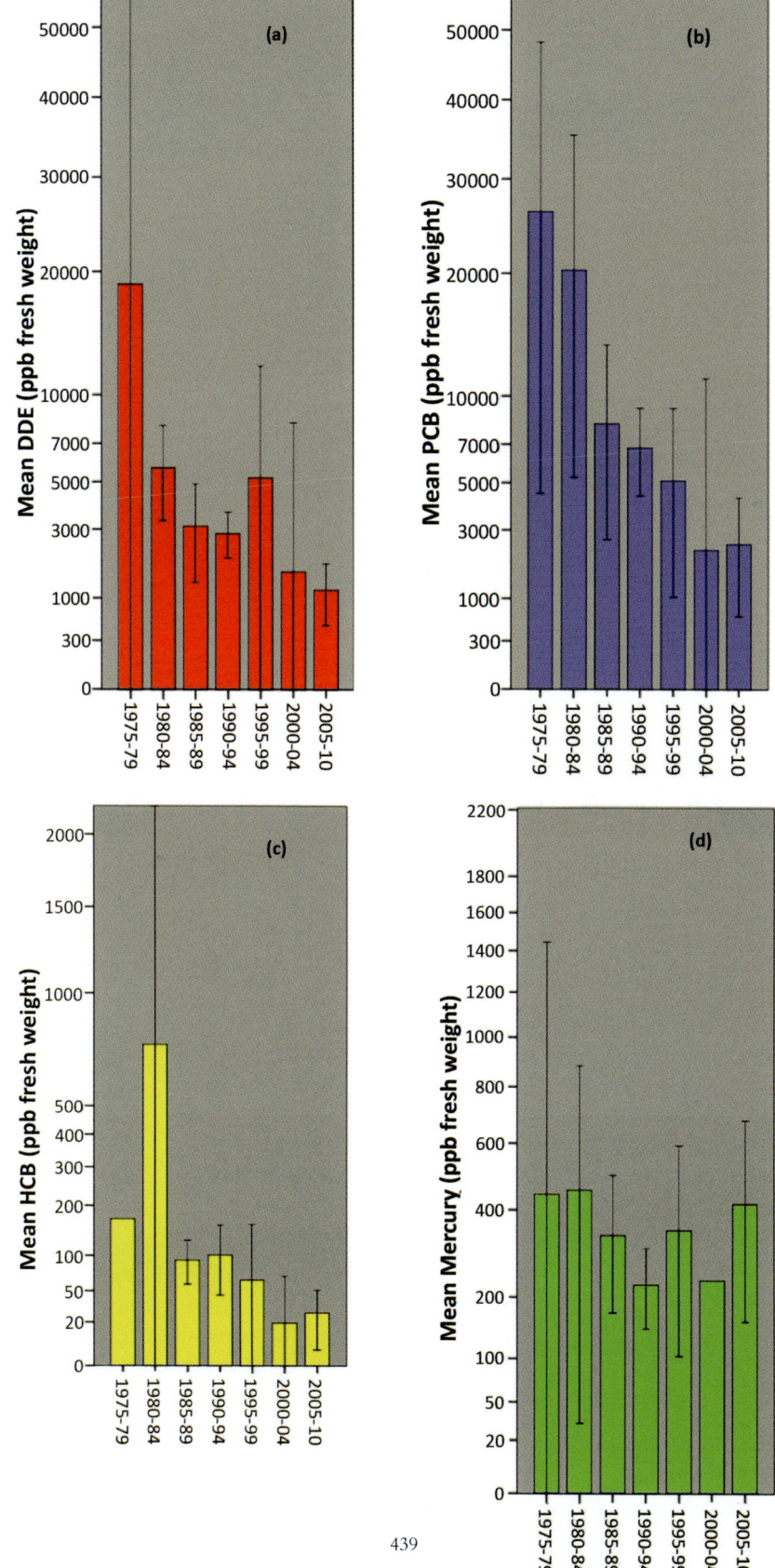

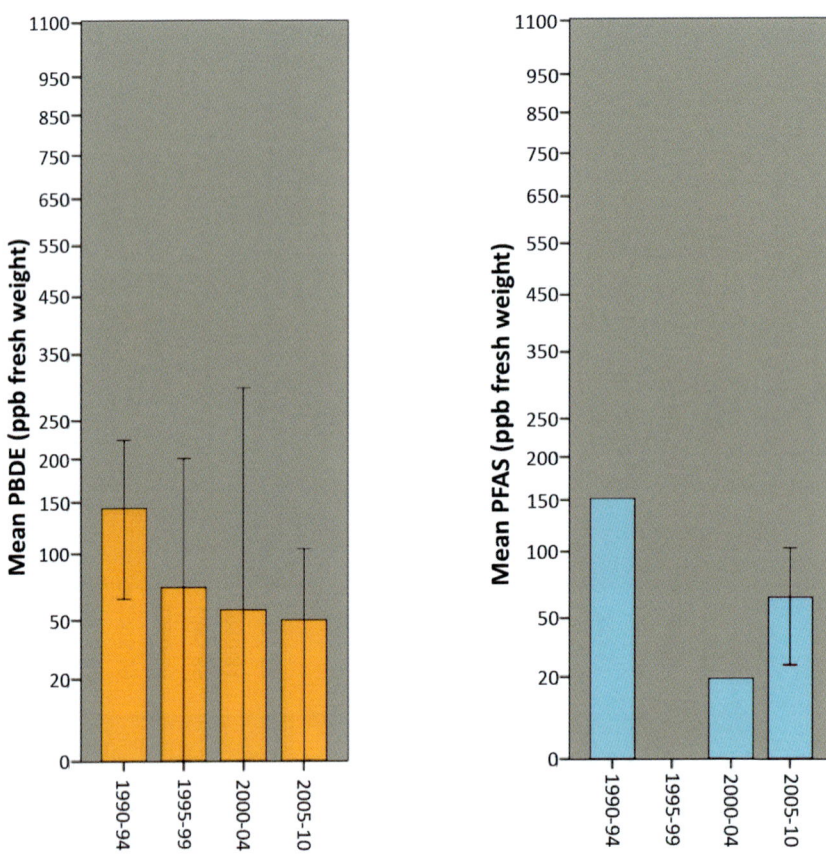

Figure 10.7 Variation of levels of contamination in Peregrine eggs in Norway for (*left*) PBDE and (*right*) PFAS. Redrawn from Nygård and Polder (2012).

Similar concerns have also recently been raised over the use of neonicotinoids, which proved very effective when introduced, but were soon subject to debate and, ultimately, bans because of their effect on bees and insectivorous birds (which may then be eaten by Peregrines). The impact of diclofenac on the Indian vulture population is yet another example of the negative effects that can occur from organic compounds developed with good intentions. Doubtless the future will see new ranges of pesticides and further unwanted side-effects for wildlife[10], and, perhaps, Peregrines will be one of the species which suffers – see, for instance, González-Rubio (2020) for a review of CECs (Contaminants of Emerging Concern) in relation to raptors. Recently a novel neurotoxin produced by cyanobacteria has been discovered in the USA which has caused the deaths of many Bald Eagles, as well as fish and other aquatic

---

[10] A particular problem identified for many of the new pesticides, as was the case for neonicotinoids, is the claim that much less pesticide, in terms of weight applied to a given area of crop/ground, was required. But as Schulz *et al.* (2021) note, for many new products this apparently useful assessment is negated by a leap in the product's toxicity.

lifeforms (Breinlinger *et al.*, 2021). It is believed that the cyanobacteria are producing the toxin from bromide-related herbicides which, if found to be the case, is another pathway from chemicals created with good intentions to potential extremely serious avian death rates.

Human Activities
The increase in the human population affects the Peregrine population in many ways. The growth of human cities and the development of previously untouched land for industry can disrupt migration routes. One very good example of this is the development of the shoreline of China's Yellow Sea where 2,000,000 birds feed and moult during migration. The coastal areas where the birds can now congregate having been significantly reduced in recent years. Doherty *et al.* (2021) analysed the results of 208 studies of 167 species and conclude that disturbance and habitat modification can alter migration pathways, this leading to negative impacts on individual fitness and survival, and, ultimately, to population viability. In a more comprehensive study, Zhang *et al.* (2022) used 'big data and machine learning' to study the populations of all 10,964 bird species. The results were not encouraging, the fragmentation of population due to human activities causing a decline in many species, particularly those in South America and south-east Asia. Most worrying, of 801 species where the population trend is currently unknown, the analysis suggested that 47% would be in decline.

In country areas where farming has become intensive interactions between humans and the falcons increase and there is little evidence to suggest that outside of urban areas Peregrines habituate to human presence (see, for instance, Pauli *et al.* (2017) for general comments on raptor tolerance of human disturbance). As well as increases in human habitation and cultivated areas, human presence in 'wild' areas is increasing as outdoor recreation becomes more popular. Monti *et al.* (2018) studied the impact of eco-tourism in a marine protected area in Corsica and found that the increase in boat and other tourist traffic had caused a drop in the breeding success of the local Osprey population, despite the increase in fish stocks within the protected area. Females left their nests more often and the levels of stress hormones in chick feathers increased. There is no reason to suppose that Peregrines will not suffer similar levels of anxiety if human traffic increases in their breeding areas[11]. Farming also brings its own hazards, with increased risks of collisions with farm traffic and buildings. There are even deaths caused by the fences erected to contain cattle herds (van der Leek *et al.*, 2019). Collision threats are heightened in urban areas with window strikes, vehicle collisions and risks of electrocution (see, for instance, Hager, 2009). The increase in high voltage power lines in areas where natural perches are sparse has led to increases in raptor deaths because of poor power line design. The growth of wind farms, a necessity in an effort to combat global warming, are a known hazard, both for

---

[11] See, also, the comment on the effect of rock climbers on Peregrine breeding success in Chapter 9 *Foes*.

soaring birds, such as White-tailed Eagles, and for hovering Kestrels which, eyes glued to the ground beneath, can be chopped in half by fast-moving blades. In a study across Europe and North Africa Gauld *et al.* (2022) note that the positioning of wind turbines inevitably conflicts with the chosen migration paths of birds enhancing the collision risk for many species. One other potential problem is the growth in electromagnetic interference as the number of radio masts for mobile phones *etc.* proliferate. In Spain, Balmori (2009) noted that microwave and radiofrequency pollution was having an effect on both White Storks (*Ciconia ciconia*) and House Sparrows. For the storks, as the electromagnetic field strength increased, the average number of chicks in nests within 100m of a phone mast decreased. For sparrows, the number of birds per hectare declined in the vicinity of phone masts. While Balmori's work does not mention Peregrines, it is possible that urban electromagnetic noise may, in the longer term, limit or reduce the falcon population in cities, either directly, or by affecting prey density.

Humans also affect Peregrines in more direct ways. Below we consider the deliberate persecution of Peregrines, but hunting, which is both necessity and sport in many areas throughout the Peregrine's range, is not only a direct cause of death, but an indirect one – fatalities as a by-product. Falcons and other raptors which feed on the bodies of waterfowl shot and killed, but not retrieved, or capture waterfowl shot, but only wounded, accumulate lead and die as a result of lead poisoning or because of its deleterious effect on the body. Lead poisoning in these circumstances has been known since early in the 20th century when an American ornithologist (Wetmore, 1919) wrote a paper which set down the evidence for the ingestion of spent shot by waterfowl. In 1999 the African-Eurasian Waterbird Agreement introduced a ban on lead shot for hunting shorebirds, ducks and geese. Although the agreement was binding across the UK, it is clear that the use of now-illegal lead shot for the shooting of ducks and geese continues (Stroud *et al.*, 2021), while attempts to outlaw the use of lead shot for the shooting of other species, such as pheasants, which might also lead to the later poisoning of raptors if wounded birds are not dispatched or carcasses not retrieved, have had little effect so far (see Pain *et al.*, 2020, Green *et al.*, 2021 and Green *et al.*, 2022). The problem of the continued use of lead shot is not confined to Britain. Andreotti *et al.* (2018) report the finding of seven lead shot balls in the digestive tract of a female Peregrine found dead on the Po Plain of northern Italy. While the shot was not the cause of death, it was probably ingested from a pigeon which, together with a Eurasian Starling, constituted the last meal of the falcon. The hunting of either prey species is illegal, but occurs extensively in northern Italy. Descalzo *et al.* (2021) studied lead poisoning in Spanish raptors. Monclús *et al.* (2020) looked at the lead poisoning of raptors across Europe, while Garvin *et al.* (2020) provide evidence for raptors across the world.

Finally, for an engrossing article on the position of raptors in general in the human world, see Donázar *et al.* (2016).

F.p. anatum downy fledgling, San Pedro, USA. *Gabriel Gruia.*

Human Persecution of Peregrines in the UK

The shooting of 'game' birds, waterfowl, pigeons (and some other species such as larger waders) for recreation, food and to prevent crop damage is seen in many areas where Peregrines breed, although it is most intensive in North America and Europe. Recreational shooting tends to be on rough land (*i.e.* land where the species are seen, but not directly cultivated for the shooters) but in some European countries, and particularly in Britain, land is managed for game birds and overstocked with them (*i.e.* the population is far higher than would exist without man management) for the benefit of the shooters (more targets) and the land owner (more income from satisfied shooters). This section of the book deals specifically with the position in Britain because the nature of 19th century industrialisation coupled with the feudal system in medieval times and its carry over into modern times has created an egregious, uniquely British, position.

In the UK the Inclosure Act of 1773 permitted landowners to enclose land which had formerly either been a shared resource – an open field system with parcels of land occupied by peasant farmers, sometimes working for their feudal lord as well as themselves – or common (*i.e.* communal) land used for grazing. Although in principle enclosure had to be agreed between all relevant parties before parliament could be petitioned and royal assent granted, unscrupulous landowners frequently avoided informing or seeking the opinion of those affected. The Act was later amended to demand that the petition be mounted on the local church door for a period to allow locals to see what was proposed and to object if they wished, but by then much of Britain's commons had been enclosed. One effect of the enclosures was to allow formerly common land to become available for private shooting. With shooting, even if it was not initially on a commercial basis, came gamekeeping, the job of the keeper being to maintain a sufficient level of game for the enjoyment of the shooters. Inevitably gamekeepers saw any species that preyed on game as an enemy, and the large-scale shooting, trapping and poisoning of raptors (and of mammalian predators), followed. On some shooting estates hundreds of raptors were killed annually. Ratcliffe (1993) quotes the inventory from the Glengarry estate in the county of Inverness which noted the killing of 1,799 raptors, including 98 Peregrines, during the four years between 1837 and 1840. Ratcliffe also notes that on the Isle of Arran the bounty for Peregrines was 2s.6d. (two shillings and sixpence: 12p in today's decimal currency) for an individual bird and 10s.6d. (52p) for a nest. In each case the amounts represented considerable sums to the locals, and it is no surprise that much effort went into the killings. Neither was the slaughter confined to those raptors which took game birds, even smaller raptors, Eurasian Sparrowhawks, Merlin *etc.*, being seen as potential threats, if not to adult game birds, then to their chicks. And, of course, a gin-trap placed on a post suitable as a perch catches whatever species is unlucky enough to land on it. For the Common Kestrel, a rodent feeder, which could not take adult game birds and rarely took chicks, indiscriminate killing seems particularly dreadful. But in its case the problem was that the rearing of Pheasants for estate shooting involved the use of surrogate hen chickens which were cooped while the young Pheasants they were helping raise ran free in wire enclosures. Such chicks were extremely vulnerable to Kestrels which were, therefore, seen to be as much of a threat as the larger raptors[12].

The most comprehensive account of the persecution of raptors, and other species deemed a risk to game birds, has been provided by McMillan (2011), who analysed data on record cards kept by the gamekeepers of the nine beats

---

[12] In their assessment of the costs and benefits of game bird management for non-game species Mustin *et al.* (2018), analysing the results from 35 European studies (though mostly from the UK), found that 63% of significant effects of game bird management were positive for non-raptor avian species, but that this percentage was dominated by agricultural habitats. In non-agricultural habitats 65% of effects were negative. Overall, legal predator control was always positive or benign, but illegal predator control (including that of Peregrines and other avian raptors) was always negative.

which covered the 56,650ha (140,000 acre) Atholl estate close to Pitlochry, Scotland. McMillan's analysis covered two periods, 1867/68 to 1910/11, when the records were collated from the returns of the nine beats, and from 1915/16 onwards, when, following the 1914–1918 war, individual records for the nine beats were available. In the first period, covering 44 years, the records show that 659,975 Red Grouse were killed (an annual mean of 14,999), together with 19,972 (annual mean 454) Black Grouse; 4,886 (annual mean 110) Capercaillie (*Tetrao urogallus*); 2,827 (annual mean 64) Ptarmigan; and 78,067 (annual mean 1,774) Pheasant. During the same period the estate's gamekeepers killed 13,272 'crows' (an annual mean of 302); 11,428 (annual mean 260) 'hawks'; 3,731 (annual mean 85) 'owls'; 1,434 (annual mean 33) Ravens; and 777 (annual mean 18) Magpies. The 'crows' were probably Hooded Crows rather than Carrion Crows. The 'hawks' are likely to include all the raptors which could have been seen on the estate – Buzzard, Golden Eagle, Hen Harrier, Common Kestrel, Merlin, Peregrine and Eurasian Sparrowhawk – which were recorded throughout the period – and, potentially, Goshawk, Osprey, Red Kite and White-tailed Eagle, which became extinct during the period. The 'owls' were likely to be Barn Owl, Long-eared Owl, Short-eared Owl and Tawny Owl. The killing of avian 'vermin' would, of course, have been accompanied by the killing of terrestrial predators (see, also, Box. 10.2 on p448). Not until 1880 was an Act of Parliament passed to protect birds from indiscriminate slaughter, but of the 86 species listed (some of which were duplicated because of the use of local names) none were raptors. Not until 1896 were raptors protected (though even then the Red Kite was excluded from the schedule), at least during the closed (breeding) season, and only then in certain areas. The effect was to allow continued persecution at other times, and the closed season was largely ignored in any case. The data presented by McMillan suggest a more or less uniform annual toll for all 'vermin' types apart from Ravens, where the numbers killed fell sharply after 1896/97, and for Magpies, where the numbers fell after 1903/04 and were zero after 1909/10: while Raven numbers have recovered somewhat, Magpies were driven virtually to extinction and are still essentially absent from the area.

It would appear, therefore, that even the first attempts at legislation against the wholesale slaughter of raptors (and owls) were ignored. Only with the passing of the Protection of Birds Act 1954 did the shooting of any raptor seen on a game bird estate become significantly less than routine, though the data which McMillan presents suggest that as far as the Atholl estate was concerned the new, stronger legislation made little difference to the nine gamekeepers: individual beat records suggesting that on some the killing continued more or less unabated until 1988/89, when someone noticed that the records were publicising the slaughter of protected species and the forms were altered to avoid that unfortunate occurrence. However, McMillan notes that the records for the individual beats show that some gamekeepers were not killing raptors and were teaching their successors to behave in the

Nominate Peregrine, Isle of Man, UK. *Peter Christian.*

same way, although others were continuing the slaughter. The fact that the gamekeepers acted entirely as they saw fit suggests that neither the landowner nor the estate manager was responsible for encouraging the law to be flouted, but the fact that there were individual differences strongly suggests there was no guidance that they should obey it either. Apparently even after the police became involved and there were meetings between the Royal Society for the Protection of Birds (RSPB) and the estate, the situation continued much as before. Most telling in McMillan's data is the indication that during the years 1980/81 to 1987/88 (when recording ceased), only in one year was the return from a single estate beat (one of nine, it must be remembered) for the number of 'hawks' killed lower than the number reported illegally killed for Scotland as a whole. Overall, over the eight-year period, the return from that single beat (controlled by a single gamekeeper) suggested twice as many hawks and owls were killed as were recorded by the RSPB for the whole of Scotland. In a study by Amar *et al.* (2012), data from 1,081 nests across northern England between 1980 and 2006 were analysed against areas of grouse shooting. Though clutch and brood sizes were the same for grouse managed areas and non-grouse areas, suggesting that neither cohort was food-resource limited, the productivity of the falcons on actively managed grouse moors was 50% lower. Analysis of wildlife crime data indicated that persecution of nominate Peregrines was more frequent on the actively managed grouse moors. Ironically the work of Redpath and Thirgood (1999) suggested that the predation of grouse chicks by Hen Harriers was positively correlated with grouse density, while Peregrine predation was inversely correlated with grouse density.

Murgatroyd *et al.* (2019) studied the loss of Hen Harriers in northern Britain. Of 58 raptors fitted with satellite tags, four were certainly killed illegally, while another 38 tags stopped working suddenly without any prior suggestion of malfunction. Suspicions were raised regarding the fate of these birds because deaths were disproportionally high for birds of that age, and final positions of the missing birds were mostly on grouse moors, particularly on managed moors: the harriers were seven-times more likely to disappear over land that was predominantly grouse moor than land that was not so managed (Fig. 10.8).

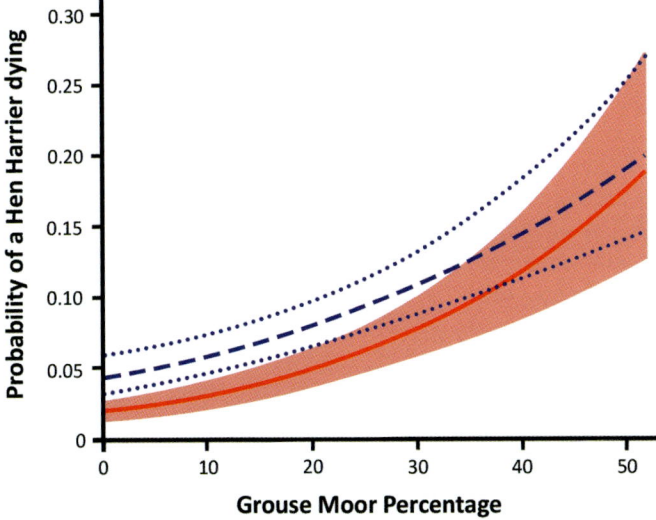

Figure 10.8 Probability of a Hen Harrier dying in relation to grouse moor percentage. The percentage of grouse moor was calculated from a 20km x 20km grid on the UK mainland and Isle of Man. The red line is the probability of dying of an individual harrier, with the pink-shaded area representing the confidence interval. The blue dashed line is for harriers known to have been illegally killed and harriers with satellite tags which stopped abruptly with no prior warning of a problem. The dotted blue lines are confidence limits. See Murgatroyd *et al.* (2019) for the definition of the confidence limits. Redrawn from Murgatroyd *et al.* (2019).

Box 10.2 The Glengarry Estate, Scotland, 1837-1840
In 1877 a letter was printed in The Times newspaper written by the Rev. Francis Orpen Morris, the incumbent of the parish of Nunburnholme in Yorkshire's East Riding (Morris, 1877). Morris (1810-1893) was born near Cork, Eire, and was, amongst many other things, an ornithologist (publishing *A History of British Birds* in six volumes between 1850 and 1857) and a passionate (and early) conservationist. His letter to The Times, starts by noting that '*Our native British birds are now more or less sufficiently protected*' before wishing that something could be done to protect species which were vagrant or migratory. After extolling the beauty of several of these – Hoopoe, Golden Oriole (*Oriolus oriolus*) etc. – he moves to the reason for his letter, the fact that he is '*continually receiving or reading accounts of the ruthless manner in which all rare birds are shot, the pleasure that many would take in seeing them alive being sacrificed to that of the few persons who shoot them for the gratification of their own.*' Morris then notes the shooting in Sussex of a Hoopoe, whose demise was announced in the 'Deaths' column of *Land and Water*, a magazine associated with *The Sporting Gazette*.

But having noted the hoopoe's shooting Rev. Morris adds a list he had received recently from a cousin in the North Riding of the birds of prey destroyed over a three-year period between 1837 and 1840 (three years?) at the Glengarry Estate in Scotland (about as far north as the Atholl Estate mentioned by McMillan, 2011), as compiled by a Mr A E Knox. The list reads:

Twenty-seven White-tailed Eagles, 15 Golden Eagles, 18 Ospreys or fishing eagles, 98 Blue Hawks, 275 Kites, 5 Marsh Harriers, 63 Goshawks, 7 Orange-legged Falcons, 11 Hobby Hawks, 285 Common Buzzards, 371 Rough-legged Buzzards, 5 Horny Buzzards, 462 Kestrels, 78 Mulen Hawks, 83 Hen Harriers, 9 Ash-coloured Harriers, 6 Ger-falcons, 1,437 Hooded or Carrion crows, 475 Ravens, 35 Horned Owls, 71 Common Fern Owls, 3 Golden, Barn or White owls, comparatively rare in Scotland, and 8 Magpies.

Knox's entry ends with the comment '*On this occasion I have omitted the quadrupeds which figured in the black list.*'

The Kites were almost certainly Red Kites. The *Blue Hawks* may have been Eurasian Sparrowhawks, though as Peregrines have sometimes been called Blue Hawks it has been suggested that might be the case here, the Orange-legged Falcons were, presumably, Red-footed Falcons, the Horny Buzzards were Honey Buzzards, the Mulen Hawks were Merlins, the Ash-coloured Harriers were Montagu's Harriers, the Horned Owls were Long-eared Owls. The puzzles are the number of Rough-legged Buzzards and the Common Fern Owls. The former suggests a remarkable number of this rare visitor to our shores, while the latter suggests the old name for the European Nightjar (*Caprimulgus europaeus*) which, of course, is insectivorous. But the biggest mystery of all is – where are the Peregrines? Were there any at Glengarry? And if the Blue Hawks were Peregrines, then where were the Sparrowhawks?

There are enlightened landowners and gamekeepers who recognised that the effect of raptors was limited and that their presence enhanced, rather than diminished, the countryside. These individuals were supported by the evidence of studies on Red Grouse ecology (Jenkins *et al.,* 1963, 1964) which showed that while the turnover of adult territorial grouse was 60–70% and mortality of first-year birds was comparable, losses to predation were small, representing about 6.5% of eggs (compared with 14.5% to other causes) and 20% of adults (of which about half were due to foxes, the remainder to avian predators, among which Golden Eagles and Hen Harriers were the most significant). However, despite the statistics, there remain landowners and gamekeepers who see the situation completely differently and choose to ensure the survival of the maximum number of game birds by ruthlessly and systematically eliminating any potential avian or mammalian predator of them by poison, gun and trap. Such eradication often includes the illegal killing of Schedule 1 (*i.e.* protected) species. The killing of Peregrines and other raptors impacts not only birds attempting to breed, but dispersing juveniles and sub-adults as well as any 'floaters' throughout the year. Overall, the story is dispiriting: the latest data on UK bird crime (RSPB, 2021) noted that the illegal killing of raptors in the UK was at an all-time high. See also Newton (2021) for an intelligent and highly readable account of the current position regarding UK grouse moors. Newton identifies six major reasons for concluding that the evidence for continued illegal is significant, these being:

1. Greater disappearance of nesting pairs, lower breeding densities or reduced occupancy of apparently suitable traditional territories on grouse moors compared with other areas.
2. Reduced nest success compared with other areas.
3. Reduced adult survival compared with other areas.
4. Reduced age of first breeding, reflecting the removal of adults from nesting territories and their replacement by birds in immature plumage.
5. Greater levels of disappearance of satellite-tracked birds on grouse moors than elsewhere.
6. The finding of poisoned baits, traps and shot or poisoned carcasses of raptors.

We believe it to be very difficult (if not impossible) to argue against this 'smoking gun' for illegal, wholesale, systematic and deliberate persecution of Peregrines and other raptors on managed grouse moors. Until the law is properly enforced, and penalties are introduced that have a real deterrent effect, then it is unlikely that attitudes and behaviours towards raptors in the UK, and, therefore, the outlook for raptors in certain areas of the UK, will improve in the short term.

Game management on grouse moors of the UK continues to be a source of extreme contention between nature lovers and shooting interests particularly on driven grouse moors[13] which are extremely lucrative for their owners. But the income from the moors comes at a cost. To satisfy their clients the moor owner must have high numbers of target birds. Environmentalists have expressed concern that the annual rearing and release of around 50 million game birds into the wild – with a biomass comparable to the total of all wild bird species – is having an impact on the ecology of the countryside (see, for instance, Neumann *et al.* (2015), who noted the change in local invertebrate communities following the release of Pheasants for shooting). There has also been important recent work on the effect of the burning of moorland heather on the local ecology, the burning being carried out because it is necessary for the moor to carry the grouse population density required for the industry to thrive. In a time of concern over global warming, deliberate moorland burning on a huge scale (see, for instance, Douglas *et al.*, 2015) is difficult to justify. The burning is also known to adversely affect moorland (peat) hydrology and chemistry, as well as the ecology of local rivers, the latter due to alterations in water chemistry which reduced invertebrate populations (see, for instance, Brown and Holden, 2020). Fig. 10.9 illustrates the density of shooting butts and burn percentage on moorland managed for grouse shooting in Scotland. The figure notes both the extensive coverage of managed moor and the relationship between shooting butt density (and, therefore, commercial income) and moorland burning[14]. Fig. 10.10 (overleaf) illustrates the extent of grouse moors in England and their position relative to National Parks (NP) and Areas of Outstanding Natural Beauty (AONB).

Moors and Heathland accounted for 18,547km^2 (7.5%) of UK land area in 2012 this comprising 2.5% of England; 15.6% of Scotland; 8% of Wales; and 4.5% of Northern Ireland. Scottish grouse moors are estimated to cover 10,000km^2 (54% of the UK total) while English grouse moors cover 2230km^2 (12% of the UK total). In total, grouse moors account for 66% of the total UK moorland. Several UK National Parks are dominated by intensively managed grouse moors, many of which are associated with the controversial burning of vegetation and the illegal persecution of birds of prey including Peregrine Falcons. Grouse moors make up 44% of the Cairngorms NP in Scotland. In England such moors account for 28% of the North York Moors NP, 20% of the Peak District NP and 2% of the Lake District (NP). The fact that illegal activities continue on National Park land, nominally held in trust for the benefit of the nation as a whole, continues to attract hand-wringing articles in British journals devoted to exploring the country's natural world. Burnside and Pamment (2020) wrote an interesting article on driven grouse moors, noting the vast number of Common Pheasants being released each year (nine times more in 2011 than had been in 1961, with an annual total now exceeding

---

[13] On driven grouse moors, a form of hunting unique to the British Isles, beaters are employed to drive grouse towards shooters standing in wait in 'butts' for the targets to appear above them. It is, of course, more expensive for the shooters, but in exchange for their cash there is no need for leg-wearying walks over rough moorland in search of an elusive target.

[14] For further information specific to Scottish grouse see Tingay and Wightman (2018).

Population

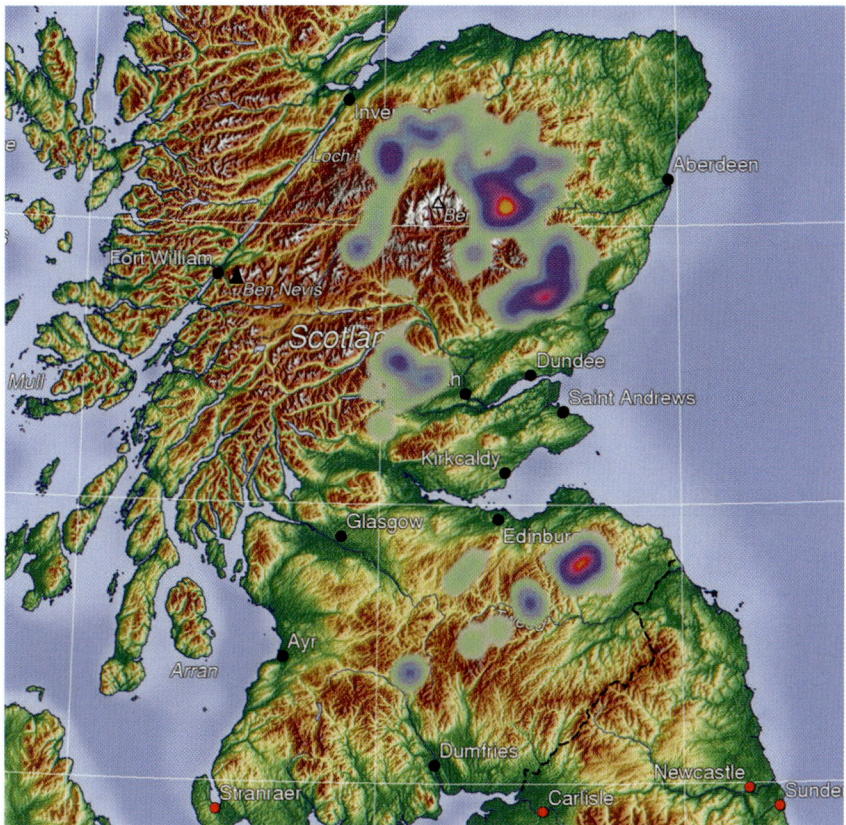

Figure 10.9 Density of Grouse shooting butts weighted by burn percentage. The overlaid colours on the map of southern Scotland are a 'heat map' which varies with the intensity of management defined by the density of shooting butts and the percentage of burned rough grazing. The colour variation is:

Low through , and to High

The authors apologise for the overlays having semi-obliterated 'Inverness' and 'Loch Ness' at the top of the map, 'Scotland' at the centre, and almost completely obliterated 'Perth', also at the centre.

The map has been drawn from Matthews *et al.* (2020), which was a major reference to Thomson *et al.* (2020).

40 million) and lamenting the fact that heather burning was intensifying and that the law in England and Wales did not reflect the vicarious liability legislation on landowners in Scotland. But the article produced an immediate response (Barker, 2020) noting that the idea of a '*shared solution*' to the driven grouse moor problem was hardly a way forward when one of the parties to the conversation continued to break the law with apparent impunity. Barker notes that the suggestion that moors will only '*thrive for wider socio-economic,*

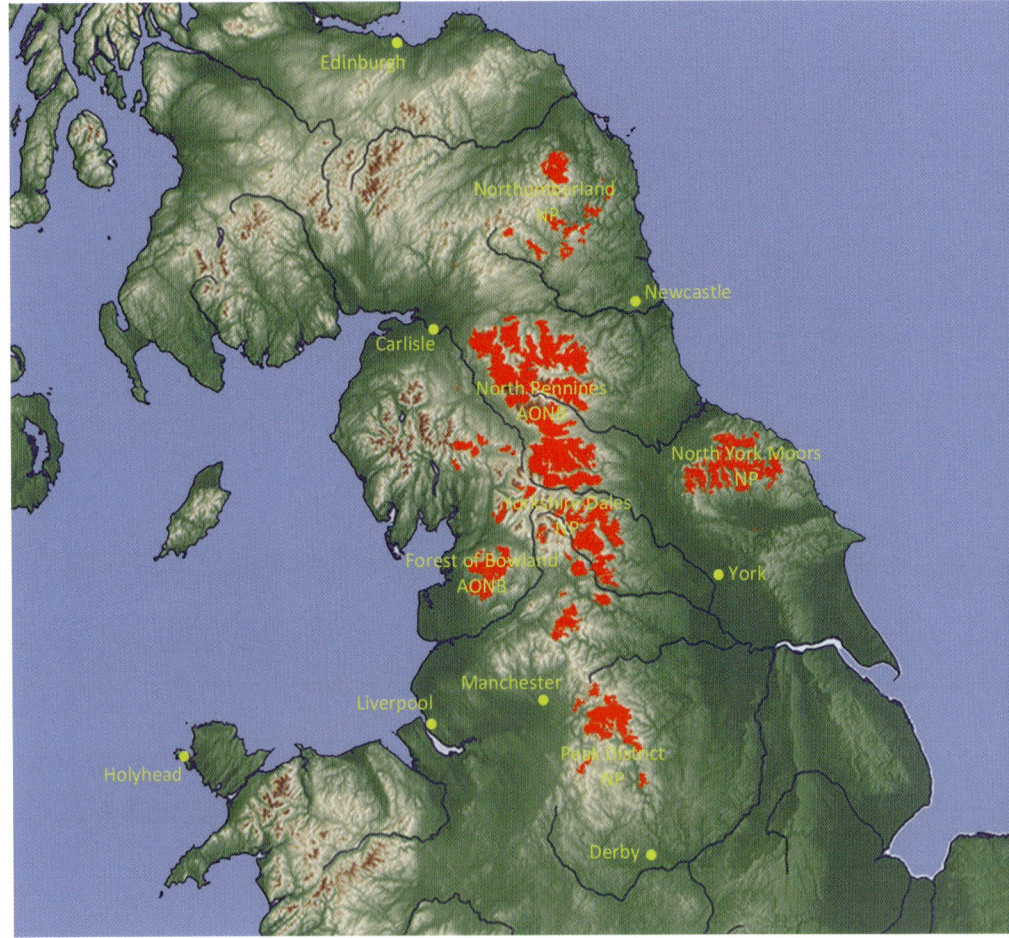

Figure 10.10 Extent of grouse moors in England (shown in red) relative to the position of major cities and National Parks (NP) and Areas of Outstanding Natural Beauty (AONB). In each case the 'NP' or 'AONB' is placed at the centre of the given area.

conservation, and biodiversity purposes' will only be possible when they cease to be incinerated grouse moors.

More recently, Crowle *et al.* (2022) have noted that the view that British uplands are only of value as land for sheep farming, forestry or grouse moors is an outdated attitude given the current understanding of carbon storage and emissions. Crowle *et al.* conclude that future diversification of upland usage would deliver multiple benefits, not only for the environment, but for society as a whole, helping to address the biodiversity crisis currently seen in the UK, as well as assisting in counteracting global warming. While the aim is laudable, it is likely Barker, (2020) and those who share his view would have a response along similar lines to those stated in the paragraph above.

While the persecution of raptors on grouse moors is clearly the most blatant human destruction, there is a long history of other forms – indiscriminate destruction of birds and nests other than on shooting estates, egg theft and the removal of chicks – which have a long history in the UK. While egg collection, which was at one-time a 'standard' hobby for both adults and children is now largely confined to a small number of people, it continues, as do the other crimes. For a specific analysis of these crimes in Wales over the last three decades, see the excellent article by Hughes *et al.* (2021). While the overall crime rate has declined for all forms of persecution, it is still the case that raptors generate a peculiar form of hatred among some humans whose activities remain a persistent problem.

Population

BirdLife International (2022) suggest a world, mature individual Peregrine population of 100,000-499,999, with a footprint of 413,000,000km^2 across the Earth[15] and a Red List category of Least Concern. But this global figure does not allow for those sub-species which have limited populations in small and isolated areas, and whose long-term survival is precarious.

Across its range the Peregrine has recovered from, or is recovering from, the catastrophic declines due to the post-war pesticide crisis, with those populations isolated from that crisis being relatively stable. Populations of each of the sub-species are given in White *et al.* (2013a). A literature search over the last decade since that book was published has revealed updates for some, but not all, sub-species, and for some, but not all, countries. Elts *et al.* (2013) suggest the population of Estonia was ≤5 breeding pairs and stable, though the data were estimated. Nygård *et al.* (2019) note that from around 2000 the Peregrine population of Norway, which was virtually extinguished in the early 1970s, had increased steadily and was probably approaching the pre-crisis level of perhaps 1000 pairs. Interestingly the rise in population was similar in three areas that define the country – the North Sea coast, fjords and inland. Sutton *et al.* (2021) suggest a population of 150-300 pairs, with a total adult population of <1000 birds for *F.p. radama* on Madagascar. Ooi *et al.* (2020) suggest that the population figure for *F.p. ernesti* in west Malaysia (70-80 pairs) quoted by Molard *et al.*, 2007 (and referenced in White *et al.*, 2013a) may be an overestimate, but admit the fact that some potential nest sites were not visited and the urban population is hard to estimate with accuracy. In South Africa, Altwegg *et al.* (2014) noted a continuing rise in the urban Peregrine population in Cape Town, the number of breeding pairs rising from 3-18 between 1997 and 2010. But arguably the most instructive study was that of Sriram and Huettmann (2017) who used data from the Global Biodiversity Information Facility (GBIF.org) and a range of biological, climatic and socio-economic predictors

---

[15] Note that the BirdLife International Peregrine footprint figure is not universally accepted. As an example, Cruz *et al.* (2021) suggest a figure of 'only' 195,000,000km^2. But each agrees that the footprint is very large.

to investigate the probable population recovery of Peregrines. The result suggested a world population (2016) of about 320,000 individuals (Figs. 10.11 and 10.12). For a specific study of Arctic breeding Peregrines (and Gyrfalcons) see Franke *et al.* (2020).

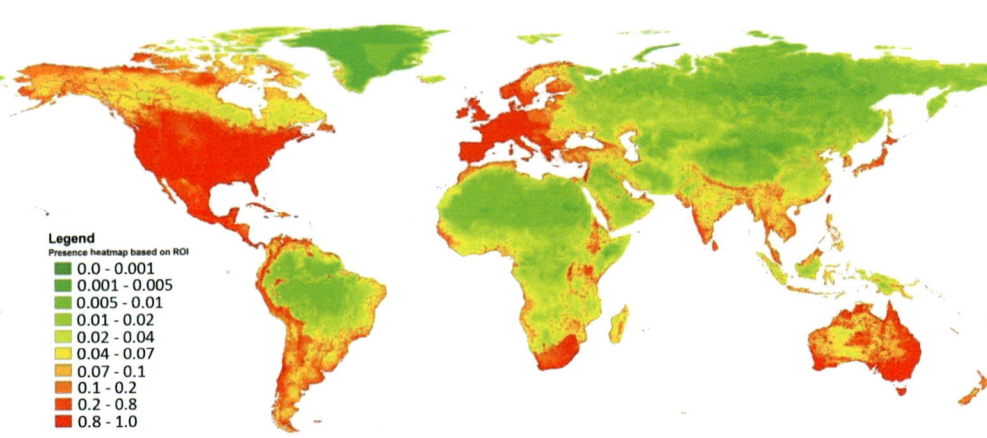

Figure 10.11 Global heat map of the predicted presence of Peregrine Falcons, showing the relative occurrence index (ROI) from 0 to 1. Reproduced, with minor modifications, from Sriram and Huettman (2017), with thanks to Falk Heuttman.

Figure 10.12 Interpreted best-available binary Presence/Absence prediction map of Peregrine Falcons, with a cut-off of 0.01. In this context 0.01 is the 1% probability that falcons will be in a given area (using the TreeNet algorithm in SPM7 provided by Salford Systems Ltd (https://www.salford-systems.com/). Reproduced from Sriram and Huettman (2017), with thanks to Falk Heuttman.

Nominate Peregrine capturing a juvenile European Herring Gull, Isle of Man, UK. *Peter Christian*.

The Peregrine Falcon

The data collected on Peregrine populations in specific areas show that the falcons establish a constant breeding density in a relatively short time. Cozic (2019) studied the rise in population in Brittany, north-west France from 1995, when the first breeding pair arrived, to 2015 by which time the département had some 60 breeding pairs. Initially the Peregrines bred in coastal cliff eyries, but as available sites became filled, pairs moved to the larger (3-40ha) islands. These too became filled and smaller (≤3ha) islands were occupied, as were quarries and other suitable inland sites. The growth of the Brittany population is shown in Fig. 10.13.

In a long-lived species, population growth rate is highly sensitive to changes in adult survival. As noted in Chapter 6 *Territory and Hunting Range*, Hunt (1998) investigated mathematically the way in which populations grow (Fig. 10.14). The two graphs (Cozic, Brittany and Hunt, theory) are remarkably similar, the Brittany growth mirroring the theoretical modelling, both for quarry eyries and all other eyries, each implying that it takes around 20 years for an influx of Peregrines into an area to create a stable breeding population. However, note the comments on the captions of both graphs for reservations on being precise regarding time periods.

F.p. anatum **adults arguing over prey remains.** *Gabriel Gruia.*

Population

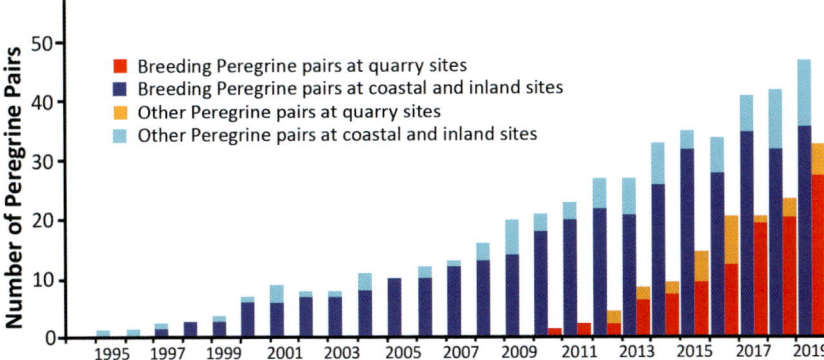

Figure 10.13 Rise in population of breeding and non-breeding Peregrines at sites in Brittany. Redrawn from Cozic (2019). Cozic distinguishes between breeding pairs and other, non-breeding, pairs. It is likely that 'other pairs' were actually a mix of floaters and juveniles though it is not possible to suggest a division between these.

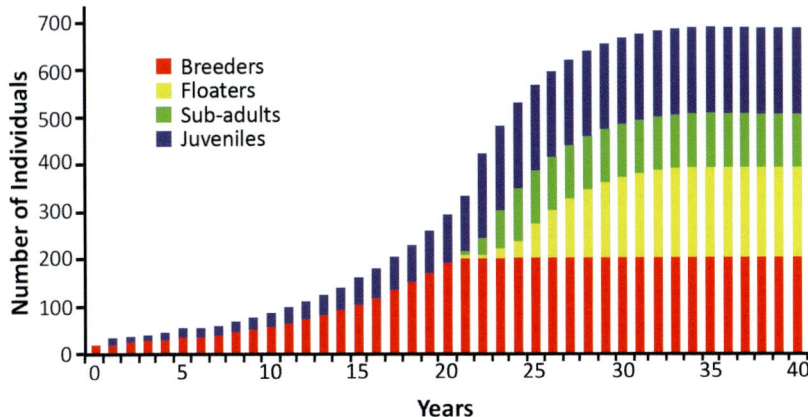

Figure 10.14 Theoretical increase in Peregrine population, including breeding pairs, floaters and juveniles. Redrawn from Hunt (1998). Hunt's basic assumptions for his model were that the initially raptor-free area had 100 usable breeding sites. In Year 1 10 breeding pairs moved in and produced 1.8 fledglings/nest. The survival rate of the first-year birds was assumed to be 60%, that of older birds to be 80%. Hunt's graph shows a stable breeding population is established after 20-22 years and that an overall stable population (*i.e.* one that includes year-on-year populations of breeding adults, floaters, sub-adults and juveniles) is established after 32 years. The survival rates used by Hunt are a reasonable approximation to the data across the Peregrine's range. The existence of floaters (adult birds capable of breeding) and sub-adults (non-juveniles, but not yet capable of breeding), as well as juveniles is also consistent with Peregrine data. However, the mathematics of the rise in the various population groups vary straightforwardly according to the survival data chosen, so the graph itself remains the same, only the number of years on the horizontal axis changing if different survival data is assumed.

UK and Isle of Man Peregrine Falcon population
The data presented here, on the number of territorial Peregrine pairs, has been derived from surveys conducted at *c.*10 year intervals by the British Trust for Ornithology (BTO) over the period 1961-2014. To create a baseline for consideration of the effect of the DDT crisis, a figure for 1930-1939 was estimated by Ratcliffe (1993) as part of his work for the 1961 survey. The stable UK population estimates for the period 1930-39 was 874 territorial pairs, having been constantly re-assessed as historical information came to light over the 30-year period to 1991. It is important to understand that the UK Peregrine population has recovered and extended its range naturally and has never been augmented by captive bred falcons, as happened in the USA.

Before considering the UK population, and how this has changed both spatially and temporally over the 75 years to 2014, we review some of the fundamental drivers of population changes.

Land use
Humans have always impacted land use and as a result have shaped the ecological environment. This process has accelerated as human populations have increased requiring more land allocation for intensive food production, industrial, commercial and leisure activity. Land use in UK was mapped in 2012 by the European Commission as part of the 'COoRdination of INformation on the Environment' programme (CORINE). A modified version of this map is shown in Fig. 10.15, together with separate maps for England, Scotland, Wales and Northern Ireland (Fig.10.16 overleaf). Note that this mapping has been updated to 2018 but the changes are insignificant in the context of this discussion.

Note that CORINE's stated UK land area of 247,517km^2 (Fig.10.15) does not equate to the addition of the four UK country data (251,627km^2 – Fig.10.16), a difference of 1.6%. We also note that UK Office of National Statistics (ONS) state that UK land area is 248,532km^2, 0.4% higher than CORINE's figure. CORINE also overstate England's area by comparison to ONS, this appearing to account for most of the difference.

Farmland
The effect of the catastrophic impact on Peregrines and other avian and mammalian predators by the use of toxic chemicals on arable farmland in the 1950/60s has been dealt with earlier in this Chapter.

In 2012 farmland comprised almost 140,000km^2 (56%) of UK land use (England, 73%; Scotland, 19%; Wales, 58% and Northern Ireland, 73%) (Figs 10.15 and 10.16). As farming has become more intensive and subject to the deposition of high concentrations of chemical applications, the land has become increasingly hostile to insects and, therefore, to the farmland birds that feed on them. Additionally, as land has been more intensively farmed commensurate with the economic requirement for ever greater yields, field

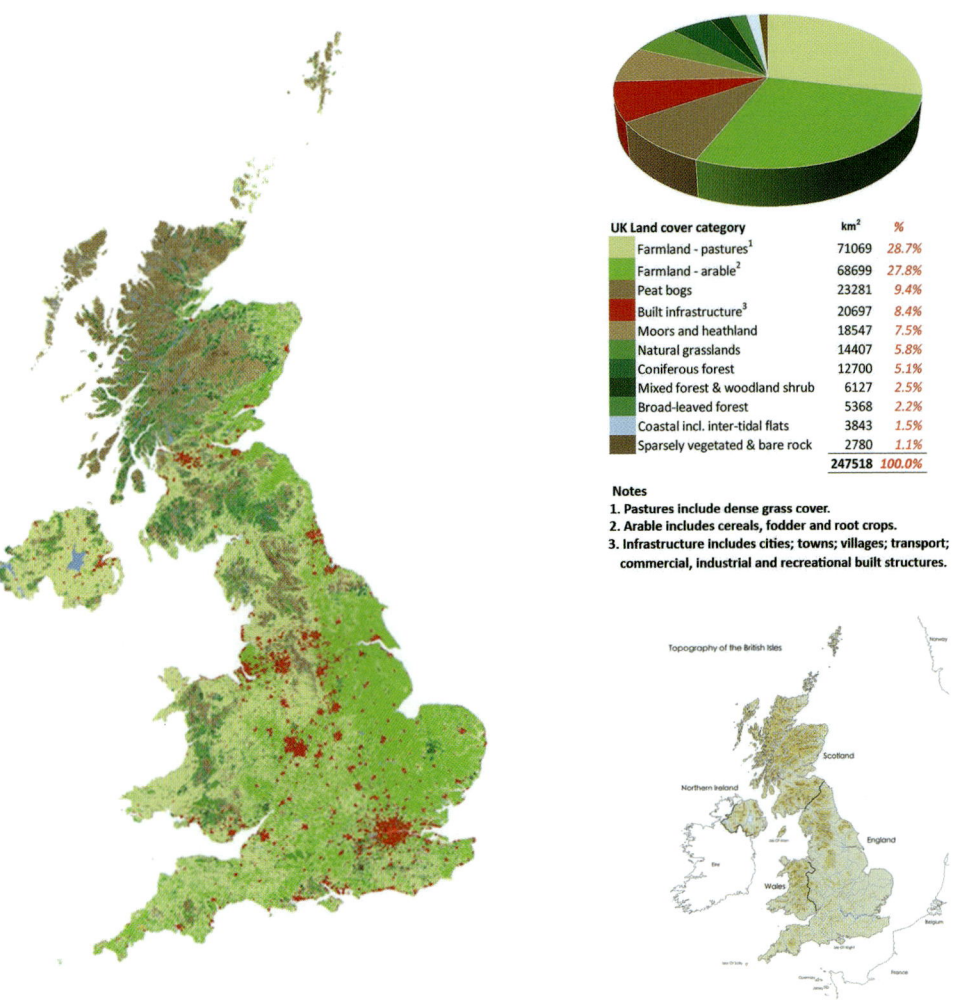

Figure 10.15 Land use in the UK as a whole constructed by the European Commission CORINE programme.

hedges have been lost or degraded, reducing the availability of weed seeds to seed-eating birds. DEFRA (2021) state that on the basis of their research that often declines in farmland bird species are associated with more than one factor impacting on a species simultaneously and since 1970 they cite loss of mixed farming and the consequent loss of habitat diversity, increased use of pesticides, adverse changes in crops grown, loss of winter fallow habitats, adverse changes in grassland management and loss of habitat resulting from

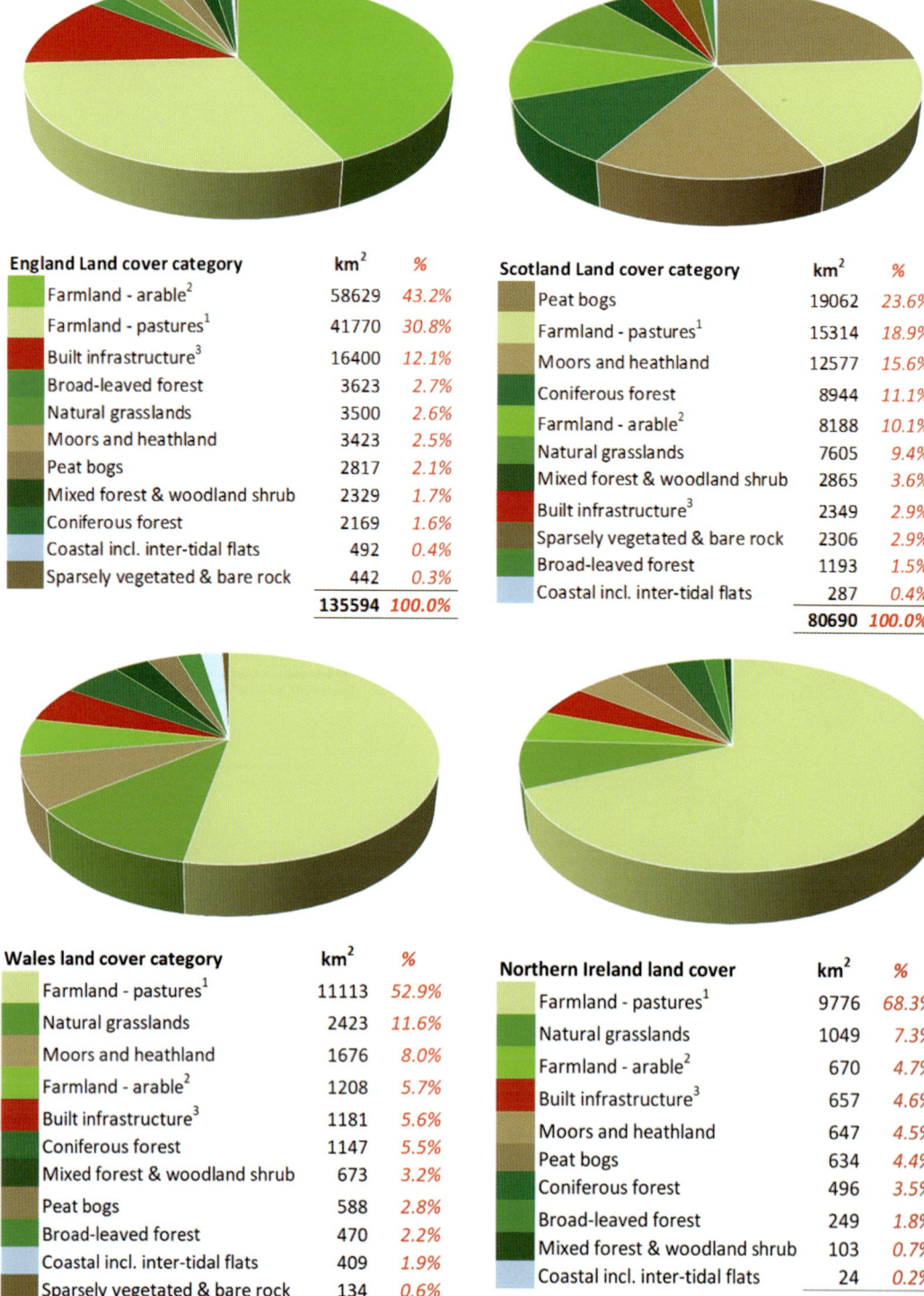

Figure 10.16 Land use in the individual countries of the UK as constructed by the European Commission CORINE programme.
The Notes on Fig. 10.15 on the previous page also apply to this figure.

increased industrial sized farms which maximise economically productive areas. The effect has been a decline in bird species (Fig. 10. DE5).

Farmland does not provide breeding habitat for Peregrines but can provide prey for 'floaters' and dispersing juveniles. Clearly, as farmland comprises 56% of UK land area then its management has serious consequences for avian ecology and the density of Peregrines that can be supported by diminished prey availability. This will be more important for non-breeding 'floaters' and for dispersing juveniles outside of the breeding season than for breeding adults which tend to stay on territory throughout the year. Fig. 10.17 notes a 57% decline in UK farmland birds from 1970 to 2019. Of particular note is the split between generalists (those species not restricted to or highly dependent on farmland habitats such as Woodpigeons, crows) and specialists (those species that are restricted to or highly dependent on farmland habitats such as Grey Partridge and Corn Bunting (*Emberiza calandra*)).

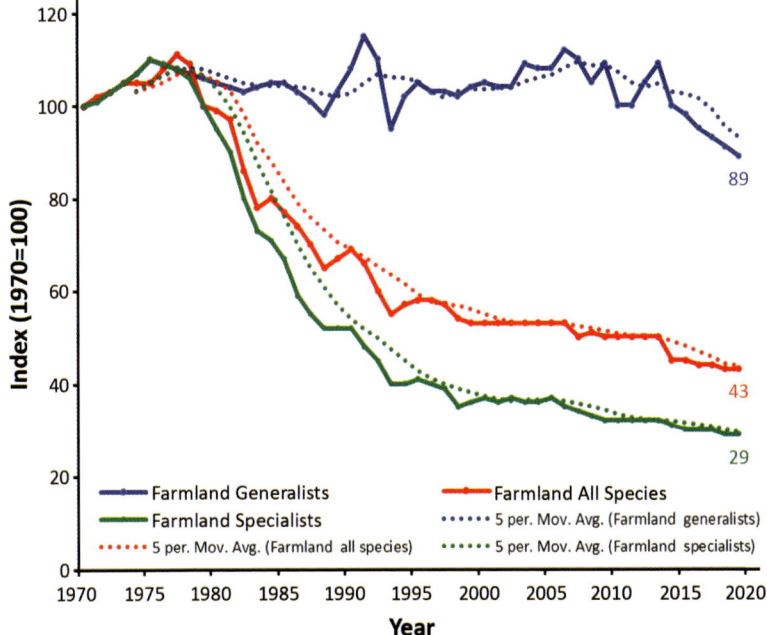

Figure 10.17 Decline of the population of breeding birds in the UK from 1970 to 2019. Redrawn from DEFRA (2021) but with a caveat. It is clear that there is a charting error in respect of the 'Generalists' as the origin on the chart starts at an Index point of 95 in 1970 not 100, which is clearly erroneous as the Y axis is titled '1970=100'. We have, therefore, displaced the Generalists chart line. This has the effect of aligning it to an indicator data sheet (DEFRA, 2016) labelled 'DE5 indices for farmland bird populations: England' for the period 1970-2015. This, of course, has the effect of raising the whole chart line up by 5 index points for the entire period 1970-2015.

Note that '5 per Mov. Avg' in the Figure indicates that the dotted lines have been compiled as a moving average of the data of the previous five years.

Forestry

In 2012 forestry comprised 24,200km² of UK land use (England 6%; Scotland 16%; Wales 11% and Northern Ireland 6%), these figures having remained essentially static to 2018. Forestry comprises almost 10% of UK land area and is not widely used by breeding Peregrines though sporadic and very occasional nesting has occurred in trees over recent years. In September 2021 the UK Forestry Commission reported woodland statistics (Table 10.2): it is interesting to note historical changes and recent country-by-country forestry development, particularly the plans of the Scottish Government to increase forest and woodland cover to 21% of the total area of Scotland by 2032, in line with the Forestry and Land Management (Scotland) Act (FLM(S)A) 2018.

Year	UK percentage historical woodland land cover				
	England	Wales	Scotland	N. Ireland	UK
1905	5.2	4.2	4.5	1.1	4.7
1924	5.1	5.0	5.6	1.0	5.0
1947	5.8	6.2	6.6	1.7	5.9
1965	6.8	9.7	8.4	3.1	7.4
1980	7.3	11.6	11.8	4.9	9.0
1995	8.4	13.8	16.4	6.0	11.3
1998	9.5	14.4	16.7	6.0	12.0
2021	10.1	15.0	19.0	8.6	13.3

Table 10.2 Historical woodland area as a percentage of total land cover in the UK, split by country. Data from the Forestry Commission, Forestry England, Scottish Forestry, Forestry and Land Scotland, the Welsh Government, Natural Resources Wales and the Forest Service and National Forestry Inventory.

Notes:
1. These figures include woodland on farmland totalling 10,690km² in 2020, this accounting for the 3.5% woodland cover difference between CORINE and the Forestry Commission.

2. Over the last century UK woodland cover has increased from 5% to 13.3%. In Wales it has tripled; in Scotland almost quadrupled; and in Northern Ireland has increased by a factor of 860%. 13.3% UK woodland land cover equates to 32,280 km² an increase of about 20,000km² over the last century, almost equivalent to the entire land area of Wales.

It is important to note that Peregrines do regularly nest in trees in European countries and there seems to be no good reason not to assume that this range extension might well occur in the UK in future, possibly 'seeded' by falcons

Nominate adult Peregrine, Germany. *Torsten Pröhl.*

crossing from Europe, with offspring remaining natal site-type-faithful when breeding in the UK, as has been known to have occurred with European urban site breeders (see Chapter 6). This could potentially provide another stimulus for UK breeding habitat expansion. However, there is a caveat – whereas human structure and quarry nesting Peregrines have few natural predators, this would not be the case for woodland nesters where Grey Squirrels (*Sciurus carolinensis*), Pine Martens (*Martes martes*), Goshawks, Buzzards and corvids are all potential breeding disrupters. Additionally, Peregrines will not nest in dense woodland, preferring isolated trees, small copses or woodland edges. This would naturally limit available woodland breeding habitat.

Special Protection Areas (SPAs)

Special Protection Areas are a network of areas listed (designated) to protect vulnerable wild bird species listed in Annex 1 to the EU Birds Directive (2009/147/EC) which naturally occur in the EU member states and the UK. They are protected areas for birds and are defined by and covered by several Acts of Parliament for the four countries that make up the United Kingdom. The acts cover not only the mainland and islands, but offshore marine habitats. Some Grouse Moors are within SPAs and these typically perform badly and as the BTO note (Wilson *et al.*, 2018) the SPAs are not performing well for Peregrines. Both the estimated number of pairs and observed breeding productivity in these areas confirm the conclusion of the latest SPA review (Stroud *et al.*, 2016) that SPA provision for Peregrines is insufficient, several confirmed instances of illegal killing of Peregrines within them having occurred. A recent study examining breeding histories of Peregrine sites in northern England, some of which were in SPAs, showed that grouse moor management is negatively associated with Peregrine breeding success (Amar *et al.*, 2012). In the Muirkirk and North Lowther Uplands SPA, no Peregrine pairs bred in 2014 (Table 10.3).

SPA	Country	No. of sites	Est.no. SPA pairs 2014	Occupancy %	Pairs No. successful	Pairs No. failed	Pairs Successful %	Fledged Min. no.	Fledged Est. no.	Fledged Est. no. /pair	Peat land %
North Pennine Moors	England	27	7	26%	0	5	0%	0	0	0.00	93%
Muirkirk and N. Lowther Hills	Scotland	15	0	0%	0	0	0%	0	0	0.00	64%
Migneint-Arenig-Dduallt	Wales	22	5	23%	0	1	0%	0	0	0.00	68%
Bowland Fells	England	14	4	29%	1	3	25%	1	1	0.25	80%
South Pennine Moors	England	25	15	60%	3	8	27%	5	6	0.40	67%
East Caithness Cliffs	Scotland	15	5	33%	–	–	0%	–	5	1.00	1%
North Caithness Cliffs	Scotland	10	3	30%	–	–	0%	–	3	1.00	1%
Rathlin Island	N. Ireland	8	5	63%	2	3	40%	6	6	1.20	0%
Hoy	Scotland	22	5	23%	2	0	100%	4	7	1.40	44%
Cairngorms	Scotland	23	7	30%	4	2	67%	9	10	1.43	60%
Berwyn	Wales	18	8	44%	5	2	71%	9	12	1.50	72%
Totals			199	64	32%	17	24	41%	34	50	0.78	71%
Comparative data for other inland natural sites				36%			57%			1.26	

Table 10.3 UK Special Protection Areas (SPAs). The table shows the estimated occupancy rate and output of fledged young/territorial pair, ranked by lowest output. Adapted from Wilson *et al.* (2018). Grouse Moor SPAs are indicated in red.

When comparing SPA data with surveyed natural counterparts all breeding metrics are worse in the former. Occupancy rates are 10% lower, the proportion of Peregrines successfully raising broods is 28% lower and 38% less young are fledged per territorial pair. It is particularly telling that where pairs were successful on SPAs the mean fledged brood size (2.0) was consistent with natural counterparts, strongly suggesting that prey availability is not limiting breeding productivity.

UK population data (including the Isle of Man)
As Ratcliffe (1993) notes, although the medieval age of falconry meant that the position of known Peregrine eyries was documented throughout the period from about 1200 onwards, the likelihood that this was a complete list is minimal, and the true British population before about 1900 cannot be computed. Once the coasts and wildernesses of Britain were more fully explored, the number of Peregrine eyries could be better established. Using data on the number of territories known to have been lost during the period 1880–1930, mainly due to disturbance by humans, Ratcliffe estimated that the population in the last years of the nineteenth century was about 1,365 pairs.

Population data 1939 – 2014
A detailed analysis of the fluctuating, but exceptionally well-documented UK Peregrine population over the period 1939-2014 allows us to identify specific spatial and temporal consequences of the impacts on the population that have been discussed earlier in this Chapter.

Our analysis will often reference Ratcliffe (1993). Ratcliffe led the Peregrine surveys for 1961, 1971 and 1981 and co-led that in 1991 (Crick and Ratcliffe, 1995). His seminal book (Ratcliffe, 1993) remains the definitive work on the subject from that era and population data within it have been used to reconstruct regional data for 1961-1991 using an amplified (2002 and 2014) categorisation. The survey data for periods to 1939, 1961, 1962, 1971, 1981 and 1991 were split into 15 different regional categories, including coastal splits where appropriate. However, the 2002 UK national census (Banks *et al.*, 2010) tabulated 24 regions and then further split each between coastal and inland where appropriate. The 2014 census (Wilson *et al.*, 2018) continued with the 2002 categorisation, but added the Scilly Isles, the Channel islands and split out the Scottish islands, Lewis and Harris, and Uist, to form 27 regional categories.

All of the data below has been tabulated and charted by reconstructing the Ratcliffe era data (1939-1991) on the basis of the 2014 revised and expanded categorisation. We have, however, excluded the Channel Isles from the analysis as we have no information on comparative data prior to 2014. The 1939-1991 data was reconstructed by re-analysing the comprehensive notes in Ratcliffe (1993) and rebuilding the data line-by-line into the updated 2014

The Peregrine Falcon

categories. For clarification, the data, charts and analyses presented here all refer to numbers of territorial pairs (as opposed to numbers of breeding pairs), unless otherwise stated[16].

In carrying out our analyses, which includes suggestions on likely future, short-term population levels based on our current understanding of habitat density levels (actual and potential) and breeding, survival and occupation level trends for multiple habitat types, we are indebted to Mark Wilson and Rob Robinson of the BTO for advice and support.

Fig. 10.18 sets down the summary country population data compiled from the surveys noted above.

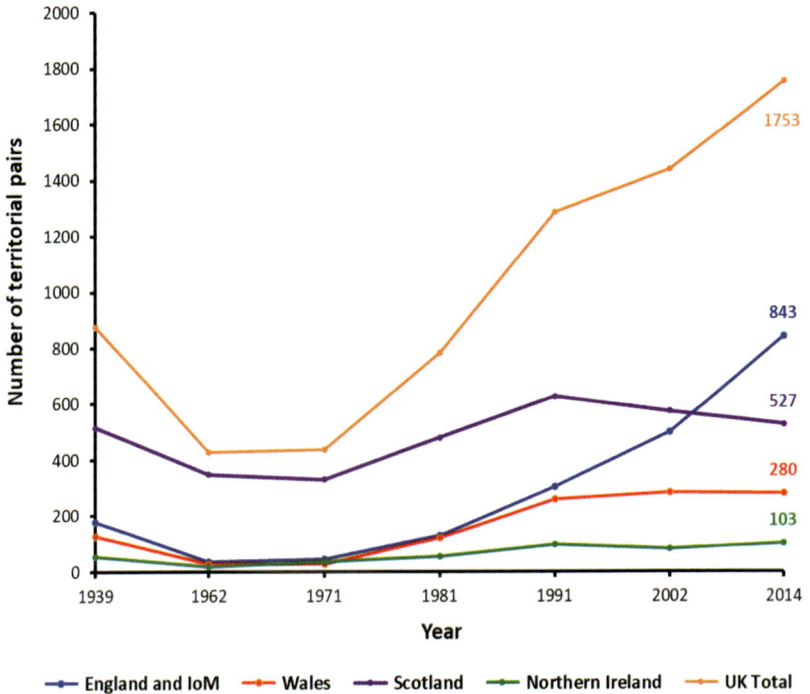

Figure 10.18 Number of nominate Peregrine territorial pairs in the UK (including the Isle of Man, but excluding the Channel Islands) 1939-2014. Having considered all available data, the authors consider the full survey low point of the falcon's population to have been in 1962. However, Ratcliffe carried out sample surveys in 1963 and considered the population low point may have been that year.

---

[16] The authors have used territorial pair data from Ratcliffe (1993) throughout in respect of all periods up to and including 1991 (as per Table 7 in that book) and we have added Northern Ireland data to Ratcliffe's GB data to obtain UK totals. We are aware that the BTO website refers to different figures for the period 1961-1991 as does a BTO report (Baillie *et al.*, 2010). However, having enquired we have received no evidence to suggest that we should amend Ratcliffe's original data.

Population

The landmark drivers and key survey conclusions and estimates for this period are summarised below;

1940/45 On 1 July 1940 the UK government enacted the Destruction of Peregrine Falcons Order to protect homing pigeons returning from mainland Europe that were carrying critical wartime information. It is estimated that some 600 Peregrines were shot during this period and countless eyries destroyed (Ferguson-Lees, 1951) The law expired in February 1946. This short-term negative impact on local territorial pair populations was to some extent reversed by c.1955.

1947 Introduction of widespread agricultural use of wartime stocks of organochlorines (*e.g.* DDT) and consequent raptor eggshell thinning with a marked negative impact on breeding metrics. Ratcliffe (1993) concluded that *'a mean level of 13.6ppm (wet weight) DDE (the DDT metabolite) corresponded with an average shell thinning of 21.6%.'*

1956/63 Serious collapse in UK Peregrine populations spreading from south to north reducing UK population to about 44% of pre-war levels. In 1956/7 cyclodiene pesticides such as dieldrin, aldrin and heptachlor were introduced and subsequently found to have acute lethal toxicity in granivorous birds. This in turn impacted apex mammalian and avian predators to the point where increased Peregrine adult mortality was recorded. Ratcliffe (1993) estimated that UK Peregrine populations were declining by 13.3% per annum between 1960-63, the worst affected areas being Wales and southern England. The pattern of population decline correlated closely with arable land usage and application intensity of the cyclodiene group of pesticides though DDT and lindane (another organochlorine) were also implicated (see Fig10.15).

1961 Incomplete voluntary ban in UK on cyclodiene pesticides for dressing spring sown cereals. This first impacted 1962 Spring sowings.

1962 The GB Peregrine population low point (44% of 1930s baseline). Concern was expressed by Ratcliffe at this point (see Ratcliffe, 1993) that the Peregrine could become extinct as a breeding bird in England and Wales by 1967 should the rate of decline observed between 1961-1963 continue. See caption for Fig. 10.18 opposite.

1971 The BTO survey results demonstrate a marked, though patchy, recovery in population numbers with mean territory occupation at 56% and successful breeding at 24% of occupied territories. UK population was estimated at 437 UK pairs.

1975 Dieldrin banned on a 'voluntary' basis in the UK.

1979   EC regulations made the ban on organochlorines mandatory rather than voluntary.

1981   All remaining agricultural uses of the cyclodiene insecticides formally discontinued. DDT was banned in 1982.

1981   The BTO survey results (Ratcliffe, 1984) demonstrate a significant and widespread increase in population numbers and breeding metrics with UK average 78% territory occupation and 49% successful breeding at those territories. The population was estimated at 783 UK pairs, an increase of 346 (79%) over 1971 levels. It was demonstrated that generally coastal populations were failing to recover at the same rate as inland populations, most notably in south-east England and south Strathclyde in Scotland. Industrial marine contaminants such as PCBs and heavy metals were implicated in the Scottish failure to recover, though arable agricultural practices were implicated in south-east England coastal population's failure to recover. However, the English Lake District and Pennines populations increased 62% and 46% respectively compared to the estimated 1939 baseline.

1980s   The population increase was, at first, merely a recovery to the pre-contamination level, but after the early 1980s numbers began to exceed the assumed stable population of the 1930s. The increase involved not only a recolonisation of ancient territories, but range expansion. DDT, being highly stable and persistent in the environment, continued to be a small component of the vast majority of Peregrine bio-material analyses and consequently were still possibly limiting breeding potential.

1985   Ratcliffe (Ratcliffe, 1993) estimates that UK peregrine population had regained its estimated pre-war baseline level of 874 pairs.

1991   BTO survey results demonstrated that population numbers were at the highest level ever recorded and breeding metrics showed 83% territory occupation with successful breeding at 61% of occupied territories where breeding outcome was recorded (Crick and Ratcliffe, 1995). The overall UK population was estimated at 1283 pairs, an increase of 500 (64%) over 1981 levels. Coastal populations had now largely recovered to 1939 baseline levels, but the south-east England coastal Peregrines were only at 25% of their 1939 levels.

Crick and Ratcliffe (1995) note that the 1991 survey revealed that 161 quarries were occupied by breeding Peregrines of which 101 were new territories since 1981. There were also seven new territories on built structures. It could not have been known at the time that this was to be the start of a dramatic population growth and habitat type extension seen over subsequent decades. Crick and Ratcliffe (1995) also noted territory size reductions because of greater density of territorial pairs in areas of increasing populations. It noted 68 instances of two breeding pairs being present in what had previously been a single territory and a remarkable 13 cases where three breeding pairs were present at territories previously holding only a single pair.

2002 UK population estimated at 1437 pairs, an increase of 154 (12%) over 1991 (Banks *et al.*, 2010). This signified a slowing in the rate of population increase and a return to traditional population stability, albeit at a higher level. The population levels at traditional inland and coastal sites had remained broadly stable compared to 1991 for Scotland, Wales and Northern Ireland, but there were gains in England, particularly in the south. English breeding habitat expansion accounted for 222 (+62%) extra territories compared to 1991. The increase in England pairs in this 11-year period therefore accounted for the whole of the UK population increase over this time.

Geographical splits and mapping were recategorized to allow for reporting in more detail.

2014 UK population estimated at 1753 pairs (excluding Channel Isles – 16 pairs), an increase of 316 (22%) over 2002 levels (Wilson *et al.*, 2018). To reflect the dramatic breeding habitat expansion due to the increased use of quarries and man-made structures for breeding the methodology of the 2014 survey was modified. Random 5km x 5km squares were searched and from the occupation data extrapolations were applied to the whole search area. This technique was deemed appropriate in a time of rapid and, to a degree, unpredictable breeding habitat expansion. In addition, traditional sites were surveyed, as in previous surveys. SPA breeding data, breeding outcomes and site types were assessed and reported for the first time.

Significant population increases were seen in southern and eastern England in the 12 years to 2014, these resulting in a doubling of the number of inland England territorial pairs from 315 to 636. For the first time on record the Peregrine breeding population in England was more than the combined populations of Wales and Scotland.

Figs 10.19 and 20 on subsequent pages show the variation of the UK Peregrine population with time.

Fig. 10.21 on the page following the population maps is a further visualisation of the spatial and temporal realtive density of the UK's Peregrine population.

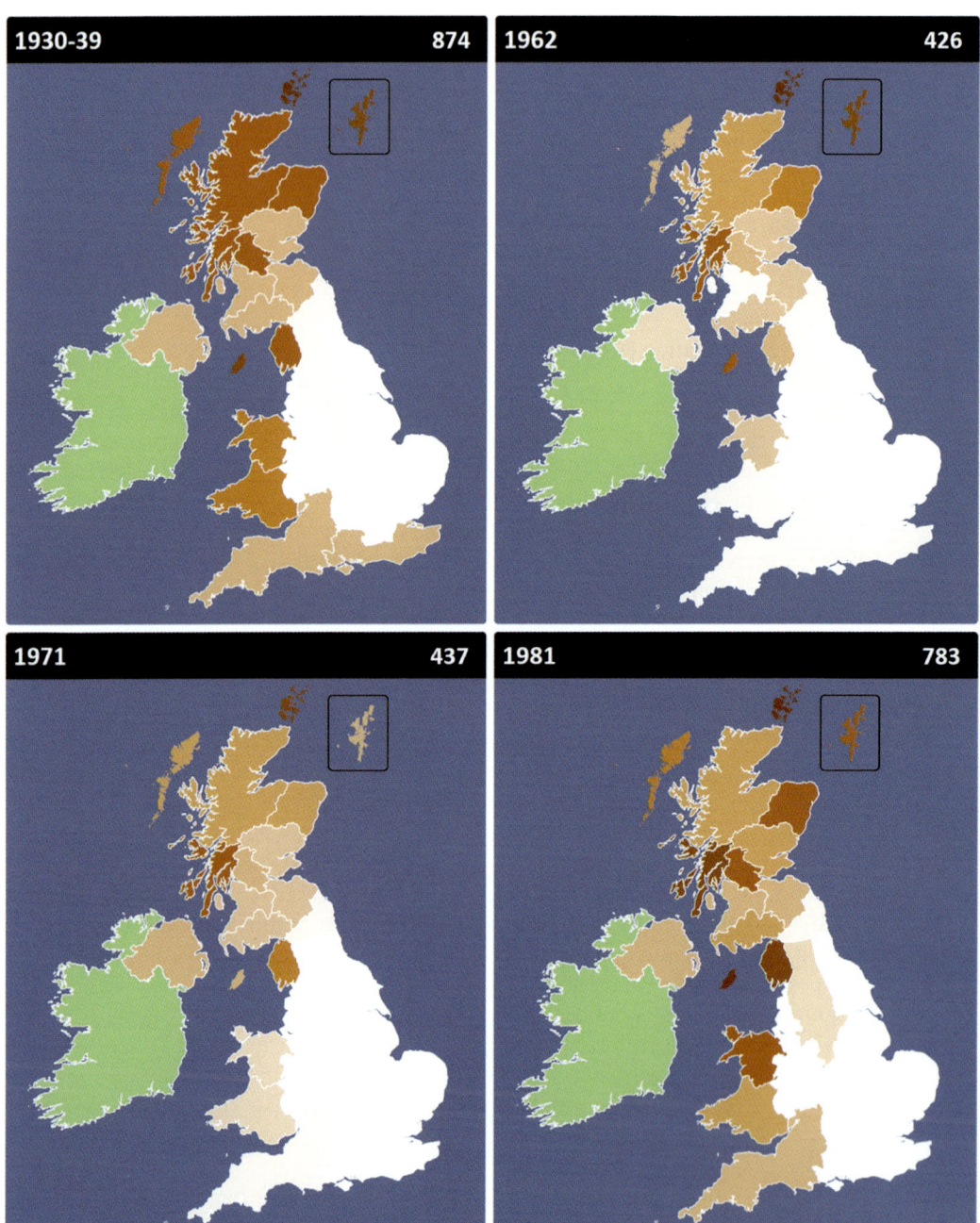

Figure 10.19 Number of nominate Peregrine territorial pairs in the UK (including the Isle of Man, but excluding the Channel Islands) 1939-2014. Having considered all available data, the authors consider the low point of the falcon's population to have been in 1962. The maps demonstrate the temporal and spatial relative density of UK Peregrine population by region over the period 1939-2014. The 1962 low population point is shown again on the opposite page side-by-side with the 2014 high population point to show a ready visual comparison.

See **Figure 10.20** opposite for map key and definition of areas and colour code.

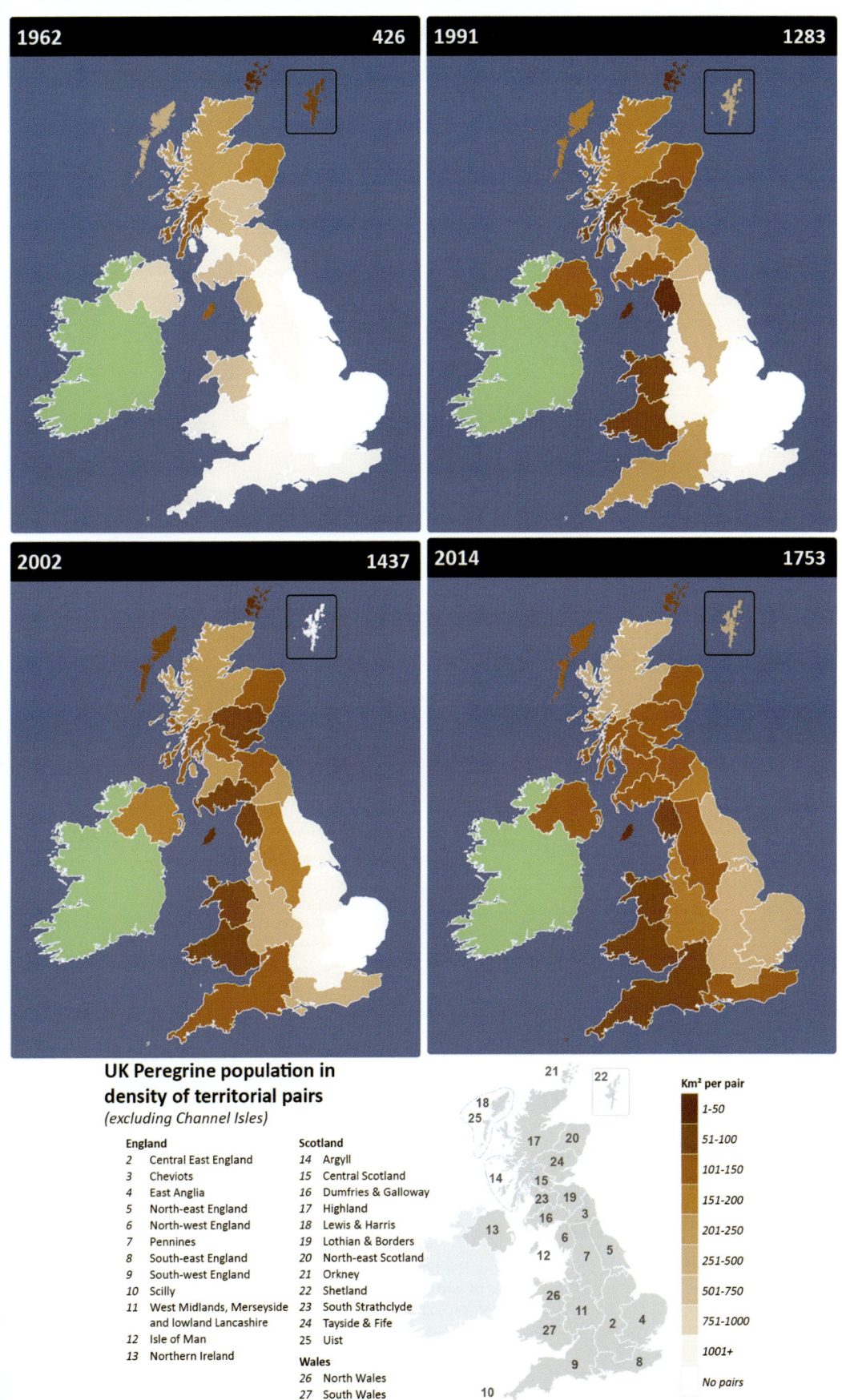

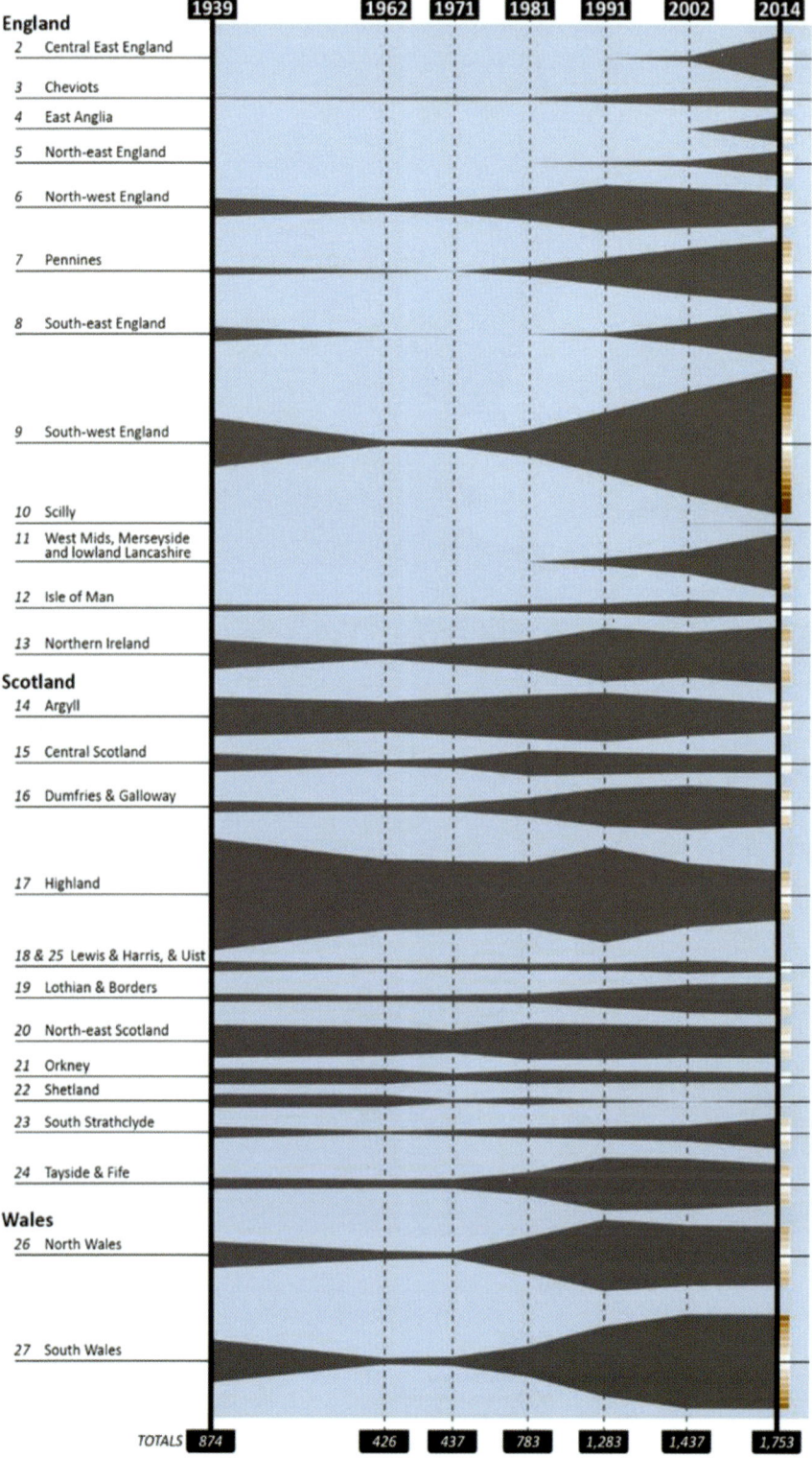

Figure 10.21 This figure spaces out the time periods between Peregrine population surveys consistent with the number of elapsed years, and visualises temporal and spatial population changes by region, providing an alternative way of visualising the trends shown by the mapping figures on the previous pages.

Nominate Peregrine food passes on the Isle of Man, UK. *Peter Christian.*

Table 10.4 (overleaf) is a 'heat map' of all the UK regional data split as per the 2014 re-categorisation, showing a summary of all the UK population BTO survey data, split by area, coastal and inland, summarised by country. The 'sparkline' at the right-hand edge gives a line chart of each line and is colour coded green for inland and blue for coastal. Note that the Channel Island data which was first included in the 2014 national census has been excluded as we have no prior comparative data.

2014 region category	BTO map key	1930/39 Baseline	1962	1971	1981	1991	2002	2014	Sparkline
England (incl. Isle of Man)									Inland / Coastal
South-east England inland	8	4	0	0	0	0	14	64	
South-west England inland	9	0	0	0	0	19	69	103	
Central East England	2	0	0	0	0	0	9	83	
East Anglia	4	0	0	0	0	0	0	45	
South-east England coastal	8	23	1	0	0	6	24	18	
South-west England coastal	9	91	9	14	48	95	121	155	
North-west England coastal	6	3	2	0	0	1	6	3	
North-east England coastal	5	0	0	0	0	3	2	8	
West Midlands, Merseyside and Lancs	11	0	0	0	0	15	42	104	
North-west England inland, Lakeland	6	31	12	24	47	83	66	58	
Pennines	7	13	4	0	19	50	84	113	
Cheviots	3	2	4	5	4	13	23	30	
North-east England inland	5	0	0	0	0	0	8	36	
Scilly Isles	10	0	0	0	0	0	0	1	
Isle of Man	12	11	4	2	12	18	31	22	
England (incl. Isle of Man) totals		178	36	45	130	303	499	843	
Wales									
South Wales coastal	27	46	3	7	29	51	64	86	
North Wales coastal	26	16	5	4	14	21	22	20	
South Wales inland	27	32	8	9	26	77	108	86	
North Wales inland	26	33	11	7	52	109	89	88	
Wales totals		127	27	27	121	258	283	280	

Table 10.4 Number of territorial pairs of nominate Peregrines in the UK and Isle of Man (excluding the Channel Islands) 1939-2014. The 'heat map' gradient varies from low (white) to high (red). The sparkline also indicates the change in population with time.

Notes:
1. The data in the above table are caveated on the basis that reconstruction of the re-categorised historic data into the 2014 categorisation required detailed analysis of the text in Chapter 4 of Ratcliffe (1993). Whereas every effort was made to allocate the text to the correct new regional category, absolute accuracy cannot be guaranteed because judgments were necessary in allocating data close to new region category borders. Nevertheless, our view is that it is worth showing the historical data in this format to identify broad regional trends over the whole survey periods 1961 – 2014 and including the retrospective Ratcliffe assessment of baseline 1930/39 territorial pair numbers.

2. The heat map identifies highs and lows for each year or region and is presented to enable visualisation of population movements at local, regional and country levels over time and for each survey period.

2014 region category	BTO map key	1930/39 Baseline	1962	1971	1981	1991	2002	2014	Sparkline Inland Coastal
Scotland									
South Strathclyde coastal	23	16	1	5	1	6	15	18	
Dumfries and Galloway coastal	16	10	4	5	20	22	25	27	
Lothian and Borders coastal	19	0	0	0	0	5	12	12	
South Strathclyde inland	23	5	4	5	17	17	15	38	
Dumfries and Galloway inland	16	10	8	9	14	47	55	40	
Lothian and Borders inland	19	15	11	13	17	29	40	46	
Argyll and Bute coastal	14	36	28	33	54	49	41	47	
Tayside inland	24	17	14	17	41	93	86	67	
Argyll and Bute inland	14	35	28	34	26	40	28	6	
Central all inland	15	33	12	17	45	40	33	32	
Highlands inland	17	118	77	88	83	120	76	59	
North-east Scotland inland	20	49	43	40	61	51	40	38	
Tayside coastal	24	3	1	0	1	2	6	7	
North-east Scotland coastal	20	11	9	1	4	12	18	20	
Orkney	21	28	26	13	24	22	20	18	
Shetland	22	25	22	5	12	5	0	4	
Highlands coastal	17	86	52	36	48	54	41	32	
Western Isles - Lewis, Harris, Uist	18/25	18	7	9	11	12	22	16	
Scotland totals		515	347	330	479	626	573	527	
Northern Ireland									
Northern Ireland coastal	13	22	6	10	14	26	23	20	
Northern Ireland inland	13	32	10	25	39	70	59	83	
Northern Ireland totals		54	16	35	53	96	82	103	
UK and Isle of Man totals		874	426	437	783	1283	1437	1753	

3. The data demonstrate temporal peaks in local and regional populations. As an example, the inland Scottish highland line shows a peak of 120 territorial pairs since the baseline of 1939, but then shows significant reductions to 2014 by which time the territorial population had halved to 59 pairs. This trend can be seen in other Scottish inland populations such as Argyll and Bute where the population has reduced even more markedly.

4. The most remarkable growth is in lowland England, this being primarily associated with the use of man-made structures and both working and disused quarries as breeding sites. This change is doubtless the lead story of the UK 21st century Peregrine population and has been facilitated by benign human attitudes, an abundance of locally available prey and high levels of occupation, coupled with strong breeding metrics and the very obvious adaptability of the bird itself.

While the population of Peregrines in the UK has increased, the total figure masks a major distributional shift away from the uplands (North-east Scotland Raptor Study Group 2015 (Wilson *et al.*, 2018) and towards lowland regions and the coast. Illegal persecution continues to limit numbers in the Scottish uplands, particularly on grouse moors and in northern England, breeding productivity on grouse moors has been 50% lower than at nests in other habitats, indicating that illegal persecution on land managed for Red Grouse shooting remains an important pressure on the population (Amar *et al.*, 2012). Sporadic persecution of Peregrines also occurs at the hands of the more militant members of the pigeon fancier community.

In summary, over the period of 75 years 1939 – 2014, UK Peregrine populations have recovered from the poisonous effects of organochlorine pesticides in the 1950s and 1960s (Newton, 2013) which had reduced the baseline 1930s population by 51% to 426 pairs by 1962. By 2014 that population low point had more than quadrupled to 1753 pairs largely because of human structure and quarry range expansion mainly in lowland England enabled by the ban on organochlorine agrochemicals and reduction of persecution during the 20th century (Wilson *et al.*, 2018). It is heart-warming to note that in the 21st century, the same human agency that poisons, traps and shoots our Peregrines on grouse moors in the name of 'sport', that shot them by government order in support of the UK war effort and then that poisoned them with organochlorine pesticides in the 1950s and 1960s in the name of agricultural industrialisation, now welcomes and supports this magnificent falcon in the very heart of our urban landscapes.

Urban Peregrines
Having noted the significant increase in urban-breeding Peregrines in the UK in the 20th century, it is worth looking in greater detail how this increase arose.

As Robbrecht *et al.* (2007) note in their excellent article, apart from populations in south-west Germany (Baden-Württemburg) and east-central France (Vosges), the consequence of the post-war agricultural pesticide crisis was the virtual extinction of Peregrines across much of continental Europe. Peregrines began to breed again in Berlin in 1986 and by the late 1980s/early 1990s they had established themselves in many cities across the country (Daniel Schmidt pers. Comm. to RS, with thanks to Remo Probst). In The Netherlands prior to the early 1990s Peregrines had bred sporadically on the ground (*e.g.* Brouwer, 1930), in stick nests in trees (*e.g.* Wigman, 1951) and on buildings, but the urban population began to expand in the early 1990s. In Belgium, Robbrecht *et al.* (2007) note that the falcons disappeared completely as a breeding species though migratory Peregrines from Sweden were seen occasionally, as were a handful of individual birds that had been ringed in Germany. Sporadic breeding attempts were noted in the late 1980s, but Eagle Owl predation kept successful breeding attempts to a minimum. Then, in

F.p. peregrinator, Gujarat, India. *Nirav Bhatt.*

F.p. minor and Hamerkop *(Scopus umbretta)*, Ethiopia. *Torsten Pröhl.*

1994 Peregrines bred in a nest box placed on a power station chimney, other pairs following in later years on other human structures (bridges, churches and power stations *etc.*) so that the number of successful breeding pairs reached 32 by 2006. In Burgundy (adjacent to the Vosges) predation by Eagle Owls also limited falcon numbers, but again from the mid-1990s an urban population began to develop and to spread, slowly, across France (see Strenna (2013) for Burgundy and Pasquier *et al.* (2018) for south-west France, from Toulouse to the Pyrénées).

The Peregrine Falcon

To further investigate the way in which urban breeding developed in the UK we have looked at the degree to which Peregrine juveniles fledged in different site types go on to breed in the same site type in which they were raised – is there evidence of site type imprinting? 95 records, based on ringing re-sightings during the period 1994-2021 were collected, and have been set down in Figs. 10.22-10.25. In each case the figure shows the site type used by a juvenile from a defined site type on its own first breeding.

As noted above, the interesting statistic is that four of the seven additions to the urban building data were European Peregrines. The increase in urban breeding in the UK, which seems to have begun in the south of England appears to have begun later than in Germany and the neighbouring countries. While it is, of course, possible that there were falcons in the UK who transferred from quarry sites to urban sites, it seems possible that the significant increase in the urban population in the UK was, at least in part, aided by an influx of European falcons finding readily available breeding sites after short northward dispersal flights. This suggestion is in line with the normal dispersal distances seen in European nominate Peregrines, and also in the continuing growth of urban Peregrines in England, particularly southern England. For an interesting, and thoroughly modern, article on the relationship that has developed between humans and Peregrines, see Searle *et al.* (2022).

F.p. pealei, north-west USA. Nick Dunlop.

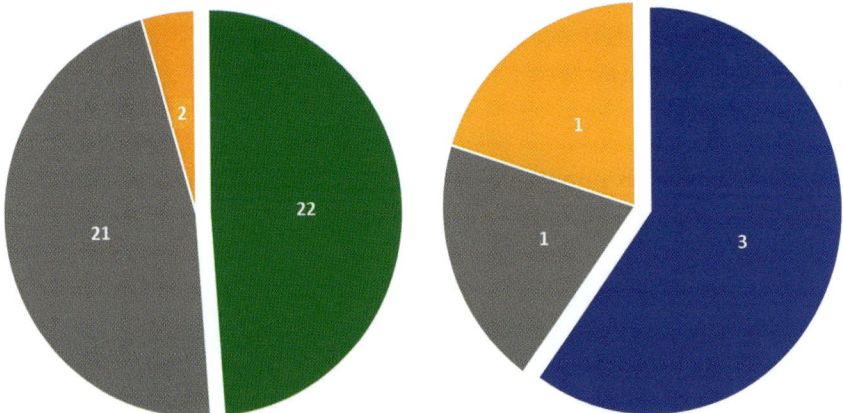

Figure 10.22 Natural Inland natal site to breeding site type. Data from 1995-2018 (45 movements recorded). The data shows that 49% of fledglings at natural inland sites remained faithful to that site type, with 47% moving to quarries and 4% to human structures.

Figure 10.23 Coastal natal site to breeding site type. Data from 1995-2018. Only five records which is insufficient to draw reliable conclusions. Data as recorded shows 60% site fidelity, with 20% moving to quarries and 20% to human structures.

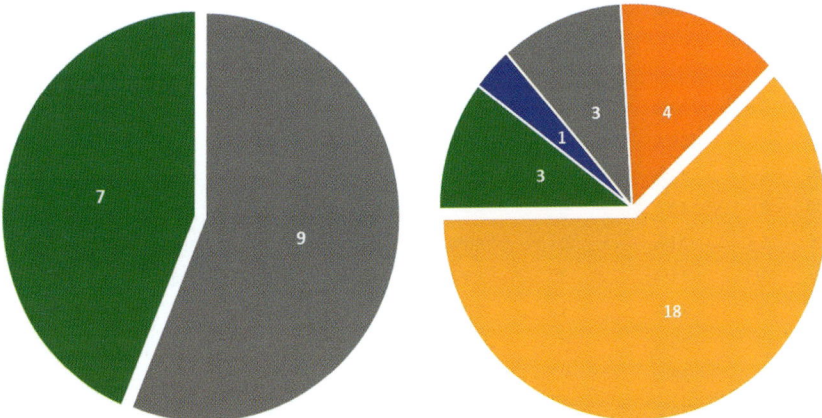

Figure 10.24 Quarry natal site to breeding site type. Data from 1995-2018 (16 movements recorded). The data shows 56% site fidelity, with 44% moving to Natural Inland sites.

Figure 10.25 Human Structures natal site to breeding site type. Data from 1995-2018 (29 movements recorded). Excluding the four European incomers, the data shows 72% site fidelity, the highest fidelity seen in this exercise. 12% of juveniles moved to natural inland sites with 12% moving to quarries. One juvenile (4%) moved to a coastal site. The interesting statistic is that four of the juveniles recorded at human structure sites had fledged in mainland Europe (three in The Netherlands and one in France). All four birds were female.

The Key to all four figures is below.

Selected sample UK population data – post 2014

London LNHS area

The population and breeding data of this area is set down in Table 10.5 and demonstrate urban and suburban breeding habitat expansion commencing 2001 (though a pair was first observed in Docklands in 1998). The London Natural History Society (LNHS) area of 3258km^2 lies within a 20-mile radius around St. Pauls Cathedral and Greater London comprises approx. 50% of that area. It overlaps the adjacent counties of Berkshire, Buckinghamshire, Essex, Hertfordshire, Kent and Surrey.

Year	Number pairs present	Cumulative population growth rate	Number pairs successful	Breeding success percent.	Number juveniles fledged	Number fledged/ successful breeding attempt	Number fledged/ territorial pair	Km^2 per pair	Average spacing Km
2001	1		1	100%	3	3.00	3.00	3258	64.39
2002	3	73%	2	67%	1	0.50	0.33	1086	37.18
2003	4	59%	1	25%	1	1.00	0.25	815	32.21
2004	4	41%	2	50%	0	0.00	0.00	815	32.21
2005	6	43%	5	83%	3	0.60	0.50	543	26.29
2006	5	31%	3	60%	2	0.67	0.40	652	28.81
2007	13	44%	3	23%	8	2.67	0.62	251	17.87
2008	15	40%	4	27%	9	2.25	0.60	217	16.62
2009	19	39%	5	26%	11	2.20	0.58	171	14.75
2010	22	36%	5	23%	8	1.60	0.36	148	13.72
2011	28	35%	9	32%	23	2.56	0.82	116	12.15
2012	23	30%	10	43%	28	2.80	1.22	142	13.44
2013	21	26%	15	71%	38	2.53	1.81	155	14.05
2014	25	26%	18	72%	51	2.83	2.04	130	12.86
2015	25	24%	16	64%	47	2.94	1.88	130	12.86
2016	25	22%	21	84%	53	2.52	2.12	130	12.86
2017	27	21%	17	63%	41	2.41	1.52	121	12.41
2018	41	23%	25	61%	63	2.52	1.54	79	10.03
2019	53	23%	26	49%	69	2.65	1.30	61	8.81
2020	51	22%	28	55%	72	2.57	1.41	64	9.03

Total number of juveniles fledged 2001-2020   531

Table 10.5 Nominate Peregrines breeding in Inner London, Greater London and into the Home Counties. This area, as defined by the London Natural History Society (LNHS) comprises a 20 mile (32km) radius circle centred on St Paul's Cathedral. With thanks to Stuart Harrington.

Population

This London population increased from 3 to 25 pairs in the 12-year period between 2002 and 2014 (the last two UK national survey dates), an 8-fold increase, whilst over the same period human structure territorial pairs increased 6-fold in England as a whole from 42 to 258 pairs. On this basis it appears that London is broadly representative of England as a whole. By 2020 the London population had doubled again in 6 years from 25 to 51 pairs at an annual growth rate of 22% since 2001 (Fig.10.26).

While these high growth rates are typical of human structure populations in the rest of England, for Scotland, Wales and Northern Ireland this trend started later and only became evident in the 2014 census data. Breeding success rates as measured by percentage of occupied territory holders fledging one or more juveniles vary with seasons, fluctuating principally with weather conditions at critical times and prey availability. The 2014 census shows that the 72% success rate in London compared to 68% for England as a whole on human structures. This compares to an overall success rate for all site types in England of 57%.

Additionally in 2014, number of fledglings produced per successful site was 2.83 in London vs 2.60 for human structure type sites in England as a whole. The combination of high breeding success rates and high output of fledged young produced a remarkable figure of 2.04 fledged young per territorial pair. It is clear that the human structure population continues to produce surplus juveniles which will add to the overall UK population and create more pressure for new and existing urban site territories.

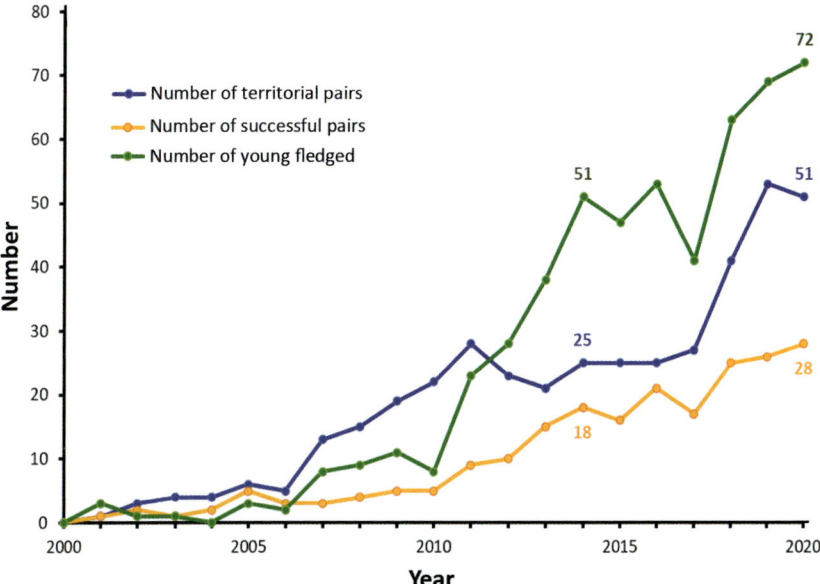

Figure 10.26 Number of territorial pairs, successful pairs and fledged young of nominate Peregrines within a 20-mile (32km) radius of St Paul's Cathedral, London.

The Peregrine Falcon

Since 2014, LNHS London breeding metrics have trended downwards with successful breeding at 55% of sites in 2020 and the number of fledglings produced per successful site being 2.57 (a success rate of 1.41 fledglings/territorial pair). As a result, the rate of population growth has slowed. By 2020 the density of territorial pairs in LNHS London stood at 1 per 64km^2 (2014 – 130km^2) an average inter-pair spacing of 9km (2014 – 12.9km). Based on available post-2014 London data, it would seem reasonable to assume that the England human structure Peregrine population will have continued to increase, but perhaps at a reduced rate, when compared to the 12 years ending in 2014. A territorial next pair average spacing of 9km in 2020 is by no means considered dense in the context of traditional UK site types, and given that nest sites and food supply are not limiting, then it seems that this population can expand further in London.

Gloucestershire

Gloucestershire is a largely rural county of 3150km^2 in the west of England which comprises 50% arable farmland, 31% Pastures and grassland, 10% mixed woodland and 9% discontinuous urban land cover. Fig.10.27 summarises Peregrine breeding metrics (unpublished data with thanks to Gloucestershire Raptor Monitoring Group (GRMG)).

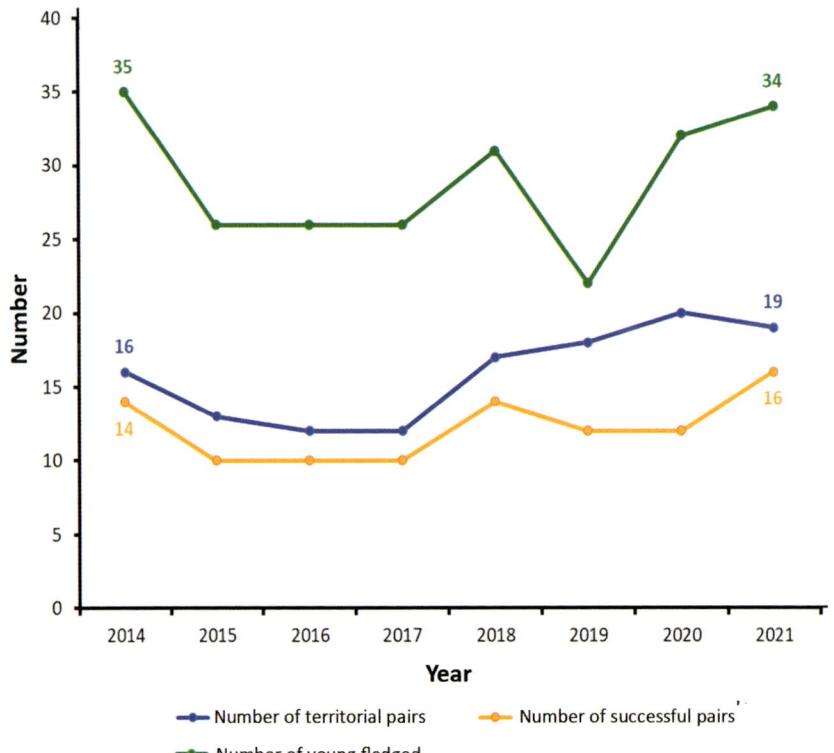

Figure 10.27 Number of territorial pairs, successful pairs and fledged young of nominate Peregrines within an area of 3150km^2 of Gloucestershire, England.

This population has maintained a stable level over the 7-year period since the last UK census in 2014. Although it has only increased overall by 20% in the 7 years ending in 2021, these data mask some underlying trends. In particular, three pylon breeding pairs have dropped out and been replaced by three pairs on buildings. Moreover, since 2020 there has been a notable increase in falcons seen prospecting site opportunities at human structures across the county, particularly on small local village churches where nestboxes have been provided. This augurs well for further population expansion in the county.

For the 8 years ending in 2021 the mean number of fledglings produced per successful site was 2.30 in Gloucestershire and the combination of good breeding success rates and good output of fledged young produced a mean figure of 1.83 fledged young/territorial pair.

North and west Cornwall 'Atlantic' coast

The authors produced a paper (Watson and Sale, 2020) which detailed a severe decline in the Peregrine Falcon population of this archetypal traditional Peregrine stronghold. In stark contrast with urban and inland rural sites, from around 2014 (the time of the last UK national survey), this population went into severe decline both in numbers of territorial pairs in occupation and in all breeding metrics Fig.10.28).

Figure 10.28 Number of territorial pairs, successful pairs and fledged young of nominate Peregrines within an area comprising *c*.225km of coastline in north and west Cornwall, south-west England.

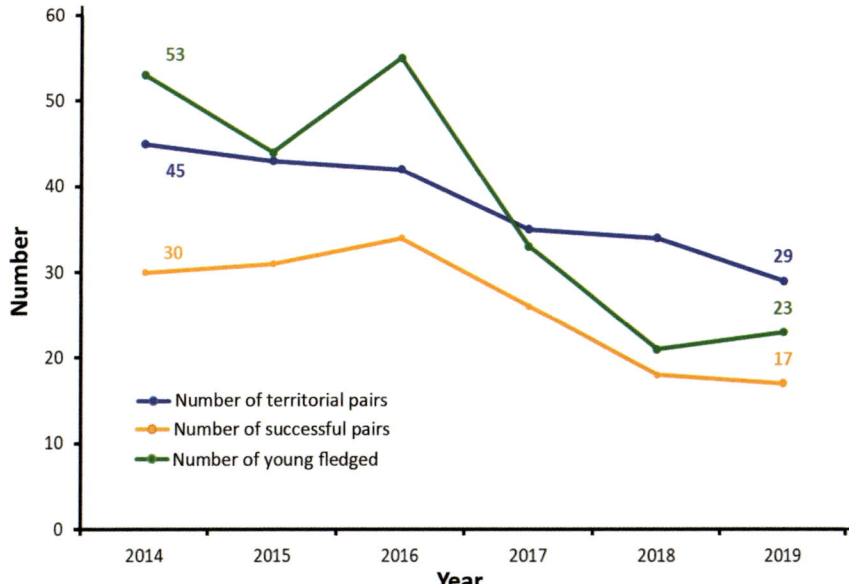

The Peregrine Falcon

Southern Scotland sample territorial pairs data 2014 -2021

Southern Scotland includes BTO mapping areas Dumfries and Galloway, Lothian and Borders, South Strathclyde and Central Scotland, $c.25,000km^2$ which comprises 13% arable farmland, 40% pasture and grassland, 21% mixed woodland, 12% moorland, 8% peat bog and 6% discontinuous urban land cover. A summary of Peregrine breeding metrics is set down in Fig.10.29.

This population has maintained a stable level over the 7-year period since the last UK census in 2014. It is notable that the 'human structure' element has not increased at the same rate as in lowland England. In fact all site type populations in the sample remained stable since the last national census in 2014. For the seven years ending in 2021, the mean number of fledglings produced per successful site was 2.72, and the combination of good breeding success rates and good output of fledged young produced a mean figure of 1.52 fledged young per territorial pair over the same period.

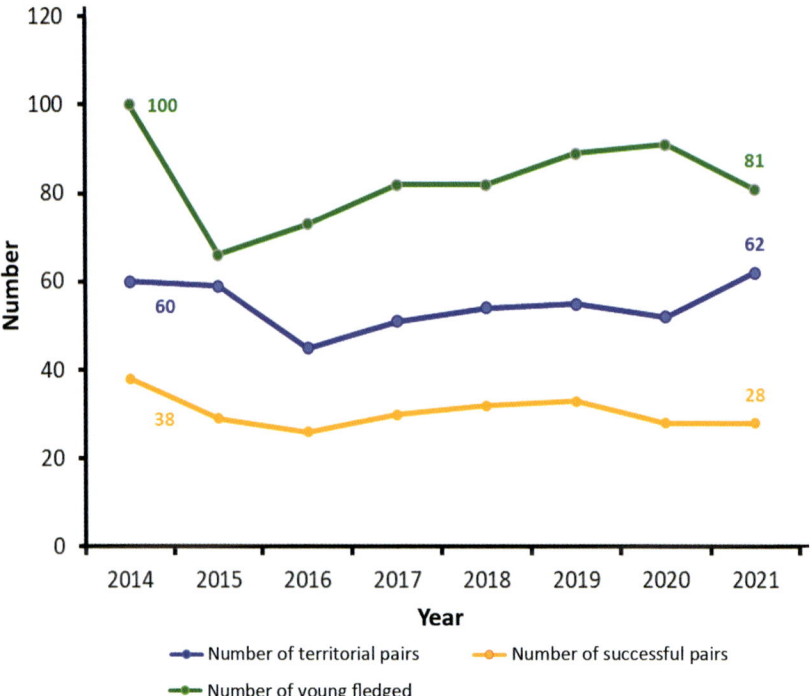

Figure 10.29 Number of territorial pairs, successful pairs and fledged young of nominate Peregrines within an area of $c.25,000km^2$ of southern Scotland.

Population

The South Dorset/Devon coast

The number of occupied territories on this 200km stretch of coast from Lyme Regis in Dorset to Plymouth in Devon has remained stable from 2014 to 2022, with 37 pairs in 2014 and 36 pairs in 2022. We have been advised by P. Johnson (pers. comm.) that in 2022, 28 (78%) of 36 territories produced 64 young being 2.25 per successful pair and 1.75 per occupied territory. This amounts to the best breeding season on this stretch of coast since the 'crash' in terms of productivity and number of young produced. The data for this area is set down in Table 10.6 which compares/contrasts the distinctly different success rates for nominate Peregrines in differing habitats across the UK.

	Principal habitat	Number of territorial pairs								8-year Percentage change
		2014	2015	2016	2017	2018	2019	2020	2021	
London LNHS area	Urban	25	25	25	27	41	53	51	51	+104
Gloucestershire	Rural	16	13	12	12	17	18	20	19	+19
South Scotland	Rural/Coast	60	59	45	51	54	55	52	62	+3
North Cornwall coast	Coast	45	43	42	35	34	29	27	25	-44
South Devon coast	Coast	37	36	35	35	32	32	32	32	-14
Sample total		183	176	159	160	178	187	182	189	
UK totals 2014 census		1753								

Table 10.6 Summary of territorial pairs population movements from 2014-2021 for nominate Peregrines at five differing habitats across the UK. The numbers in the grey boxes are minimum estimates for the given years, provided by S. Harrington for LNHS and P. Johnson for the south Devon coast. P. Johnson also advised that the number of coastal territorial pairs in south Devon had risen to 36 in 2022, with 64 young fledged, the highest number since 2000.

These five, largely unpublished, snapshots represent a sample of 10.4% of the 2014 census UK population. Populations have remained stable from 2014-2021 in Gloucestershire and southern Scotland, but have decreased by 44% on the coast of north Cornwall. However, we believe that the decline on this coast is limited to Cornwall as adjacent coastlines in Devon have remained broadly at stable 2014 levels over the same period (M. Darlaston pers. comm. to authors for north Devon, data above for south Devon). In stark contrast the London LNHS area has seen a doubling of its population from 26 pairs to 51 pairs in just seven years. This highlights a continuation of the increase in Peregrine breeding density and range in lowland England urban and semi-urban environments which have been recorded since the 1990s. One of the authors (SW) monitors Gloucestershire Peregrines and has noted significant increases in sightings of 'third birds' at local 'human structure' territories since the 2014 census in the breeding season. Additionally, many more low buildings, principally churches, have been prospected and occupied over the past several years. Such breeding sites are low when compared to city sites but stand high in their local town or village environment.

This, albeit small, data sample suggests a continuation of the recent trend of population stability for Peregrines breeding at traditional sites in the UK in contrast to the apparent continuation in growth of the lowland England populations breeding on human structures and in quarries. The 11% annual growth in LNHS area Peregrine population since 2014 represents a slowing down in the previous annual growth rate of 18% between 2002 and 2014. It will be interesting to see at what level UK lowland urban Peregrine populations become stable, given that in the LNHS area average linear spacing per territory holding pair was 9km by 2021.

Global Climate Change
Over the course of the first twenty years of the new millennium UK Peregrines have dramatically expanded their range into all areas of human habitation from the smallest village church to the highest tower in the middle of our urban landscapes. A combination of a warming world and the fact that cities are warmer than rural environments has encouraged the falcons to breed earlier, allowing their offspring a longer period to hone their hunting skills before winter arrives – see Fig. 10.30.

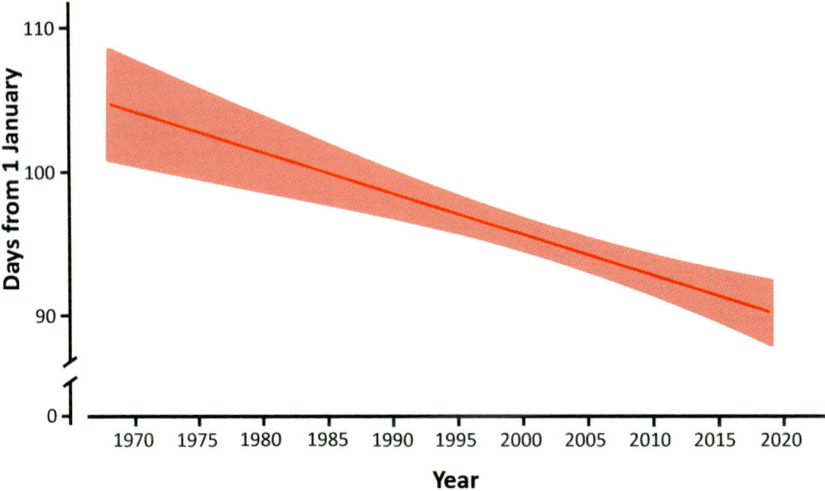

Figure 10.30 The change in mean first egg laying date for UK Peregrine Falcons. The change is consistent with the findings of McLean *et al.* (2022) who investigated the life history and phenotypical changes potentially attributable to climate change in 60 European bird species. Peregrines were not one of the 60, but McLean *et al.* note that temporal changes in laying date, together with body condition and offspring number were consistent with rising temperatures. However, McLean *et al.* also noted that non-temperature related factors also influenced behaviours, so it was not solely climate change that was influencing some species, as these were, for instance, very sensitive to human impacts. In the case of Peregrines and their move to urban breeding, it seems that both climate change and human impacts are affecting breeding behaviour. The figure is redrawn from data on the website of the British Trust for Ornithology.

But this change has not been replicated in 'wild' country breeding Peregrines. Studying *F.p. brookei* on Sicily, Sarà *et al.* (2022) found that although spring temperatures had increased on the island, global warming had also brought an increase in rainfall in February, and that had delayed the start date of the falcons' incubation by seven days between 1979 and 2019: a similar delay was also seen in the island's Lanner Falcon population. As noted in Chapter 8, Potvin *et al.* (2016) also noted changes in wintering and breeding ranges as a result of global warming[17].

As the data in Chapter 4 note, the Peregrine has an extremely plastic diet, its menu at any given time and place essentially matching the available prey spectrum, including mammalian prey in times of peak rodent years (as occurs regularly at northern latitudes). In the northern hemisphere the effect of global warming has been to allow a northward expansion of the range of some species, usually at the expense of the southern range limit as temperatures increase. There has been a significantly less southern expansion as easily reached and usable land masses become much rarer as higher latitudes are reached. It is anticipated that Peregrines of all sub-species will follow the prey as long as this is an option, although the possibility that island species may eventually become extinct cannot be discounted. On islands, healthy (perhaps young) adults may move to nearby land masses to breed, the remaining island (perhaps mostly elderly) population dying out. If no young birds move, then inbreeding may hasten population collapse as prey and Peregrine numbers reduce.

As noted in Chapter 8, Gu *et al.* (2021) looked at the potential change in both breeding and wintering areas of *F.p. calidus* as a consequence of global warming using model simulations to predict the position in 2070. As regards breeding areas, Gu *et al.* suggest the northward expansion of the tundra will probably see Peregrines abandoning Kolguev Island and that there will be reductions in the populations of all the other five current breeding areas, these varying from 37.2% on the Yamal Peninsula to 92.7% on the Kola Peninsula. Whether nominate Peregrines from further south in Russia will move north to replace their Arctic cousins probably rests on whether adequate prey fills in the gaps left by the northern-movement of current *F.p. calidus* prey, and whether human expansion northwards occurs. If the latter occurs, will it leave room for both Peregrines and Peregrine prey species? Gu *et al.* also predict changes in migration timing and in the position of wintering areas for *F.p. calidus*, something which was also noted by Therrien *et al.* (2017) in their study of eastern North American migratory raptors: Peregrines were not mentioned specifically in the work, though were obviously covered by it.

What is clear from the data on global warming is that the main driver is the increase in 'greenhouse' gases – primarily CO_2 (carbon dioxide), but also

---

[17] For a very interesting review of the effect of local warming across species, see Antão *et al.* (2022) who used a unique dataset covering almost 1500 species of birds, mammals, insects and plants in Finland to study the changes brought about by global warming over the period 1978-2017. The data showed hundreds of examples of distribution changes, with local extinctions. In general, the changes saw species moving to higher latitudes or elevations.

methane (CH_4) – in the atmosphere. The polar regions are warming faster than more southerly areas because of the reduction in solar albedo. Albedo is a measure of the reflection of incident solar radiation by any surface. The lighter (whiter) the surface, the higher the albedo and, correspondingly, the darker (blacker) the surface, the lower the albedo. As the albedo of ice and snow is higher than that of dark ocean water, as the sea ice of the polar regions reduces, the more of the Sun's heat is absorbed by the replacement dark ocean waters and the local area warms. Sea ice coverage during Arctic winters has been declining for many years, and during the winter 2021-22 in the southern hemisphere Antarctic sea ice coverage was a record low (Raphael and Handcock, 2022). An additional, and potentially worrying, feedback mechanism was reported recently, Duan *et al.* (2022) noting that sea ice loss in the Arctic (the Barents and Kara seas) was enhancing winter warming on the Tibetan Plateau. The plateau covers 2,500,000km^2 and is home to thousands of glaciers that feed Asia's great rivers. The plateau provides water for 40% of the world's population so any threat to it caused by warming is a threat to the world order. To add to the concern over this linkage, Jeong *et al.* (2022) have noted another, warming of the Indian Ocean being a driver of increased temperatures in the Arctic: the Earth is not parcelled as neatly as mankind would wish. As John Donne wrote (in the 17th century), '*no man is an island*'.

Snow also has a higher albedo than dark earth, so as snow melts and the underlying darker tundra soils are exposed, warming occurs. This affects the Arctic more than the Antarctic because of the thick snow/ice covering of the southern continent. In the Arctic the warming is also enhanced by melting of the permafrost that underlies the tundra. As the permafrost thaws, the organic material trapped within it inevitably warms, leading to increased microbial activity and the release of both CO_2 and methane, increasing the atmospheric content of both, and creating a vicious circle of global warming. Not surprisingly, the Arctic is warming at twice the rate of the rest of the Earth. The effect of this warming is to increase surface run-off from Greenland's covering ice sheet (Tedstone and Machguth, 2022), raising sea levels and so endangering low-lying coastal areas: such increased levels will have serious implications for countries such as Bangladesh.

Global warming is also changing the weather, bringing more and fiercer storms. Rainfall is increasing as are much heavier rainfall rates. Franke *et al.* (2011), studying *F.p. tundrius* in northern Canada, noted a lower survival rate in adult autumnal migrating falcons due to higher precipitation, *i.e.* the local weather which, of course, is being influenced by global warming, was reducing an adult Peregrine's chances of surviving. In a later study in the same area of the Canadian Arctic, Anctil *et al.* (2014) noted that heavy rainfall during the breeding season was also increasing the mortality of *F.p. tundrius* chicks. Death due to prolonged rainfall has always been the case, with as most tundra nests are open to the elements, but during the last 30 years, the period during which global warming has been clearly noticeable, with the number of times annually when heavy rain (defined as ≥8mm/day) has occurred increasing (Fig. 10.31), with a corresponding increase in chick deaths (Fig. 10.32).

While, again, range and prey plasticity is in the favour of the Peregrine as the Earth warms, the most recent IPCC report (IPCC, 2022) on global warming paints a bleak picture relative to the attempt to limit the Earth's temperature rise to below 1.5°C. Whether the Peregrine's future remains as secure as it currently seems will depend on how this result is received by world governments and how they respond to its message.

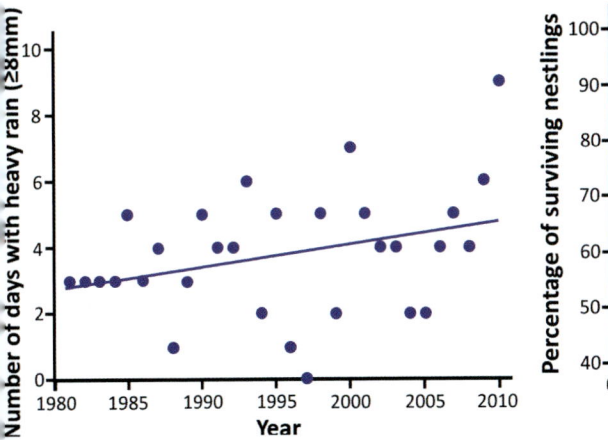

Figure 10.31 Number of days with heavy rain (≥8mm/day) recorded by the Canadian Rankin Inlet Airport Weather Station in the months of July and August during the period 1981-2010. The line is a best fit to the data points. Redrawn from Anctil *et al.* (2014).

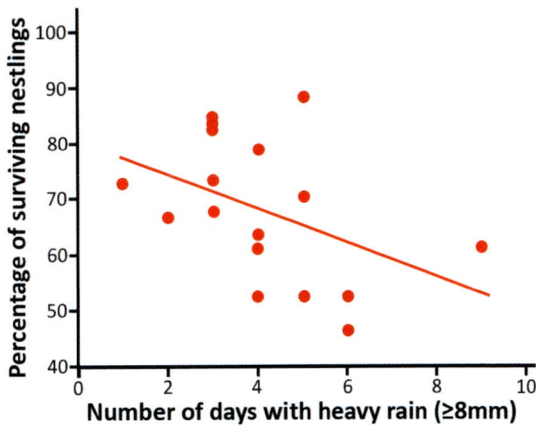

Figure 10.32 Percentage of Peregrine Falcon nestlings surviving to 25 days of age in the months of July and August during the period 1981-2010 in relation to the Number of days with heavy rain (≥8mm/day) recorded by the Canadian Rankin Inlet Airport Weather Station. The annual number of nestlings at hatch varied from 36 to 77, with a mean of 44. The line is a best fit to the data points. Redrawn from Anctil *et al.* (2014).

The effect of severe climate change will not only be the alteration of the temperature pattern across the Earth which has created environmental stability over the years since the last Ice Age and, as far as species other than mankind are concerned over millennia. Bowgen *et al.* (2022) reviewed and analysed hundreds of reports describing the response of individual species to environmental shifts brought about by climatic changes – see, also, Bosco *et al.* with extremely interesting data specifically for avian species in Finland. The researchers noted that conservation interventions to counter the negative effect of changes were best targeted at individual species rather than being generic. That is an important finding, and an encouraging one. But unfortunately may be overwhelmed if climatic changes induce human migration on a potentially massive scale as high temperatures causing crop failures and the lack of water in the sub-continental areas accelerates. Such migrations will have potentially catastrophic effects on vulnerable species whatever interventions are applied. As noted above the range footprint of Peregrines is enormous, but for many other species it is very much smaller. Cruz *et al.* (2021) note that disruption by human migration and/or resettlement will have a much more significant effect on species with smaller ranges – the examples Cruz *et al.* choose for comparison are the Peregrine (range 195 million km^2) – and the Annobón Scops-owl (*Otus feae*), where a population of no more than 250 owls occupy the 16km^2 island of Annobón in the Atlantic: the island is part of Equatorial Guinea. As Gaüzère and Devictor (2021) note, most terrestrial species of birds and mammals cannot react to either temporal or spatial changes as fast as humans can.

While writing this book on the Peregrine, one of the planet's more charismatic species, both authors have been acutely aware that we now live at a potential turning point for our Earth's rich bounty of species. In our lifetimes, human agency has brought our fragile planet to a critical juncture where it is now clear that we must radically change our priorities, give primacy to reversing biodiversity and habitat losses, while preserving and expanding our planet's ecosystems for the benefit of all the species that thrive on this, our one and only Earth.

F.p.anatum, California, USA taking a timeout from hunting for the chicks to have a meal. *Gabriel Gruia.*

'I remember in my childhood seeing or rather recognising my first Peregrine as he glided through immeasurable space among the clouds and never in all my life can I recall having witnessed anything in Wild Nature which left an impression so indelible as that small black cross against the sky.' *H. Mortimer Batten, 1923.*

The painting is by Martin Buckley from a photograph taken by Bill Mayo at the Evesham Bell Tower, UK.

REFERENCES
Titles in parentheses are translations of the originals.

Adams, B., Sinayskiy, I. and Petruccione, F. (2018). An open quantum system approach to the radical pair mechanism, *Sci Rep.* 8, 15719. https://doi.org/10.1038/s41598-018-34007-4

Albuquerque, J.L.B. (1982). Observations on the use of rangle by the Peregrine Falcon (*Falco peregrinus* tundrius) wintering in southern Brasil. *Raptor Res.*, 16, 91–92.

Alerstam, T. (1987). Radar observations of the stoop of the Peregrine Falcon *Falco peregrinus* and the Goshawk *Accipiter gentilis*. *Ibis*, 129, 267–273.

Altwegg, R., Jenkins, A. and Abadi, F. (2014). Nestboxes and immigration drive the growth of an urban Peregrine Falcon *Falco peregrinus* population. *Ibis*, 156, 107–115.

Álvarez, J.C., Meseguer, J., Meseguer, E. and Pérez, A. (2001). On the role of the alula in the steady flight of birds. *Ardeola*, 48, 161-173.

Amar, A., Court, I.R., Davison, M., Downing, S., Grimshaw, T., Pickford, T. and Raw, D. (2012). Linking nest histories, remotely sensed land use data and wildlife crime records to explore the impact of grouse moor management on Peregrine Falcon populations. *Biol. Conserv.*, 145, 86–94.

Ambrose, R.E. and Riddle, K.E. (1988). Population dispersal, turnover, and migration of Alaska Peregrines. In *Peregrine Falcon Populations: Their Management and Recovery* (ed. T.J. Cade et al.). The Peregrine Fund, Boise, Idaho, pp. 677–684.

Anctil, A., Franke, A. and Bêty, J. (2014). Heavy rainfall increases chick mortality of an Arctic top predator: experimental evidence and long-term trend in Peregrine Falcons. *Oecologia*, doi 10.1007/s00442-013-2800-y

Anderson, C.M. and White, C.M. (2000). Recent observations on Peregrine Falcons *Falco peregrinus* of the Cape Verde Islands, Atlantic Ocean. In *Raptors at Risk* (ed. R.D. Chancellor and B.-U. Meyburg). World Working Group on Birds of Prey and Owls/ Hancock House, pp. 685-689.

Andersson, M. and Norberg, R.Å. (1981). Evolution of reversed sexual size dimorphism and role partitioning among predatory birds with a size scaling of flight performance. *Biol. J. Linne. Soc.*, 15, 105–130.

Andreotti, A., Fabbri, I., Menotta, S. and Borghesi, F. (2018). Lead gunshot ingestion by a Peregrine Falcon, *Ardeola*, 65, 53-58.

Antão, L.H., and 25 others. (2022). Climate change reshuffles northern species within their niches. *Nature Climate Change*, doi:org/10.1038/s41558-022-01381-xhttps://doi.org/10.1038/s41558-022-01381-xtps://doi.org/10.1038/s41558-022-01381-x

Applegate, J.R., Van Wettere, A., Christiansen, E.F. and Degernes, L.A. (2017). Management and case outcone of gastric impaction in four raptors: a case series. *J. Avian Med. Surg.*, 31, 62-69.

Artukhin, Y. (2009). Peregrine Falcon *Falco peregrinus japonensis* on the Kurile Islands. In *Peregrine Falcon Populations* (Sielicki, J. and Mizera, T. eds), Turul Publishing and Poznań University of Life Sciences, pp. 311-322.

Aschoff, J. and Pohl, H. (1970). Rhythmic variations in energy metabolism. *Fed. Proc.*, 29, 1541-1552.

Askew, G. N. and Ellerby, D. J., (2007). The mechanical power requirements of avian flight. *Biol. Lett.* 3, 445-448.

Augspurger, T. and Boynton, A. (1998). Organochlorines and mercury in Peregrine Falcons eggs from western North Carolina. *J. Raptor Res.*, 32, 251-254.

Ayala-Perez, V., Carmona, R., Nallely, A. and Castillo, F. (2021). Effect of nesting substrate on the breeding performance of Peregrine Falcons in Guerro Negro, Baja Califonria Sur, Mexico, 2012-2016. *J. Raptor Res.*, 55, 181-189.

Baillie, S.R., Marchant, J.H., Leech, D.I., Joys, A.C., Noble, D.G., Barimore, C., Downie, I.S., Grantham, M. J., Risely, K. and Robinson, R.A. (2010). Breeding Birds in the wider countryside: their conservation status 2009. *BTO Research Report No. 541*, BTO, Thetford, UK. www.bto.org/birdtrends.

Baker, J.A. (1967). *The Peregrine*. Collins, London.

Ballard, J.W.O. and Whitlock, M.C. (2004). The incomplete natural history of mitochondria. *Mol. Ecol.*, 13, 729–744.

Balmori, A. (2009). Electromagnetic pollution from phone masts. Effects on wildlife. *Pathophysiology*, doi:10.1016/j.pathophys.2009.01.007

Banks, A.N., Crick, H.Q.P., Coombes, R., Benn, S., Ratcliffe, D.A. and Humphreys, E.M. (2010). The breeding status of Peregrine Falcons *Falco peregrinus* in the UK and Isle of Man in 2002. *Bird Study*, 57, 421–436.

Barboza, T., Beaufrére, H. and Moens, N. (2020). Effects of perching surfaces and foot bandaging on central metatarsal foot pad weight loading of the Peregrine Falcon (*Falco peregrinus*). *J. Avian Med. Surg.* 34, 9-16.

Barker, A. (2020). Letter in response to Burnside and Pamment article 'If it flies, it dies'. *Brit. Birds*, 113, 425.

Barlow, H.B. and Ostwald, T.J. (1972). Pecten of the Pigeon's eye as an inter-ocular eye shade. *Nature New Biology*, 236, 88-90.

Barnaby, W. (1975), Fight for the falcon. *Nature*, 258, 473.

Barnes, J.G. and Gerstenberger, S.L. (2015). Using feathers to determine mercury contamination in Peregrine Falcons and their prey. *J. Raptor Res.*, 49, 43–58.

Barnes, J.G., Varland, D.E., Fleming, T.L., Buchanan, J.B. and Gerstenberger, S.L. (2018). Mercury contamination in Peregrine Falcons (*Falco peregrinus*) in coastal Washington, 2001-2016. *Wilson J. Ornith.*, 130, 958-968.

Barreau, D. and Bergier, P. (2001). (The birds of of the Marrakech region. Part 2: Non-passerines). *Alauda*, 69, 167-202.

Barton, N.W.H. and Houston, D.C. (1992). Post-mortem changes in gross intestinal morphology. *Can. J. Zool.*, 70, 1849-1851.

Barton, N.W.H. and Houston, D.C. (1993). A comparison of digestive efficiency in birds of prey. *Ibis*, 135, 363–371.

Barton, N.W.H. and Houston, D.C. (1994). Morphological adaptation of the digestive tract in relation to feeding ecology of raptors. *J. Zool. Lond.*, 232, 133-150.

Barton, N.W.H. and Houston, D.C. (1996). Factors influencing the size of some internal organs in raptors. *J. Raptor Res.* 20, 219-223.

Barton, N.W.H. and Houston, D.C. (2001). The incidence of intestinal parasites in British birds of prey. *J. Raptor Res.*, 35, 71–73.

Basso, E., Drever, M.C., Fonseca, J. and Navedo, J.G. (2021). Semi-intensive shrips farms as experimental areans for the study of predation risk from falcon to shorebirds. *Ecol. and Evol.*, 11, 13379-13389.

Baudat-Franceschi, J., Waka, Y. and Duval, T. (2013). (Breeding range and biology of the Peregrine Falcon *Falco peregrinus ernestii* in New Caledonia). *Alauda*, 81, 97-114.

Beauchamp, G. (2016a). Timing of attacks by a predator at a prey hotspot. *Behav. Eco. Sociobiol.* 70, 269-276.

Beauchamp, G. (2016b). Does sun glare increase antipredator behaviour in prey? *J. Avian Biol.*. 48, 591-595.

Beaupre, E. (1922). The Duck Hawk. *Can. Field-Nat.*, 36, 33–35.

Becker, J.J. (1987). Revision of 'Falco' Ramenta Wetmore and the Neogene evolution of the Falconidae. *Auk*, 104, 270-276.

Becking, J.H. (1989). Henri Jacob Victor Sody) 1892-1959): *His Life and Work*. E.J. Brill, Leiden, The Netherlands.

Beebe, F.L. (1960). The marine Peregrines of the northwest Pacific coast. *Condor*, 62, 145–189.

Beingolea, O. and Arcilla, N. (2021). Linking Peregrine Falcon's (*Falco peregrinus*) wintering areas in Peru with their North American natal and breeding grounds. *J. Raptor Res.*, 54, 222-232.

Bélisle, E., Gahbauer, M.A. and Bird, D.M. (2012). Unusual behaviour by a juvenile Peregrine Falcon: interference, siblicide and incest. *J. Raptor Res.*, 46, 324–326.

Bell, D.A., Griffiths, C.S., Caballero, I.C., Hartley, R.R. and Lawson, R.H. (2014). Genetic evidence for global dispersion in the Peregrine Falcon (*Falco peregrinus*) and affinity with the Taita Falcon (*Falco fasciinucha*). *J. Raptor Res.*, 48, 44–53.

Bennett, A.T.D. and Cuthill, I.C. (1994). Ultraviolet vision in birds: what is it for? *Vision Res.*, 34, 1471–1478.

Benson, C.W. (1960). The Birds of the Comoro Islands: Results of the British Ornithologists' Union Centenary Expedition 1958. *Ibis*, 130b, 1-106.

Bequaert, J.C. (1954). The Hippoboscidae or Louse-flies (Diptera) of mammals and birds: Part II. Taxonomy, evolution and revision of American genera and species. *Entomologica Americana*, 34, 1-611.

Bergulla, H.L. (1992). *Birds of Vanuatu*. Anthony Nelson, Oswestry, England.

Bertran, K., Busquets, N., Abad, F.X., Garcia de la Fuente, J., Solanes, D., Cordón, I., Costa, T., Dolz, R. and Majó, N. (2012). Highly (H5N1) and low (H7N2) pathogenic avian influenza virus infection in falcons via nasochoanal route and ingestion of experimentally infected prey. *PLoS ONE*, 7(3): e32107. doi:10.1371/journal.pone.0032107

Bhatt, N. and Ganpule, P. (2017). The identification of the Red-naped Shaheen *Falco peregrinus babylonicus*, its separation from *F.p. calidus* in the field, and its status and distribution in north-western India. *Indian Birds*, 13, 85-92.

Bird D.M. and Aubry, Y. (1982). Reproductive and hunting behaviour in Peregrine Falcons, *Falco peregrinus*, in southern Quebec. *Can. Field-Nat.*, 96, 167–171.

BirdLife International (2022) Species factsheet: *Falco peregrinus*. Downloaded from http://www.birdlife.org on 06/03/2022.

Blank, M., Kiger, L., Thielebein, A., Gerlach, F., Hankeln, T., Marden, M.C. and Burmester, T. (2011) Oxygen supply from the bird's eye perspective: Globin E is a respiratory protein in the chicken retina. *J. Biol. Chem.*, 286, 26507-26515.

Blezard, E., Garnett, M., Graham, R. and Johnson, T.L. (1943). The Birds of Lakeland. *Transactions of the Carlisle Natural History Society*, 6. Carlisle Natural History Society, Carlisle.

Boal, C.W. and Dykstra, C.R. (eds). (2018). *Urban Raptors: ecology and conservation of birds of prey in cities*. Island Press, Washington DC, USA.

Bondí, S., Guzzo, E., Vitale, E., Baragona, A., Grancagnolo, D. and Sarà, M. (2016). Factors affecting the diet of Peregrine Falcon in Italy. *Avocetta*, 40, 33-42.

Booms, T.L. and Fuller, M.R. (2003). Gyrfalcon diet in central west Greenland during the nesting period. *Condor*, 105, 528–537.

Bosco, L., Xu, Y., Deshpandee, P. and Lehikoinen, A. (2022). Range shifts of overwintering birds depend on habitat type, snow conditions and habitat specialization. *Oecologia*, 199, 725-736.doi.org/10.1007/s00442-022-05209-5

Bossu, C.M., Heath, J.A., Kaltrenecker, G.S., Helm, B. and Ruegg, K.C. (2021). Clock-linked genes underlie seasonal migratory timing in a diurnal raptor. *Proc. Roy. Soc. B*, 289, 2021507, doi.otg/10.1098/rspb.2021.2507

Boulet, M., Olsen, P., Cockburn, A. and Newgrian, K. (2001). Parental investment in male and female offspring by the Peregrine Falcon *Falco peregrinus*. *Emu*, 101, 95-103.

Bowgen, K.M., Kettel, E.F., Butchart, S.H.M., Carr, J.A., Foden, W.B., Magin, G., Morecroft, M.D., Smith, R.K., Stein, B.A., Sutherland, W.J., Thaxter, C.B. and Pearce-Higgins, J.W. (2022). Conservation interventions can benefit species impacted by climate change. *Biol. Conserv.*, 269, 109524, doi.org/10.1016/j.biocon.2022.109524

Boyce, D.A. (1985). Merlins and the behaviour of wintering shorebirds. *Raptor Res.*, 19, 95–96.

Bradley, M. and Oliphant, L.W. (1991). The diet of Peregrine Falcons in Rankin Inlet, Northwest Territories: an unusually high proportion of mammalian prey. *Condor*, 93, 193–197.

Bradley, M., Johnstone, R., Court, G. and Duncan T. (1997). Influence of weather on breeding success of Peregrine Falcons in the Arctic. *Auk*, 114, 786–791.

Brambilla, M., Rubolini, D. and Guidali, F. (2004). Rock climbing and Raven *Corvus corax* occurrence depress breeding success of cliff-nesting Peregrines *Falco peregrinus*. *Ardeola*, 51, 425–430.

Brambilla, M., Rubolini, D. and Guidali, F. (2006a). Factors affecting breeding habitat selection in a cliff-nesting Peregrine *Falco peregrinus* population. *J. Orn.*, 147, 428–435.

Brambilla, M., Rubolini, D. and Guidali, F. (2006b). Eagle Owl *Bubo bubo* proximity can lower productivity of cliff-nesting Peregrines *Falco peregrinus*. *Ornis Fennica*, 83, 20–26.

Brazil, M.A. and Hanawa, S. (1991). The status and distribution of diurnal raptors in Japan. Birds of Prey Bulletin No. 4.

Bregulla, H.L. (1992). *Birds of Vanuatu*. Anthony Nelson, Oswestry, UK.

Breinlinger, S., Phillips, T.J., Haram, B.N., Mareš, J., Martínez Yerena, J.A., Hrouzek, P., Sobotka, R., Henderson, W.M., Schmieder, P., Williams, S.M., Lauderdale, J.D., Wilde, H.D., Gerrin, W., Kust, A., Washington, J.W., Wagner, C., Geier, B., Liebeke, M., Enke, H., Niedermeyer, T.H. and Wilde, S.B. (2021). Hunting the eagle killer: A cyanobacteria neurotoxin causes vacuolar myelinopathy. *Science*, 371, eaax9050 (2021), 26 March 2021.

Bright, J.A., Marugán-Lobón, J., Cobb, S.N. and Rayfield, E.J. (2016). The shapes of bird beaks are highly controlled by nondietary factors. *Proc. Nat. Acad. Sci. USA*, 113, 5352-5357.

Brighton, C.H., Zusi, L., McGowan, K.A., Kinniry, M., Kloepper, L.N. and Taylor, G.K. (2021). Aerial attack strategies of hawks hunting bats and the adaptive benefits of swarming. *Behav. Ecol.* 32, 464-476.

Brighton, C.H., Kloepper, L.N., Harding, C.D., Larkman, L., McGowan, K., Zusi, L. and Taylor, G.K. (2022). Raptors avoid the confusion effect by targeting fixed points in dense aerial prey aggregations. *Nature Communications*, 13:4778, doi.org/10.1038/s41467-022-32354-5.

Bringmann, A. (2019). Structure and function of the bird fovea. *Anat. Histol. Embryol.*, 48:177–200.

Brodkorb, P., (1964). Catalogue of fossil birds: Part 2 (Anseriformes through Galliformes). *Bull. Florida State Mus. Biol. Sci.*, 8.

Brooke, M. de L. (1995). The modern avifauna of the Pitcairn Islands, *Biol. J. Linne. Soc.*, 56, 199-212.

Brouwer, G.A. (1930). (A second case of breeding *Falco peregrinus* (Tunst). in the Netherlands). *Ardea*, 16, 66-67.

Brown, J.W., van Coeverden de Groot, P.J., Birt, T.P., Seutin, G., Boag, P.T. and Friesen, V.L. (2007). *Mol. Ecol.*, 16, 327-343.

Brown, L.E. and Holden, J. (2020). Contextualizing UK moorland burning studies with geographical variables and sponsor identity. *J. App. Ecol.* 57, 2121-2131, https://doi.org/10.1111/1365-2664.13708

Bruce-Mitford, R.L.S. (1975). *The Sutton Hoo Ship Burial*. Trustees of the British Museum, London.

Bruderer, B. and Boldt, A. (2001). Flight characteristics of birds: 1. radar measurements of speeds. *Ibis*, 143, 178–204.

Bruderer, B., Peter, D., Boldt, A. and Liechti, F. (2010). Wing-beat characteristics of birds recorded with tracking radar and cine camera. *Ibis*, 152, 272–291.

Bruggeman, J., Swem, T., Andersen, D.E., Kennedy, P.L. and Nigro, D. (2021). Incorporating productivity as a measure of fitness into models of breeding area quality of Arctic Peregrine Falcons. *Wildlife Biology*, wlb.00475 (doi: 10.2981/wlb.00475).

Bruinzeel, L.W. and van de Pol, M. (2004). Site attachment of floaters predicts success in territory acquisition. *Behav. Ecol.*, 15, 290–296.

Buchanan, J.B. (1991). Two cases of carrion-feeding by Peregrine Falcons in western Washington. *Northwestern Naturalist*, 72, 28–29.

Buchanan, J.B. (1996). A comparison of behaviour and success rates of Merlins and Peregrine Falcons when hunting Dunlin in two coastal habitats. *J. Raptor Res.*, 30, 93–98.

Buchanan, J.B. (2012). Change in Merlin hunting behaviour following recovery of Peregrine Falcon populations suggests mesopredator suppression. *J. Raptor Res.*, 46, 349-356.

Buchanan, J.B., Herman, S.G. and Johnson, T.M. (1986). Success rate of the Peregrine Falcon (*Falco peregrinus*) hunting Dunlin (*Calidris alpina*) during winter. *Raptor Res.*, 20, 130–131.

Buchanan, J.B., Schick, C.T., Brennan, L.A. and Herman, S.G. (1988). Merlin predation on wintering Dunlins: hunting success and Dunlin escape tactics. *Wilson Bull.*, 100, 108–118.

Buchanan, J.B., Hamm, K.A., Salzer, L.J., Diller, L.V. and Chinnici, S.J. (2014). Tree-nesting by Peregrine Falcons in North America: historical and additional records. *J. Raptor Res.*, 48, 61–67.

Bundle, M.W., Hansen, K.S. and Dial, K.P., (2007). Does the metabolic rate–flight speed relationship vary among geometrically similar birds of different mass?, *J. Exp. Biol.*, 210, 1075-1083.

Burnham, K.K. and Burnham, W.A. (2011). Ecology and biology of Gyrfalcons in Greenland. In *Gyrfalcons and Ptarmigan in a Changing World, Volume II* (Watson, R.T., Cade, T.J., Fuller, M., Hunt, G. and Potapov, E. (Eds.), pp1-20. The Peregrine Fund, Boise, Idaho, USA. http://dx.doi.org/ 10.4080/gpcw.2011.0209

Burnham, W.A., Enderson, J.H. and Boardman, T.J. (1984). Variation in Peregrine Falcon eggs. *Auk*, 101, 578–583.

Burnside, E. and Pamment, N. (2020). 'If it flies, it dies'. *Brit. Birds*, 113, 190-192.

Burnside, R.J., Salliss, D., Collar, N.J. and Dolman, P.M. (2021). Birds use individually consistent temperature cues to time their migration departure. *PNAS*, 118 (28), e2026378118

Byre, V.J. (1990). A group of young Peregrine Falcons prey on migrating bats. *Wilson Bull.* 102, 728-730.

Caballero, I.C., Bates, J.M., Hennen, M. and Ashley, M.V. (2016). Sex in the City: Breeding Behavior of Urban Peregrine Falcons in the Midwestern US. *PLoS ONE*, 11(7): e0159054. doi:10.1371/journal.pone.0159054.

Cade, T.J. (1960). Ecology of the Peregrine and Gyrfalcon populations in Alaska, *University of California Publications in Zoology*, 63, 151–290.

Cade, T.J. (1982). *The Falcons of the World*. Cornell University Press, Ithaca; Collins, London.

Cade, T.J. and Bird, D.M. (1990) Peregrine Falcons, *Falco peregrinus*, nesting in an urban environment: a review. *Can. Field-Nat.*, 104, 209–218.

Cadena, C.D. and Zapata F. (2021). The genomic revolution and species delimitation in birds (and other organisms). Why phenotypes should not be overlooked. *Ornithology*, 138, 1-18.

Cameron, M., and Olsen, P. (1993). Significance of caching in Falco: Evidence from a nesting pair of Peregrine Falcons *Falco peregrinus*. In *Australian Raptor Studies: Proceedings of the 10th Anniversary Australasian Raptor Association*, Canberra, September 21-22, 1993 (P. Olsen, ed.), pp43-54.

Campbell, K.E. (1976). The late Pleistocene avifauna of La Corolina, southwestern Ecuador. *Smithsonian Contributions to Paleobiology*, 27, 155-168.

Campioni, L., Delgado, M.D.M. and Penteriani, V. (2010). Social status influences microhabitat selection: breeder and floater Eagle Owls *Bubo bubo* use different post sites. *Ibis*, 152, 569–579.

Canby, J.V. (2002). Falconry (Hawking) in Hittite lands. *Journal of Near Eastern Studies*, 61, 161-201.

Carere, C., Montanino, S., Moreschini, F., Zoratto, F., Chiarotti, F., Santucci, D. and Alleva, E.. (2009). Aerial flocking patterns of wintering Starlings, *Sturna vulgaris*, under different predation risk. *Anim Behav.*, 77, 101–107.

Carlier, P. (1993). Sex differences in nesting site attendance by Peregrine Falcons (*Falco peregrinus brookei*). *J. Raptor Res.*, 27, 31–34.

Carlier, P. (1995). Vocal communications in Peregrine Falcons *Falco peregrinus* during breeding. *Ibis*, 137, 582–585.

Carter, K.M., Lacki, M.J., Dzialak, M.R., Burford, L.S. and Bethany, R.O. (2003). Food habits of Peregrine Falcons in Kentucky. *J. Raptor Res.*, 37, 344-349.

Carver, M. (1998). *Sutton Hoo:Burial Ground of Kings?* British Museum Press, London.

Casini, L. and Morelli, F. (2008). A Peregrine Falcon in flight retrieves nestling falling from a cliff. *J. Raptor Res.*, 42, 225.

Castellanos, A., Argüelles, C., Salinas, F., Rodríguez, A. and Ortega-Rubio, A. (2006), Diet of breeding Peregrine Falcons at a coastal lagoon, Baja California Sur, Mexico. *J. Raptor Res.*, 40, 241-244.

Castilla, A.M., Van Dongen, S., Herrel, A., Francesch, A., Martinez de Aragón. J., Malone, J. and Negro.J.J. (2010). Increase in membrane thickness during development compensates for eggshell thinning due to calcium uptake by the embryo in falcons, *Naturwissenschften*, 97, 143-151.

Chen, D., La Guardia, M.J., Harvey, E., Amaral, M., Wohlfort, K. and Hale, R.C. (2008). Polybrominated Diphenyl Ethers in Peregrine Falcon (*Falco peregrinus*) eggs from northeastern US. *Environ. Sci. Tech.*, 42, 7594-7600.

Chesser, R.T., Banks, R.C., Barker, F.K., Cicero, C., Dunn, J.L., Kratter, A.W., Lovette, I.J., Rasmussen, P.C., Remsen, J.V., Rising, J.D., Stotz, D.F. and Winker, K. (2012). Fifty-third supplement to the American Ornithologists' Union check-list of North American birds. *Auk*, 129, 573–588.

Choi, C-Y. and Nam, H-Y. (2012). Migrating dragonflies: famine relief for resident Peregrine Falcons *Falco peregrinus* on islands. *Forktail*, 28. 149-151.

Choi, C-Y. and Nam, H-Y. (2015). Diet of Peregrine Falcons (*Falco peregrinus*) in Korea: food items and seasonal changes. *J. Raptor Res.* 49, 376-388.

Clarke, A. (1977). Contamination of Peregrine Falcons (*Falco peregrinus*) with Fulmar stomach oil. *J. Zool. Lond.* 181, 11-20.

Clarke, A., Croxall, J.p., Poncet, S., Martin, A.R. and Burton, R. (2012). Important bird areas: South Georgia. *Brit. Birds*, 105, 118-144.

Clevinger, A.P. (1987). Atypical incubation rates at a New Mexico Peregrine Falcon eyrie. *J. Raptor Res.*, 21, 33–35.

Clum, N.J. (1995). Effects of aging and mate retention on reproductive success of captive female Peregrine Falcons. *Amer. Zool.*, 35, 329–339.

Clunie, F. (1972). A contribution to the natural history of the Fiji Peregrine. *Notornis*, 19, 302-322.

Clunie, F. (1976a). Long-tailed Fruit Bats as Peregrine prey. *Notornis*, 23, 245.

Clunie, F. (1976b). A Fiji Peregrine Falcon (*Falco peregrinus*) in an urban-marine environment. Notornis, 23, 8-28.

Cochrane, W.W. and Applegate, R.D. (1986). Speed of flapping flight of Merlins and Peregrine Falcons. *Condor*, 88, 397–398.

Cocker, M. (2007). Peregrine Falcon retrieving prey from a flock of Carrion Crows. *Brit. Birds*, 100, 307.

Collar, N.J. (2002). Insectivory and kleptoparasitism in Peregrine Falcons. *Brit. Birds*, 95, 142.

Combridge, P. (2008). Peregrine Falcon defending prey from flock of Carrion Crows. *Brit. Birds*, 101, 383–384.

Cooper, J. (2020). Sacred specimens. *Evolve*, 42, 56-61.

Cooper, J.E. (1984). Developmental abnormalities in two British falcons (Falco *spp.*). *Avian Pathology*, 13, 639–645.

Cooper, J.E. (1993). Diseases in the Peregrine. In Ratcliffe, D., *The Peregrine Falcon*. Poyser, London, pp. 360–361 and Table 31.

Cooper, J.E., Redig, P.T. and Burnham, W. (1980). Bacterial isolates from the pharynx cloaca of the Peregrine Falcon (*Falco peregrinus*) and Gyrfalcon (*F. rusticolus*) (Bacteria from Falcons). *Raptor Res.*, 14, 6–9.

Court, G.S. (1986). Some aspects of the reproductive biology of tundra Peregrine Falcons. MSc thesis, University of Alberta.

Court, G.S., Gates, C.G. and Boag, D.A. (1988). Natural history of the Peregrine Falcon in the Keewatin District of the Northwest Territories. *Arctic*, 41, 17–30.

Court, G.S., Bradley, D.M., Gates, C.G. and Boag, D.A. (1989). Turnover and recruitment in a tundra population of Peregrine Falcons *Falco peregrinus*. *Ibis*, 131, 487–496.

Court, G.S., Gates, C.G., Boag, D.A., MacNeil, J.D., Bradley, D.M., Fesser, A.C., Patterson, J.R., Stenhouse, G.B. and Oliphant, L.W. (1990). A toxicological assessment of Peregrine Falcons, *Falco peregrinus tundrius*, breeding in the Keewatin District of the Northwest Territories, Canada. *Can. Field-Nat.*, 104, 255-272.

Cozic, E. (2019). (The Peregrine Falcon in Brittany: status of the breeding population, nesting habitat, breeding performance and diet), *Ornithos*, 26, 273-291.

Craig, G.R., White, G.C. and Enderson, J.H. (2004). Survival, recruitment, and rate of population change of the Peregrine Falcon population in Colorado. *J. Wild. Mgt.*, 68, 1032-1038.

Cramp, S. and Simmons, K.E.L. (eds.) (1980). *Handbook of the Birds of Europe, the Middle East and North Africa: The Birds of the Western Palearctic, Vol. 2.* Oxford University Press, Oxford.

Cresswell, W. (1993). Escape responses by Redshanks, *Tringa totanus*, on attack by avian predators. *Anim. Behav.*, 46, 609–611.

Cresswell, W. (1994). Flocking is an effective anti-predator strategy in Redshanks, Tringa totanus. *Anim. Behav.*, 47, 433-442.

Cresswell, W. (1996). Surprise as a winter hunting strategy in Sparrowhawks *Accipiter nisus*, Peregrines *Falco peregrinus* and Merlins *F. columbarius*. *Ibis*, 138, 684–692.

Cresswell, W. and Quinn, J.J. (2013). Contrasting risks from different predators change the overall nonlethal effects of predation risk. *Behav. Ecol.* 24, 871-876.

Cresswell, W. and Whitfield, D.P. (1994). The effects of raptor predation on wintering wader populations at the Tyninghame estuary, southeast Scotland. *Ibis*, 136, 223–232.

Cresswell, W., Lind, J. and Quinn, J.L. (2010). Predator-huting success and prey vulnerability: quantifying the spatial scale over which lethal and non-lethal effects of predation occur. *J. Anim. Ecol.*, 79, 556-562.

Crick, H.Q.P. and Ratcliffe, D.A. (1995). The Peregrine *Falco peregrinus* breeding population of the United Kingdom in 1991. *Bird Study*, 42, 1-19.

Crombach, M. (2020). (Female Peregrine Falcon *Falco peregrinus* with GPS sender from Kolguev, Russia, found dead in Nijmegen, The Netherlands, *De Takkeling*, 28, 86-91.

Cronin, T.W. and Marshall, J. (2011) Patterns and poperties of polarized light in air and water. *Phil. Trans. R. Soc. B.* 366, 619-626.

Crowle, A.J.W., Glaves, D.J., Oakley, N., Drewitt, A.L. and Denmark-Melvin, M.E. (2022). Alternative future land use options in the British uplands. *Ibis*, doi: 10.1111/ibi.13041

Cruz, C., Santulli-Sanzo, G. and Ceballos, G. (2021). Global patterns of raptor distribution and protected areas optimal selection to reduce the extinction crisis. *PNAS*, 118 (37), e2018203118

Culver, E. (1919). Duck hawks wintering in the centre of Philadelphia. *Auk*, 36, 108–109.

Curk, T., Kulikova, O., Fufachev, I., Wikelski, M. Safi, K. and Pokrovsky, I. (2022). Arctic migratory raptor selects nesting area during the previous breeding season. *Front. Ecol. Evol.*, 10:865482, doi:10.3389/fevo.2022.865482.

Curtis, O.E. and Jenkins, A.R. (2002). Shell thickness and size of Peregrine Falcon, *Falco peregrinus minor*, eggs from two areas in South Africa. *Ostrich*, 73, 64-66.

Czechura, G.V. (1984). The Peregrine Falcon (*Falco peregrinus macropus*) Swainson in southeastern Queensland. *Raptor Res.*, 18, 81-91.

Daan, S., Dijkstra, C. and Tinbergen, J.M. (1990). Family planning in the Kestrel (*Falco tinnunculus*): the ultimate control of covariation of laying date and clutch size, *Behaviour* 14, 83-116.

Danner, R.M., Gulson-Castillo, E.R., James, H.F., Dzielski, S.A., Frank, D.C., Sibbaled, E.T. and Winkler, D.W. (2017). Habitat-specific divergence of air conditioning structures in bird bills. *Auk*, 134, 65-75.

Danter, L. (2018). Aerodynamic effect and biomimetic potential of the bony tubercle inside the falcon nostril. MSc. thesis, Hochschule Bremen City University of Applied Science.

Davis, K. (2008). *Falcons of North America*. Mountain Press Publishing Company, Missoula, Montana, USA.

Dawson, R.D., Mossop, D H. and Boukall, B. (2011) Prey use and selection in relation to reproduction by Peregrine Falcons breeding along the Yukon river, Canada. *J. Raptor Res.*, 45, 27-37.

De Rosa, D., Di Febbraro, M., De Lisio, L., De Sanctis, A. and Loy, A. (2019). The decline of the Lanner Falcon in Mediterranean landscapes: competition displacement or habitat loss? *Anim. Conserv.* 22, 24-34.

DeCandido, R. and Allen, D. (2006). Nocturnal hunting by Peregrine Falcons at the Empire State Building, New York City. *Wilson J. Ornith.*, 118, 53-58.

DEFRA (2016). Observatory monitoring framework – indicator data sheet. UK Government Department for Environment and Rural Affairs.

DEFRA (2021). Wild Bird Populations in the UK, 1970 to 2019 – updated for wintering waterbirds. UK Government Department for Environment and Rural Affairs.

Dekker, D. (1980). Hunting success rates, foraging habits, and prey selection of Peregrine Falcons migrating through central Alberta. *Can. Field-Nat.*, 94, 371–382.

Dekker, D. (1988). Peregrine Falcon and Merlin predation of small shorebirds and passerines in Alberta. *Canadian J. Zoo.*, 66, 925–928.

Dekker, D. (1995). Prey capture by Peregrine Falcons wintering on southern Vancouver Island, British Columbia. *J. Raptor Res.*, 29, 26–29.

Dekker, D. (1998). Over-ocean flocking by Dunlins, Calidris alpina, and the effect of raptor predation at Boundary Bay, British Columbia. *Can. Field-Nat.*, 112, 694–697.

Dekker, D. (2003). Peregrine Falcon predation on Dunlins and ducks and kleptoparasitic interference from Bald Eagles wintering at Boundary Bay, British Columbia. *J. Raptor Res.*, 37, 91–97.

Dekker, D. and Bogaert, L. (1997). Over-ocean hunting by Peregrine Falcons in British Columbia. *J. Raptor Res.*, 31, 381–383.

Dekker, D. and Drever, M.C. (2016). Interactions of Peregrine Falcons (*Falco peregrinus*) and Dunlin (*Calidis alpina*) wintering in British Columbia, 1994-2015. *J. Raptor Res.*, 50, 363-369.

Dekker, D. and Ydenburg, R. (2004). Raptor predation on wintering Dunlins in relation to the tidal cycle. *Condor*, 106, 415–419.

Dekker, D., Out, M., Tabak, M. and Ydenberg, R. (2012). The effect of kleptoparasitic Bald Eagles and Gyrfalcons on the kill rate of Peregrine Falcons hunting Dunlins wintering in British Columbia. *Condor*, 114, 290–294.

Delmore, K.E., Toews, D.P.L., Germain, R.R., Owens, G.L. and Irwin, D.E. (2016). The genetics of seasonal migration and plumage color, *Current Biology*, 26, 2167–2173.

Dement'ev G.P. (1940). (*Key to the birds of the Kirghiza SSR, Volume 2 (diurnal birds of prey)*). Frunze (now Bishkek, Kyrgyzstan).

Dement'ev, G.P. (1957). On the Shaheen *Falco peregrinus babylonicus*. *Ibis*, 99, 477-482.

Dennhardt, A.J. and Wakamiya, S.M. (2013). Effective dispersal of Peregrine Falcons (*Falco peregrinus*) in the midwest USA. *J. Raptor Res.*, 262-270.

Dennis, R.H. (1970). The oiling of large raptors by Fulmars. *Scottish Birds*, 6, 198–199.

Descalzo, E., Canarero, P.R., Sánchez-Barbudo, I.S., Martinez-Haro, M., Ortiz-Santaliestra, M.E. Moreno-Opo, R. and Mateo, R. (2021). Integrating active and passive monitoring to assess sublethal effects and mortality from lead poisoning on brids of prey. *Sci. Total Environ.*, 750, 142260, doi.org/10.1016/j.scitotenv.2020.142260

Dial, K. P., Biewener A. A., Tobalske B. W. and Warrick D. R., (1997). Mechanical power output of bird flight. Nature, 390, 67-70.

Diaz del Campo, F. (1974). (Some comments on the diet of the Peregrine Falcon (*Falco peregrinus*)), *Ardeola*, 19, 351-357.

Dixon, A. and Richards, C. (2003). Estimating the number of racing pigeons killed at Peregrine (*Falco peregrinus*) territories in south Wales. *Welsh Birds*, 3, 344–353.

Dixon, A., Richards, C., Haffield, P. *et al.* (2010). Population decline of Peregrines *Falco peregrinus* in central Wales associated with a reduction in racing pigeon availability. *Birds in Wales*, 7, 3–11.

Dixon, A., Rahman, L., Sokolov, A. and Sokolo, V. (2017). Peregrine Falcons crossing the 'Roof of the World'. In *Bird Migration across the Himalayas* (ed. H.H.T Prins and T. Namgail), Cambridge University Press, pp. 128-141.

Dixon, N. 2013. Peregrine aggression towards Buzzards at St. Michaels church, Exeter. *Devon Birds,* 66, 30-31.

Dixon, N. (2021). The value of webcams for monitoring nesting Peregrine Falcons. *Brit. Birds*, 114, 352-357.

Dixon, N and Drewitt, E.J.A. (2018). A 20-year study investigating the diet of Peregrines, *Falco peregrinus*, at an urban site in south-west England (1997-2017). *Ornis Hungarica*, 26, 177-187.

Dixon, N. and Gibbs, A. 2015a. Cooperative attacks by urban Peregrines on Common Buzzards. *Brit. Birds*, 108, 253-263.

Dixon, N. and Gibbs, A. 2015b. Territorial aggression shown by urban Peregrine falcons towards Common Buzzards. *Devon Birds*, 68, 13-20.

Dixon, N. and Gibbs, A. 2016. Cooperative attacks by urban Peregrines on Common Buzzards: an update. *Brit.Birds*, 109, 411-415.

Djorgova, N., Ragyov, D., Biserkov, V., Biserkov, J. and Nikolov, B.P. (2021). Habitat preferences of diurnal raptors in relation to human access to their breeding territories in the Balkan Mountain range, *Bulgarian Avian Res.*, 12:29, https://doi.org/10.1186/s40657-021-00265-6

Doherty, T.S., Hays, G.C. and Driscoll, D.A. (2021). Human disturbance causes widespread disruption of animal movements. *Nature Ecol. Evol.*, 5, 513-519.

Dolnik, V.R. (1995). (*Energy and Time Resources in Birds in Nature*). Nauka, St. Petersburg (in Russian).

Donázar, J.A., Cortés-Avizanda, A., Fargallo, J.A., Margalida, A., Moleón, M. Morales-reyes, Z., Moreno-Opo, R., Pérez-Garciá, J.M., Sánchez-Zapata, J.A., Zuberogoitia, I. and Serrano, D. (2016). Roles of raptors in a changing world: from flagships to providers of key ecosystem services. *Ardeola*, 63, 181-234.

Doolittle, T.C.J., Berger, D.D. and Van Stappen, J.F. (2013). Easternmost recovery in Europe of a Peregrine Falcon banded in North America. *J. Raptor Res.*, 47, 75–76.

Dorogoi, I. (1988). (Data on the biology of rare birds of the Kolyma lowlands). In *Rare Birds of the Far East and their Conservation*, Nauka, Vladivostok.

Douglas, D.J.T., Buchanan, G.M., Thompson, P., Amar, A., Fielding, D.A., Redpath, S.M. and Wilson, J.D. (2015). Vegetation burning for game management in the UK uplands is increasing and overlaps spatially with soil carbon and protected areas. *Biol. Conserv.*, 191, 243-250.

Dreiss, A.N., Ruppli, C.A., Faller, C. and Roulin, A. (2015). Social rules govern vocal competition in the Barn Owl. *Anim. Behav.*, 102, 95–107.

Drewitt, E.J.A. and Dixon, N. (2008). Diet and prey selection of urban-dwelling Peregrine Falcons in southwest England. *Brit. Birds*, 101, 58–67.

Drewitt, E. and Sutton, L.J. (2021). Trans-continental dispersal of a UK-bred juvenile Peregrine Falcon. *Brit. Birds*, 114, 46.

Driver, R.J. and Balakrishnan, C.N. (2021). Highly contiguous genomes improve the understanding of avian olfactory receptor repertoires from the Symposium presented at the annual meeting of the Society for Integrative and Comparative Biology virtual meeting January 3-February 28, 2021, doi.org/10.1093/icb/icab150.

Duan, A., Peng, X., Liu, J, Chen, Y., Wu, G., Holland, D.M., He, B., Hu, W., Tang, Y and Li, X. (2022). Sea ice loss of the Barents-Kara Sea enhances the winter warming over the Tibetan Plateau. *Npj Climate and Atmospheric Science*, 5:26, doi.org/10.1038/s41612-022-00245-7

Duquet, M. and Nadal, R. (2012). The capture of bats by raptors. *Ornithos*, 19, 184–195.

Durston, N.E., Wan, X., Liu, J.G. and Windsor, S.P. (2019). Avian surface reconstruction in free flight with application to flight stability analysis of a Barn Owl and Peregrine Falcon. *J. Exp. Biol.*, 222, jeb185488. doi:10.1242/jeb.185488

Dutson, G. (2001). New distributional ranges for Melanesian birds. *Emu*, 101, 237-248.

Dvorak, D., Mark, R. and Reymond, L. (1983). Factors underlying falcon grating acuity. *Nature*, 303, 729-730.

Eastham, C.P., Quinn, J.L. and Fox, N.C. (2000). Saker *Falco cherrug* and Peregrine *Falco peregrinus* Falcons in Asia: Determining migration routes and trapping pressure. In *Raptors at Risk* (ed. R.D. Chancellor and B.-U. Meyburg). World Working Group on Birds of Prey and Owls/Hancock House, pp. 247-258.

EC. (2009). Directive on the conservation of wild birds. Directive 2009/147/EC. The English language version of the Directive is available at:http://eur-lex.europa.eu/legal-content/EN/TXT/?uri=CELEX:32009L0147

Einwich, A., Seth, P.K., Bartölke, R, Botle, P., Feederle, R., Dedek, K. and Mouritsen, H. (2022). Localisation of cryptochrome 2 in the avian retina. *J. Comp. Physiol. A*, 208, 69-81.

Ellis, D.H. (2006). Swift-hunting behavior of the Peregrine Falcon in Arizona. *Western Birds*, 37, 149-155.

Ellis, D.H. and Groat, D.L. (1982). A Prairie Falcon intrudes at a Peregrine Falcon eyrie and pirates prey. *Raptor Res.*, 16, 89–91.

Ellis, D.H., Rohwer, V.G. and Rohwer, S. (2016). Experimental evidence that a large raptor can detect and replace heavily damaged flight feathers long before their scheduled moult dates. *Ibis*, 159, 217-220.

Ellis, D.H., Sabo, B.A., Fackler, J.K. and Millsap, B.A. (2002). Prey of Peregrine Falcon (*Falco peregrinus cassini*) in southern Argentina and Chile. *J. Raptor Res.*, 36, 315-319.

Ellis, D.H., LaRue, C.T., Fackler, J.K. and Nelson, R.W. (2007). Insects predominate in Peregrine Falcon predation attempts in Arizona. *Western Birds*, 38, 261-267.

Ellis, D.H., Saggese, M.D., Nelson, R.W. and Quaglia, A.I.E. (2019). Robbing Ibis nests as a foraging strategy of Peregrine Falcons in Patagonia, Argentina. *J. Raptor Res.*, 53, 438-440.

Ellis, D.H., Saggese, M.D., Franke, A. and Nelson, R.W. (2020). Extreme color variation in the Peregrine Falcon (*Falco peregrinus*) in Patagonia, *El Hornero*, 35, 65-76.

Elts, J., Leito, A., Leivits, A., Luigujõ, L., Mägi, E., Nellis, Rein, Nellis, Renno, Ots, M. and Pehlak, H. (2013). (Status and numbers of Estonian birds, 2008–2012), *Hirundo*, 26, 80-112.

Emison, W.B., Bren, W.M. and White, C.M. (1993). Influence of weather on the breeding of the Peregrine Falcon *Falco peregrinus* near Melbourne. In *Australian Raptor Studies*: *Proceedings of the 10th Anniversary Australasian Raptor Association*, Canberra, September 21-22, 1993 (P. Olsen, ed.), pp.26-32.

Emslie, S.D., (1985). The late Pleistocene (Rancholabrean) avifauna of Little Box Elder Cave, Wyoming. *Univ. Wyoming. Contrib. Geol.*, 23, 63-82.

Enderson, J.H. (1969). Peregrine and Prairie Falcon Life Tables based on band-recovery data. In *Peregrine Falcon Populations: their Biology and Decline* (ed. J.J. Hickey). University of Wisconsin Press, Madison, pp.505–509.

Enderson, J.H. and Craig, G.R. (1988). Population turnover in Colorado Peregrines. In *Peregrine Falcon Populations: their management and recovery*, The Peregrine Fund, Boise, Indiana, USA. pp.685-688

Enderson, J.H. and Craig, G.R. (1997). Wide ranging by nesting Peregrine Falcons (*Falco peregrinus*) determined by radiotelemetry. *J. Raptor Res.*, 31, 333–338.

Enderson, J.H. and Kirven, M.N. (1983). Flights of nesting Peregrine Falcons recorded by telemetry. *Raptor Res.*, 17, 33–37.

Enderson, J.H., Temple, S.A. and Swartz L.G. (1973). Time-lapse photographic records of nesting Peregrine Falcons. *Living Bird*, 11, 113–128.

Environment and Climate Change Canada. (2017). Management plan for the Peregrine Falcon *pealei* subspecies (*Falco peregrinus pealei*).In *Canada. Species at Risk Act Management Plan Series*. Environment and Climate Change Canada, Ottawa. Parts 1 and 2.

Ericson, P.G.P., Anderson, C.L., Britton, T., Elzanowski, A., Johansson, U.S., Källersjö, M.,Ohlson, J.I., Parsons, T.J., Zuccon, D. and Mayr, G., (2006). Diversification of Neoaves: integration of molecular sequence data and fossils. *Biology Letters*, 2, 543-547.

Faccio, S.D., Amaral, M., Martin, C.J., Lloyd, J.D., French, T.W. and Tur, A. (2013). Movement patterns, natal dispersal, and survival of Peregrine Falcons banded in New England. *J. Raptor Res.*, 47, 246–261.

Falk, K., Møller, S. and Mattox, W.G. (2006). A long-term increase in eggshell thickness of Greenlandic Peregrine Falcons *Falco peregrinus tundrius. Science Total Environ.*, 355, 127-134.

Feenders, G. and Bateson, M. (2013). Hand rearing affects emotional responses but not basic cognitive performance in European Starlings. *Anim. Behav.*, 86, 127–138.

Fenech, N. and Sammut, M. (2017). Prey taken and other specific behaviour by breeding Peregrine Falcons in the Maltese Islands. *Brit. Birds*, 110, 51-53.

Feng, S. and 149 others, (2020). Dense sampling of bird diversity increases power of comparative genomics. *Nature*, 587, 252-257 plus supplementary information.

Ferguson-Lees, I.J. (1951). The Peregrine population of Britain, Parts 1 and 2, Bird Notes, 24, 200-205.

Fisher, D. (1978). Peregrine retrieving prey from sea. *Brit. Birds*, 71, 461.

Fiuczynski, K.D. and Sömmer, P. (2000). Adaptation of two falcon species *Falco femoralis* and *Falco subbuteo* to an urban environment. In *Raptors at Risk* (ed. R.D. Chancellor and B.-U. Meyburg). World Working Group on Birds of Prey and Owls/Hancock House, pp. 463-467.

Ford, B. (2007). Recovery of a Peregrine Falcon after hitting the sea. *Brit. Birds*, 100, 304–305.

Formon, A. (1969). (A contribution to the study of Peregrine Falcons *Falco peregrinus* in eastern France). *Nos Oiseaux*, 30, 109–139.

Fox, R., Lehmkuhle, S.W. and Westendorf, D.H. (1976). Falcon visual acuity. *Science*, 192, 263–265.

Franke, A., Therrien, J-F., Deschamps, S. and Bêty, J. (2011). Climatic conditions during outward migration affect apparent survival of an Arctic top predator, the Peregrine Falcon, *Falco peregrinus*. *J. Avian Biol.*, 42, 544-551.

Franke, A. and Bêty, J. (2014). Heavy rainfall increases nestling mortality of an Arctic top predator: experimental evidence and long-term trend in Peregrine Falcons. *Oecologia*, 174, 1033-1043.

Franke, A., Galipeau, P. and Nikolaiczuk, L. (2013). Brood reduction by infanticide by Peregrine Falcons. *Arctic*, 66, 226-229.

Franke. A. + 34 others. (2020). Status and trends of circumpolar Peregrine Falcon and Gyrfalcon populations. *Ambio*, 49, 762-783, doi.org/10.1007/s13280-019-01300-z

Franklin, K. (1999). Vertical flight. *Journal of the North American Falconers Association*, 38, 68–72.

Fransson, T., Jakobsson, S., Johansson, P., Kullberg, C., Lind, J. and Vallinn, A. (2001). Magnetic cues trigger extensive refuelling, *Nature*, 414, 35-36.

Fretwell, S.D. and Lucas, H.L. (1969). On territorial behavior and other factors influencing habitat distribution in birds: I. Theoretical development. *Acta Theoretica*, 19, 16-36.

Fuchs, J., Johnson, J.A. and Mindell, D.P., (2015). Rapid diversification of falcons (Aves: Falconidae) due to expansion of open habitats in the Late Miocene. *Mol. Phylogenet. Evol.*, 82, 166-182.

Fuller, M.R., Seegar, W.S. and Schueck, L.S. (1998) Routes and travel rates of migrating Peregrine Falcons *Falco peregrinus* and Swainson's Hawk *Buteo swainsoni* in the western hemisphere. *J. Avian Biol.*, 29, 433-440.

Fyfe, R.W., Temple, S.A. and Cade, T.J. (1976). The 1975 North American Peregrine Falcon Survey. *Can. Field-Nat.*, 228-273.

Gaffney, M.F. and Hodos, W. (2003). The visual acuity and refractive state of the American Kestrel (*Falco sparverius*). *Vision Res.*, 2053-2059.

Gahbauer, M.A., Bird, D.M., Clark, K.E., French. T., Brauning, D.W. and McMorris, F.A. (2015). Productivity, mortality, and management of urban Peregrine Falcons in northeastern North America. *J. Wildlife Manage.*, 79, 10–19.

Galtier, N., Nabholz B., Glémin S. and Hurst G.D. (2009). Mitochondrial DNA as a marker of molecular diversity: a reappraisal. *Mol. Ecol.*, 18, 4541–4550.

Galván, I., Rodríguez-Martinez, S. and Carrascal, L.M. (2018) Dark pigmentation limits thermal niche position in birds. *Func. Ecol.*, 32, 1531-1540.

Ganusevich, S.A., Maechtle, T.L., Seegar, W.S., Yates, M.A., McGrady, M.J., Fuller, M., Schueck, L., Dayton, J. and Henny, C.J. (2004). Autumn migration and wintering areas of Peregrine Falcons *Falco peregrinus* nesting on the Kola Peninsula, northern Russia. *Ibis*, 146, 291–297.

Garcia, E.F.J. (1978). Persecution of raptors by Peregrines at Gibraltar. *Brit. Birds*, 71, 460-461.

Gargiulo, A., Fioretti, A., Russo, T.P., Varriale, L., Rampa, L., De Luca Bossa, L.M., Raia, P. and Dipineto, L. (2017). Occurrence of enteropathogenic bacteria in birds of prey in Italy. *Letters in Applied Microbiolgy*, 66, 202-206.

Garvin, J.C., Slabe, V.A., and Cuadros Díaz, S.F. (2020). Conservation Letter: Lead poisoning of raptors. *J. Raptor Res.*, 54, 473-479.

Gauld, J.G. + 50 others. (2022). Hotspots in the grid: Avian sensitivity and vulnerability to collision risk from energy infrastructure interactions in Europe and North Africa. *J. App. Ecol.*, 59, 1496-1512. DOI: 10.1111/1365-2664.14160

Gaüzère, P. and Devictor, V. (2021). Mismatches between birds' spatial and temporal dynamics reflect their delayed response to global changes. *Oikos*, 130, 1284-1296.

Gerson, A.R. and Guglielmo, C. (2016). Avian osmoregulation in flight: unique metabolic adaptations present novel challenges, *FASEB J.*, 30 (S1), 976.

Glasser, A. and Howland, H.C. (1996). A history of studies of visual accommodation in birds. *Quarterly Review of Biology*, 71, 475-509.

Glasser, A., Pardue, M.T., Andison, M.E. and Sivak, J.G. (1997). A behavioral study of refraction, corneal curvature, and accommodation in raptor eyes. *Can. J. Zool.* 75, 2010-2020.

González-Martín-Moro, J., Hernández-Verdejo, J.L. and Clement-Corral, A. (2017). The visual system of diurnal raptors: update review. *Arch. Soc. Esp. Oftalmol.* 92, 225-232.

González-Rubio, S., Ballesteros-Gómez, A., Asimakopoulos, A.G. and Jaspers, V.L.B. (2020). A review of contaminants of emerging concern in European raptors (2002-2020). *Sci. Total Environ.*, 760, 14337.

Goslow, G.E. (1971). The attack and strike of some North American raptors. *Auk*, 88, 815–827.

Goss, N.S. (1878). Breeding of duck hawks in trees. *Bulletin of theNuttall Ornithological Club*, 3, 32-34.

Gould, W.R. and Fuller, M.R. (1995). Survival and population studies estimation in raptor studies: a comparison of two methods. *J. Raptor Res.*, 29, 256-264.

Gowree, E.R., Jagadeesh, C., Talboys, E., Lagemann, C. and Brücker. (2018). Vortices enable the complex aerobatics of Peregrine Falcons. *Communications Biology*, 1.27, DOI: 10.1038/s42003-018-0029-3

Green, R.E., Taggart, M.A., Pain, D.J., Clark, N.A., Clewley, L., Cromie, R., Elliot, B., Green, R.M.W., Huntley, B., Huntley, J., Leslie, R., Porter, R., Robinson, J.A., Smith, K.W., Smith, L., Spencer, J. and Stroud, D. (2021). Effect of a joint policy statement by nine UK shooting and rural organisations on the use of lead shotgun ammunition for hunting Common Pheasants *Phasianus colchicus* in Britain. *Conservation Evidence J.*, 18, 1-9.

Green, R.E., Taggart, M.A., Pain, D.J., Clark, N.A., Clewley, L., Cromie, R., Dodd, S.G., Elliot, B., Green, R.M.W., Huntley, B., Huntley, J., Pap, S., Porter, R., Robinson, J.A., Sheldon, R., Smith, K.W., Smith, L., Spencer, J. and Stroud, D. (2022). Effectiveness of actions intended to achieve a voluntary transition from the use of lead to non-lead shotgun ammunition for hunting in Britain. *Conservation Evidence J.*, 19, 8-14.

Greenwood, P.J. (1980). Mating systems, philopatry and dispersal in birds and mammals. *Anim. Behav.*, 28, 1140–1162.

Greenewalt, C.H. (1962). Dimensional relationships for some flying animals. *Smithsonian Miscellaneous Collection*, 144, No. 2.

Griffiths, C.S. (1994a). Syringeal morphology and the phylogeny of the Falconidae. *Condor*, 96, 127–140.

Griffiths, C.S. (1994b). Monophyly of the Falconiformes based on syringeal morphology. *Auk*, 111, 787–805.

Griffiths, C.S. (1997). Correlation of functional domains and rates of nucleotide substitution in cytochrome b. *Mol. Phylogenet. Evol.*, 7, 352–365.

Griffiths, C.S. (1999). Phylogeny of the Falconidae inferred from molecular and morphological data. *Auk*, 116, 116–130.

Grimmett, R., Inskipp, C. and Inskipp, T. (2000). *Birds of Nepal*. Christopher Helm, London (Helm Field Guide).

Groombridge, J.J., Jones, C., Bayes, M.K., van Zyl, A.J., Carillo, J., Nichols, R.A. and Bruford, M.W. (2002). A molecular phylogeny of African kestrels with reference to divergence across the Indian Ocean. *Mol. Phylogenet. Evol.*, 25, 267–277.

Gu, Z., Pan, S., Lin, Z., Hu, L,. Dai, X., Chang, J., Xue, Y., Su, H., Long, J., Sun, M., Ganusevich, S., Sokolov, V., Sokolov, A., Pokrovsky, I., Ji, F., Bruford, M.W., Dixon, A. and Zhan, X. (2021). Climate-driven flyway changes and memory-based long-distance migration, *Nature*, 251- 264, https://doi.org/10.1038/s41586-021-03265-0

Guglielmo, C.G. (2018). Obese super athletes: fat-fuelled migration in birds and bats. *J. Exp. Biol.*, jeb165753. Doi:10.1242/jeb.165753.

Hackett, S.J. and 17 others. (2008). A phylogenomic study of birds reveals their evolutionary history. *Science*, 320, 1763–1768.

Hagar, J.A. (1938). In Bent, A.C., *Life Histories of North American Birds of Prey (Part 2)*. Smithsonian Institution, United States National Museum Bulletin 170. US Government Printing Office, Washington.

Hager, S.B. (2009). Human-related threats to urban raptors. *J. Raptor Res.*, 43, 210–226.

Hahmann, U. and Güntürkün, O. (1993). The visual acuity for the lateral visual field of the pigeon (*Columba livia*). *Vision Res.*, 33, 1659-1664.

Hall, G.H. (1955). *Great Moments in Action: the Story of the Sun Life Peregrines*. Mercury Press, Montreal.

Hambly, C., Harper, E.J. and Speakman, J.R. (2004). The energy cost of loaded flight is substantially lower than expected due to alterations in flight kinematics. *J. Exp. Biol.*, 207, 3969–3976.

Hancock, R. and Martin, J.R. (2015). Predation of Rose-ringed Parakeets by raptors and owls in Inner London. *Brit. Birds*, 108, 349-353.

Hanna, W.C. (1940). Siberian Peregrine Falcon in North America. *Condor*, 42, 166-167.

Hart, L.A., Wreford, E.P., Brown, M. and Downs, C.T. (2018). Hunting speeds of five southern African raptors, *Ostrich*, 89, 251-258.

Hart, N.S., Mountford, J.K., Davies, W.I.L., Collin, S.P. and Hunt, D.M., (2016). Visual pigments in a palaeognath bird, the *Emu Dromaius novaehollandiae*: implications for spectral sensitivity and the origin of ultraviolet vision, *Proc. R. Soc. B*, 283, 20161063.

Hartley, R. (1992). Status and productivity of Peregrines in low latitudes. *Gabar*, 7, 38-40.

Hartley, R.R. (2000). Ecology of the Taita *Falco fasciinucha*, Peregrinus *F. peregrinus minor* and Lanner *F. biarmicus* Falcons in Zimbabwe. In *Raptors at Risk* (ed. R.D. Chancellor and B.-U. Meyburg). World Working Group on Birds of Prey and Owls/Hancock House, pp. 87-105.

Hartley, R.R., Newton. I. and Robertson, M. (1995). Organochlorine residues and eggshell thinning in the Peregrine Falcon *Falco peregrinus minor* in Zimbabwe. *Ostrich*, 69-73.

Hays, L.L. (1987). Peregrine Falcon nest defense against a Golden Eagle. *J. Raptor Res.*, 21, 67.

Heerenbrink, M.K., Johansson, L.C. and Hedenström, A. (2015). Power of the wingbeat: modelling the effects of flapping wings in vertebrate flight. *Proc. Roy. Soc. A*, 471, 20140952.

Helbig, A.J. (2000). Contributions of the molecular studies to the phylogeny and systematics of African birds. *Ostrich*, 71, 40, doi:10.1080/00306525.2000.9639863

Hemelrijk, C.K., van Zuidam, L. and Hildenbrandt, H. (2015). What underlies waves of agitation in Starling flocks? *Behavioral Ecology and Sociobiology*, 69, 755–764.

Henderson, M.T., Booms, T.L., Robinson, B.W., Johnson, D.L. and Anderson, D.L. (2021) Direct and indirect effects of nesting site characteristics for a cliff-breeding raptor in western Alaska. *J. Raptor Res.*, 55, 17-32.

Henny, C.J., Galushin, V.M., Dudin, P.I., Khrustov, A.V., Mischenko, A.L., Moseikin, V.N., Sarychev, V.S. and Turchin, V.G. (1998). Organochlorine pesticides, PCBs and mercury in hawk, falcon, eagle and owl eggs from the Lipetsk, Voronezh, Novgorod and Sartov regions, Russia, 1992-1993. *J. Raptor Res.*, 32, 143-150.

References

Henny, C.J., Seegar, W.S., Yates, M.A., Maechtle, T.L., Galushin, and Fuller, M.R. (2000). Contaminants and wintering areas of Peregrine Falcons, *Falco peregrinus*, from the Kola Peninsula, Russia. In *Raptors at Risk* (ed. R.D. Chancellor and B.-U. Meyburg). World Working Group on Birds of Prey and Owls/Hancock House, pp. 871-878.

Herbert, R.A. and Herbert, K.G.S. (1965). Behavior of Peregrine Falcons in the New York City region. *Auk*, 82, 62–94.

Hewitt, S. (2013). Avian drop catch play. *Brit. Birds*, 106, 206–216.

Heyers, D., Manns, M., Luksch, H., Güntürkün, O. and Mouritsen, H. (2007). A Visual Pathway Links Brain Structures Active during Magnetic Compass Orientation in Migratory Birds. *PLoS ONE* 2(9): e937. doi:10.1371/journal.pone.0000937

Hickey, J.J. (Ed). (1969). Peregrine Falcon Populations: their biology and decline. *Proccedings of an International Conference*, 1965, University of Wisconsin Press, Madison, Milwaukee.

Hickey, J. J. and Anderson, D.W. (1968). Chlorinated hydrocarbons and eggshell changes in raptorial and fish-eating birds. *Science*, 162, 271-273.

Hickey, J. J. and Anderson, D.W. (1969). The Peregrine Falcon: life history and population literature. In *Peregrine Falcon Populations: their Biology and Decline* (ed. J.J. Hickey). University of Wisconsin Press, Madison, pp. 3–44.

Hilbers, J.P., Huijbregts, M.A. and Schipper, A.M. (2018). Predicting reintroduction costs for wildlife populations under anthropogenic stress. *J. App. Ecol.*, 57, 192-201.

Hilgert, N. (1988). Aspects of breeding and feeding behavior of Peregrine Falcons in Guayllabamba, Ecuador. In *Peregrine Falcon Populations: their management and recovery*, The Peregrine Fund, Boise, Indiana, USA, pp. 749-755.

Hill, D.S. (1962). A study of the distribution and host preferences of three species of Ornithomyia (Diptera: Hippoboscidae) in the British Isles, *Proc Roy. Ent. Soc. Lond. A*, 37, 37-48.

Hirsch, J. (1982). Falcon visual sensitivity to grating contrast. *Nature*, 300, 57–58.

Hirsch, J. (1983). Response to Dvorak *et al.* 1983. *Nature*, 303, 730.

Holland, D.C. (1989). An instance of carrion-feeding by the Peregrine Falcon (*Falco peregrinus*). *J. Raptor Res.*, 23, 184.

Homberger, B., Jenni, L., Duplain, J., Lanz, M. and Schaub, M. (2021). Strong effects of radio-tags, social groups and release date on survival of reintroduced Grey Partridges. *Anim. Conserv.*, 24, 677-688.

Horváth, G., Takács, P., Kretzer, B., Szilasi, S., Száz, D., Farkas, A. and Barta, A. (2017). Celestial polarization patterns sufficient for Viking navigation with the naked eye: detectability of Haidinger's brushes on the sky versus meteorological conditions. *R. Soc. Open Sci.*, 4:160688, http://dx.doi.org/10.1098/rsos.160688

Hovis, J., Snowman, T.D., Cox, V.L., Fay, R. and Bildstein, K.L. (1985). Nesting behavior of Peregrine Falcons in west Greenland during the nestling period. *J. Raptor Res.*, 19, 15-19.

Howland, H.C., (1974). Optimal strategies for predator avoidance: the relative importance of speed and manoeuvrability. *J. Theor. Biol.*, 47, 333-350.

Hoyt, D.F. (1979). Practical methods of estimating volume and fresh weight of bird eggs. *Auk*, 96, 73–77.

Hudson, D.M. and Bernstein, M.H., (1983). Gas exchange and energy costs in the White-necked Raven, *Corvus cryptoleucus*. *J. Exp. Biol.*, 103, 121-130.

Hughes, J., Mason, H., Bruce, M. and Sharrock, G. (2021) Crimes against raptors in Wales 1990-2019. *Birds in Wales*, 18, 3-19.

Hunt, B.S. (2012). Greenshank seeking protection during Peregrine Falcon attack. *Brit. Birds*, 105, 279.

Hunt, W.G., Driscoll, D.E., Bianchi E.W. and JacKman, R.E. (1992). Ecology of Bald Eagles in Arizona. Report to U.S. Bureau of Reclamation, Nevada, Contract 6-CS-30-04470. BioSystems Analysis Inc., Santa Cruz. (The report is in four volumes; the information on the Bald Eagle attack is in Volume 2, pp. C-131–C-132.)

Hunt, W.G. (1998). Raptor floaters at Moffat's equilibrium. *Oikos*, 82, 191–197.

Hurley, V.G. (2009). An assessment of nest site imprinting in Peregrine Falcons in Australia. In *Peregrine Falcon Populations* (Sielicki, J. and Mizera, T. eds), Turul Publishing and Poznań University of Life Sciences, pp. 463-478.

Hustler, K. (1983). Breeding biology of the Peregrine Falcon in Zimbabwe. *Ostrich*, 54, 161-171.

Ibarra, J., Mera y Sierra, R.L., Neira, G., Ibaceta, D.E.J., Saggese, M.D. (2019). Air sac nematode (Serratospiculum tendo) infection in an austral Peregrine Falcon (*Falco peregrinus* cassini) in Argentina. *J. Wild. Dis.*, 55, 179-182.

IPCC (Intergovernmental Panel on Climate Change) (2022). Climate Change 2022: Impacts, Adaptation and Vulnerability. Contribution of Working Group 11 to the Sixth Assessment Report of the Intergovernmental Panel on Climate Change. Cambridge University Press.

Ishizawa, J. and Chiba, S. (1967). (Stomach analysis of 12 species of Japanese hawks). *Journal of the Yamashina Institute for Ornithology*, 5, 13-33.

Jacobsen, F., Nesje, M., Bachmann, L. and Lifjeld, J.T. (2008). Significant genetic admixture after reintroductionof Peregrine Falcon (*Falco peregrinus*) in southern Scandinavia. Conserv. Genet., 9, 581-591.

Jarvis, E.D. and 104 others (2014). Whole-genome analyses resolve early branches in the tree of life of modern birds. *Science*, 15 December, 1320–1331. (This article was the first in a series of ten which filled that issue of *Science* and which were all related to the work of the Avian Phylogenomics Consortium.)

Jenkins, A.R. (1995). Morphometrics and flight performance of southern African Peregrine and Lanner falcons. *J. Avian Biol.* 26, 49-58.

Jenkins, A.R. (1998). Behavioural ecology of Peregrine Falcons and Lanner Falcons in South Africa. PhD thesis, University of Cape Town, South Africa.

Jenkins, A.R. (2000a). Hunting mode and success of African Peregrines *Falco peregrinus minor*: does nesting habitat quality affect foraging efficiency? *Ibis*, 142, 235–246.

Jenkins, A.R. (2000b). Characteristics of Peregrine and Lanner Falcons nesting habitats in South Africa. *Ostrich*, 71, 416-424.

Jenkins, A.R. (2000c). Factors affecting breeding success of Peregrine and Lanner Falcons in South Africa. *Ostrich*, 71, 385-392.

Jenkins, A.R. and Benn, G.A. (1998). Home range and habitat requirements of Peregrine Falcons on the Cape Peninsula, South Africa. *J. Raptor Res.*, 32, 90–97.

Jenkins, A.R. and Avery, G.M. (1999). Diets of breeding Peregrine and Lanner falcons in South Africa, *J. Raptor Res.*, 33, 190-206.

Jenkins, A.R. and Hockey, P.A.R. (2001). Prey availability influences habitat tolerance: an explanation for the rarity of Peregrine Falcons in the tropics. *Ecography*, 24, 359-367.

Jenkins, D., Watson, A. and Miller G.R. (1963). Population studies on Red Grouse, Lagopus lagopus scoticus (Lath.) in north-east Scotland. *J. Anim. Ecol.*, 32, 317–376.

Jenkins, D., Watson, A. and Miller G.R. (1964). Predation and Red Grouse populations. *J. Appl. Ecol.*, 1, 183–195.

Jennings, A.R. (1961). An analysis of 1,000 deaths in wild birds. *Bird Study*, 8, 25-31.

Jeong, Y.-C., Yeh, S.-W., lim, Y.-K., Santose, A. and Wang, G. (2022). Indian Ocean warming as key driver of long-term positive trend of Arctic Oscillation. *Nature Climate and Atmospheric Science*, 5:56, doi.org/10.1038/s41612-022-00279-x

Jiguet, F., Robert, A., Micol, T. and Barbraud, C. (2007). Quantifying stochastic and deterministic threats to island seabirds: the last endemic prions face extinction from falcon peregrinations. *Anim. Conserv.*, 10, 245-253.

Johnson, D. (2011). Peregrine Falcon nest relief at night. *Brit.Birds*, 104, 217.

Johnson, P. (2008). Peregrine Falcons feeding Common Kestrel chicks. *Brit. Birds*, 101, 327.

Johnstone, R.M. (1998). Aspects of the population biology of tundra Peregrine Falcons (*Falco peregrinus tundrius*). PhD thesis, University of Saskatchewan, Canada.

Johnstone, R.M., Court, G.S., Fesser, A.C., Bradley, D.M., Oliphant, L.W. and MacNeil, J.D. (1996). Long-term trends and sources of organochlorine contamination in Canadian tundra Peregrine Falcons, *Falco peregrinus tundrius*. *Environ. Poll.*, 93, 109-120.

Johnstone, S. and Earl, T. (2018). Common Starlings forming a 'baitball' in response to an attack by Peregrine Falcon. *Brit. Birds*, 111, 765.

Jones, M.P., Pierce, K.E. and Ward, D. (2007). Avian vision: a review of form and function with special consideration to birds of prey. *J. Exot. Pet Med.*, 16, 69–87.

Kane, S.A. and Zamani, M. (2014). Falcons pursue prey using visual motion cues: new perspectives from animal-borne cameras. *J. Exp. Biol.*, 217, 225–234.

Kane, S.A., Fulton, A.H. and Rosenthal, L.J. (2015). When hawks attack: animal-borne video studies of Goshawk pursuit and prey-evasion strategies. *J. Exp. Biol.*, 218, 212–222.

Kano, F., Walker, J., Sasaki, T. and Biro, D. (2018). Head-mounted sensors reveal visual attention of free-flying homing pigeons. *J. Exp. Biol.*, 221, doi:10.1242/jeb.183475

Katzner, T., Winton, J.D., McMorris, F.A. and Brauning, D. (2012). Dispersal, band encounters, and causes of death in a reintroduced and rapidly growing population of Peregrine Falcons. *J. Raptor Res.*, 46, 75-83.

Kauffman, M.J., Frick, W.F. and Linthicum, J. (2003). Estimation of habitat-specific demography and population growth for Peregrine Falcons in California. Ecol. Appl., 13, 1802-1816.

Kelly, G.M. and Thorpe, J.P. (1993). A communal roost of Peregrine Falcons and other raptors. *Brit. Birds*, 86, 49–52.

Kerlinger, P. (1985). Water-crossing behavior of raptors during migration. *Wilson Bull.* 97, 109-113.

Kerlinger, P., Cherry, J.D. and Powers, K.D. (1983). Records of migrant hawks from the North Atlantic Ocean. *Auk*, 100, 488–490.

Kéry, M., Banderet, G., Müller, C., Pinaud, D., Savioz, J., Schmid, H., Werner, S. and Monneret, R-J. (2022). Spatio-temporal variation in post-recovery dynamics in a large Peregrine Falcon (*Falco peregrinus*) population in the Jura Mountains 2000-2020. *Ibis*, 164, 217-239.

Kesel, A.B., Hoffmann, F. Baars, A. and Danter, L. (2019). Flow phenomena in the falcon nose - an interpretation approach. In *Bionik aus der Natur: Innovationspoten für Tecnologieanwendungen 9, Bremer Bionik-Kongress*, Bremen, Germany, pp. 224-227.

Kettel, E.F., Gentle, L.K., and Yarnell, R.W. (2016). Evidence of an urban Peregrine Falcon (*Falco peregrinus*) feeding young at night. *J. Raptor Res.*, 50, 321-323.

Kettel, E.F., Gentle, L.K., Quinn, J.L. and Yarnell, R.W. (2018). The breeding performance of raptors in urban landscapes: a review and meta-analysis. *J. Orn.*, 159, 1-18.

Kettel, E.F., Gentle, L.K., Yarnell, R.W. and Quinn, J.L. (2019). Breeding performance of an apex predator, the Peregrine Falcon, across urban and rural landscapes. *Urban Ecosystems*, 22, 117-125.

Khlopotova, A.V. (2013). Study on the ecology of the Peregrine Falcon (*Falco peregrinus* Tunstall, 1771) in the Chusovaya River Nature Park. *Russsian J. Ecology*, 44, 358-360.

Kimball, R.T., Parker, P.G. and Bednarz, J.C. (2003). Occurrence and evolution of cooperative breeding among the diurnal raptors (Accipitridae and Falconidae). *Auk*, 120, 717–729.

King, S.S. (2008). Peregrine Falcon egg breakage. *Brit. Birds*, 101, 326–327.

King, S.S. (2009). Peregrine Falcon robbing Hobby of prey. *Brit. Birds*, 102, 406.

Kirmse, W. (2001). Notes on restricted exchange between nest site types in the Peregrine (*Falco peregrinus*). *Buteo*, 12, 85–88.

Kjellén, N. (1992). Differential timing of autumn migration between sex and age groups in raptors at Falsterbo, Sweden. *Ornis Scand.*, 23, 420–434.

Klaassen, R.H.G., Hake, M., Strandberg, R., Koks, B.J., Trierweiler, C., Exo, K-M., Birlein, F. and Alerstam, T. (2014). When and where does mortality occur in migratory birds? Direct evidence from long-term satellite tracking of raptors. *J. Anim. Ecol.*, 83, 176-184, doi: 10.1111/1365-2656.12135.

Kleinstäuber, G., Kirmse, W. and Sömmer, P. (2009). The return of the Peregrine to eastern Germany – re-colonisation in the west and east; the formation of an isolated tree-nesting subpopulation and further management. In *Peregrine Falcon Populations* (Sielicki, J. and Mizera, T.

Korbel, R.T., Redig, P.T., Walde, I. and Reese, S. (2000). Video fluorescein angiography in the eyes of various raptors and mammals. *Proc. Ann. Conf. Association Avian Veterinarians*, 89-90.

Korpimäki, E. (1986a). Reversed size dimorphism in birds of prey, especially Tengmalm's Owl *Aegolius funereus*: a test of the 'starvation hypothesis'. *Ornis Scand.*, 17, 326–332.

Krekels, R. (2017). (Peregrine chicks *Falco peregrinus* predated by Eagle Owl *Bubo bubo* near Maastricht). De Takkeling, 25, 258-262.

Krone, O. (2007). Pathology: endoparasites. In *Raptor Research and Management Techniques* (ed. D.L. Bird and D.R. Bildstein). Hancock House, Blaine, WA, pp. 318–328.

Krüger, O. (2005). The evolution of reversed sexual size dimorphism in hawks, falcons and owls: a comparative study. *Evol. Ecol.*, 19, 467–486.

Kurosawa, T. and Kurosawa, R. (2003). A helper at the nest of Peregrine Falcons in northern Japan. *J. Raptor Res.*, 37, 340–342.

Kuyt, E. (1967). Two banding returns for Golden Eagle and Peregrine Falcon, *Bird-Banding*, 38, 78-79.

Lagemann, C., Gowree, E.R., Jagadeesh, C., Talboys, E. and Brücker, C. (2018). Experimental and numerical analysis of the aerodynamics of Peregrine Falcons during stoop flight. In *Deutscher Luft- und Raumfahrtknogress (German Aerospace Congress)* 480311, Bonn, doi:10.25967/483011.

Lamarre, V., Franke, A., Love, O.P., Legagneux, P. and Bêty, J. (2017). Linking pre-laying energy allocation and timing of breeding in a migratory Arctic raptor. *Oecologica*, 183, 653-666.

Langgemach, T., Sömmer, P., Kirmse, W., Saar, C. (1997). (First record of tree-nesting Peregrine Falcons *Falco p. peregrinus*, in Brandenburg, Germany twenty years after the extinction of the European tree-nesting population). *Vogelwelt*, 118, 79–94.

Lank, D.B., Butler, R.W., Ireland, J. and Ydenberg, R.C. (2003). Effects of predation danger on migration strategies of sandpipers. *Oikos*, 103, 303-319.

Laybourne, R.C. (1974). Collision between a vulture and an aircraft at an altitude of 37,000 feet. *Wilson Bull.*, 86, 461–462.

Leckie, F. and Campbell, S. (2000). A new record of successful tree nesting Peregrines. *Scottish Birds*, 21, 45–46.

Lee, Y.-F. and Kuo, Y.-M. (2001). Predation of Mexican Free-tailed Bats by Peregrine Falcons and Red-tailed Hawks. *J. Raptor Res.*, 35, 115–123.

Leveque-Beaudin, V. and Sinclair, B.J. (2021). Louse fly (Diptera, Hippoboscidae) associations with raptors in southern Canada, with new North American and European records. *Int. J. Parasitology: Parasite and Wildlife*, 16, 168-174.

Lindberg, P. (1977). (Peregrine ring recoveries in Sweden). In *Pilgrimsfalk: Report of a conference on Peregrine Falcons at Grimsö, Sweden* (ed. P. Lindberg), pp. 39-42.

Lindberg, P. (1983). Relationship between the diet of Fennoscandian Peregrines *Falco peregrinus* and organochlorines and mercury in their eggs and feathers, with a comparison to the Gyrfalcon *Falco rusticolus*. PhD thesis, University of Gothenburg.

Lindberg, P. and Sjöberg, U. (2009). Captive breeding and restocking of the Peregrine Falcon in Sweden. In *Peregrine Falcon Populations* (Sielicki, J. and Mizera, T. eds), Turul Publishing and Poznań University of Life Sciences, pp. 677-694.

Lindner, M. (2018). Influence of the Eagle Owl (*Bubo bubo*) on the Peregrine Falcon (*Falco peregrinus*) population in Germany. *Ornis Hungarica*, 26, 243-253.

Lindström, Å., Alerstam, A., Andersson, A., Bäckman, J., Bahlenberg, P., Bom, R., Ekblom, R., Klaassen, R.H.G., Korniluk, M., Sjöberg, S. and Weber, J.K.M. (2021). Extreme altitude changes between night and day during marathon flights of Great Snipes. *Current Biology*, 31, 3433-3439.

Lockhart, K., Irvine, C., MacLaurin, J., Dolgova, S. and Hebert, C.E. (2020). Peregrine Falcon scavenges adult Herring Gull at nest site on Lake Superior, Ontario, Canada. *J. Raptor Res.*, 54, 470-472.

Łukasiewicz, M., Wnęk, K and Boruc, K. (2014). Biology of embryo development in pigeon *Columba livia domesticus*, in conditions of artificial incubation. *Adv. Anim. Vet. Sci.*, 2, 401-406.

Maa, T.C. (1969). A revised checklist and concise host index of Hippoboscidae (Dipteria). *Pacific Insects Monographs*, 2, 261–299.

Macdonald, H. (2006). *Falcons*. Reaktion Books, London.

MacNally, L. (1979). Peregrine apparently killed by Golden Eagle. *Scottish Birds*, 10, 234.

Mahr, K. and Hoi, H. (2018). Red-legged Partridges perceive the scent of predators and alarm scents of an avian heterospecific. *Anim. Behav.*, 144, 109-114.

Mak, B., Francis, R.A. and Chadwick, M.A. (2021). Breeding habitat selection of urban Peregrine Falcons (*Falco peregrinus*) in London. *J. Urban Ecol.*, 7, 1-9, doi: 10.1093/jue/juab017

Marconot, B. (2003). (Nocturnal hunting by Peregrine Falcons in eastern France). *Ornithos*, 10, 207-211.

Martin, A.P. (1980). A study of a pair of breeding Peregrine Falcons (*Falco peregrinus peregrinus*) during part of the nesting period. BSc dissertation, University of Durham.

Martin, G.R. (1986). Shortcomings of an eagle's eye. *Nature*, 319, 357.

Martinez, J.E., Martínez, J.A., Zuberogoitia, I., Zabala, J., Redpath, S.M. and Calvo, J.F. (2008). The effect of intra- and inter-specific interactions on the large-scale distribution of cliff-nesting raptors. *Ornis Fennica*, 85, 13-21.

Masman, D., Gordijn, M., Daan, S. and Dijkstra, C. (1986). Ecological energetics of the Kestrel: field estimates of energy intake throughout the year. *Ardea*, 74, 24–39.

Masman, D., Daan, S. and Beldhuis, J.A. (1988). Ecological energetics of the Kestrel: daily energy expenditure throughout the year based on the time-energy budget, food intake and doubly labelled water methods. *Ardea*, 76, 64–81.

Matthews, K.B., Fielding, D., Miller, D.G., Gandossi, G., Newey, S., Thomson, S. (2020) Mapping the areas of moorland that are actively managed for grouse and the intensity of current management regimes. Part 3 – Research to assess socioeconomic and biodiversity impacts of driven grouse moors and to understand the rights of gamekeepers. *Commissioned Report for Scottish Government*. https://sefari.scot/document/part-3-mapping-the-areas-and-management-intensity-of-moorland-actively-managed-for-grouse

Matsyna, A.I., Matsyna, E.I. and Matsyna, A.A. (2010). Observations (of) Peregrine Falcon hunting habits along the coast of Iturp Island, South Kuril islands, Russia. *Raptors Conservation*, 18, 184–185.

Mäntylä, E., Kipper, S. and Hilker, M. (2020). Insectivorous birds can see and smell systemically herbivore-induced pines. *Eco. Evol.*, 10, 9358-9370.

McFadden, S.A. and Reymond, L. (1985). A further look at the binocular visual field of the pigeon (*Columba livia*). *Vision Res.*, 25, 1741-1746.

McGrady, M.J., Maechtle, T.L., Vargas, J.J., Seegar, W.S. and Catalina Porras Peña, M. (2002). Migration and ranging of Peregrine Falcons wintering on the Gulf of Mexico coast, Tamaulipas, Mexico. *Condor*, 14, 39–48.

McLean, N., Kruuk, L.E.B., van der Jeugd, H.P., Leech, D., van Turnhout, C.A.M. and van de Pol, M. (2022). Warming temperatures drive at least half of the magnitude of long-term trait changes in European birds. *PNAS*, 119, 10e2105416119

McMillan, R.L. (2011). Raptor persecution on a large Perthshire estate: a historical study. *Scottish Birds*, 31, 195–205.

McPherson, S.C., Sumasgutner, P. and Downs, C.T. (2021). South African raptors in urban landscapes: a review. *Ostrich*, 92, 41-57.

Mearns, R. (1982). Winter occupation of breeding territories and winter diet of Peregrines in south Scotland. *Ornis Scand.*, 13, 79–83.

Mearns, R. (1983). Breeding Peregrines oiled by Fulmars. *Bird Study*, 30, 243–244.

Mearns, R. (1985). The hunting ranges of two female Peregrines towards the end of the breeding season. *J. Raptor Res.*, 19, 20–26.

Mearns, R. and Newton, I. (1984). Turnover and dispersal in a Peregrine *Falco peregrinus* population. *Ibis*, 347–355.

Mearns, R. and Newton, I. (1988). Factors affecting breeding success of Peregrines in south Scotland. *J. Anim. Ecol.*, 57, 903–916.

Mebs, T. (1960). (Investigations of the moult order of wing and tail feathers in large falcons.) *J. Orn.*, 101, 175–194.

Mebs, T. (1971). (Causes of death and mortality rates in German and Finnish Peregrines derived from ring recoveries.) *Vogelwarte*, 26, 98–105.

Mebs. T. (2001). Ground-nesting Peregrine Falcons (*Falco peregrinus*) in Europe: situation in past and today. *Buteo*, 12, 81–84.

Mebs, T. (2009). (The nocturnal hunting of Peregrine Falcons *Falco peregrinus* at buildings illuminated by powerful halogen floodlights: a review of documented cases in Europe.) *Vogelwelt*, 129, 107–113.

Meese, R.J. and Fuller, M.R. (1989). Distribution and behaviour of passerines around Peregrine *Falco peregrinus* eyries in western Greenland. *Ibis*,131, 27–32.

Meier, A.J., Noble, R.E. and McKenzie, P.M. (1989). Observations of autumnal courtship behavior in Peregrine Falcons. *J. Raptor Res.*, 23, 121–122.

Mendelsohn, J.M., Kemp, A.C., Biggs, H.C., Biggs, R. and Brown, C.J. (1989). Wing areas, wing loadings and wing spans of 66 species of African raptors. *Ostrich*, 60, 35-42.

Meyburg, B-U., Paillat, P., Meyburg, C. and McGrady, M. (2018). Migration of a Peregrine Falcon *Falco peregrinus calidus* from Saudi Arabia to Cape Town as revealed by satellite telemetry. *Ostrich*, 89, 93-96.

Mikkola, H. (1983). *Owls of Europe*. Poyser, London.

Mills, R., Hildenbrandt, H., Taylor, G.K. and Hemelrijk, C.K. (2018). Physics-based simulations of aerial attacks by Peregrine Falcons reveal that stooping at high speed maximises catch success against agile prey. *PLoS Comput. Biol.*, 14, e1006044.

Mills, R., Taylor, G.K. and Hemelrijk, C.K. (2019) Sexual size dimorphism, prey morphology and catch success in relation to flight mechanics in the Peregrine Falcon: a simulation study. *J. Avian. Biol.*, e01979, doi: 10.1111/jav.01979

Mitkus, M., Olsson, P., Toomey, M.B., Corbo, J.C. and Kelber, A. (2017). Specialized photoreceptor composition in the raptor fovea. *J. Comp. Neurol.*, 525, 2152-2163.

Mlíkovský, J. and Hruška, J. (2000). Food of the Peregrine Falcon (*Falco peregrinus*) in Plzeň, Czech Republic. *Buteo*, 11, 125-128.

Mizera, T. and Sielicki, J. (2009). *Peregrine Falcon Populations* (Sielicki, J. and Mizera, T. eds), Turul Publishing and Poznań University of Life Sciences.

Molard, L., Kéry, M. and White, C.M. (2007). Estimating the resident population size of Peregrine Falcon *Falco peregrinus* in Peninsular Malaysia. *Forktail*, 23, 87-91.

Molard, L. (2009). Behaviour and ecology of resident Peninsular Malaysian Peregrine Falcons. In *Peregrine Falcon Populations* (Sielicki, J. and Mizera, T. eds), Turul Publishing and Poznań University of Life Sciences, pp. 479-492.

Monclús, L., Shore, R.F. and Krone, O. (2020). Lead contamination in raptors in Europe: a systematic review and meta-analysis. *Sci. Total Environ.*, 748, 141437, doi.org/10.1016/j.scitotenv.2020.141437

Monneret, R-J. (1974). (The behavioural display repertoire of Peregrine Falcons). *Alauda*, 42, 407–428.

Monneret, R-J. (1983). (Helpers at the nests of Peregrine Falcons). *Alauda*, 51, 241–250.

Monneret, R-J. (2010). (The spread of Eurasian Eagle Owl *Bubo bubo* and its consequences for Peregrine Falcon *Falco peregrinus* in the French Jura Mountains). *Alauda*, 78, 81-91.

Monneret, R-J., Kéry, M., Couerdassier, M., Cretin, J-Y., Parish, D., Herold, J-P. Ruffinoni, R. and members of the Jura Peregrine Group. (2015). (Consequences of artificial nest use by the Peregrine Falcon *Falco peregrinus* in the Jura range). *Alauda*, 83, 133–142.

Monti. F., Duriez, O., Dominici, J-M., Sforzi, A., Robert, A., Fusani, L. and Grémillet, D. (2018). The price of success: integrative long-term study reveals ecotourism impacts on a flagship species at a UNESCO site. *Anim. Conserv.*, 21, 448-458.

Mooney, N. and Brothers, N. (1993). Dispersion, nest and pair fidelity of Peregrine Falcons *Falco peregrinus* in Tasmania. In *Australian Raptor Studies: Proceedings of the 10th Anniversary Australasian Raptor Association*, Canberra, September 21-22, 1993 (P. Olsen, ed.), pp. 33-42.

Moore, B.A., Murphy, C.J., Marlar, A., Dubielzig, R.R., Teixeira, L.B.C., Ferrier, W.T. and Hollingsworth, S.R. (2019). Presumed photoreceptor dysplasias in Peregrine Falcons (*Falco peregrinus*) and Peregrine Falcon hybrids. *J. Wild. Dis.*, 55, 325-334.

Morris, F.O. (1877). Letter to The Times newspaper, 11 November.

Morrison, J.L., Terry, M. and Kennedy, P.L. (2006). Potential factors influencing nest defence in diurnal North American raptors. *J. Raptor Res.*, 40, 98–110.

Mouze, M. (2022). (Exceptional survival of a one-eyed Peregrine). *Ornithos*, 29, 66-68.

Mueller, H.C., Mueller, N.S., Berger, D.D., Allez, G., Robichaud, W. and Kaspar, J.L. (2000). Age and sex differences in the timing of the fall migration of hawks and falcons. *Wilson Bull.*, 112, 214–224.

Müller, M.G., Wernery, U. and Kösters, J. (2000). Bumblefoot and lack of exercise among wild and captive-bred falcons tested in the United Arab Emirates. *Avian Dis.*, 44, 676-680.

Murgatroyd, M., Redpath, S.M., Murphy, S.G., Douglas, D.J.T., Saunders, R. and Amar, A. (2019). Patterns of satellite tagged Hen Harriers disappearances suggest widespread illegal killing on British grouse moors. *Nature Communications*, 10:1094, doi.org/10.1038/s41467-019-09044-w

Mustin, K., Arroyo, B., Beja, P., Newey, S., Irvine, R.J., Kestler, J. and Redpath, S.M. (2018). Consequence of game bird management for non-game species in Europe. *J. App. Ecol.*, 55, 2285-2295.

Naoroji, R. (2006). *Birds of Prey of the Indian Subcontinent*, Helm, London.

Negro, J.J., Bortolotti, G.R. and Sarasola, J.H. (2007). Deceptive plumage signals in birds: manipulation of predators or prey? *Biol. J. Linne. Soc.*, 90, 467-477.

Negro, J.J., Grande, J.M. and Sarasola, J.H. (2004). Do Eurasian Hobbies (*Falco subbuteo*) have 'false eyes' on the nape?. *J. Raptor Res.*, 38, 287-288.

Nelson, R.W. (1970). Some aspects of the breeding behaviour of Peregrine Falcons on Langara Island, British Columbia. MSc thesis, University of Calgary.

Nelson, R.W. (1972). The incubation period in Peale's Peregrine. *Raptor Res.*, 6, 11-15.

Nelson, R.W. (1988). Do large natural broods increase mortality of parent Peregrine Falcons? In *Peregrine Falcon Populations: Their Management and Recovery* (ed. T.J. Cade et al.). The Peregrine Fund, Boise, Idaho, pp. 719-728.

Nelson, R.W. (1990). Status of the Peregrine Falcon, *Falco peregrinus pealei*, on Langara Island, Queen Charlotte Islands, British Columbia, 1968-1989. *Can. Field-Nat.*, 104, 193-199.

Nelson, R.W. and Myres, M.T. (1976). Declines in population of Peregrine Falcons and their seabird prey at Langara Island, British Columbia. *Condor*, 78, 281-293.

Nesje, M., Røed, K.H., Bell, A.D., Lindberg, P. and Lifjeld, J.T. (2000) Microsatellite analysis of population structure and genetic variability in Peregrine Falcons (*Falco peregrinus*). *Anim. Conserv.*, 3, 267-275.

Nethersole-Thompson, D. (1931). Observations on the Peregrine Falcon (*Falco peregrinus peregrinus*). *The Oologists' Record*, 11, 73-80.

Nethersole-Thompson, C. and Nethersole-Thompson, D. (1943). Nest site selection by birds. *Brit. Birds*, 37, 108–113.

Neumann, J.L., Holloway, G.J., Sage, R.B. and Hoodless, A.N. (2015). Releasing of Pheasants for shooting in the UK alters woodland invertebrate communities. *Biol. Conserv.*, 191, 50–59.

Newton, I. (1979). *Population Ecology of Raptors*. Poyser, London.

Newton, I. (2021). Killing of raptors on grouse moors: evidence and effects. *Ibis*, 163, 1-19.

Newton, I. (2013). Organochlorine pesticides and birds. *Brit. Birds*, 106, 189-205.

Newton, I. and Mearns, R. (1988). Population ecology of Peregrines in southern Scotland. In *Peregrine Falcon Populations: Their Management and Recovery* (ed. T.J. Cade et al.). The Peregrine Fund, Boise, Idaho, pp. 651–665.

Newton, I, McGrady, M.J. and Oli, M.K. (2016). A review of survival estimates for raptors and owls. *Ibis*, 158, 227-248.

Newton, I., Bogan, J.A. and Haas, M.B. (1989). Organochlorines and mercury in the eggs of British Peregrines *Falco peregrinus*. *Ibis*, 131, 355-376.

Ng, C.S. and Li, W-H. (2018). Genetic and molecular basis of feather diversity in birds. *Genome Biol.*, 10, 2572-2586.

Nisbet, I.C.T. (1988). The relative importance of DDE and Dieldrin in the decline of Peregrine Falcon populations. In *Peregrine Falcon Populations: Their Management and Recovery* (T.J. Cade et al., eds.). The Peregrine Fund, Boise, Idaho, pp. 351-375.

Nittinger, F., Haring, E., Pinsker, W., Wink, M. and Gamauf, A. (2005). Out of Africa? Phylogenetic relationships between *Falco biarmicus* and the other Hierofalcons (Aves: Falconidae). *J. Zool. Syst. Evol. Res.*, 43, 321–331.

Nittinger, F., Gamauf, A., Pinsker, W., Wink, M. and Haring, E. (2007). Phylogeography and population structure of the Saker Falcon (*Falco cherrug*) and the influence of hybridisation: mitochondrial and microsatellite data. *Mol. Ecol.*, 16, 1497–1517.

Noguera, J.C. and Velando, A. (2019). Bird embryos perceive vibratory cues of predation risk from clutch mates, *Nature Ecol.and Evol.*, 3, 1225-1232.

Norriss, D.W. (1995). The 1991 survey and weather impacts on the Peregrine Falcon *Falco peregrinus* breeding population in the Republic of Ireland. *Bird Study*, 42, 20–30.

Nygård, T. (1983). Pesticide residues and shell thinning in eggs of Peregrines in Norway. *Ornis Scand.*, 14, 161-166.

Nygård, T. (1999). Correcting eggshell indices of raptor eggs for hole size and eccentricity. *Ibis*, 141, 85–90.

Nygård, T. and Polder, A. (2012). Environmental Pollutants in Eggs of Birds of Prey in Norway: Current Situation and Time Trends. *Norsk Institutt for Naturforskning (NINA) Report 834.*

Nygård, T., Sandercock, B.K., Reinsborg, T. and Einvik, K. (2019) Population recovery of Peregrine Falcons in central Norway in the 4 decades aince the DDT ban. *Ecotoxcology*, 28, 1160-1168.

O'Connor, J.P. and Sleeman, D.P. (1987). A review of Irish Hippoboscidae (Insecta: Diptera). *The Irish Naturlaists' Journal.* 22, 236-239.

Ochs, M.F., Zamani, M., Gomes, G.M.R., Neto, R.C.de N. and Kane, S.A. (2017). Sneak peak: Raptors search for prey using stochastic head turns. *Auk*, 134, 104-115.

Oehme, (1964). (Comparative investigations of the eyes of birds of prey). *Z. Morph. Ökol. Tiere*, 53, 618-635.

Oliphant, L.W. (1991). Hybridization between a Peregrine Falcon and a Prairie Falcon in the wild. *J. Raptor Res.*, 25, 36–39.

Olsen, J. (2013). Reversed sexual dimorphism and prey size taken by male and female raptors: a comment on Pande and Dahanukar. *J. Raptor Res.* 47, 79–81.

Olsen, J. and Georges, A. (1993). Do Peregrine Falcon fledglings reach independence during peak abundance of their main prey? *J. Raptor Res.*, 27, 149-153.

Olsen, J. and Tucker, A.D. (2003). A brood-size manipulation experiment with Peregrine Falcons, *Falco peregrinus*, near Canberra. *Emu*, 103, 127-132.

Olsen, J., Fuentes, E., Bird, D.M., Rose, A.B. and Judge, D. (2008). Dietary shifts based upon prey availability in Peregrine Falcons and Australian Hobbies breeding near Canberra, Australia. *J. Raptor Res.*, 42, 125-137.

Olsen, P.D. (1982). Ecogeographic and temporal variation in the eggs and nests of the Peregrine, *Falco peregrinus*, (Aves: Falconidae) in Australia. *Aust. Wildl. Res.*, 9, 277-291.

Olsen, P.D. and Cockburn, A. (1991). Female-biased sex allocation in Peregrine Falcons and other raptors. *Behav. Eco. Sociobiol.*, 28, 417-423.

Olsen, P.D. and Olsen, J. (1988). Population trends, distribution and status of the Peregrine Falcon in Australia. In *Peregrine Falcon Populations: Their Management and Recovery* (ed. T.J. Cade et al.), The Peregrine Fund, Boise, Idaho, pp. 255–274.

Olsen, P.D, Morris, A. and Bennett, C. (1993a). Observations on the reproductive diet of the Peregrine Falcon *Falco peregrinus* in New South Wales. In *Australian Raptor Studies: Proceedings of the 10th Anniversary Australasian Raptor Association*, Canberra, September 21-22, 1993 (P. Olsen, ed.), pp. 78-80.

Olsen, J., Olsen, P., Billett, T. and Jolly, J. (1993b). Observations on the reproductive diet of the Peregrine Falcon *Falco peregrinus* in South Australia. *In Australian Raptor Studies: Proceedings of the 10th Anniversary Australasian Raptor Association*, Canberra, September 21-22, 1993 (P. Olsen, ed.), pp. 81-83.

Olsen, P.D., Olsen, J. and Mason, I. (1993c). Breeding and non-breeding season diet of the Peregrine Falcon *Falco peregrinus* near Canberra, prey selection, and the relationship between diet and reproductive success. In *Australian Raptor Studies: Proceedings of the 10th Anniversary Australasian Raptor Association*, Canberra, September 21-22, 1993 (P. Olsen, ed.), pp. 55-77.

Olsen, P., Marshall, R.C. and Gaal, A. (1989). Relationships within the genus Falco: a comparison of the electrophoretic patterns of feather proteins. *Emu*, 89, 193–203.

Olsen, P., Barry, S., Baker, G.B., Mooney, N., Cam, G. and Cam, A. (1998a). Assortative mating in falcons: do big females pair with big males? *J. Avian Biol.*, 29, 197–200.

Olsen, P., Doyle, V. and Boulet, M. (1998b). Variation in male provisioning in relation to brood size in Peregrine Falcon *Falco peregrinus*. *Emu*, 98, 297-304.

Ooi, B.Y., Kéry, M., Percival, R., Lee, Z.H. and Chu, S.C. (2020). A population study of Peregrine Falcons (*Falco peregrinus ernesti*) on West Malaysia. *Ornis Hungarica*, 28, 11-27.

Oro, D. (1996). Colonial seabird breeding in dense and small sub-colonies: an advantage against aerial predation. *Condor*, 98, 848–850.

Orton, D.A. (1975). The speed of a Peregrine's dive: Putting flights of fancy over the velocity of the stoop in mathematical perspective. *The Field*, 25 Septmeber, 588-560.

O'Rourke, C.T., Hall, M.I., Pitlik, T. and Fernández-Juricic, E. (2010). Hawk Eyes I: Diurnal Raptors Differ in Visual Fields and Degree of Eye Movement. *PLoS ONE* 5(9): e12802. doi:10.1371/journal.pone.0012802

Osmolovskaya, B.I. (1948). (The ecology of raptors of the Yamel Peninsula). *Proceedings of Geography Institute of AS of USSR*, 41, 5-77.

Paganini, S.T., Stafford, A., Hirschheydt, J. and Kéry, M. (2018). A large aggregation of 50 Peregrine Falcons (*Falco peregrinus*) during migration in the western Gulf of Mexico. *J. Raptor Res.*, 52, 500-502.

Pagel. J.E. (1989). Use of explosives to enhance a Peregrine Falcon eyrie. *J. Raptor Res.*, 23, 176-178.

Pagel, J.E. and Schmitt, N.J. (2013). American Marten remains with Peregrine Falcon prey sample in Yellowstone National Park. *J. Raptor Res.*, 47, 419-420.

Pagel, J.E. and Sipple, J. (2011). Incident of full sibling mating in Peregrine Falcons (*Falco peregrinus*). *J. Raptor Res.*, 45, 97–98.

Pagel, J.E., Patton, R.T. and Latta, B. (2010). Ground nesting of Peregrine Falcons (*Falco peregrinus*) near San Diego, California. *J. Raptor Res.*, 44, 323-325.

Pain, D.J., Swift, J., Green, R. and Cromie, R. (2020). The tide is turning for lead ammunition. *Brit. Birds*, 113, 110-118.

Paine, R.T., Wootton, J.T. and Boersma, P.D. (1990). Direct and indirect effects of Peregrine Falcon predation on seabird abundance. *Auk*, 107, 1–9.

Palacios, A. and Vareal, F.J. (1992). Color mixing in the pigeon (*Colomba livia*) II: A psychophysical determination in the middle, short and near-UV wavelength region. *Vision Res.* 32, 1947-1953.

Palleroni, A., Miller, C.T., Hauser, M. and Marler, P. (2005). Prey plumage adaptation against falcon attack. *Nature* 434, 973–974.

Pande, S. and Dahanukar, N. (2012). Reversed sexual dimorphism and differential prey delivery in Barn Owls (Tyto alba). *J. Raptor Res.*, 46, 184–189.

Pande, S. and Dahanukar, N. (2013). Reversed sexual dimorphism and prey delivery: response to Olsen. *J. Raptor Res.*, 47, 81–82.

Pande, S., Zduniak, P. and Yosef, R. (2017). Nest occupancy and reproductive success of a subspecies of the Peregrine Falcon, the Black Shaheen (*Falco peregrinus peregrinator*) in western India, *J. Raptor Res.*, 51, 470-475.

Park, J-S., Holden, A., Chu, V., Kim, M., Rhee, A., Patel, P., Shi, Y., Linthicum, J., Walton, B.J., McKeown, K., Jewell, N.P. and Hooper, K. (2009). Time-trends and congener profiles of PBDEs and PCBs in California Peregrine Falcons (*Falco peregrinus*). *Environ. Sci. Tech.*, 43, 8744-8751.

Parker, A. (1979). Peregrines at a Welsh coastal eyrie. *Brit. Birds*, 72, 104–114.

Paskhalny, S.P. and Golovatin, M.G. (2009). The current status of the Peregrine population in Yamal and Lower Ob region. In *Peregrine Falcon Populations* (Sielicki, J. and Mizera, T. eds), Turul Publishing and Poznań University of Life Sciences, pp. 373-394.

Pasquier, C., Fehlmann, M., Bresson, C., Fremaux, S., Jean, S. and Thelliez, J.-P. (2018). Monitoring of Peregrine Falcons in the Ariège Pyrenees and Toulouse, Midi-Pyrénées region, France. *Ornis Hungarica*, 26, 143-158.

Pauli, B.P., Spaul, R.J. and Heath, J.A. (2017). Forecasting disturbance effects on wildlife: tolerance does not mitigate effects of increased recreation on wildlands. *Anim. Conserv.*, 20, 251-260.

Pazhenkov, A., Karyakin, I., Afanasyev, D., Krivopalova, A. and Pazhenkova, E. (2018). Restoration of the tree-nesting Peregrine Falcon (*Falco peregrinus*) population in the Volga-Ural region. *Ornis Hungarica*, 26, 254-258.

Peakall, D.B. and Kiff, L.F. (1988). DDE contamination in Peregrines and American Kestrels and its effects on reproduction. In *Peregrine Falcon Populations: Their Management and Recovery* (T.J. Cade et al., eds.). The Peregrine Fund, Boise, Idaho, pp. 337-350.

Pennycuick, C. J., (1988). Empirical estimates of body drag of large waterfowl and raptors. *J. Exp. Biol.*, 135, 253–264.

Pennycuick, C.J. (2008). *Modelling the Flying Bird*. Academic Press, London.

Pennycuick, C.J., Fuller, M.R., Oar, J.J. and Kirkpatrick, S.J. (1994). Falcon versus Grouse: flight adaptations of a predator and its prey. *J. Avian Biol.*, 25, 39–49.

Pennycuick, C.J., Fast, P.L.F., Ballerstädt, N. and Rattenborg, N. (2012). The effect of an external transmitter on the drag coefficient of a bird's body, and hence on migration range, and energy reserves after migration. *J. Orn.*, 153, 633–644.

Pérez-Camacho, L., Martínez-Hesterkamp, S., Rebollo, S., García-Salgado G. and Morales-Castilla, I., (2018), Structural complexity of hunting habitat and territoriality increase the reversed sexual size dimorphism in diurnal raptors, *J. Avian Biol.*, 49 (10): e01745

Perkins, M., Ferguson, L., Lanctot, R.B., Stenhouse, I.J., Kendall, S., Brown, S., Gates, H.R., Hall, J.O., Regan, K. and Evers,D.C. (2016). Mercury exposure and risk in breeding and staging Alaskan shorebirds. *Condor*, 118, 571-582.

Peter, D. and Kestenholz, M. (1998). (Stoops of Peregrine Falcon *Falco peregrinus* and Barbary Falcon *Falco pelegrinoides*). *Ornithologische Beobachter*, 95, 107–112.

Philips, J.R. (2000). A review and checklist of the parasitic mites (Acarina) of the Falconiformes and Strigiformes. *J. Raptor Res.*, 34, 210–231.

Philips, J.R. and Dindal, D.L. (1977). Raptor nests as a habitat for invertebrates: a review. *Raptor Res.*, 11, 86–96.

Pictor, G. (2013). Tree nesting Peregrines in Wiltshire. *Hobby (Wiltshire Bird Report No. 40), Wiltshire Ornithological Society*, 101-104.

Pinzon-Rodriguez, A. and Muheim, R. (2021). Cryptochrome expression in avian UV cones: revisiting the role of CRY1 as magnetoreceptor. *Nature Portfolio Scientific Reports*, (2021) 11.12683, dpi.org/10/1038/s41598-021-92056-8

Pokrovsky, I., Lecomte, N., Sokolov A., Sokolov, V. and Yoccoz, N.G. (2010). Peregrine Falcons kill a Gyrfalcon feeding on their nestling. *J. Raptor Res.*, 44, 66–69.

Pokrovsky, I., Ehrich, D., Fufachev, I., Ims, R.A., Kulikova, O., Sokolov A., Sokolova, N., Sokolov, V. and Yoccoz, N.G. (2020), Nest association between two predators as a behavioural response to the low density of rodents. *Auk*, 137. 1-13.

Ponnikas, S., Ollila, T. and Kvist, L. (2017). Turnover and post-bottleneck genetic structure in a recovering population of Peregrine Falcons *Falco peregrinus*. *Ibis*, 159, 311-323.

Ponitz, B., Triep, M. and Brücker, C. (2014a). Aerodynamics of the cupped wings during Peregrine Falcon's diving flight. *Open Journal of Fluid Dynamics*, 4, 363-372. http://dx.doi.org/10.4236/ojfd.2014.44027

Ponitz, B., Schmitz, A., Fischer, D., Bleckmann, H. and Brücker, C. (2014b). Diving-flight aerodynamics of a Peregrine Falcon. *PLoS ONE*, 9(2): e86506.

Ponton, D.A. (1983). Nest site selection by Peregrine Falcons. *Raptor Res.*, 17, 27–28.

Potapov, E. and Sale R. (2005). *The Gyrfalcon*. T&AD Poyser, London.
Potapov, E. and Sale R. (2012). *The Snowy Owl*. T&AD Poyser, London.
Potier, S. (2020). Olfaction in raptors. Zoo. J. Linn. Soc., 189, 7snyder13-721.
Potier, S., Lieuvin, M., Pfaff, M. and Kelber, A., (2020a). How fast can raptors see?, *J. Exp. Biol.* 223, jeb.209031.
Potier, S., Mitkus, M. and Kelber, A. (2000b). Visual adaptations of diurnal and nocturnal raptors. *Seminars in Cell and Developmental Biology*, 106, 116-126.
Potvin, D.A., Välimäki, K. and Lehikoinen, A. (2016). Differences in shifts of wintering and breeding ranges lead to changing migration distances in European birds. *J. Avian Biol.*, 47, 619-628.
Price-Waldman, R. and Stoddard, M.C. (2021). Avian coloration genetics: recent advances and emerging questions. *J. Heredity,* 112, 395-416.
Probst, R., Pavlicev, M. and Schmid, R. (2007). Differences in the diet of three Peregrine Falcon *Falco peregrinus* pairs nesting in Chukotka, north-east Russia. *Forktail*, 175-177.
Pruett-Jones, S.G., White, C.M. and Devine, W.R. (1981). Breeding of the Peregrine Falcon in Victoria, Australia. *Emu*, 80, 253-269.
Puchała, K.O., Nowak-Życzyńska, Z., Sielicki, S. and Olech, W. (2021). Assessment of the genetic potential of the Peregrine Falcon (*Falco peregrinus peregrinus*) population used in the reintroduction program in Poland. *Genes*, 12, 666, doi.org/10.3390/genes1205066
Pulido, F. (2007). The genetics and evolution of avian migration, *Bioscience*, 57, 165-174.
Pyle, P. (2005). First-cycle molts in North American Falconiformes. *J. Raptor Res.*, 39, 378–385.
Pyle, P. (2013). Evolutionary implications of synapomorphic wing-molt sequences among falcons (Falconiformes) and parrots (Psittaciformes). *Condor*, 115, 593–602.
Quinn, J.L. (1997). The effects of hunting Peregrines *Falco peregrinus* on the foraging behaviour and efficiency of the Oystercatcher *Haematopus ostralegus*. *Ibis*, 139, 170–173.
Quinn, J.L., Kokorev, Y., Prop, J. and Black, J.M. (2000). Are Peregrine Falcons in northern Siberia still affected by organochlorines? In *Raptors at Risk* (ed. R.D. Chancellor and B.-U. Meyburg). World Working Group on Birds of Prey and Owls/Hancock House, pp. 279-294.
Quinn, J.L., Prop, J., Kokorev, Y. and Black, J.M. (2003). Predator protection or similar habitat selection in Red-Breasted Goose nesting associations: extremes along a continuum. *Anim. Behav.*, 65, 297–307.
Rabøl, J. (1993). The orientation systems of long-distance passerine migrants displaced in Autumn from Denmark to Kenya, Ornis Scand., 24, 183-196.
Raidal, S.R., Jaensch, S.M. and Ende, J. (1999). Preliminary report of a parasitic infection of the brain and eyes of a Peregrine Falcon *Falco peregrinus* and Nankeen Kestrels *Falco cenchroides* in western Australia. *Emu*, 99, 291-292.
Rand, A.L. and Gilliard, E.T. (1967). *Handbook of New Guinea Birds*, Weidenfield and Nicolson, London, UK.
Raphael, M.N. and Handcock, M.S. (2022). A new record minimum for Antarctic sea ice. *Nature Reveiws: Earth and Environment*, 3, 215-216, doi.org/10.1038/s43017-022-00281-0
Ratcliffe, D.A. (1958). Broken eggs in Peregrine eyries. *Brit. Birds*, 51, 23-26.
Ratcliffe, D.A. (1962). Peregrine incubating Kestrel eggs. *Brit. Birds*, 55, 131–132.
Ratcliffe, D.A. (1963). Peregrine rearing young Kestrels. *Brit. Birds*, 56, 457–460.
Ratcliffe, D.A. (1970). Changes attributable to pesticides in egg breakage frequency and eggshell thickness in some British birds. *J. App. Ecol.*, 7, 67–115.
Ratcliffe, D.A. (1984). The Peregrine Falcon breeding population of the United Kingdom in 1981. *Bird Study*, 31. 1-18.
Ratcliffe, D.A. (1993). *The Peregrine Falcon* (second edition). Poyser, London. (First edition published 1980.)

Ravi, S., Noda, R., Gagliardi, S., Kolomenskiy, D., Coombes, S., Liu, H., Biewener, A.A. and Konow, N. (2020). Modulation of flight muscle recruitment and wing rotation enables hummingbirds to mitigate aerial roll perturbations. *Curr. Biol.* 30, 187-195.

Razafimanjato, G., Rene de Roland, L-A., Rabearivony, J. and Thorstrom, R. (2007). Nesting biology and food habits of the Peregrine Falcon *Falco peregrinus radama* in south-west and central plateau of Madagascar. *Ostrich*, 78, 7-12.

Redpath, S.M. and Thirgood, S.J. (1999). Numerical and functional responses in generalist predators: Hen Harriers and Peregrines on Scottish grouse moors. *J. App. Ecol.*, 68, 879–892.

Rees, G. (2009). More Peregrine kleptoparasitism. *Brit. Birds*, 102, 511.

Rejt, Ł. (2001). Feeding activity and seasonal changes in prey composition of urban Peregrine Falcons *Falco peregrinus*. *Acta Ornithologica*, 36, 165-169.

Rejt, Ł. (2004). Nocturnal feeding of young by urban Peregrine Falcons (*Falco peregrinus*) in Warsaw (Poland). *Polish J. Ecology*. 52, 63-68.

Rejt, Ł. And Sielicki, S. (2007). Feeding activity and seaonal changes in prey composition of Peregrines in Poland. In *Peregrine Falcon Populations* (Sielicki, J. and Mizera, T. eds), Turul Publishing and Poznań University of Life Sciences, pp. 555-566.

Reymond, L. (1985). Spatial visual acuity of the eagle *Aquila audax*: a behavioural, optical and anatomical investigation. *Vision Res.*, 25, 1477–1491.

Reymond, L. (1987). Spatial visual acuity of the falcon *Falco berigora*: a behavioural.optical and anatomical investiagtion. *Vision Res.*, 27, 1859-1874.

Reynolds, M.H., Nash, S.A.B. and Courtot, K.N. (2015). Peregrine Falcon predation of endangered Laysan Teal and Laysen Finches on remote Hawaiian atolls. *Hawai'i Cooperative Studies Unit, Universtiy of Hawai'i, Technical Report HCSU-065*.

Ritz, T., Adem, S. and Schulten, K. (2000). A model for photoreceptor-based magnetoreception in birds. *Biophysical J.*, 78, 707-718.

Rivas-Salvador, J., Aguilera-Alcalá, N., Tella, J.L. and Carrete, M. (2021). Assessing the introduction of exotic raptors into the wild from falconry. *Biol. Invasions*, 23, 1131-1140.

Roalkvam, R. (1985). How effective are hunting Peregrines? *Raptor Res.*, 27-29.

Roberts, E.L. (1946). Unusual hunting behaviour of Peregrine Falcon. *Brit. Birds*, 39, 318–319.

Robinson, B.G., Franke, A. and Derocher, A.E. (2017). Weather-mediated decline in prey delivery rates causes food-limitation in a top avian predator. *J. Avian Biol.*, 48, 748-758.

Robinson, B.G., Franke, A. and Derocher, A.E. (2018). Stable isotope mixing models fail to estimate the diet of an avian predator. *Auk*, 135, 60-70.

Robinson, J.A. (2018). Kleptoparasitism of food-passing Marsh Harriers by a Peregrine Falcon. *Brit. Birds*, 111, 402-403.

Robinson, R.A. and Wilson, M. (2021) Contrasting trends in age-specific survival of Peregrine Falcons (*Falco peregrinus*) in Britain using smoothed estimates of recovery probabilities. *Ibis*, 163, 890-898.

Robinson, R.A., Leech, D.L. and Clark, J.A. (2021). The Online Demography Report: Bird ringing and nest recording in Britain & Ireland in 2010. BTO, Thetford (http://www.bto.org/ringing-report, created on 27 June 2020). Report retrieved June 2022.

Robbrecht G., Bekaert, M., Van Nieuwenhuyse, D., Vangeluwe, D., Louette, M. and Lens, L. (2007), The Peregrine Falcon *Falco peregrinus* back in Belgium, the story of a successful nest box action. *Nature.oriolus*, 73, 4-18.

Rodríguez, B., Siverio, F., Siverio, M. and Rodríguez, A. (2019). Falconry threatens Barbary Falcons in the Canary Islands through genetic admixture and illegal harvesting of nestlings. *J. Raptor Res.*, 53, 189-197.

Rogers, W. and Leatherwood, S. (1981). Observations of feeding at sea by a Peregrine Falcon and Osprey. *Condor*, 83, 89–90.

Rollie, C. and Christie, G. (2006). Peregrine Falcon catching and killing a bat in daylight, *Scottish Birds*, 26, 45–46.

Rosenfield, R.N., Schneider, J.W., Papp, J.M. and Seegar, W.S. (1995). Prey of Peregrine Falcons in west Greenland. *Condor*, 97, 763–770.

Rounsley, K.J. and McFadden, S.A. (2005). Limits of visual acuity in the frontal field of the Rock Pigeon (*Columba livia*). *Perception*, 34, 983-993.

RSPB (2021). Birdcrime 2020: Exposing bird of prey persecution in the UK. Royal Society for the Protection of Birds, Sandy.

Rudebeck, G. (1951). The choice of prey and modes of hunting of predatory birds with special reference to their selective effect. *Oikos*, 3, 200–231.

Ruthven, G. (2013). Peregrine retrieving prey from water. *Scottish Birds*, 33, 125.

Saggese, M.D., Ellis, D.H., Trejo, A., Nelson, R.W., Quaglia, A.I.E., Caballero, I.C., Ellis, C.H. and Amoros, M.B. (2019). Avian prey remains at eyries of the Austral Peregrine Falcon (*Falco peregrinus cassini*) in southern Atlantic Argentine Patagonia during the breeding season. *J. Raptor Res.*, 53, 207-211.

Sale. R. (2020). *The Common Kestrel*, Snowfinch Publishing, Coberley, UK.

Sale, R. and Messenger, A. (2021). *The Eurasian Hobby*. Snowfinch Publishing, Coberley, UK.

Salminen, P. and Wikman, M. (1977). (Peregrine population trend in Finland). In *Pilgrimsfalk: Report of a conference on Peregrine Falcons at Grimsö, Sweden* (ed. P. Lindberg), pp. 24-30.

Samson, A., Ramakrishnan, B., Santhoshkumar, P. and Karthick, S. (2017). Observation of Shaheen Falcon *Falco peregrinus peregrinator* (Aves: Falconiformes: Falconidae) in the Nilgiris, Tamil Nadu, India. *J. Threatened Taxa*, 9, 10850-10852.

Sankey, D.W.E., Shepard, E.L.C., Biro, D. and Portugal, S.J. (2019). Speed consensus and the 'Goldilocks principle' in flocking birds (*Columba livia*). *Anim. Behav.*, 157, 105-119.

Santema, P., Schlicht, L., Beck, K.B., Sheldon, B.C. and Kempenaers, B. (2021a). Why do nestlings fledge early in the day? *Anim. Behav.*, 174, 79-86.

Santema, P., Schlicht, L., Beck, K.B., Sheldon, B.C. and Kempenaers, B. (2021b). Experimental evidence that nestlings adjust their fledging time to each other in a multiparous bird. *Anim. Behav.*, 180, 143-150.

Santos, C.D., Muñoz, S.J.P., Onrubia, A-R. and Wikelski, M. (2020). The gateway to Africa: what determines sea crossing performance of a migratory soaring bird at the strait of Gibraltar? *J. Anim. Ecol.*, 89, 1317-1328.

Sarà, M., Mascara, R., Nardo, A. and Zanca, L. (2022). Climate effects on breeding phenology of Peregrine and Lanner falcons in the Mediterranean. *Ardea*, 110, 1-18.

Saxby, H.L. (1874). *The Birds of Shetland* (ed. S.H. Saxby). MacLachlan and Stewart, Edinburgh.

Schick, C.T., Brennan, L.A., Buchanan, J.B., Finger, M.A., Johnson, T.M. and Herman, S.G. (1987). Organochlorine contamination in shorebirds from Washington State and the significance for their falcon predators. *Environ. Mon. Assess.*, 9, 115-131.

Schmidl, D. (1988). Dusting in falcons. *J. Raptor Res.*, 22, 59–61.

Schmitz, A., Ondreka, N., Poleschinski, J., Fischer, D., Schmitz, H., Klein, A., Bleckmann, H. and Bruecker, C., (2018). The Peregrine Falcon's rapid dive: on the adaptedness of the arm skeleton and shoulder girdle, *J. Comp. Physiol. A*, 204, 747-759.

Schoenjahn, J., Pavey, C.R. and Walter, G.H., (2020). Why female birds of prey are larger than males, *Biol. J. Linne. Soc.*, 129, 532-542.

Schoonmaker, P.K., Wallace, M.P. and Temple, S.A. (1985). Migrant and breeding Peregrine Falcons in northwestern Peru. *Condor*, 87, 423-424

Schulten, K., Swenberg, C.E. and Weller, A. (1978). A biomagnetic sensory mechanism based on magnetic field modulated coherent electron spin motion. *Z. Phys. Chem.*, 111, S1, 1-5.

Schulz, R., Bub, S., Petschick, L.L., Stehle, S. and Wolfram J., (2021). Applied pesticide toxicity shifts towards plants and invertebrates, even in GM crops, *Science*, 372, 81-84.

Schulze, C., (2016): (On the morphology of the physiological foveae at the fundus of birds of prey and owls using optical coherence tomography (OCT)). Dissertation, Ludwig Maximilian University of Munich: Faculty of Veterinary Medicine.

Schwab, I.R. and Maggs, D. (2004). The falcon's stoop. *Br. J. Opthalmol.* 88 (1), 4.

Searle, A., Turnbull, J. and Adams, W.M. (2022). The digital Peregrine: a technonatural history of a cosmopolitan raptor. Trans. Inst. British Geog., doi.org/10.1111/tran.12566

Serjeantson, D. (2009). *Birds (Cambridge Manuals in Archaeology)*. Cambrige University Press, England.

Serra, G., Lucentini, M. and Romano, S. (2001). Diet and prey selection of non-breeding Peregrine Falcons in an urban habitat of Italy. *J. Raptor Res.*, 35, 61-64.

Shafaeipour, A. (2014). Prey selection of the Barbary Falcon (*Falco peregrinoides*) in south-western Iran. *Zoology in the Middle East*, 60, 13-19.

Shafaeipour, A., Siverio, M. and Siverio, F. (2016). Data on habitat and breeding biology of the Barbary Falcon, *Falco peregrinus pelegrinoides*, Temminck, 1829, from south-western Iran. *Acta Zoologica Bulgarica*, 68, 85-88.

Sherrod, S.K. (1988). *Behaviour of Fledgling Peregrines*, The Peregrine Fund, Ithaca, New York.

Sherrod, S.K. (1988). Behavioural differences in Peregrine Falcons. In *Peregrine Falcon Populations: their management and recovery*, The Peregrine Fund, Boise, Indiana, USA, p741-748.

Sielicki, J. and Mizera, J. (2009). Peregrine Falcon Population: Status and Perspectives in the 21st Century: Proceedings of the 2nd International Peregrine Conference, 19-23 September 2007, Piotrowo/Poznań, Poland. TURUL, Warsaw.

Sielicki, S. and Sielicki, J. (2009). Restoration of Peregrine Falcon in Poland 1989-2007. In *Peregrine Falcon Populations* (Sielicki, J. and Mizera, T. eds), Turul Publishing and Poznań University of Life Sciences, pp. 699-722.

Siverio, F., Rodríguez, A. and Padilla, D.P. (2007). Kleptoparasitism by Eurasian Buzzard (*Buteo buteo*) on two Falco species. *J. Raptor Res.*, 42, 67-68.

Siverio, M., Siverio, F., Rodríguez, B and Rodríguez, A. (2011). Long-term monitoring of an insular population of Barbary Falcon *Falco peregrinus pelegrinoides*. *Ostrich*, 82, 225-230.

Slagsvold, T. and Sonerud, G.A. (2007). Prey size and ingestion rate in raptors: importance for sex roles and reversed sexual size dimorphism. *J. of Avian Biol.*, 38, 650–661.

Smith, G.D., Murillo-García, O.E., Hostetler, J.A., Mearns, R., Rollie, C., Newton I., McGrady, M.J. and Oli, M.K. (2015). Demographic of population recovery: survival and fidelity of Peregrine Falcons at various stages of population recovery. *Oecologia*, 178, 391-401.

Smith, J.M., Stauber, E. and Bechard, M.J. (1993). Identification of Peregrine Falcons using a computerized classification system of toe-scale pattern analysis. *J. Raptor Res.*, 27, 191-195.

Smith, M.B. (1992). Peregrines nesting beside Kestrels on urban chimney. *Brit. Birds*, 85, 498.

Snyder, A.W. and Miller, W.H., (1978). Telephoto lens system of falconiform eyes. *Nature*, 275, 127-129.

Sodhi, N.S., Warkentin, I.G., James, P.C. and Oliphant, L.W. (1991). Effects of radiotagging on breeding Merlins. *J. Wild. Mgt.*, 55, 613–616.

Sokolov, V., Lecomte, N., Sokolov, A., Rahman, Md.L. and Dixon, A. (2014). Site fidelity and home range variation during the breeding season of Peregrine Falcons in Yamal, Russia. *Polar Bio.* 37, 1621-1631.

Sonerud, G.A., Steen, R., Løw, L.M., Røed, L.T., Skar, K., Selås, V. and Slagsvold, T., (2013). Size-based allocation of prey from male to offspring via female: family conflicts, prey selection, and evolution of sexual size dimorphism in raptors, *Oecologia*, 172, 93-107.

Sonerud, G.A., Steen, R., Løw, L.M., Røed, L.T., Skar, K., Selås, V. and Slagsvold, T., (2014a). Evolution of parental roles in raptors: prey type determines role asymmetry in the Eurasian Kestrel, *Anim. Behav.* 96, 31-38.

Sonerud, G.A., Steen, R., Selås, V., Aanonsen, O.M., Aasen, G-H., Fagerland, K.L., Fosså, A., Kristiansen, L., Løw, L.M., Rønning, M.E., Skouen, S.K., Asakskogen, E., Johansen, H.M., Johnsen, J.T., Karlsen, L.I., Nyhus, G.C., Røed, L.T., Skar, K., Sveen, B-A., Tveiten, R. and Slagsvold, T., (2014b). Evolution of parental roles in provisioning birds: diet determines role asymmetry in raptors, *Behav. Ecol.*, 25, 762-772.

Spofford, W.R. (1947). A successful nesting of the Peregrine Falcon with three adults present, *Migrant*, 18, 56–58.

Spofford, W.R. and Amadon, D. (1993). Live prey to young raptors – incidental or adaptive? *J. Raptor Res.*, 27, 180–184.

Sprunt, A. (1951). Aerial feeding of Duck Hawk, *Falco p. anatum. Auk*, 68, 372–373.

Sriram, S. and Huettman, F. (2017). A global model of predicted Peregrine Falcon (*Falco peregrinus*) distribution with Open Source GAS Code and 104 Open Access Layers for use by the global public. *Earth Syst. Sci. Data Discuss.*, doi:10.1594/essd-2016-65, 2017

Stager, K.E. (1941). A group of bat-eating Duck Hawks. *Condor*, 43, 137–139.

Stepanyan, L.S. (1969a). (Observations of the Red-naped Shaheen in Central Asia). *Falke*, 16, 124-130.

Stepanyan, L.S. (1969b). (Observations of the falcon (*Falco pelegrinoides babylonicus*) in Central Asia). *Bull. Moscow Naturalist's Society*, 74(6), 37-48.

Stephan, C. and Bugnyar, T. (2013). Pigeons integrate past knowledge across sensory modalities. *Anim. Behav.*, 85, 605–613.

Stirrup, S. (2021). Peregrine Falocn 'leaf bathing'. *Brit. Birds*, 114, 424.

Stoddard, M.C., Eyster, H.N., Hogan, B.G., Morris, Soucy, E.R. and Inouye, D.W., (2020). Wild hummingbirds discriminate nonspectral colours, *PNAS*, 117, 15112-15122.

Storms, R.F., Carere, C., Zoratto, F. and Hemelrijk, C.K. (2019). *Behav. Ecol. Sociobiol.*, 73: 10, https://doi.org/10.1007/s00265-018-2609-0

Strenna, L. (2013). The evolution of the population of Peregrine Falcons *Falco peregrinus* in Burgundy). *Alauda*, 81, 17-38.

Stroud, D.A., Bainbridge, I.P., Maddock, A., Anthony, S., Baker, H., Buxton, N., Chambers, D., Enlander, I., Hearn, R.D., Jennings, K.R., Whitehead, S. on behalf of the UK SPA and Ramsar Scientific Working Group (eds.). (2016). *The Status of UK SPAs in the 2000s: the Third Network Analysis*. Joint Nature Conservation Committee, Peterborough, UK.

Stroud, D.A., Pain, D.J. and Green, R.E. (2021). Evidence of widespread illegal hunting of waterfowl in England, despite partial regulation of the use of lead shotgun ammunition. *Conservation Evidence J.*, 18, 18-24.

Stutchbury, B.J. (1991). Floater behaviour and territory acquisition in male Purple Martins. *Animal Behaviour*, 42, 435–443.

Suh, A., Paus, M., Kiefmann, M., Churakov, G., Franke, F.A., Brosius, J., Kriegs, J.O. and Schmitz, J., (2011). Mesozoic retroposons reveal parrots as closest living relatives of passerine birds. *Nat. Comms.*, 2, 443, do1: 10.1038/ncomms1448.

Sulkava, S. (1968). A study on the food of the Peregrine, *Falco p. peregrinus* Tunstall, in Finland. *Aquilo*, 6, 19–31.

Sullivan, T.N., Wang, B., Espinosa, H.D. and Meyers, M.A. (2017). Extreme lightweight structures: avian feathers and bones. *Materials Today*, 20, 377-391.

Sutton, L.J., Réné de Roland, L-A., Thorstrom, R. and McClure, C.J.W. (2021). Distribution and habitat use of the Madagascar Peregrine Falcon: first estimates of the area of habitat and population size. *BioR xiv preprint* https://doi.org/10.1101/2021.08.30.458216

Swann, R.L. (1998). Baby Stoat in Peregrine nest, Scottish Birds, 19, 180.

Sweeney, S.J., Redig, P.T. and Tordoff, H.B. (1997). Morbidity and productivity of rehabilitated Peregrine Falcons in the upper Midwestern USA. *J. Raptor Res.*, 31, 347–352.

Talbot, S.L., Sage, G.K., Sonsthagen, S.A., Gravley, M.C., Swem, T., Wlliams, J.C., Longmire, J.L., Ambrose, S., Flamme, M.J., Lewis, S.B., Phillips, L., Anderson, C. and White, C.M. (2017). Intraspecific evolutionary relationship among Peregrine Falcons in western North America high latitudes. *PLoS ONE*, 12(11): e0188185. doi.org/10.1371/journal.pone.0188185

Tamini, L.T., Chavez, L.N. and Shigihara, Y.A. (2016). (Record of a Peregrine Falcon (*Falco peregrinus*) preying on two species of petrels from a fishing vessel in the Argentine Sea. *El Hornero*,31, 117-120.

Tarboton, W. (1984). Behavior of the African Peregrine during incubation. *Raptor Res.*, 18, 131-136.

Taylor, L., Slankard, K. and Stump, A. (2020). Home range of an adult female Peregrine Falcon (*Falco peregrinus*) during the breeding and non-breeding seasons in Kentucky, USA. *Journal of the Kentucky Academy of Science*, 81, 1-10.

Tedstone, A.J. and Machguth, H. (2022). Increasing surface runoff from Greenland's firn areas. *Nature Climate Change*, 12, 672-676.

Telford, E.A. (1996). Peregrine Falcons in the north-eastern United States: sonographic analysis of the defense call, population turnover, and dispersal. MSc thesis, Boise State University.

Therrien, J-F., Lecomte, N., Zgirski, T., Jaffré, M., Beardsell, A., Goodrich, L.J., Bêty, J., Franke, A., Zlonis, E. and Bildstein, K.L. (2017). Long-term phenological shifts in migration and breeding-area residency in eastern North American raptors. *Auk*, 134, 871-881.

Thiollay, J-M. (1988). Prey availability limiting an island population of Peregrine Falcons in Tunisia. In *Peregrine Falcon Populations: Their Management and Recovery*, p701-710, The Peregrine Fund, Boise, Indiana, USA.

Thiollay, J-M. (1983). (Current position of diurnal raptors in northern Borneo). *Alauda*, 51, 109-123.

Thomas, A.R.L. and Balmford, A., (1995). How Natural Selection Shapes Birds' Tails. *Am. Nat.*, 146, 848-868.

Thompson, G.B. (1954). V. Contributions toward a study of the ectoparasites of British birds and mammals – No. 2. *Annals and Magazine of Natural History*, 7:73, 17-39.

Thomson, S., McMorran, R., Newey, S., Matthews, K.B., Fielding, D., Miller, D.G., Glass, J., Gandossi, G., McMillan, J. and Spencer, M. (2020). Summary Report - Socio-economic and biodiversity impacts of driven grouse moors in Scotland. Commissioned report for Scottish Government (CR/2019/01), https://www.gov.scot/ISBN/978-1-80004-212-4

Thomsett, S. (1988). Distribution and status of the Peregrine in Kenya. In *Peregrine Falcon Populations: Their Management and Recovery* (T.J. Cade et al. eds). The Peregrine Fund, Boise, Idaho, pp. 289–295.

Thorstrom, R., Rene de Roald, L.-A. and Watson, R.T. (2003) Falconiformes and Strigiformes: Ecology and status of raptors. In *The Natural History of Madagascar* (Goodman, S.J. and Benstead, J.P., eds.), University of Chicago Press.

Tingay, R.E. and Wightman, A. (2018). The case for reforming Scotland's driven grouse moors. *The Revive Coalition*, Edinburgh, Scotland.

Tobalske B. W., Hedrick, T. L., Dial, K. P. and Biewener A. A., (2003). Comparative power curves in bird flight. *Nature*, 421, 363-366, 2003.

Tordoff, H.B. and Redig, P.T. (1997). Midwest Peregrine Falcon demography. *J. Raptor Res.*, 31, 339-346.

Tordoff, H.B. and Redig, P.T. (1998). Apparent siblicide in Peregrine Falcons. *J. Raptor Res.*, 32, 184.

Tordoff, H.B. and Redig, P.T. (1999a). Two fatal Peregrine Falcon territorial fights. *Loon*, 71, 182–186.

Tordoff, H.B. and Redig, P.T. (1999b). Close inbreeding in Peregrine Falcons in midwestern United States. *J. Raptor Res.*, 33, 326-328.

Tornberg, R., Korpimäki, V-M., Rauhala, P. and Rytkönen, S. (2016). Peregrine Falcon (*Falco peregrinus*) may affect local demographic trends in wetland bird species. *Ornis Fennica*, 93, 172-185.

Trainer, D.O. (1969). Diseases in raptors: a review of the literature. In P*eregrine Falcon Populations: their Biology and Decline* (ed. J.J. Hickey). University of Wisconsin Press, Madison, pp. 425–433.

Treleaven, R.B. (1977). *Peregrine: the Private Life of the Peregrine Falcon*. Headline Publications, Penzance, England.

Treleaven, R.B. (1980). High and low intensity hunting in raptors. *Z. Tierpsychol.* (now *Ethology*), 54, 339–345.

Treleaven, R.B. (1998). *In Pursuit of the Peregrine*. Tiercel Publishing, Wheathampstead, England.

Tucker, V.A., (1968). Respiratory exchange and evaporative water loss in the flying Budgerigar. *J. Exp. Biol.*, 48, 67-87.

Tucker V.A., (1972). Metabolism during flight in the Laughing Gull (*Larus atricilla*). *Am. J. Physiology*, 222, 237-245.

Tucker. V.A. (1998) Gliding flight: speed and acceleration of ideal falcons during diving and pull out. *J. of Exp. Biol.*, 201, 403–414.

Tucker, V.A. (1999). The stoop of a falcon: how fast, how steep, how high? *Hawk Chalk*, 38, 58–63.

Tucker, V.A. (2000a). The deep fovea, sideways vision and spiral flight paths in raptors. *J. Exp. Biol.*, 203, 3745–3754.

Tucker, V.A. (2000b). Gliding flight: drag and torque of a hawk and a falcon with straight and turned heads, and a lower value for the parasite drag coefficient. *J. Exp. Biol.*, 203, 3733–3744.

Tucker, V.A., Cade, T.J. and Tucker, A.E. (1998). Diving speeds and angles of a Gyrfalcon (*Falco rusticolus*). *J. Exp. Biol.*, 201, 2061–2070.

Tucker, V.A., Tucker, A.E., Akers, K. and Enderson, J.H. (2000). Curved flight paths and sideways vision in Peregrine Falcons (*Falco peregrinus*). *J. Exp. Biol.*, 203, 3755–3763.

Tyler, S.J. and Ormerod, S.J. (1990). Mammals taken by Peregrines in mid-Wales. *Welsh Bird Report*, 4, 57.

Tyrberg, T., (2008). *Pleistocene birds of the Palearctic*. web.telia.com/~u11502098/Pleistocene.pdf.

Uttendörfer, O. (1952). (*New results on the diet of birds of prey and owls*). Eugen Ulmer, Stuttgart.

van der Leek, S.A., DeMerritt, S.L.A. and Kasner, A.C. (2019). Description of a Peregrine Falcon (*Falco peregrinus*) fence mortality in the southern high plains of Texas. *The Southwestern Naturalist*, 64, 228-230.

Varland, D.E., Buchanan, J.B., Fleming, T.L., Kenney, M.K. and Vanier, C. (2009). Scavenging as a food-acquisition strategy by Peregrine Falcons. *J. Raptor Res.*, 52, 291-308.

Varland, D.E., Powell, L.A., Buchanan, J.B., Fleming, T.L. and Vanier, C. (2020). Peregrine Falcon survival rates derived from long-term study at a migratory and overwintering area in coastal Washington State. *J. Raptor Res.*, 54, 207-221.

Vasina, W.G. and Straneck, R.J. (1984). Biological and ethological notes on *Falco peregrinus cassini* in Central Argentina. *Raptor Res.* 18, 123-130.

Velarde, E. (1993). Predation of nesting alcids by Peregrine Falcons at Rasa Island, Gulf of California, Mexico, *Condor*, 95, 706-708.

Vickers-Rich, P., Monaghan, J.P., Baird, R.F. and Rich, T.H. (1991). *Vertebrate Palaeontology of Australasia.* Pioneer Design Studio/Monash University Publications, Melbourne, Australia.

Victory, N., Segovia, Y. and Garcia, M. (2021). Foveal shape, ultrastructure and photoreceptor composition in Yellow-legged Gull, *Larus michahellis* (Naumann, 1840). *Zoomorph.*, 140, 151-167.

Vidal, A., Baldomà, L., Molina-López, R.A., Martin, M. and Darwich, L. (2017). Microbiological diagnosis and antimicrobial sensitivity profiles in diseased free-living raptors. *Avian Pathology*, 46, 442-450.

Videler, J.J. (2005). *Avian Flight*. Oxford University Press, Oxford.

Videler, J.J., Vossebelt, G., Gnodde, M. and Groenewegen, A. (1988a). Indoor flight experiments with trained Kestrels: I. Flight strategies in still air with and without added weight. *J. Exp. Biol.*, 134, 173–183.

Videler, J.J., Groenewegen, A., Gnodde, M. and Vossebelt, G. (1988b). Indoor flight experiments with trained Kestrels: II. The effect of added weight on flapping flight kinematics. *J Exp. Biol.*, 134, 185–199.

Village, A. (1990). The Kestrel. Poyser, London.

Vitovich, O.A., Tkachenko, I.V. and Akbaev, I.M. (2000). (The diet and hunting behaviour of the Caucasian Peregrine Falcon (*Falco peregrinus* brookei). *Russian J. Orn.*, 114, 17-20.

Voous, K.H. (1961). Records of Peregrine Falcons on the Atlantic Ocean. *Ardea*, 49, 176–177.

Vorkamp K., Møller, S., Falk, K., Rigét, F.F., Thomsen, M. and Sørenson, P.B. (2014). Levels and trends of toxaphene and chlordane-related pesticides in Peregrine Falcon eggs from south Greenland. *Science Total Environ.*, 468-469, 614-621.

Vrettos, M., Reynolds, C. and Amar, A. (2021). Malar stripe size and prominence in Peregrine Falcons vary positively with solar radiation: support fo the solar glare hypothesis. *Biol. Lett.* 17, 20210116

Walker, D. (1988). Peregrine taking Leach's Petrel. *Brit. Birds*, 81, 395.

Walker, K.S. (2019). The influence of urbanisation on the diet of breeding Peregrine Falcons (*Falco peregrinus*) in Cape Town, South Africa. BSc thesis, University of Cape Town.

Walker, L.A., Chaplow, J.S., Grant, H.K., Lawlor, A.J., Pereira M.G., Potter, E.D. and Shore, R.F. (2016). Mercury (Hg) concentrations in predatory bird livers and eggs as an indicator of changing environmental concentrations: a *Predatory Bird Monitoring Scheme (PBMS) report*. Centre for Ecology & Hydrology, Lancaster, UK.

Wallen, M.S. (1992). Talon-locking between Merlin and Peregrine. *Brit. Birds*, 85, 496.

Wallis, R. J., (2017). 'As the Falcon her Bells' at Sutton Hoo? Falconry in Early Anglo-Saxon England, *Arch. J.*, 174, 409-436.

Walls, G.L. (1942). *The vertebrate eye and its adaptive radiation*. Cranbrook Press, Bloomfield Hills, Maryland,

Watson, S.J. and Sale R.G. (2020). A significant decline of the breeding Peregrine Falcon in coastal north and west Cornwall. *Brit. Birds*, 113, 354-362.

Watts, B.D. and Truitt, B.R. (2021). Influence of introduced Peregrine Falcons on the distribution of Red Knots within a spring staging site. *PLoS ONE*, 16(1): e0244459. https://doi.org/10.1371/journal.pone.0244459

Wegner, P., Kleinstäuber, G., Baum, F. and Schilling, F. (2005). Long-term investigation of the degree of exposure of German Peregrine Falcons (*Falco peregrinus*) to damaging chemicals from the environment. *J. Orn.*, 146, 34-54.

Weir, D.N. (1977). The Peregrine in N.E. Scotland in relation to food and pesticides. In *Pilgrimsfalk: Report of a conference on Peregrine Falcons at Grimsö, Sweden* (ed. P. Lindberg), pp. 56-58.

Weir, D.N. (1978). Wild Peregrines and grouse. *Falconer*, 7, 98–102.

Wells, D.R. (1998). *The Birds of the Thai-Malay Peninsula, Vol. 1, Non-Passerines*, Academic Press, London.

Wendt, A.M. and Septon, G.A. (1991). Notes on a successful nesting by a pair of yearling Peregrine Falcons (*Falco peregrinus*). *J. Raptor Res.*, 25, 21–22.

Wernham, C., Toms, M., Marchant, J.H., Clark, J., Siriwardena, G. and Baillie, S.R. (eds.) (2002). *The Migration Atlas: Movements of the Birds of Britain and Ireland*. Poyser, London.

Wetmore, A. (1919). Lead poisoning in waterfowl. US Department of Agriculture Bulletin 793.

White, C.M. and Brimm, D.J. (1990). Insect hawking by a Peregrine Falcon (*Falco peregrinus*) in Fiji. *Notornis*, 37, 140.

White, C.M. and Nelson, R.W. (1991). Hunting range and strategies in a tundra breeding Peregrine and Gyrfalcon observed from a helicopter. *J. Raptor Res.*, 25, 49-62.

White, C.M., Brimm, D.J. and Clunie, F. (1988a). A study of Peregrines in the Fiji islands, south Pacific Ocean. In *Peregrine Falcon Populations: Their Management and Recovery* (T.J. Cade et al. eds). The Peregrine Fund, Boise, Idaho, pp. 275–287.

White, C.M., Emison, W.B. and Bren, W.M. (1988b). Atypical nesting habitat of the Peregrine Falcon (*Falco peregrinus*) in Victoria, Australia. *J. Raptor Res.*, 22, 37-43.

White, C.M., Brimm, D.J. and Wetton, J.H. (2000). The Peregrine Falcon *Falco peregrinus* in Fiji and Vanuatu. In *Raptors at Risk* (ed. R.D. Chancellor and B.-U. Meyburg). World Working Group on Birds of Prey and Owls/Hancock House, pp. 707-720.

White, C.M., Emison, W.B. and Williamson, F.S.L. (1973). DDE in a resident Aleutian Island Peregrine population. *Condor*, 75, 306-311.

White, C.M., Clum, N.J., Cade, T.J. and Hunt, W.G. (2020). Peregrine Falcon (*Falco peregrinus*), version 1. In *The Birds of North America Online* (S.M. Billerman, ed.). Cornell Laboratory of Ornithology, Ithaca, NY, USA. https://doi.org/10.2173/bow.perfal.01.

White, C.M., Cade, T.J. and Enderson, J.H. (2013a). *Peregrine Falcons of the World*. Lynx, Barcelona.

White, C.M., Sonsthagen, S.A., Sage, G.K., Anderson, C. and Talbot, S.L. (2013b). Genetic relationships among some subspecies of the Peregrine Falcon (*Falco peregrinus* L.), inferred from mitochondrial DNA control-region sequences. *Auk*, 130, 78–87.

Whitman, J.S. and Caikoski, J.R. (2008). Peregrine Falcon nesting in tree stick-nest in Alaska. *J. Raptor Res.*, 42, 300-302.

Whittaker, D.J. and Hagelin, J.C. (2021). Female-based patterns and social function in avian chemical communication. *J. Chem. Eco.*, 47, 43-62.

Wiebe, K.L., Wiehn, J., and Korpimäki, E., (1998). The onset of incubation in birds: can females control hatching patterns?, *Anim. Behav.*, 55, 1043-1052.

Wightman, C.S. and Fuller, M.R. (2006). Influence of habitat heterogeneity on distribution, occupancy patterns, and productivity of Peregrine Falcons in central west Greenland. *Condor*, 108, 270–281.

Wigman, A.B. (1951). (Breeding Peregrine Falcons, *Falco peregrinus* Tunst. in the 1950 season). *Limosa*, 24, 8-10.

Wikelski, M., Quetting, M., Cheng, Y., Fiedler, W., Flack, A., Gagliardo, A., Salas, R., Zannoni, N. and Williams, J. (2021). Smell of green leaf volatiles attracts White Sorks to freshly cut meadows. *Scientific Reports*, 11, 129129.

Wilcox, J.J.S., Boissinot, S. and Idaghdour, Y. (2019). Falcon genomics in the context of conservation, speciation and human culture. *Ecol. and Evol.* 9, 14523-14537.

Williamson, J.L. and Witt, C.C. (2021). Elevational niche-shift migration: Why the degree of elevational change matters for the ecology, evolution and physiology of migratory birds. *Ornithology*, 138, 1-26.

Wilson, M.D., Balmer, D.E., Jones, K., King, V.A., Raw, D., Rollie, C.J., Rooney, E., Ruddock, M., Smith, G.D., Stevenson, A., Stirling-Aird, P.K., Wernham, C.V., Weston, J.M. and Noble, D.G. (2018). The breeding population of Peregrine Falcon *Falco peregrinus* in the United Kngdom, Isle of Man and Channel Islands in 2014. *Bird Study*, 65, 1-19, doi: 10.1080/00063657.2017.1421610

Wilson, U.W., McMillan, A. and Dobler, F.C. (2000). Nesting, population trend and breeding success of Peregrine Falcons on the Washington outer coast, 1980-1998. *J. Raptor Res.* 34, 67-74.

Wiltschko, R. and Wiltschko, W. (2022). The discovery of the use of magnetic navigational information. *J. Comp. Physiol. A.* 208, 9-18.

Wiltschko, W. (1968). (On the influence of static magnetic fields on the orientation of the Robin (*Erithacus rubecula*)), *Z. Tierpsychol.* (now *Ethology*), 25, 537-558.

Wiltschko, W. and Wiltschko, R. (1999). The effect of yellow and blue light on magnetic compass orientation in European Robins, *Erithacus rubecula*. *J. Comp. Physiol. A*, 184, 295-299.

Wiltschko, W. and Wiltschko, R. (2007). Magnetoreception in birds: two receptors for two different tasks. *J. Orn.*, 148 (Supp. 1), S61-S76.

Wiltschko, W., Traudt, J., Güntürkün, O., Prior, H. and Wiltschko, R. (2002). Lateralization of magnetic compass in a migratory bird. *Nature*, 419, 467-470.

Wimsatt, W.A. (1940a). Early nesting of the Duck Hawk in Maryland. *Auk*, 57, 109.

Wimsatt, W.A. (1940b). Homing instinct and profligacy in the Duck Hawk. *Auk*, 57, 107–109.

Winger, B.M, Lovette, I.J. and Winkler, D.W. (2012), Ancestry and evolution of seasonal migration in the Parulidae, *Proc. Roy. Soc. B*, 279. 610-618.

Wink, M. and Sauer-Gürth, H. (2000). Advances in the molecular systematics of African raptors. In *Raptors at Risk* (ed. R.D. Chancellor and B.-U. Meyburg). World Working Group on Birds of Prey and Owls/Hancock House, pp. 135–147.

Wink, M., Seibold, I., Lotfikhah, F. and Bednarek, W. (1998). Molecular systematic of Holarctic raptors (Order Falconiformes). In *Holarctic Birds of Prey* (ed. R.D. Chancellor and B.-U. Meyburg). World Working Group on Birds of Prey and Owls/ADENEX, pp. 29–48.

Wink, M., Sauer-Gürth, H., Ellis, D. and Kenward R. (2004). Phylogenetic relationships in the Hierofalco complex (Saker-, Gyr-, Lanner-, Lagger Falcon). In *Raptors Worldwide* (ed. R.D. Chancellor and B.-U. Meyburg). World Working Group on Birds of Prey and Owls/ MME, pp. 499–504.

Wink, M., Sauer-Gürth, H., El-Sayed, A.A. and Gonzalez, J., (2007). (A look through the magnifying glass of genetics: birds of prey from the DNA perspective). In *Greifvögel und Falknerei 2005/2006 (Yearbook of the German Falcon Order).* J. Neumann-Neudamm Verlag, Melsungen.

Wolfe, L.R. (1938). Birds of central Luzon. *Auk,* 55, 198-224.

Woods, R.W. and Woods, A. (1997). *Atlas of Breeding Birds of the Falkland Island*, Anthony Nelson, London.

Woźniakowski, G.J., Samorek-Salamonowicz, E., Szymański, P., Wencel, P. and Houszka, M. (2013). Phylogenetic analysis of Columbid herpesvirus-1 on Rock Pigeons, birds of prey and non-raptorial birds in Poland. *BMC Veterinary Research,* 9:52, http://www.biomedcentral.com/1746-6148/9/52

Wrege, P.H. and Cade, T.J. (1977). Courtship behavior of large falcons in captivity. *J. Raptor Res.,* 11, 1-46.

Wright, T.F., Schirtzinger, E.E., Matsumoto, T., Eberhard, J.R., Graves, G.R., Sanchez, J.J., Capelli, S., Müller, H., Scharpegge, G.K., Chambers, G.K. and Fleischer, R.C., (2008). A multi-locus molecular phylogeny of the parrots (Psittaciformes): support for a Gondwanan origin during the Cretaceous, *Mol. Biol. Evol.,* 25, 2141-2156.

Wygnanski-Jaffe, T., Murphy, C.J., Smith, C., Kubai, M., Christopherson, P., Ethier, C.R. and Levin, A.V. (2007). Protective ocular mechanisms in woodpeckers. *Eye,* 21, 83-89.

Xing, L., Niu, K., Ma, W., Zelenitsky, D.K., Yang, T.-R. and Brusatte, S.L. (2022). An exquisitely preserved in-ovo theropod dinosaur embryo sheds light on avian-like prehatching postures. *iScience,* 25, 103516, doi.org/10.1016/j.sci.2021.103516

Yamada, I. (2001). (The status of Peregrine Falcons in the Setouchi District of western Japan). *Goshawk,* 3, 4–8.

Yanuschevich, A.I., Tyurin, P.S., Yakovleva, I.D., Kydyraliev, A. and Semeonva, N.I. (1959). (*Birds of Kirghiza, Volume 1*). Academy of Kirghiza Soviet Socialist Republic, Bishkek (formerly Frunze).

Ydenberg, R.C., Butler, R.W., Lank, D.B., Smith, B.D. and Ireland, J. (2004). Western Sandpipers have altered migration tactics as Peregrine Falcon populations have recovered. *Proc. Roy. Soc. B.,* 271, 1263-1269.

Zabala, J. and Zuberogoitia, I. (2015). Breeding performance and survival in the Peregrine Falcon *Falco peregrinus* support an age-related competence improvement hypothesis mediated by an age threshold. *J. Avian Biol.,* 45, 141–150.

Zhan, X. and 24 others. (2013). Peregrine and Saker falcon genome sequences provide insights into evolution of a predatory lifestyle. *Nature Genetics,* 45, 563-566.

Zhang, X., Campomizzi, A.J. and Lebrun-Southcott, Z.M. (2022). Predicting population trends of birds worldwide with big data and machine learning. *Ibis,* 164, 750-770.

Zhao, H., Xu, W., Zhang, Y., Li, X., Zhang, H., Xuan, J. and Jia, B. (2018). Polarization patterns under different sky conditions and a navigation method based on the symmetry of the AOP map of skylight. *Optics Express,* 26, 28589-28603, https://doi.org/10.1364

Zhao, X., Zhang, M., Che, X and Zou, F. (2020). Blue light attracts nocturnally migrating birds. *Ornithological Applications,* 122, 1-12.

Zuberogoitia, I., Iraeta, A. and Martínez, J.A. (2002). Kleptoparasitism by Peregrine Falcons on Carrion Crows. *Ardeola,* 49, 103–104.

Zuberogoitia, I., Martínez, J.E., Larrea, M. and Zabala, J. (2018). Parental investment of male Peregrine Falcons during incubation: influence of experience and weather. *J. Orn.,* 159, 275-282.

Zuberogoitia, I., Zabala, J., Martínez, J.E. and Olsen, J. (2015) Alternative eyrie use in Peregrine Falcons: is it a female choice? *J. Zoo.,* 296, 6-14.

Zuberogoitia, I., Martínez, J.A., Azkona, A., Martínez, J.E., Castillo, I. and Zabala, J. (2009). Using recruitment age, territorial fidelity and dispersal as decisive tools in the conservation and management of Peregrine Falcon (*Falco peregrinus*) populations: the case for a healthy population in northern Spain. *J. Orn.,* 150, 95–101.

Above   Nominate Peregrines in Germany. The male, to the left, although the size difference makes the difference between the two obvious. The male is an adult, the female is in her second calendar year. *Torsten Pröhl.*

Below   F.p. pealei, north-west USA. *Nick Dunlop.*

Index

Albertus Magnus/Albert the Great, 35-37
Avian associations with Peregrines, 394-399

Bald Eagle, 405
Book of St Albans, 48-49

Canada Goose, 410-411
Common Buzzard, 401-403
Courtship feeding, 201

Dame Juliana Barnes, 48-49
DDT and related pesticides, 428-437

Eagle Owl, 408-410
Egyptian mummies, 34-36
Evolution of birds, 10-16

Falconidae, origins of, 10-16
Falconry
  origins of 40, 42-44
  nomenclature of, 40, 48-49
  in Anglo-Saxon UK, 44-48
  diplomatic and cultural importance of, 50, 51
Falcons
  relationship to hawks, 11-12,
  live prey, 27
  method of killing, 26
  'True Falcons', 16-25
  general characteristics, 25-29
Food caching, 242-244
Frederick II, 37-39
Fulmars, 399-400

Global climate change and Peregrines, 487-491
Golden Eagle, 404-405
Goshawk, 404
Grouse moors, 445-454
Gyrfalcon, 406

Hierofalcons, 17-18
Hittites, 43

Inertial Measuring Units (IMUs), 56, 101-102

Jackdaw, 314, 316

Kleptoparasitism, 242

Lifetime reproductive success (LRS), 292-294
Linnaeus, 39

Neonicotinoids, 440

Organochlorine contamination, 428-437
Osprey, 413-415

Pellets, 164, 172-173

Peregrine Falcon
  age at first breeding, 269
  breeding
    decisions regarding, 292-294
    studies in the UK, 294-319
      London, 296-302
      Evesham, Worcestershire, 303-313
      Scottish Borders, 314-318
    causes of failure, 345
    nearest neighbour distance, 265, 275, 276, 277
  breeding success, 342-353
    effect of weather, 344-349
    effect of parental age, 350-353
  cannibalism, 326
  causes of death 423-454,
    collisions, 423
    disease and parasites, 423-428
    pesticides 428-438
    other contaminates, 438-442
    human activities, 442-443
    human persecution, 444-454
  chick rearing and growth, 329-342
    prey deliveries to chicks, 299-302, 309-313, 314-318, 319, 335-337
      dependency on weather, 315
    variation of weight, 338, 347
    development of feathers, 339-340
    early flights, 340-342
  clutch size, 323-324
    variation with weather, 347
    second clutches, 324
  copulation, 289-291
  courtship, 272-273
  courtship feeding, 263, 272
  diet, 162-209
    prey spectrum, 162-169
    carrion feeding, 166-168
    composition of diet
      preference for pigeons, 169-172
      by nominate Peregrines, 174-176, 183
      by sub-species 177-202 (see sub-species below for exact page numbers)
      pellet formation, 172-173
    of urban birds across the range, 203
    seasonal variation of, 204-205
    annual variation of, 205-207
    variation with sub-species, 207-209
    food caching, 242-244
  dimensions and weight, nominate, 52
    see also individual sub-species
  dispersal, 361-362
  displays, 272-275
  distribution of, 82-84
  egg laying, 296-297, 304-305, 320-321
    period/interval/timing, 296-297, 304-305, 321-322
    incubation, 298, 305-309
  eggs, 222-323
    clutch size, 323-324
    false eyes – see ocelli
    family of, 18-20
    female fecundity, 350-353

Index

fiidelity, 353-361
  to site, 353-260
  territorial residency, 360
  to mate, 361
flight
  characteristics, 92-161
  cruising speed, 104-108
  effect of attaching IMUs on, 101-103
  flapping flight, 100-104
  landing, 130-131
  migration, 158-159
  soaring and gliding, 108-112
  stooping, 112-128
    adaption of nasal structure for, 114-116
    speeds, verified and claimed, 113, 116-117
    theories and practicalities of, 128
    g-forces imposed by, 121, 126, 132
    shapes during, 122-123
    adaptation of bone structure for, 126-128
  take-off, 129
  wing beat frequency, 144
  wing loading, 94-99
food consumption and energy balance, 246-255
  weight loss during breeding, 251-253
  during chick raising, 256
  digestive efficiency, 253-255
floaters, 266-267
global climate change and, 487-491
habitat, 86-88,
home range see territory/home range
humans and, 396, 412-413
hunting,
  flights observed by IMU, 131-135, 156-157
  strategies, 209-230
    method of capture, 214-215
    retrieval of prey from water, 216-218
    use of twilight and urban lighting, 218-220
    cooperative hunting, 220-221
    defensive techniques of prey, 222-230
      flocking, 222-227
      other techniques, 228-230
    success of, 230-241
    kleptoparasitism, 242
incest, 270-271
incubation, 324-329
internal organs of, 29-32
life expectancy, 422-423
migration, 364-393
  and navigational clues, 366-374
    celestial, 366-368
    magnetic, 370-374
  preparation for, 374
  curious distances of, 375-376
  timing of, 376-377, 379
  differential migration adults and juveniles, 378
  of northern Peregrines, 379-392
  effect of climate change on, 393
  inherent nature of, 364-365
  use of satellite tags to study, 383-389
  impact of tags on birds, 101-103
moult, sequence and timing, 84-86
name, 37, 39-40

nest sites, 275-289
  choice of, 275-280
  traditional sites, 271, 275, 277-279
  orientation, 276
  ground nesting, 282
  man-made structures (other than urban buildings), 282-283
  urban nesting, 284-288
ocelli, 52-53
olfaction, 160
origin of name, 35, 37
pair formation, 267-272
parasites of, 424-428
pellets, 164, 172-173
plumage,
  nominate adults, 50-51
  nominate juvenile, 52
  nominate nestling, 53
population
  world, 454-458
  UK, 459-487
    UK land use, 459-466
    population data by year, 466-474
    by area, 475-477
    in urban areas, 477-480
    by natal site, 479-480
    in London, 481-483
    in Gloucestershire, 483-484
    on Cornwall's Atlantic coast, 484
    in southern Scotland, 485
    in south Dorset/Devon, 486-487
Reverse Sexual Size Dimorphism of, 53-59
second clutches, 322
sex ratio of chicks, 331
survival, 419-422
sub-species, 59-61
  African Peregrine (*F.p. minor*), 71-72, diet, 188-190
  American Peregrine (*F.p. anatum*), 62-64, diet, 179-181
  Australian Peregrine (*F.p. macropus*), 80, diet, 198-200
  Barbary Falcon (*F.p. pelegrinoides*), 70-71, diet, 186
  Black Shaheen (*F.p. peregrinator*), 75-76, diet, 194
  Cape Verde Peregrine (*F.p. madens*), 70, diet, 188
  Ernest's Peregrine (*F.p. ernesti*), 78-79, diet, 197-198
  Izu Peregrine (*F.p. furuitii*), 78, diet, 197
  Japanese Peregrine (*F.p. japonensis*), 77, diet, 195-197
  Madagascan Peregrine (*F.p. radama*), 61, diet, 190
  Mediterranean Peregrine (*F.p. brookei*), 69, diet, 183-186
  Melanesian Peregrine (*F.p. nesiotes*), 81, diet, 201-202
  Peale's Peregrine (*F.p. pealei*), 64-65, diet, 178-179

The Peregrine Falcon

Red-naped Shaheen (*F.p. babylonicus*), 73-75, diet, 192-193
Siberian Peregrine (*F.p. calidus*), 61, diet, 191-192
South American Peregrine (*F.p. cassini*), 66-67, diet, 181-182
Tundra Peregrine (*F.p. tundrius*), 61, diet, 177-178
Sub-species are specifically dealt with in Chapters 2 and 4, but are also mentioned in other chapters.

territory/home range, 256-265
vision of, 25-26, 135-157
  vision field of, 136-138
  eye structure of, 138-140
  frequency spectrum of, 140-143
  receptors, 141-142
  flicker-fusion frequency (FFF) of, 143
  spatial discrimination of, 143-147
  twin fovea of, 148-153, 156-157
  use in hunting, 155-157
  one-eyed Peregrine, 158
voice, 88-91
wintering grounds, 380-389

Raven, 400
Reverse Sexual Size Dimorphism (RSD), 53-59
Rough-legged Buzzard (Rough-legged Hawk), 398-399

Shamanism, 35, 37

The Pesticide Crisis, 428-437
Third birds at nest, 266-267
Tomia/Tomial tooth, 26-27
'True' Falcons, 11-12, 16-25
Tunstall, Marmaduke, 39-4

Nominate Peregrine in transitional plumage. Prague, Czechia. *Ondře Prosicky*.

Taf. XXV.

Verlag u. Chromo-Lith. von Th. Fischer, Cassel.

$\frac{1}{3}$

FALCO PEREGRINUS, LINN.
Wanderfalke.

Mittelaltes Weibchen.   Altes Männchen.